THE **NUTR** SOLUTION

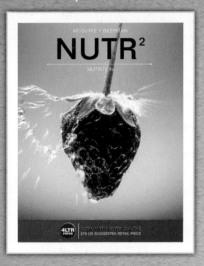

Print + Online

NUTR² delivers all the key terms and core concepts for the **Nutrition** course.

NUTR Online provides the complete narrative from the printed text with additional interactive media and the unique functionality of **StudyBits**—all available on nearly any device!

What is a StudyBit™? Created through a deep investigation of students' challenges and workflows, the StudyBit™ functionality of **NUTR Online** enables students of different generations and learning styles to study more effectively by allowing them to learn their way. Here's how they work:

COLLECT WHAT'S IMPORTANT
Create StudyBits as you highlight text, images or take notes!

WEAK

FAIR

STRONG

UNASSIGNED

RATE AND ORGANIZE STUDYBITS
Rate your understanding and use the color-coding to quickly organize your study time and personalize your flashcards and quizzes.

StudyBit™

TRACK/MONITOR PROGRESS
Use Concept Tracker to decide how you'll spend study time and study YOUR way!

85%

PERSONALIZE QUIZZES
Filter by your StudyBits to personalize quizzes or just take chapter quizzes off-the-shelf.

CORRECT

INCORRECT

INCORRECT

INCORRECT

NUTR²

Michelle Kay McGuire
Kathy A. Beerman

Vice President, General Manager:
  Liz Covello
Product Director, 4LTR Press: Steven E. Joos
Content/Media Developer:
  Victoria Castrucci
Product Assistant: Lauren Dame
Marketing Manager: Tom Ziolkowski
Marketing Coordinator: Lorreen Towle
Content Project Manager: Darrell E. Frye
Manufacturing Planner: Ron Montgomery
Production Service: SPi Global
Sr. Art Director: Bethany Casey
Internal Designer: Ke Design, Mason, OH
Cover Designer: Lisa Kuhn/Curio Press, LLC
Cover Image: Yuji Saka/Getty Images
Intellectual Property Analyst: Amber Hill
Intellectual Property Project Manager:
  Nick Barrows

For product information and technology assistance, contact us at
Cengage Learning Customer & Sales Support, 1-800-354-9706

For permission to use material from this text or product,
submit all requests online at **www.cengage.com/permissions**
Further permissions questions can be emailed to
permissionrequest@cengage.com

Library of Congress Control Number: 2016941409

Student Edition ISBN: 978-1-337-09748-2

Student Edition with Online ISBN: 978-1-337-09747-5

Cengage Learning
20 Channel Center Street
Boston, MA 02210
USA

Cengage Learning is a leading provider of customized learning solutions with employees residing in nearly 40 different countries and sales in more than 125 countries around the world. Find your local representative at **www.cengage.com.**

Cengage Learning products are represented in Canada by Nelson Education, Ltd.

To learn more about Cengage Learning Solutions, visit **www.cengage.com**

Purchase any of our products at your local college store or at our preferred online store **www.cengagebrain.com**

Printed in the United States of America
Print Number: 01   Print Year: 2016

McGUIRE / BEERMAN
NUTR²

BRIEF CONTENTS

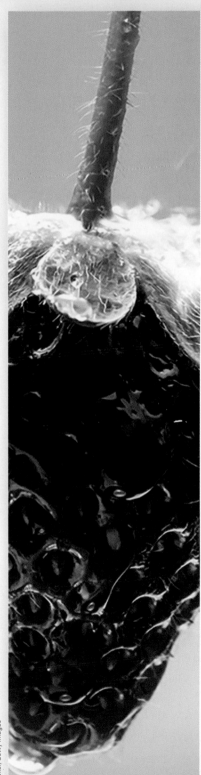

Yuji Sakai/Getty Images

CONTENTS

© mythja/Shutterstock.com

© Daxiao Productions/Shutterstock.com

© Kapu/Shutterstock.com

NUTR
ONLINE
STUDY YOUR WAY
WITH STUDYBITS!

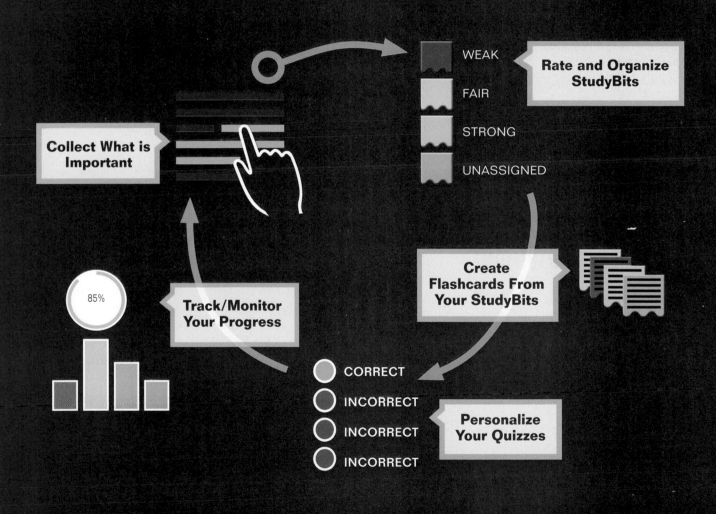

WEAK

FAIR

STRONG

UNASSIGNED

Rate and Organize StudyBits

Collect What is Important

Create Flashcards From Your StudyBits

Track/Monitor Your Progress

85%

CORRECT

INCORRECT

INCORRECT

INCORRECT

Personalize Your Quizzes

1 | Why Does Nutrition Matter?

LEARNING OUTCOMES

1-1 Define and understand the meaning of nutrition.

1-2 Understand the main purposes of nutrients and how they are classified.

1-3 Differentiate among the major groups of nutrients.

1-4 Define *calorie* and explain the concept of *energy*.

1-5 Outline the scientific method.

1-6 Evaluate the validity of a nutritional claim.

1-7 Understand the connection between nutrition and health.

1-8 Appreciate the importance of nutrition.

After finishing this chapter go to **PAGE 19** for **STUDY TOOLS.**

1-1 WHAT IS NUTRITION?

Life would not be possible without the nourishment of food, and your quality of life depends greatly on which foods you choose to eat. Hopefully, you are interested in making sure that your nutrition is as health promoting as possible. If you apply the information you learn in this course to your life, you may very well be on your way to reaching this goal. In this first chapter, you will learn many fundamental concepts necessary to understand the importance of good nutrition. You will also learn how scientists study nutrition, how national health is assessed, and how you can use scientific reason—not rumor—to select a healthy diet for years to come.

You have probably heard the terms *nutrition* and *nutrient*, but you may not know exactly what these words really mean. The term **nutrition** refers to how living organisms obtain and use food to support all the processes required for their existence. Because this process is complex, the study of nutrition incorporates a wide variety of scientific fields. Scientists who study nutrition, *nutritional scientists*, work in many disciplines, such as immunology, medicine, genetics, biology, physiology, biochemistry, education, psychology, sociology, and of course nutrition. A **dietitian** is a

nutrition The science of how living organisms obtain and use food to support processes required for existence.

dietitian A nutrition professional who helps people make dietary changes and food choices to support a healthy lifestyle.

The foods you choose now will influence both your immediate and long-term health.

nutrition professional who helps people make dietary changes and food choices to support a healthy lifestyle. Some dietitians are also involved in scientific research. Thus, the science of nutrition reflects a broad spectrum of academic and social disciplines.

1-2 WHAT ARE NUTRIENTS, AND WHAT DO THEY DO?

But what are nutrients, and why do you need them? In other words, why do people actually need to eat? A **nutrient** is a substance found in food that is required by the body and used for energy, maintenance of body structure, and/or regulation of chemical processes. For example, fats and carbohydrates provide the energy needed to fuel your body, calcium and phosphorus build and strengthen your teeth and bones, and many vitamins facilitate the chemical reactions that protect your cells from the damaging effects of excessive sunlight and pollution.

Not all nutrients in foods are naturally occurring. Sometimes foods are fortified with nutrients to make the products more nutritious. When a food has been fortified in a way that meets rigorous national standards, as put forth by the U.S. Food and Drug Administration (FDA), it can be labeled as being enriched. For instance, refined wheat flour is typically available in two forms: plain and enriched. **Fortification** and **enrichment** of foods is completely voluntary in the United States. These concepts are described in more detail in Chapter 7.

There are many ways to classify nutrients, foods, and food components. These classifications help nutritionists and other scientists distinguish the source, purpose, chemical composition, and importance to sustaining life of any given substance found in food. Although a multitude of nutrient classification systems are important to scientific research, only three are needed to understand basic nutrition. Nutrients can be classified as essential, nonessential, or conditionally essential; as organic or inorganic; and as macronutrients or micronutrients. Each of these classification systems will be introduced in the following sections.

nutrient A substance found in food that is used by the body for energy, maintenance of body structure, or regulation of chemical processes.

fortification The intentional addition of nutrients to a food.

enrichment A type of fortification whereby specific amounts of selected nutrients are added to certain foods.

1-2a Essential, Nonessential, and Conditionally Essential Nutrients

One way nutritionists classify a nutrient is by whether it must be obtained from the foods you eat to support life. Although your body can theoretically use all the nutrients found in foods, you only *need* to consume some of them. A substance that must be obtained from the diet to sustain life is referred to as an **essential nutrient**. Your body needs essential nutrients, but it either cannot make them at all or cannot make them in adequate amounts. A **nonessential nutrient** is a substance that your body needs, but if necessary, can produce in amounts needed to satisfy its requirements. Therefore, you do not actually need to consume nonessential nutrients. Most foods contain a mixture of essential and nonessential nutrients. For example, milk contains a variety of essential vitamins and minerals, such as vitamin A and calcium, as well as several nonessential nutrients, such as cholesterol.

There are, however, circumstances under which a normally nonessential nutrient becomes essential. In these situations, the nutrient is called a **conditionally essential nutrient**. For example, older children and adults must obtain two essential lipids through the diet, whereas babies require at least four that they are unable to produce because their physiologic systems are too immature. The additional lipids are therefore conditionally essential during early life. Certain diseases can also cause normally nonessential nutrients to become conditionally essential. You will learn about some of these in later chapters.

1-2b Organic Nutrients, Inorganic Nutrients, and Organic Foods

essential nutrient A substance that must be obtained from the diet to sustain life.

nonessential nutrient A substance that sustains life but does not necessarily need to be obtained from the diet.

conditionally essential nutrient A normally nonessential nutrient that, under certain circumstances, becomes essential.

organic compound A substance that contains carbon and hydrogen atoms.

inorganic compound A substance that does not contain carbon.

Nutrients can also be distinguished as organic or inorganic. By definition, a substance that contains carbon and hydrogen atoms is an **organic compound**. Carbohydrates, proteins, lipids, and vitamins are therefore chemically organic nutrients because they all contain carbon and hydrogen atoms. An **inorganic compound** does not contain carbon. Because neither water nor minerals contain carbon, they are

Are Organic Foods Healthier?

Because the U.S. Department of Agriculture (USDA) makes no claims that organically produced food is safer or more nutritious than conventionally produced food, the *organic* label is not meant to suggest superior nutritional quality or food safety. Furthermore, most scientific studies on the properties of organic food have not found that organic foods contain higher levels of nutrients than their nonorganic counterparts.[1] The only appreciable differences between organic and conventionally produced foods are the methods used to grow, handle, and process them. Whether these alternative agricultural practices promote enhanced environmental integrity and ecological balance is another area of active debate.

examples of inorganic compounds. In this way, all foods are considered organic—at least in the chemical sense of the term.

The term *organic* also has an additional and very different meaning when a person uses it to describe how a food is grown, harvested, or manufactured. When a food is labeled *certified organic*, it has been grown and processed according to U.S. Department of Agriculture (USDA) national organic standards. These foods are usually identified by the USDA's organic foods seal, which is displayed on the foods' packaging (see Figure 1.1). There are many rules and regulations that must be followed for

FIGURE 1.1 THE USDA ORGANIC FOODS SEAL

USDA, National Organic Program

Certified organic foods can be identified by this seal.

a crop or food to be certified as organic by the USDA. For example, whereas a farmer growing conventional crops can use all pesticides and herbicides that have been approved by the USDA, farmers growing organic crops can only use a subset of them.

Not all organic foods are made entirely with organic ingredients. To find out the percentage of organic ingredients that a product contains, you can examine its food label (see Figure 1.2). Foods that carry the USDA organic seal and are labeled *100% organic* must have at least 95 percent organically produced ingredients. Foods labeled *organic* must have at least 70 percent organic ingredients. Products that contain less than 70 percent organic ingredients may list specific organically produced ingredients on the side panel of the package, but may not make any organic claims on the front of the package.

1-2c Macronutrients and Micronutrients

Finally, nutrients are classified based on the quantity a person must consume to maintain health. Nutrients consumed in relatively large amounts (more than a gram per day) are classified as **macronutrients**. The macronutrients include carbohydrates, proteins, lipids, and water. Because you need only very small amounts of vitamins and minerals (often micrograms or milligrams each day), these substances are called **micronutrients**. Over the course of a lifetime, a typical adult requires about 2,700 pounds (1,200 kilograms) of protein, a macronutrient, but only about 0.3 pounds (0.14 kilograms) of iron, a micronutrient.

1-2d Other Health-Promoting Substances

As scientists research nutrition, they learn more and more about the relationship between diet and health. Not too long ago, scientists discovered that, in addition to the traditional established nutrients, foods contain other substances that likely benefit health. Scores of these compounds have only recently been uncovered and are therefore less understood than traditional nutrients. In fact, because scientific technology and nutritional knowledge have advanced so much during the last few decades, the definition of *nutrient* is evolving. Consequently, the list of recognized nutrients will likely grow as researchers identify how the myriad substances found in the foods we eat work together to promote health and well-being.

When a health-promoting compound is found in plants, it is called a **phytochemical** (or **phytonutrient**). When a health-promoting compound is found in animal-based food, it is called a **zoochemical** (or **zoonutrient**).[2] As scientists learn more about these compounds, some may be reclassified as nutrients.

FIGURE 1.2 UNDERSTANDING COMPOSITION OF "ORGANIC" FOODS

100% Organic Cereal

Must have 95–100 percent certified organic ingredients

Organic Cereal

Must have at least 70 percent certified organic ingredients

Cereal
Made with organic grains

Organic ingredients can be listed on side panel

Cereal

No organic claim is being made

When listed on food labels, the term *organic* can have different meanings.

macronutrients A class of nutrients that humans need to consume in relatively large quantities (more than a gram per day).

micronutrients A class of nutrients that humans need to consume in relatively small quantities.

phytochemical (or **phytonutrient**) A compound found in plants that likely benefits human health beyond the provision of essential nutrients and energy.

zoochemical (or **zoonutrient**) A compound found in animal-based foods that likely benefits human health beyond the provision of essential nutrients and energy.

You may have heard of functional foods or even seen them advertised. A **functional food** (or **super food**) is a food or product that likely promotes optimal health beyond simply helping the body meet its basic nutritional needs. Functional foods contain either:

1. a high concentration of traditional nutrients,
2. phytonutrients, and/or
3. zoonutrients.[3]

For example, soymilk is often referred to as a functional food because it contains phytochemicals that are believed to decrease the risk of some cancers. Other examples are cow's milk, which is rich in zoochemicals that may lower the risk of cancer and high blood pressure, and tomatoes, which may promote cardiovascular health and lower risk for prostate cancer. Although consuming functional foods may improve your health, the processes by which this occurs are often poorly understood.

Macronutrients	Micronutrients
Carbohydrates	Vitamins
Proteins	Minerals
Lipids	
Water	

1-3 HOW ARE MACRONUTRIENTS AND MICRONUTRIENTS CLASSIFIED?

Scientists organize macronutrients and micronutrients into six general groups based on their chemical natures (see Table 1.1). Each major group or *class* of nutrients consists of many different compounds, and each contributes to the structure and/or function of your body in one way or another. In this section, each of the six macro- and micronutrient classes will be introduced. This is not the last time you will see them, however; each will be addressed in much greater detail in subsequent chapters.

1-3a Carbohydrates

Carbohydrates, consisting of carbon, hydrogen, and oxygen atoms, serve a variety of functions in the body. There are many different types of carbohydrates. For example, those found in grain-based foods such as rice and pasta are quite different from those found in fruits and sweet desserts. Of the many carbohydrates that exist, perhaps the

Grains and cereals are good sources of carbohydrates. Experts recommend that you choose whole-grain products at least half the time.

© White78/Shutterstock.com

functional food (or **super food**) A food that likely benefits human health by providing a high concentration of nutrients, phytochemicals, or zoochemicals.

most important is glucose, which you may know as *blood sugar*. Glucose is important because most of your body's cells use it as their primary source of energy. However, your body uses other carbohydrates for many other purposes as well. For instance, some make up your genetic material (DNA). Others, such as dietary fiber, play roles in maintaining the health of your digestive tract and decreasing the risk of certain conditions such as heart disease and type 2 diabetes.

1-3b Proteins

Protein is abundant in many foods, including meat, legumes (such as dried peas and beans), dairy foods, and some cereal products. Although most proteins consist primarily of carbon, oxygen, nitrogen, and hydrogen atoms, some also contain sulfur or selenium atoms. The thousands of different proteins in your body serve many varied roles. For instance, proteins comprise the major structural materials of the body, including muscle, bone, and skin. Proteins allow you to move, support your complex internal communication systems, keep you healthy by fulfilling their roles in the immune system (which protects against infection and disease), and regulate many of the chemical reactions needed for life. When needed to do so, proteins can also serve as a source of energy.

Most dairy products are excellent sources of protein.

© Mike Flippo/Shutterstock.com

1-3c Lipids

Lipids, referred to as oils when they are liquid and fats when they are solid, are the third major macronutrient class. Lipids generally consist

Olives are often recommended because they are sources of healthy oils.

© Rafa Irusta/Shutterstock.com

of carbon, oxygen, and hydrogen atoms. They provide large amounts of energy, are important to the structure of cell membranes, and are necessary for the development and maintenance of your nervous and reproductive systems. Lipids also regulate a variety of processes that happen within cells. Many foods contain lipids, although the types of lipids found in plant-based foods, such as nuts and corn oil, are typically quite different from those found in animal-based foods, such as meat, fish, eggs, and milk.

1-3d Water

The water consumed through both foods and beverages is essential to many bodily functions.

© Horiyan/Shutterstock.com

Without water, the fourth macronutrient class, there would be no life. Water, which is comprised of oxygen and hydrogen atoms, makes up approximately 60 percent of your total body weight. Humans typically consume water every day, whether as a beverage or in the foods we eat. The functions of water are varied and vital. More specifically, water functions as the transporter of nutrients, gases, and waste products; as the fluid in which chemical reactions occur; and as a partner in many chemical reactions needed for your body to function. Water is also important in regulating body temperature and protecting internal organs from damage.

1-3e Vitamins

Vitamins, which are micronutrients, have a variety of chemical structures. Although they all contain carbon, oxygen, and hydrogen atoms, some vitamins also contain substances such as phosphorus and sulfur. Vitamins are abundant in most naturally occurring foods—especially fruits, vegetables, and grains. Your body requires vitamins to control hundreds of chemical reactions needed for its function. Vitamins also promote healthy and appropriate growth and development. Some vitamins, called antioxidants, protect your body from the damaging effects of harmful compounds such as air pollution. Unlike carbohydrates, proteins, and lipids, vitamins are not used directly for structure or energy. However, they play important roles in the chemical processes that build and maintain tissue and in the utilization of energy obtained from macronutrients.

© Stephen VanHorn/Shutterstock.com

Vitamins can be subdivided based on how they interact with water. Water-soluble vitamins (vitamin C and the B vitamins) dissolve easily in water, while fat-soluble vitamins (vitamins A, D, E, and K) do not. Much contemporary nutritional research focuses on the role of vitamins in the prevention and management of diseases such as heart disease and certain types of cancer.

1-3f Minerals

The mineral sodium chloride (salt) is essential in small quantities, but harmful for some when consumed in excess.

© Mega Pixel/Shutterstock.com

At least 15 minerals, each of which serves a specific purpose, are considered to be essential nutrients. For example, calcium, abundant in dairy products, provides the matrix for various structural components in your body, such as bone. Other minerals, such as sodium, help regulate a variety of body processes, such as water balance. Still other minerals, such as selenium, which is abundant in many seeds and nuts, facilitate chemical reactions. Like vitamins, minerals are not themselves used for energy, though many drive energy-yielding reactions involving the macronutrients. Scientists are still discovering new ways that minerals prevent and perhaps even treat various diseases.

1-4 HOW IS THE ENERGY IN FOOD MEASURED?

As you have just learned, the macronutrients (with the exception of water) supply the body with energy. But what exactly is *energy*? And how can foods contain this important commodity? **Energy** is the capacity of a physical system to do work. In other words, if something has energy, it can cause something else to happen. Energy is not a nutrient, but in terms of nutrition, the body uses energy found in foods to grow, develop, move, and fuel the many chemical reactions required for life. Carbohydrates, proteins, and lipids all contain energy your body can utilize and are therefore classified as **energy-yielding nutrients**. After you eat an energy-yielding nutrient, cells in your body transfer the energy that it contains into a special

energy The capacity to do work.

energy-yielding nutrient A nutrient that the body can use for energy.

FIGURE 1.3 NUTRITIONAL ENERGY

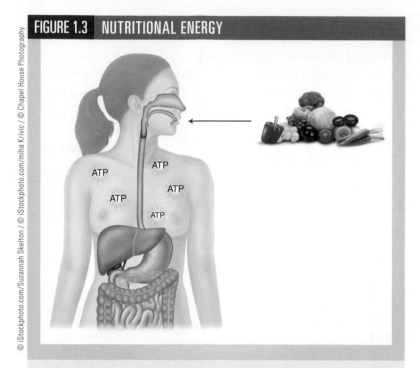

The body transfers energy found in carbohydrates, proteins, and lipids into AT P—the only form of energy used by cells.

substance called **adenosine triphosphate (ATP)**, which stores energy somewhat like a molecular battery. Your body can then use the energy stored in ATP to power its many processes (see Figure 1.3).

1-4a Calories Represent the Amount of Energy in a Food

The amount of energy in the food you eat varies, and the unit of measurement used to express the amount of energy in a food is called the **calorie**. The more calories a food has, the more ATP the body can make from it. Because a single calorie represents a very small amount of energy, the energy content of foods is typically expressed in a unit representing 1,000 calories—a *kilocalorie*. The kilocalorie is often abbreviated as *kcalorie* or *kcal*. Further, a kilocalorie is sometimes referred to as a *Calorie* (note the capital *C*) outside of scientific research, as on food labels. Therefore, 1 Calorie is equivalent to 1 kilocalorie, or 1,000 calories. A slice of cherry pie contains

1,000 calories = 1 kilocalorie = 1 Calorie

adenosine triphosphate (ATP) A chemical that provides energy to cells in the body.

calorie The unit of measurement used to express the amount of energy in a food.

approximately 480 Calories, which is equivalent to 480 kilocalories, or 480,000 calories.

Carbohydrates and proteins provide approximately 4 kcal/g. That is, they provide 4 kilocalories of energy for each gram (g) of substance consumed. Lipids provide approximately 9 kcal/g. Thus, 10 grams of a pure carbohydrate or protein contain 40 kilocalories (4 kcal/g × 10 g), whereas 10 grams of a pure lipid contain 90 kilocalories (9 kcal/g × 10 g). Although alcohol is not considered a nutrient, it provides 7 kcal/g. To practice figuring out how many calories are in a meal, try calculating the caloric content of a breakfast consisting of oatmeal, low-fat milk, brown sugar, raisins, and orange juice. The amount of each food's energy-yielding nutrients—carbohydrates, proteins, and lipids—can be found on the food's label or in any food composition table. By multiplying the weight (in grams) of each energy-yielding nutrient by its caloric content (kcal/gram) and then adding up these values, you can easily determine the number

Is There Actually Energy in Your Energy Drink?

Have you ever looked at the ingredient list on the label of an "energy" drink or power shot? You might be surprised to learn that many of these products—especially those labeled *diet* or *sugar free*—do not actually contain energy-yielding nutrients, and therefore do not really contain energy. Instead, they are chock full of vitamins and other substances (like caffeine) loosely associated with increased mental "energy" and enhanced wakefulness.[4] You can easily tell if a product contains true energy by checking to see if it provides calories in the form of carbohydrates, protein, or fat.

Some energy drinks do not actually contain energy (calories).

	Kilocalories from Energy-Yielding Nutrients			
Food	**Carbohydrates (4 kcal/g)**	**Protein (4 kcal/g)**	**Lipids (9 kcal/g)**	**Total Kilocalories**
Oatmeal, 1 cup • Carbohydrates: 25 g • Protein: 6 g • Lipids: 2 g	25 g × 4 kcal/g = 100 kcal	6 g × 4 kcal/g = 24 kcal	2 g × 9 kcal/g = 18 kcal	142 kcal
Milk, 1 cup • Carbohydrates: 12 g • Protein: 8 g • Lipids: 2 g	12 g × 4 kcal/g = 48 kcal	8 g × 4 kcal/g = 32 kcal	2 g × 9 kcal/g = 18 kcal	98 kcal
Brown Sugar, 2 tablespoons • Carbohydrates: 24 g • Protein: 0 g • Lipids: 0 g	24 g × 4 kcal/g = 96 kcal	0 g × 4 kcal/g = 0 kcal	0 g × 9 kcal/g = 0 kcal	96 kcal
Raisins, 1/2 ounce • Carbohydrates: 11 g • Protein: 0 g • Lipids: 0 g	11 g × 4 kcal/g = 44 kcal	0 g × 4 kcal/g = 0 kcal	0 g × 9 kcal/g = 0 kcal	44 kcal
Orange Juice, 1 cup • Carbohydrates: 27 g • Protein: 2 g • Lipids: 0 g	27 g × 4 kcal/g = 108 kcal	2 g × 4 kcal/g = 8 kcal	0 g × 9 kcal/g = 0 kcal	116 kcal
Total	396 kcal	64 kcal	36 kcal	496 kcal

TABLE 1.2 CALCULATING THE CALORIC CONTENT OF A TYPICAL BREAKFAST

of kilocalories provided by the meal. As illustrated in Table 1.2, the total caloric content of this breakfast is 496 kilocalories, or 496 Calories. Because of rounding errors and other factors, the total number of kilocalories (or Calories) listed in a food composition table or on a label may differ slightly from the value obtained from calculations. However, these differences are usually very small.

1-5 HOW DO NUTRITIONAL SCIENTISTS CONDUCT THEIR RESEARCH?

Few things are more significant or imperative to people than food. Food is also an important frontier of scientific research. Many questions about how food interacts with our bodies and ultimately our health are yet to be answered. But how do scientists carry out this research, and how can you distinguish reliable information from false or exaggerated claims? For centuries, scientists have explained observations (including those related to nutrition) using a series of steps collectively called the **scientific method**. There are three basic steps in the scientific method:

1. making an observation,
2. proposing a hypothesis, and
3. collecting data.[5]

1-5a Step 1: The Observation Must Be Accurate

An appropriate and accurate observation about an event or phenomenon serves as both the framework and foundation for the rest of the scientific method. If the observation is flawed, any resulting conclusions will likely be flawed as well. For example, consider the observation that there has been an alarming rise in childhood obesity over the past few decades. Before a researcher can develop an explanation for

scientific method A series of steps used by scientists to explain observations.

this observation, he or she should first consider several questions about it. Are girls more likely to be obese than boys? At what age do the rates of obesity increase? Are adult obesity rates also rising? Answering these questions will help ensure that the observation is complete and accurate and provides a solid base on which the rest of the scientific method can be applied.

1-5b Step 2: A Hypothesis Explains the Observation

Once the researcher makes an observation and understands the details associated with it, the next step is to explain why the event occurred. At this point, the scientist must propose a **hypothesis**, a prediction about the relationship between variables, to explain the observation. For example, the researcher might hypothesize that the increase in childhood obesity is due to a lack of exercise. Alternatively, the hypothesis might focus on excess calorie consumption. Scientists make two general types of hypotheses: those that predict cause-and-effect (causal) relationships and those that predict correlations. A **cause-and-effect relationship** (or **causal relationship**) is a relationship whereby an alteration to one variable causes a change in another variable. When two variables are clearly related, but one cannot be shown to cause the other, one can only say that there is a **correlation** (or **association**). For example, if your alarm clock goes off every morning around the time the sun rises, the two events are correlated, but neither actually causes the other. This situation is very different from a cause-and-effect relationship, because when two factors are simply correlated with each other, altering one will not change the other. Clearly, setting your alarm to sound earlier will not cause the sun to come up earlier. In a true cause-and-effect relationship, if you change one variable (the cause), you will necessarily change the other (the effect).

Understanding the difference between cause-and-effect relationships and correlations is important in all scientific disciplines, not just nutrition. Although

Although there is probably an association between when your alarm sounds and the time the sun rises, changing your wake-up alarm will not alter the sunrise. In other words, correlation does not infer causality.

many studies are designed to test for correlations, their results are unfortunately interpreted or reported as proving causal relationships. In the next section, you will learn about what types of experiments scientists can use to test hypotheses related to correlations versus causal relationships.

1-5c Step 3: Experimentation

It is important to realize that if the study design is flawed, a good observation and/or hypothesis can be completely wasted. A basic understanding of sound experimental practices will help you discern which nutrition claims are unfounded and which are valid.

Epidemiologic Studies When a researcher wants to determine whether one variable is simply correlated with another variable, an **epidemiologic study** can be conducted. In this type of study, a researcher examines the relationship between variables in a group of people simply by recording information and then examining how various factors are statistically related to each other. In epidemiologic studies, researchers do not actually ask people to change their behaviors, alter their food intake patterns, or undergo any sort of treatment. Consequently, epidemiologic studies should not be used to test hypotheses predicting causal relationships—only correlations.

As an example, a research team might want to examine the relationship between maternal caffeine intake and infant birth weight. This might involve interviewing pregnant women to estimate caffeine intake and then mathematically relating this to the weight of their newborn infants. However, it is possible that other associated factors known as **confounding variables** might similarly impact birth weight, and high-quality research usually

hypothesis A prediction about the relationship between variables.

cause-and-effect relationship (or **causal relationship**) A relationship whereby an alteration to one variable causes a change in another variable.

correlation (or **association**) A relationship whereby an alteration to one variable is related to a change in another variable.

epidemiologic study A study in which data are collected from a group of people who are not asked to change their behaviors in any way.

confounding variable A coincidental factor related to a study's outcome but not of primary interest to the research hypothesis; this should be accounted for in the analysis.

accounts for such factors using statistics. For instance, it is possible that women who consume the most caffeine also tend to be smokers, so an apparent relationship between caffeine and birth weight could actually be due to the confounding effect of smoking. As such, it is important to recognize and control for any known confounding variables.

Intervention Studies In contrast to an epidemiologic study, an **intervention study** requires participants—regardless of whether they are humans, animals, or simply cells—to undergo a treatment or intervention (see Figure 1.4). In most studies, some of the participants receive the treatment, while others do not.

Participants who do not receive a treatment or intervention are considered to be in the **control group**.

FIGURE 1.4 INTERVENTION STUDIES

Nutrition intervention studies can involve humans, animal models, or cell-culture systems, depending on the hypothesis that the researcher wants to test.

A control group is needed to determine whether the effects witnessed in the treatment group are caused by the treatment, are due to chance, or are a result of some other aspect of the study. For example, a researcher interested in understanding the childhood obesity epidemic might test the hypothesis that nutrition education can decrease obesity in children. To do so, the researcher might have some children attend a nutrition education class (the treatment group), and others not (the control group). The researcher could then measure whether the children who received the nutrition education gained less weight than those who received no extra education.

However, the presence of a control group does not necessarily ensure that the researcher's conclusions are accurate. This is because of a well-known phenomenon called the **placebo effect**, which occurs when a study participant experiences an apparent effect of the treatment just because he or she believes that the treatment will work. Indeed, many studies have shown that taking an unmedicated sugar pill actually influences blood pressure in some people if they believe that the pill is medicated.[6] How the placebo effect works remains mysterious to scientists, but it clearly connotes a strong mind–body connection. Although the placebo effect cannot be prevented, researchers can account for it by requiring control group participants to consume or experience an inert treatment that looks, smells, tastes, and/or seems just like the actual treatment. This inert treatment is called a **placebo**. In this way, any differences between the intervention and control groups can be attributed to the actual treatment and not to placebo effect.

It is also important to note that a researcher can inadvertently influence a study's outcome simply by knowing which subjects received the actual treatment and which received the placebo. This is a type of **researcher bias**. For example, if a researcher knows that certain participants received a treatment thought to improve memory, he might unconsciously score a memory test more favorably for those individuals. Because this type of bias may affect a study's final results, it should be avoided at all costs.

intervention study An experiment in which a variable is altered to determine its effect on another variable.

control group Study participants that do not receive a treatment or intervention.

placebo effect A phenomenon whereby a study participant experiences an apparent effect of the treatment just because the participant believes that the treatment will work.

placebo An inert treatment given to the control group that cannot be distinguished from the actual treatment.

researcher bias A phenomenon whereby the researcher, usually inadvertently, influences the results of a study.

An important technique that minimizes the placebo effect and researcher bias is designing a study so that the researchers and/or participants are unaware of who has been assigned to the intervention and control groups. When researchers know who is in the treatment group and who is in the placebo group, but participants do not, the study is called a **single-blind study**. When neither the researchers nor the participants know, the study is called a **double-blind study**.

Another technique commonly used to avoid additional types of bias is the **random assignment** of participants to the treatment and control groups. For example, consider the study designed to test whether nutrition education influences childhood obesity. The random assignment of children to either the treatment or the control group would help ensure that important confounding variables, such as usual intake of sweets and exercise patterns, are distributed evenly between the two groups. It is important to remember, though, that even when researchers take all precautions to minimize potential sources of error such as bias, it is possible that other factors can also influence a study's outcome—or what is known as the **dependent variable**. In addition, even when researchers eliminate or control for all known confounding variables, it remains possible that undiscovered ones exist. Randomization distributes all known and unknown confounding variables equally in both groups, allowing researchers to be more confident that a change in the dependent variable (in this case, childhood weight gain) is actually due to the treatment (in this case, nutrition education). Every intervention study report should state whether participants were randomly assigned to study groups (see Figure 1.5).

As you now know, nutrition research takes many shapes and forms, and careful use of the scientific method helps ensure that conclusions are valid and appropriate no matter what the format of the research. Because nutrition research continues to

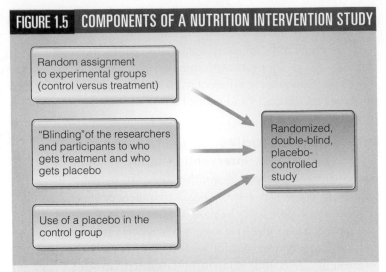

FIGURE 1.5 COMPONENTS OF A NUTRITION INTERVENTION STUDY

Random assignment to experimental groups (control versus treatment)

"Blinding" of the researchers and participants to who gets treatment and who gets placebo

Use of a placebo in the control group

→ Randomized, double-blind, placebo-controlled study

Randomized, double-blind, placebo-controlled intervention studies are considered the gold standard of nutrition research.

reveal new findings, scientists' understanding of the relationship between diet and health is continually shifting. Therefore, you should not be surprised as dietary recommendations change over time.

1-6 ARE ALL NUTRITION CLAIMS BELIEVABLE?

Without a doubt, it is the nature of science to be a long and winding road of discovery. Although the emergence of new theories, devaluation of previously accepted hypotheses, and shifts in popular opinion are to be expected, it is crucial that new nutritional claims be critically evaluated as they emerge. Sensational studies captivate public interest, but their findings are not always based on sound research. In addition, it is important that new findings are replicated before they are put into everyday practice. If unfounded nutrition claims go unchecked, they may come to be believed and communicated as fact. To help distinguish hearsay from science, keep the following questions in mind when you make nutrition-related decisions.

1-6a Where Was the Study Published?

The first important consideration when determining whether a nutrition claim is reputable is where the study was published. As you can imagine, not all sources of information are equally credible. This often makes it difficult to judge the validity of nutrition-related claims. As such, it is important to determine the *primary source* of any nutrition claim that you may see or hear of. In other

single-blind study A human experiment in which the participants do not know to which group they have been assigned.

double-blind study A human experiment in which neither the participants nor the scientists know to which group the participants have been assigned.

random assignment A research strategy whereby study participants have equal chance of being assigned to the treatment and control group.

dependent variable The outcome of interest in a research study.

© SpeedKingz/Shutterstock.com

1-6c Who Paid for the Research?

Good research is often expensive. There are many ways that scientists obtain the money needed to fund their studies, but most acquire funding by applying for grants from private companies, foundations, and state and federal agencies. Note that, just because a researcher obtained money from a private source such as a food commodity group (like the dairy industry) or a food manufacturer, this does not mean that the research and its results are somehow biased. Indeed, credible researchers go to substantial lengths and take important steps to ensure that their funding sources do not bias or influence the outcomes of their studies—especially if the funding agencies have something to gain or lose from the studies' results.

1-6d Did the Researchers Use the Right Study Design?

Once you have determined that a nutrition claim has been published in a reputable journal or report and was conducted by qualified researchers, you are ready to consider the research itself. Was the research conducted in a way that was appropriate to test the hypothesis? And do the conclusions fit the study design?

One of the best resources for finding information about a given research study is the U.S. National Library of Medicine, which hosts a searchable biomedical database called PubMed (www.ncbi.nlm.nih.gov/PubMed). The PubMed database allows easy access to more than 11 million biomedical journal citations. You can use details from PubMed to answer important questions about the study, such as:

- Was it an epidemiologic study or an intervention trial?
- Do the results suggest an association or causal relationship?
- Was an appropriate control group used?
- If it was an intervention, was the study double-blinded and was a placebo used in the control group?

With your knowledge of experimental design, you can evaluate the research and determine whether the nutrition claim is likely to be valid.

words, where was this information first published or reported? In general, a **peer-reviewed journal** is a trustworthy primary source of information. Studies published in peer-reviewed journals have been read and approved by a group of experts (peers) knowledgeable in the study's specific area of nutrition. Many private and governmental organizations, such as the National Institutes of Health and the Institute of Medicine, also publish credible studies and information concerning diet and health. These agencies are considered some of the most reliable and unbiased sources of information available. Table 1.3 lists a selection of both reputable peer-reviewed journals and organizations that publish nutrition-related articles. As a rule, you should question nutrition claims that have not first been published in a peer-reviewed journal or other highly regarded publication. Often, it is not necessary to go any further than this step to determine whether a nutrition claim is even worth considering.

> Determining fact from fiction when it comes to nutrition claims can sometimes be difficult.

1-6b Who Conducted the Study?

Next, it is important to ask, "Who conducted the research?" In general, reliable nutrition research is conducted by scientists at universities and medical schools. Researchers at private and public institutions and organizations also conduct sound nutrition research. It is important that the individuals conducting the research are qualified and knowledgeable. Finding out where the researchers conducting the study work and what their qualifications are can help you make this determination.

peer-reviewed journal A publication that requires a group of scientists to read and approve a study before it is published.

1-6e Do Public Health Organizations Concur?

Even the best, most thorough experiment does not always provide conclusive evidence that a particular nutrient influences health in a certain way. Consequently, public health experts usually wait for several studies to produce similar results before they begin to make overall claims about the effect of a certain nutrient on health. Thus, before believing a nutritional claim to be true, determine whether it is supported by major public health organizations. For example, you might check whether the American Heart Association supports a study's claim that a certain phytochemical decreases one's risk of heart disease or whether the American Cancer Society supports a study's claim that a particular nutrient decreases one's risk of cancer.

1-7 NUTRITION AND HEALTH: WHAT IS THE CONNECTION?

Nutritional scientists continue to study the impact of nutrition, because experts have very good reason to believe that it plays a dominant role in health and disease. Indeed, consuming either too little or too much of a nutrient can cause illness. But what do experts really know about nutrition and health? And how do scientists and organizations track the overall health of a nation? Although extensive answers to

TABLE 1.3 SOME RELIABLE SOURCES OF NUTRITION INFORMATION

Peer-Reviewed Journals	Government and Private Agencies and Their Websites
American Journal of Clinical Nutrition	Academy of Nutrition and Dietetics (http://www.eatright.org)
Annals of Nutrition and Metabolism	American Cancer Society (http://www.cancer.org)
Annual Review of Nutrition	American Diabetes Association (http://www.diabetes.org)
Appetite	American Heart Association (http://www.americanheart.org)
British Journal of Nutrition	American Institute for Cancer Research (http://www.aicr.org)
Clinical Nutrition	American Medical Association (http://www.ama-assn.org)
European Journal of Nutrition	American Society for Nutrition (http://www.nutrition.org)
Journal of Human Nutrition and Dietetics	Centers for Disease Control and Prevention (http://www.cdc.gov)
Journal of Nutrition	Institute of Medicine (http://www.iom.edu)
Journal of the American College of Nutrition	Mayo Clinic (http://www.mayoclinic.org)
Journal of the Academy of Nutrition and Dietetics	National Academy of Sciences (http://www.nas.edu)
Journal of the American Medical Association (JAMA)	National Institutes of Health (http://www.nih.gov)
Journal of Pediatrics Gastroenterology and Nutrition	National Institutes of Health National Center for Complementary and Alternative Medicine (http://nccam.nih.gov)
Lancet	National Institutes of Health Office of Dietary Supplements (http://dietary-supplements.info.nih.gov/)
Nature	U.S. Department of Agriculture (http://www.usda.gov)
New England Journal of Medicine	U.S. Food and Drug Administration (http://www.fda.gov)
Nutrition	U.S. Food and Nutrition Information Center (http://fnic.nal.usda.gov)
Nutrition Research	
Public Health Nutrition	
Science	
Scientific American	

Annals of Nutrition and Metabolism Vol. 57, No. 3-4, 2010, reprinted with permission of S. Karger AG, Basel.

these questions are beyond the scope of this chapter, it is important to understand some basic concepts about the relationship between nutrition and health, how this relationship is assessed on a national scale, and how it evolves.

1-7a Public Health Agencies

To appreciate the complex relationship between nutrition and health, it is important to understand how researchers and public health experts assess the health of a community or nation. In other words, what do scientists and other health professionals measure when they want to determine if a population is becoming more or less healthy? And who keeps track of all this information? In the United States, the organization responsible for monitoring health trends is the U.S. Centers for Disease Control and Prevention (CDC). The CDC surveys and monitors national and international health to prevent disease outbreaks, implement disease prevention strategies, and maintain national health statistics.

1-7b Mortality and Morbidity Rates

The CDC monitors many aspects of societal health, but the most frequently cited are morbidity and mortality rates. A **rate** is a measure of some event, disease, or condition within a specific time span. For instance, speed is expressed as a rate, such as miles per hour. One example of a health-related rate is **mortality rate**, which assesses the number of deaths in a given period of time. For example, cancer experts monitor cancer mortality rates to help determine whether deaths from this disease are increasing or decreasing over time.

An important health-related mortality rate is **infant mortality rate**, which is the number of infant deaths (occurring within the first year of life) per 1,000 live births in a given year. Infant mortality rate is often used to assess the well-being of a society, as it reflects a complex and interrelated web of environmental, social, economic, medical, and technological factors that influence overall health. The 2013 U.S. infant mortality rate of 5.96 per 1,000 live births represented a historical low, and was 13 percent lower than that in 2003. The United States' infant mortality rate has declined dramatically in the past century, as illustrated in Figure 1.6.[7] Nonetheless, the most recent data places the United States 44th internationally behind countries like Hong Kong (1st) and Sweden (2nd). International infant mortality rates are illustrated in Figure 1.7. Many factors influence infant mortality rate; genetics, access to medical care, substance abuse, maternal nutrition, and weight gain during pregnancy all affect this statistic.

Whereas mortality rates assess the number of deaths in a given period of time, a **morbidity rate** reflects illness or disease in a given period of time. Like mortality rates, morbidity rates for certain

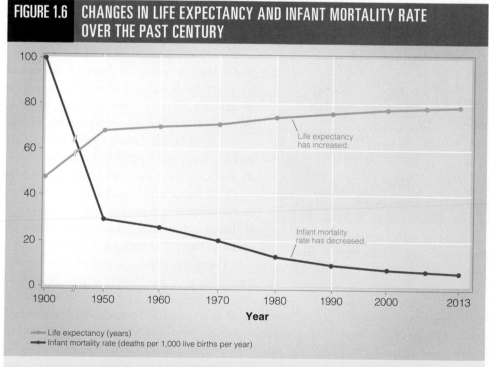

FIGURE 1.6 CHANGES IN LIFE EXPECTANCY AND INFANT MORTALITY RATE OVER THE PAST CENTURY

Life expectancy has increased.

Infant mortality rate has decreased.

- Life expectancy (years)
- Infant mortality rate (deaths per 1,000 live births per year)

Since 1900, life expectancy has increased while infant mortality rate has decreased. These shifts indicate greater societal health.

Source: Xu J. Q., Kochanek K. D., Murphy S. L., Tejada-Vera B. Deaths: Final data for 2007. National vital statistics reports. Hyattsville, MD: National Center for Health Statistics. 2010; 58:19. Available from: http://www.cdc.gov/NCHS/data/nvsr/nvsr58/nvsr58_19.pdf; National Center for Health Statistics. Health, United States, 2014: With Special Feature on Adults Aged 55–64. Hyattsville, MD. 2015. Available from: http://www.cdc.gov/nchs/data/hus/hus14.pdf.

rate A measure of some event, disease, or condition within a specific time span.

mortality rate The number of deaths that occur in a certain population group in a given period of time.

infant mortality rate The number of infant deaths per 1,000 live births in a given year.

morbidity rate The number of illnesses or diseases in a given period of time.

FIGURE 1.7   LIFE EXPECTANCY AROUND THE GLOBE

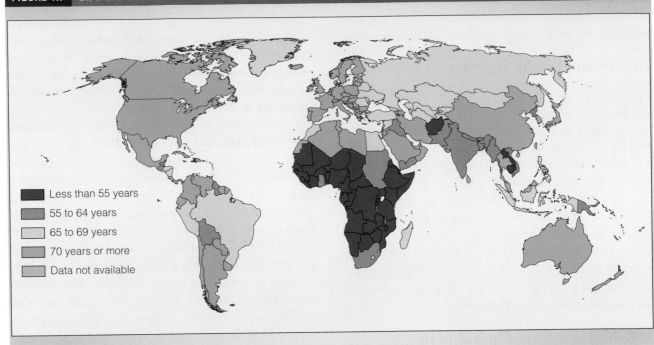

Less than 55 years
55 to 64 years
65 to 69 years
70 years or more
Data not available

The Centers for Disease Control and Prevention (CDC) is a federal agency charged with monitoring health and disease rates in the U.S. population.

1-7c   Life Expectancy

Another indicator of societal health is **life expectancy**, the average number of years of life remaining for a person at a particular age. In 2013, life expectancy at birth in the United States was 79 years—76 years for males and 81 years for females. If American men and women are still alive at 65 years of age, they can expect to live to the ages of 83 and 86, respectively. As you might have guessed, average life expectancy in the United States has increased dramatically over the past century (see Figure 1.6). For a broad constellation of reasons, including better nutrition, life expectancy is higher in industrialized than nonindustrialized nations. This is illustrated in Figure 1.7.

There are many reasons why life expectancy has increased over the past century—one of which is better nutrition.[9] Regardless of its cause, an increasing life expectancy has resulted in a phenomenon often referred to as the graying of America. The **graying of America**, characterized by an increasing proportion of the population over the age of 65, continues to significantly influence America's health.[10] For example, the graying of America has contributed to a rise in the prevalence of diseases such as heart disease and cancer, which are now the nation's leading causes of disability and death.

life expectancy A statistical prediction of the average number of years of life remaining for a person at a particular age.

graying of America A phenomenon occurring in the United States characterized by an increasing proportion of the population over the age of 65.

diseases have changed drastically over the last few decades in the United States. For example, tuberculosis rates have continuously declined over the last century, whereas type 2 diabetes rates continue to increase.[8]

1-7d Diseases Are Either Infectious or Noninfectious

You have just learned how scientists and public health experts use indices such as morbidity rates, mortality rates, and life expectancy to assess the health of a nation. But how do they categorize disease, and what is actually meant by this term? A **disease** is any abnormal condition of the body or mind that causes discomfort, dysfunction, or distress but is not the direct result of a physical injury. In other words, a broken arm due to a fall is not considered a disease. An **infectious disease** is caused by a pathogen, such as a bacterium, virus, fungus, parasite, or other microorganism, is contagious (meaning that it can be passed from one person to another person), and tends to be short-lived. An example of an infectious childhood disease is chicken pox, which is caused by a virus. Many serious infectious diseases so common during childhood in past decades, such as chicken pox and whooping cough, are now preventable via the use of vaccines. Conversely, a noninfectious disease is not spread from one person to another, does not involve an infectious agent, and tends to be long term and chronic in nature. **Noninfectious diseases** tend to be treatable rather than curable, and are often preventable. An example of a noninfectious disease is type 2 diabetes.

A **chronic degenerative disease** is a noninfectious disease that develops slowly, persists over a long period of time (chronic), and tends to result in progressive breakdown of tissues and loss of function (degenerative). Type 2 diabetes, heart disease, osteoporosis, and cancer are chronic degenerative diseases that have been linked to poor nutrition. These diseases are among the most common and costly health problems that people face today—they are also among the most preventable. The adoption of healthy behaviors such as eating nutritious foods, being physically active, and avoiding tobacco use can prevent or control the devastating effects of many of these diseases. Scientists believe that most chronic degenerative diseases are caused by a combination of genetic and lifestyle factors. Although poor nutrition is a leading cause, the exact processes by which chronic degenerative diseases develop are often poorly understood.

Another cause of illness is an **autoimmune disease**. Typically, your body's immune system protects you from disease and infection. But if you have an autoimmune disease, your immune system attacks healthy cells by mistake, causing sometimes serious consequences. Examples of autoimmune diseases that you will learn about in this book are celiac disease and type 1 diabetes. This is just too redundant.

Chronic Degenerative Diseases Have Largely Replaced Infectious Diseases

In the early 1900s, infectious diseases such as pneumonia, tuberculosis, and influenza were rampant throughout America. Infectious diseases were the leading causes of death, accounting for one-third of the nation's mortality rate. Severe nutritional deficiencies contributed to the high number of deaths and illnesses that occurred throughout this period. It may surprise you to learn that today, chronic degenerative diseases have generally replaced infectious diseases as the leading causes of death in America (see Figure 1.8).[11] Indeed, today's leading causes of death are heart disease, cancer, respiratory diseases, unintentional injuries, stroke, Alzheimer's disease, diabetes, influenza and pneumonia, nephritis, and suicide. These account for 74 percent of the 2.6 million deaths each year.

Public health efforts such as investment in water treatment and sewage disposal facilities, development of antibiotics, and implementation of childhood vaccination programs have helped reduce the incidence of many infectious diseases. As a combined result of decreased infectious disease rates, better nutrition, and other improvements in health care, life expectancy rates have increased faster than at any other time in United States' history. Although it seems like a testament to America's health, this trend is something of a double-edged sword because it has brought with it increases in the chronic degenerative diseases so common to today's society.

Risk Factors of Chronic Diseases

As previously mentioned, the specific causes (or *etiologies*) of today's most common chronic degenerative diseases are complex and often poorly understood. However, scientists do know that many lifestyle, environmental, and genetic factors are related to a person's risk of developing these diseases. Such a characteristic is called a **risk factor**. For example, the major lifestyle-related risk factors that are associated with heart disease,

---

disease An abnormal condition of the body or mind that causes discomfort, dysfunction, or distress.

infectious disease An illness that is contagious, caused by a pathogen, and often times short-lived.

noninfectious diseases An illness that is not contagious, does not involve an infectious agent, and often times long term and chronic.

chronic degenerative disease A noninfectious disease that develops slowly, persists over a long period of time, and tends to result in progressive breakdown of tissues and loss of function.

autoimmune disease An illness resulting from an inappropriate immune response that attacks the body's cells.

risk factor A lifestyle, environmental, or genetic factor related to a person's chances of developing a disease. Just because something is a risk factor does not mean that it is causative in nature.

FIGURE 1.8 FIVE LEADING CAUSES OF DEATH OVER THE PAST CENTURY

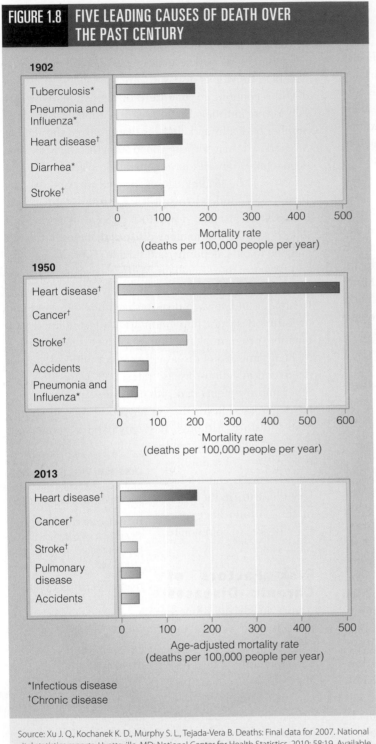

1902

Tuberculosis*
Pneumonia and Influenza*
Heart disease†
Diarrhea*
Stroke†

Mortality rate
(deaths per 100,000 people per year)
0 100 200 300 400 500

1950

Heart disease†
Cancer†
Stroke†
Accidents
Pneumonia and Influenza*

Mortality rate
(deaths per 100,000 people per year)
0 100 200 300 400 500 600

2013

Heart disease†
Cancer†
Stroke†
Pulmonary disease
Accidents

Age-adjusted mortality rate
(deaths per 100,000 people per year)
0 100 200 300 400 500

*Infectious disease
†Chronic disease

Source: Xu J. Q., Kochanek K. D., Murphy S. L., Tejada-Vera B. Deaths: Final data for 2007. National vital statistics reports. Hyattsville, MD: National Center for Health Statistics. 2010; 58:19. Available from: http://www.cdc.gov/NCHS/data/nvsr/nvsr58/nvsr58_19.pdf; National Center for Health Statistics. Health, United States, 2014: With Special Feature on Adults Aged 55–64. Hyattsville, MD. 2015. Available from: http://www.cdc.gov/nchs/data/hus/hus14.pdf.

nutrition transition A shift from undernutrition to overnutrition or unbalanced nutrition that often occurs as a society transitions to a more industrialized economy.

Many unhealthy eating patterns are considered risk factors for chronic degenerative diseases.

cancer, and stroke—the top three leading causes of death in the United States—include tobacco use, lack of physical activity, and a range of poor dietary habits. Risk factors for cancer that are related to genetics and environment include a family history of cancer, asbestos- and lead-based housing, and exposure to pollution and excessive sunlight. Although some of these risk factors may play a role in causing chronic degenerative diseases, many are simply predictive—they help doctors and nutritionists know who is at the greatest risk. *Remember that correlation does not always infer causality.*

Obesity is a major risk factor for many chronic degenerative diseases because being obese can predispose a person to life-threatening conditions such as heart disease, stroke, and type 2 diabetes. Obesity is becoming a health crisis of epidemic proportions both nationally and globally. As such, the dramatic shift from undernutrition to overnutrition or unbalanced nutrition that often occurs as a society transitions to a more industrialized economy—referred to as the **nutrition transition**—is strongly related to many of the chronic degenerative diseases facing humankind today. You will learn much more about obesity and its negative health consequences throughout this book.

1-7e Assessing the Nutritional Health of the Nation

As you have learned, nutritional concerns have changed dramatically over the past century. Specifically, Americans have transitioned from being vulnerable to micronutrient deficiencies to being over nourished and overweight. To help identify and tackle health challenges—including those related directly to nutrition—the U.S. Department of Health and Human Services (DHHS) has for three decades published a document called *Healthy People*. The most recent version, Healthy People 2020, outlines a set of overall health objectives for the nation to accomplish by 2020.[12]

Recognizing that chronic conditions such as obesity, type 2 diabetes, and osteoporosis are currently some of the most pressing health concerns throughout America, DHHS designed Healthy People 2020 to achieve four overarching health-related goals:

Regular exercise is an important component of a healthy lifestyle.

- Encourage long, high-quality lives that are free of preventable disease, disability, injury, and premature death.

- Achieve health equity, eliminate disparities, and improve the health of all groups.

- Create social and physical environments that promote good health for all.

- Promote quality of life, healthy development, and healthy behaviors across all life stages.

These four broad goals are subdivided into 39 different topic areas. The topics of diabetes; food safety; nutrition/weight status; and maternal, infant, and child health each have corresponding nationwide goals. The goal for nutrition/weight status, for example, is the promotion of health and reduction of chronic disease risk through the consumption of healthful diets and achievement and maintenance of healthy body weights. Although discussing each of Healthy People 2020's topics and goals is beyond the scope of this book, you can learn more at the Healthy People website (http://www.healthypeople.gov).

1-8 WHY STUDY NUTRITION?

Experts agree that chronic degenerative diseases and other health concerns caused, at least in part, by poor diet represent some of the United States' most serious and pressing public health issues. Researchers estimate that 69 percent of adults over the age of 20 are either overweight or obese, and that more than 280,000 deaths occur each year because of obesity alone.[13] More than 13 percent of Americans now have cardiovascular disease (the leading cause of death), 6 percent have cancer, 30 percent have high blood pressure, and 12 percent have type 2 diabetes. Fortunately, consuming a healthy balance of nutrients, phytonutrients, and zoonutrients can decrease your risk of developing all these conditions. Indeed, as the occurrence of chronic diseases increases, it is ever more important to pay attention to what you eat throughout your entire life. As the old saying goes, "An ounce of prevention is worth a pound of cure." Or perhaps, stated most succinctly and accurately by Sir Francis Bacon more than 400 years ago, "Knowledge is power."

2 | Choosing Foods Wisely

LEARNING OUTCOMES

2-1 Describe the continuum of *nutritional status*.

2-2 Differentiate the methods by which nutritional status is assessed.

2-3 Explain how the Dietary Reference Intakes are used to assess dietary adequacy.

2-4 Utilize contemporary food guidance systems.

2-5 Use food labels to plan a healthy diet.

2-6 Develop an action plan that demonstrates the four steps of dietary assessment.

After finishing this chapter go to **PAGE 47** for **STUDY TOOLS.**

2-1 WHAT IS NUTRITIONAL STATUS?

Food and eating are important components of all cultures around the world. On a more basic level, however, food is simply a vehicle for nutrients and energy that your body needs to function. But how do you know if you are consuming enough—but not too much—food? And how can you choose foods in order to meet your nutritional needs? To answer these questions, you need to understand several important concepts about the nutrients in foods and how much of them you need. In this chapter, you will learn the fundamentals of these concepts and how to apply them to your own food choices. You will also learn about important dietary regulations and guidelines that have been developed to help in this endeavor. This basic information and these helpful dietary tools will enable you to determine which foods—and how much of them—to choose to optimize your health.

Most people know that it is important for our diets to provide specific nutrients and energy in the right amounts. **Malnutrition** can result both when a person's diet does not provide enough nutrients and/or energy and when it provides too much. We often think of malnutrition in terms of starving, emaciated people, and in some cases this is correct. However, malnutrition means more than this. The term **undernutrition** is used to describe impaired health that results when a person's diet does not provide enough nutrients and/or energy to satisfy physiological needs. When this occurs, a person is said to have one or more **nutritional deficiencies**. Conversely, the term **overnutrition** refers to impaired health that results when a person's diet provides an excess of nutrients and/or energy. In most cases, an excess intake of nutrients from food is not harmful. However, with certain disease states or when some nutrients are taken as supplements in very high doses, overnutrition can cause nutrients to accumulate in the body and result

malnutrition A state of poor nutritional status caused by an imbalance between the body's nutrient requirements and nutrient availability.

undernutrition The inadequate intake of one or more nutrients and/or energy.

nutritional deficiency A condition caused by inadequate intake of one or more essential nutrients.

overnutrition A state of poor nutrition that occurs when the diet provides excess nutrients and/or energy.

in **nutritional toxicity**. When this happens, physiological function is impaired, and can sometimes result in death. When excess energy is consumed, obesity can occur. Undernutrition and overnutrition are both examples of malnutrition and make up the extreme ends of the nutritional status continuum. The term **nutritional status** refers to where a person is along this continuum. Figure 2.1 illustrates the relationships between nutrient availability (or dietary intake), health, and nutritional status. As you can see, both undernutrition and overnutrition can increase the chance that you have suboptimal health. Thus, the center of the nutritional status continuum is where you want to be.

2-1a Primary and Secondary Malnutrition

Regardless of whether it causes undernutrition or overnutrition, malnutrition can be precipitated by a variety of underlying causes. **Primary malnutrition** is a condition whereby poor nutritional status is caused strictly by inadequate diet, while **secondary malnutrition** is caused by factors other than diet, such as illness. For example, a person who is deficient in one of the B vitamins because his diet lacks vitamin-rich fruits and vegetables is experiencing primary malnutrition. If his vitamin B deficiency is instead caused by an illness that interferes with vitamin B absorption, he is experiencing secondary malnutrition. It is important to distinguish whether primary or secondary malnutrition is causing poor nutritional status because their treatments are very different. Nutritional deficiencies caused by primary malnutrition can often be cured by consuming certain foods, whereas secondary malnutrition requires addressing the underlying cause.

2-1b Adequate Nutrient Intake

How can a health care provider know if a person is malnourished? And how can you know if your diet is meeting your nutritional needs? The answers to these questions are complex because different people have different nutritional needs for optimal health.

nutritional toxicity Overconsumption of a nutrient that results in dangerous toxic effects.

nutritional status The extent to which a person's diet meets his or her individual nutrient requirements.

primary malnutrition A condition whereby poor nutritional status is caused strictly by inadequate diet.

secondary malnutrition A condition whereby poor nutritional status is caused by factors other than diet, such as illness.

Both undernutrition (left) and overnutrition (right) are forms of malnutrition.

Factors such as sex, age, physical activity, and genetics all influence a person's nutritional needs. For example, a physically active person, with all other factors (e.g., sex, age, weight, height, and so on) being the same, is likely to have different nutritional requirements than someone who is sedentary. No matter the specific nutritional needs, a person who regularly consumes the required amount of nutrients and energy to meet his or her physiologic needs has achieved **nutritional adequacy**.

Dietitians and other health professionals use several tools to determine whether a person's nutritional status is optimal, or if it could benefit from different dietary choices. As you will learn in the following sections, you can use many of these same tools to assess your own nutritional status and dietary adequacy.

nutritional adequacy A condition whereby a person regularly consumes the required amount of a nutrient to meet his or her physiological needs.

FIGURE 2.1 DIETARY INTAKE INFLUENCES NUTRITIONAL STATUS AND HEALTH

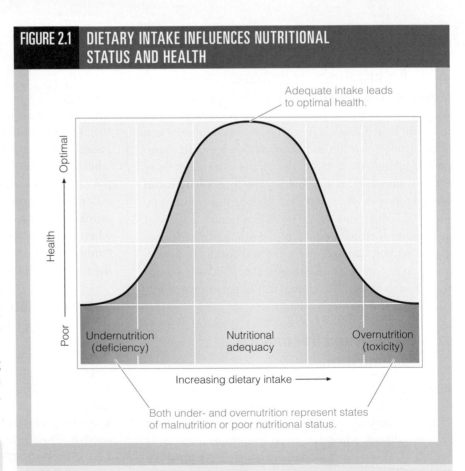

Adequate intake leads to optimal health.

Undernutrition (deficiency)

Nutritional adequacy

Overnutrition (toxicity)

Increasing dietary intake ⟶

Both under- and overnutrition represent states of malnutrition or poor nutritional status.

Both under- and overnutrition can result in malnutrition and poor health over time.

HOW IS NUTRITIONAL STATUS ASSESSED?

Because adequate nutrition is required for optimal health, it is important for health care providers to be able to assess a person's nutritional status. This is especially critical during periods of growth and development, such as infancy, when nutrient requirements are high and poor nutrition can have long-lasting consequences. In general, there are four ways by which nutritional status can be assessed. These tools are sometimes referred to as the ABCDs of nutritional assessment:

- **A**nthropometric measurement,
- **B**iochemical measurement,
- **C**linical assessment, and
- **D**ietary assessment.

Although each of these nutritional assessment methods can provide some information about nutritional status, each one by itself cannot tell you everything. It is therefore important to use multiple approaches when determining nutritional adequacy.

2-2a Anthropometric Measurement

An **anthropometric measurement** assesses the physical dimensions, such as height, or composition, such as fat mass of the body. Derived from Greek roots, the term *anthropometry* means literally "to measure the human body." Because most anthropometric measurements are easy and inexpensive to conduct, anthropometry is routinely used in clinical and research environments. Even stepping on the scale to weigh yourself is an example of an anthropometric measurement. Anthropometric measurement cannot confirm deficiency of any particular essential nutrient, but it can give a clinician clues that nutritional inadequacies might be present.

Height, Weight, and Circumference Because obesity can increase a person's risk of developing chronic degenerative diseases such as heart disease and type 2 diabetes, height and body weight are frequently used anthropometric measures. These measures can also provide important information about the progression of other diseases. For example, a loss of height might indicate a decline in bone density, and a significant loss of body weight might indicate an eating disorder. Height and weight are also commonly used to assess nutritional

status during infancy, childhood, and pregnancy. Various circumferences, such as those of the waist, hips, and head, are also sometimes measured to assess health. Differences in waist and hip circumferences reflect variations in body fat distribution patterns, and head circumference is frequently measured to monitor brain growth during infancy.

Body Composition Estimates of **body composition**—the proportions of fat, water, lean tissue, and mineral (bone) mass that make up the body—are also anthropometric measurements. The amount and distribution of these components can important indicators of one's nutritional status and overall health. For instance, adequate hydration status (water content) is important for optimal athletic performance; alterations in lean tissue can indicate advanced disease in cancer patients; too much body fat can lead to cardiovascular disease; and loss of bone mass is a major risk factor for osteoporosis. Many campus recreation centers and health clinics offer free body composition testing. Knowing this information about yourself might be useful as you learn more about nutrition and health.

2-2b Biochemical Measurement

Because anthropometric measurements are not diagnostic (i.e., they cannot be used to diagnose a problem definitively), they must be supported by other measures of nutritional status. To assess health and nutritional status beyond anthropometric measurements, biochemical measurements are often used. A **biochemical measurement** is a laboratory analysis of a biological sample such as blood or

> **anthropometric measurement** A measurement of the body's physical dimensions or composition.
>
> **body composition** The proportions of fat, water, lean tissue, and mineral (bone) mass that make up the body.
>
> **biochemical measurement** Laboratory analysis of a biological sample, such as blood or urine.

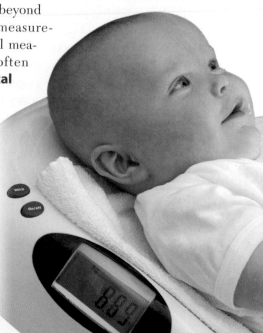

Body weight is one example of a simple anthropometric measurement.

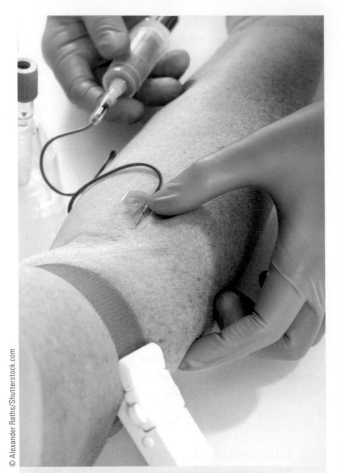

Blood is often drawn and analyzed to assess biochemical measurements of nutritional status.

Many nutrient deficiencies have distinct signs that can be seen by a trained eye. The disfiguration of this fingernail signals a vitamin B deficiency. Note, however, that fingernail disfiguration can also be caused by other nutrient deficiencies and conditions.

urine. In some cases, the sample is analyzed for a specific nutrient or a substance related to the nutrient's function. For example, although the amount of iron in the blood can be measured, a more commonly used method to assess iron status is measuring the amount of hemoglobin (an iron-containing protein that carries oxygen) in the blood. The percent of blood made up of red blood cells (hematocrit) is another commonly used test to screen for iron deficiency. Biochemical measurements are integral to nutritional assessment because they can help diagnose a specific nutrient deficiency or excess. However, because the collection and analysis of biological samples often necessitate technical expertise and sometimes costly procedures, biochemical analyses are often done only when malnutrition is already suspected.

sign A physical indicator of disease that can be seen by others, such as pale skin and skin rashes.

symptom A subjective manifestation of disease that generally cannot be observed by other people.

2-2c Clinical Assessment

Another way to evaluate nutritional status is to conduct a face-to-face assessment. During a clinical assessment, a clinician may ask questions about whether a patient has had previous diseases, unusual weight loss or weight gain, surgeries, or is taking any medications. The clinician may also ask about other relevant information such as family history. This process, called *taking a medical history*, can be helpful in determining one's overall health and risk for disease. A clinician may also take anthropometric measurements. During a clinical assessment, the clinician will likely note each visible or measurable **sign** of illness—a physical indicator of disease that can be observed or assessed objectively by someone else. For example, signs of an iron deficiency might include pale skin and shortness of breath. Other observable signs that may indicate poor nutritional status include skin rashes and swollen ankles (edema). These signs may suggest vitamin B and protein deficiencies, respectively.

Patients may also be asked during a clinical assessment if they are experiencing anything unusual in regard to their health. This enables the clinician to gather information about symptoms. A symptom is different from a sign in that a **symptom** is a subjective manifestation of disease that generally cannot be observed by other people. For example, fatigue is a symptom commonly associated with iron deficiency, and loss of appetite is a symptom of zinc toxicity. Because they generally go unnoticed by clinicians, symptoms must be reported by the patient. Although nutrient deficiencies caused by primary malnutrition were once common in the United States, deficiencies are more likely to be caused by secondary malnutrition today. This is why it is important for a clinician to gather as much information as possible about a person's health prior to treating a nutritional deficiency.

Clinical assessment has many advantages over other forms of nutritional assessment. For one, it is the only way that health care providers can determine if patients are experiencing symptoms of malnutrition. Furthermore, physical signs associated with extreme forms of malnutrition can be very distinct, and observing them can make diagnosis of a particular nutrient deficiency or toxicity easier.

2-2d Dietary Assessment

Although anthropometric, biochemical, and clinical assessments can be used to assess some aspects of nutritional status, it is also very important to evaluate the adequacy of one's dietary intake. This is called **dietary assessment**. There are three general types of dietary assessment. The first, **diet recall**, requires a person to remember, record, and evaluate every food and drink consumed over a given time span—typically 24 hours. Because diet recalls are generally based on a single day and require a person to remember information, they often are not representative of overall dietary patterns. The second type of dietary assessment, the **food frequency questionnaire**, assesses food selection patterns over an extended period of time. For example, a food frequency questionnaire used to determine overall patterns of fruit intake would ask questions about which fruits the patient typically eats and how much she normally consumes in a serving. Because it does not assess nutrient intake (only food intake patterns), information from a food frequency questionnaire is limited in accuracy and completeness.

Although diet recalls and food frequency questionnaires are relatively simple to conduct, their usefulness depends on a person's memory. To assess nutrient intake more accurately, it is best to keep a list of all foods and beverages consumed over a given period of time. This method, called a **diet record** (or **food record**) requires a person to either weigh all of the food and beverages consumed or estimate portion sizes. Portion sizes can be estimated using standard household measurements (cups, scales, etc.).

Clinical assessments are valuable in nutritional assessment because they can uncover signs and symptoms of malnutrition.

© Denise Kappa/Shutterstock.com

© Stuart Jenner/Shutterstock.com

dietary assessment The evaluation of adequacy of a person's dietary intake.

diet recall A retrospective dietary assessment method whereby a person records and analyzes every food and drink consumed over a given time span.

food frequency questionnaire A retrospective dietary assessment method whereby food selection patterns are assessed over an extended period of time.

diet record (or **food record**) A prospective dietary assessment method whereby a person records and analyzes every food and drink as it is consumed over a given time span.

To get an accurate assessment of nutrient intake, food records should be kept for at least three days—ideally two weekdays and one weekend day. When the diet record is complete, its information can be analyzed to estimate nutrient intake. Though the diet record method is usually more time- and labor-intensive than diet recall and the food frequency questionnaire, it is the preferred method of dietary assessment because of its accuracy.

Food Composition Tables and Dietary Analysis Software After completing a diet record, the next step is to analyze the information to estimate both nutrient and energy (calorie) intakes. There are two sources of information concerning the nutrient composition of foods: food composition tables and computerized nutrient databases. Food composition tables can be purchased or accessed free of charge on the website of the U.S. Department of Agriculture (USDA; http://www.ars.usda.gov/nutrientdata). Using printed food composition tables to calculate nutrient intake can be tedious. Fortunately, easy-to-use computerized nutrient databases are also available. For example, an online dietary analysis tool accompanies the MyPlate food guidance system. The MyPlate system is described in more detail later in this chapter.

Although other aspects of nutritional assessment (anthropometric, biochemical, and clinical assessment) are important, dietary assessment is one of the most effective methods to assess the adequacy of a person's diet. However, after you have determined how much of each nutrient you consume on a regular basis, how can you know if your intake is adequate? To answer this question, you must refer to a variety of dietary intake reference standards and recommendations, several of which are described next.

2-3 HOW MUCH OF A NUTRIENT IS ADEQUATE?

Dietary Reference Intakes (DRIs) A set of four dietary reference standards used to assess and plan dietary intake: Estimated Average Requirement, Recommended Dietary Allowance, Adequate Intake level, and Tolerable Upper Intake Level.

nutrient requirement The amount of a nutrient that a person must consume to promote optimal health.

Recall that different people require different amounts of nutrients depending on their sex, age, and other variables. Because individuals' bodies and activities are so diverse, there is no simple or definitive way to know what a person's nutrient needs actually are. However, to help assess dietary adequacy, the

Institute of Medicine developed a set of usable nutritional standards. These standards can help people judge whether their typical dietary intake is likely to provide too little, too much, or the right amount of the essential micronutrients, macronutrients, and calories.

2-3a Dietary Reference Intakes

Because humans are all so different, establishing concrete dietary recommendations proved an enormous hurdle for the scientific community. Indeed, the task required the input and analysis of an assembly of researchers organized by the Institute of Medicine, a division of the National Academy of Sciences. In 1994, the Institute of Medicine began developing the **Dietary Reference Intakes (DRIs)**, a set of four dietary assessment standards used to assess and plan dietary intake. The four reference values that comprise the DRIs are as follows.

- Estimated Average Requirement (EAR)
- Recommended Dietary Allowance (RDA)
- Adequate Intake (AI) level
- Tolerable Upper Intake Level (UL)

These four dietary assessment standards are illustrated in Figure 2.2, outlined in Table 2.1 and described below. Using all four types of DRI values provides a comprehensive assessment of the nutritional adequacy of individuals and populations. To utilize the DRIs, it is critical to understand some fundamental concepts underlying them. For example, because nutrient requirements differ by sex and stage of life, the DRIs provide different values for different groups of people. DRI values for the various life-stage groups take into account both age and physiologic condition, such as pregnancy and lactation. In all, there are 16 life-stage groups for females and 10 life-stage groups for males. When utilizing the DRIs, it is important to be careful to use the correct values for one's appropriate sex and life-stage group. Using incorrect values may lead to misinterpretation of intake adequacy.

It is also important to understand the term *nutrient requirement*. A **nutrient requirement** is the amount of a nutrient that a person must consume to promote optimal health. Individuals each have their own unique nutrient requirements. In general, the nutrient requirements of a population are distributed in a bell-shaped curve meaning that the vast majority of people have requirements at some mid-level amount, while equally few require much less and much more. Many

FIGURE 2.2   DIETARY REFERENCE INTAKE STANDARDS

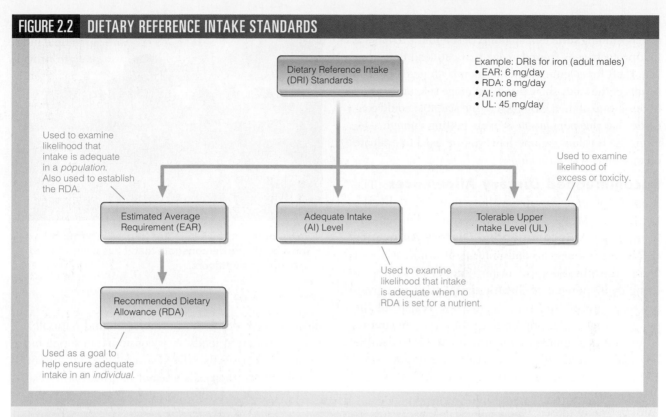

There are four sets of DRI reference values: Estimated Average Requirement, Recommended Dietary Allowance, Adequate Intake Level, and Tolerable Upper Intake Level.

factors influence a person's nutrient requirements—these factors are partially accounted for in the DRI sex and life-stage groups. However, there are many other factors, such as genetics, certain diseases, medications, lifestyle practices (e.g., smoking, exercise, alcohol consumption), and other environmental influences that can also influence nutrient requirements. For instance, tobacco use can increase one's vitamin C requirement.[1] Because of these other factors, it is virtually impossible to know how much of each essential nutrient a person actually needs; the DRIs are only *estimates* of nutrient requirements and intake goals in a healthy population. Your personal level may actually be less or more.

Nonetheless, the DRI values can be powerful tools for assessing your nutritional status and helping you plan a healthy diet. How can you use the DRIs for these purposes? The answer to this question requires a basic understanding of the standards that comprise the DRIs. In the following sections, you will learn about each of the four main dietary reference standards in detail and how to use them to determine the adequacy of your diet; in other words if you are getting too little or too much of each important nutrient.

Estimated Average Requirements When the Institute of Medicine first set out to establish nutrient requirements in the 1990s and early 2000s, it developed Estimated Average Requirements for each essential micronutrient and macronutrient. An **Estimated Average Requirement (EAR)** value represents the daily intake of a nutrient that meets the physiological requirements of half the healthy individuals in a given life-stage group and sex (see Figure 2.3). For example, the EAR for iron in breastfeeding women is 6.5 milligrams/day, meaning that half of the women in this life-stage group require less than 6.5 milligrams/day iron, and half need more. It is important to an EAR value meets or exceeds the needs of half the population, but is inadequate for the other half. Also, EAR values are only available for nutrients for which there is sufficient science-based information.

EARs are useful in research and public health settings to evaluate whether population groups are likely consuming an adequate amount of a nutrient. For instance, say a researcher

Estimated Average Requirement (EAR)
The daily intake of a nutrient that meets the physiological requirements of half the healthy individuals in a given age, physiologic state, and sex.

conducts a study to determine whether American men are consuming sufficient amounts of selenium, a mineral that protects cells from damage that can lead to cancer. The EAR for selenium in adult men is 45 µg/day. If the results of the study show that the average selenium intake in this group of men is 50 µg/day, the scientist could conclude that this *population* is likely getting enough selenium. EAR values cannot, however, be used to evaluate dietary intakes of *individuals*.

Recommended Dietary Allowances The second set of reference values developed by the Institute of Medicine is the Recommended Dietary Allowances. A **Recommended Dietary Allowance (RDA)** represents the daily intake of a nutrient that meets the physiological requirements of nearly all (roughly 97 percent of) healthy individuals in a given life-stage group and sex. In other words, if all the people in a certain population group are consuming a nutrient in amounts that approximate or exceed the RDA, 97 percent will have satisfied their requirement. RDA values are derived directly from the EARs; Figure 2.3 illustrates the relationship between these two sets of standards. In contrast to the population-oriented EARs, however, the RDAs can be utilized as nutrient-intake goals for individuals. This is because each RDA has a built-in safety margin that ensures adequate intake by 97 percent of the population if recommended levels are consumed.

Unless specifically noted, an RDA value does not

To claim that a chicken is "free range," a farmer must be able to demonstrate that it has been allowed access to the outdoors.

distinguish whether the nutrient is found naturally in foods, is added to foods, or is consumed in supplement form. Also, like EARs, RDAs are available only for nutrients for which there is sufficient information. Because

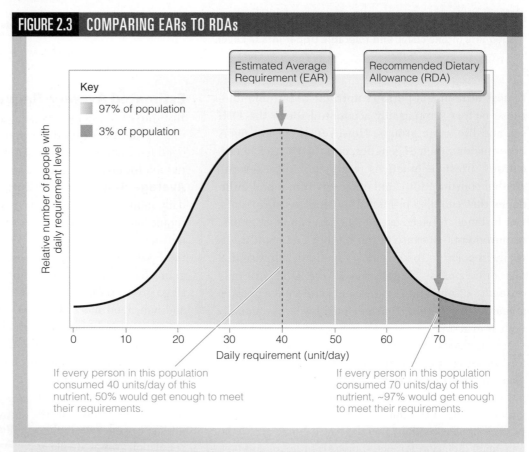

FIGURE 2.3 COMPARING EARs TO RDAs

Key
- 97% of population
- 3% of population

Estimated Average Requirement (EAR)

Recommended Dietary Allowance (RDA)

Relative number of people with daily requirement level

Daily requirement (unit/day)

If every person in this population consumed 40 units/day of this nutrient, 50% would get enough to meet their requirements.

If every person in this population consumed 70 units/day of this nutrient, ~97% would get enough to meet their requirements.

EAR values represent average requirements for a given population, whereas RDAs are intake goals for individuals.

Recommended Dietary Allowance (RDA) The daily intake of a nutrient that meets the physiological requirements of nearly all (roughly 97 percent) healthy individuals in a given life-stage and sex.

of this, another set of standards was developed to provide provisional intake goals for nutrients without EARs or RDAs until more rigorous scientific studies are available. These standards are called Adequate Intake levels.

Adequate Intake Levels When scientific evidence was insufficient to establish an EAR (and thus accurately set an RDA), the DRI committee developed an **Adequate Intake (AI)** level instead. As such, if a nutrient has an AI instead of an RDA, you can safely assume that more research is needed before a conclusive dietary intake goal can be set. An example of a nutrient with an AI instead of an RDA is potassium. This may surprise you, as adequate potassium intake is essential to heart, brain, kidney, and muscle function. More research is required before EARs and RDAs can be established for this mineral. Sometimes, further research is impossible. For example, because rigorous studies cannot ethically be conducted on infants younger than six months of age, there are no RDAs for this life-stage group—only AIs.

Like an RDA, an AI represents a daily nutrient intake goal for healthy individuals. Unlike the RDAs, which are based on rigorous scientific studies, the AIs are based on intake levels that appear to support adequate nutritional status. Continuing the above example, AIs for infants were developed by measuring average daily nutrient intakes by breastfed babies. Because breastfeeding is considered the ideal way to feed infants, these nutrient intakes are assumed to be adequate. You can learn

The Adequate Intake (AI) values for babies are based on average nutrient intakes of breastfed infants.

The Tolerable Upper Intake Levels (ULs) were developed to help consumers know how much of a nutrient supplement is safe to consume.

which nutrients have AI values and which have RDA values for each life-stage group by examining the DRI tables on the Dietary Reference Intakes card located at the back of the book as well as in Table 2.1.

Together, RDA and AI values can be used to set goals to help you consume nutrients in sufficient quantities, allowing you to maintain nutritional health. Conversely, another set of values can help you avoid the consumption of nutrients in such large quantities that they can actually do you harm. These values are called Tolerable Upper Intake Levels.

Tolerable Upper Intake Levels The RDA and AI values have been established to prevent deficiencies and optimize health. However, avoiding nutrient overconsumption and toxicity is also important. To help avoid such excesses, a **Tolerable Upper Intake Level (UL)** is established as the highest level of usual daily nutrient intake likely to be safe. For instance, the UL for selenium in men is 400 µg/day. This means that for a typical college-aged male, consuming 400 µg of this mineral a day is not likely to cause problems, but exceeding this amount might lead to selenium toxicity. The ULs are not meant to be used as goals for dietary intake. Instead, they provide limits for those who take supplements or consume large amounts of nutrient-fortified foods.

2-3b Using EARs, RDAs, AIs, and ULs to Assess Nutrient Intake

The four sets of standards that comprise the DRIs should be used in coordination to assess nutrient intake. Although this process may seem complex at first, relatively simple guidelines can be used to make inferences about your diet—simply compare the results of your dietary

Adequate Intake (AI) The daily intake of a nutrient that appears to support adequate nutritional status; established when RDAs cannot be determined.

Tolerable Upper Intake Level (UL) The highest level of usual daily nutrient intake likely to be safe.

TABLE 2.1 AVAILABLE DIETARY REFERENCE INTAKE (DRI) STANDARDS FOR ADULTS

Nutrient	EAR[b]	RDA[c]	AI[d]	UL[e]
Macronutrients[a]				
Water			●	
Linoleic acid			●	
Linolenic acid			●	
Carbohydrates	●	●		
Fats				
Protein	●	●		
Vitamins				
Thiamin	●	●		
Riboflavin	●	●		
Niacin	●	●		●
Biotin			●	
Pantothenic Acid			●	
Vitamin B$_6$	●	●		●
Folate	●	●		●
Vitamin B$_{12}$	●	●		
Vitamin C	●	●		●
Choline			●	●
Vitamin A	●	●		●

Nutrient	EAR[b]	RDA[c]	AI[d]	UL[e]
Vitamin D	●	●		●
Vitamin E	●	●		●
Vitamin K			●	
Minerals				
Sodium			●	●
Chloride			●	●
Potassium			●	
Calcium	●	●		●
Phosphorus	●	●		●
Magnesium	●	●		●
Iron	●	●		●
Zinc	●	●		●
Iodine	●	●		●
Selenium	●	●		●
Copper	●	●		●
Manganese			●	●
Fluoride			●	●
Chromium			●	
Molybdenum	●	●		●

[a]Note that there are no DRIs for energy, per se. Instead, you can estimate your caloric needs by using the Estimated Energy Requirement (EER) calculations.
[b]Estimated Average Requirement.
[c]Recommended Dietary Allowance.
[d]Adequate Intake.
[e]Tolerable Upper Intake Level.

assessment to your EAR, RDA, AI, and UL values.[2] These concepts are illustrated in Figure 2.4 and summarized as follows.

- If your intake of a nutrient is much less than the EAR, then it is likely to be inadequate, increasing your risk of nutrient deficiency.
- If your intake is between the EAR and the RDA, then you may need to increase your intake.
- If your intake is between the RDA and the UL, then it is probably adequate.
- If your intake is above the UL, then it is probably too high.

- If your intake of a nutrient falls between the AI and UL, then it is probably adequate.
- If your intake is below your AI, no conclusion can be made about the adequacy of your intake.

Consider a 20-year-old female who, upon completing a food record and dietary assessment, learns that her vitamin A intake is 1,500 µg/day. Because this value falls between the RDA (700 µg/day) and the UL (3,000 µg/day), her vitamin A intake is probably adequate. Had her vitamin A intake been 600 µg day, it would have been between her EAR (500 µg/day) and RDA (700 µg/day), indicating that she might benefit from consuming more. Had it been 3,500 μg/day (above her UL), then she could conclude that her intake of vitamin A is too high.

2-3c Energy Intake Can Also Be Assessed

So far, you have learned about using the DRIs to assess your nutritional status in terms of micronutrients and macronutrients. However, it is also important to determine whether you are consuming the right amount of calories and whether the sources of these calories are balanced. To address this issue, the Institute of Medicine developed two types of standards that you can use to assess your energy intake: Estimated Energy Requirements and the Acceptable Macronutrient Distribution Ranges.[3]

Estimated Energy Requirements The Estimated Energy Requirements are similar in theory and application to the EARs. An **Estimated Energy Requirement (EER)** value represents the average energy intakes needed for a person to maintain current weight. Unlike the reference intakes for nutrients that are based on a person's sex and life-stage, the EERs not only consider these factors but also take into account body size (weight and height) and physical activity level.

EERs are calculated using relatively simple mathematical equations. Using the EER equations for adult men and women of healthy weight below, you can calculate your own EER in kilocalories per day (kcal/day). Additional EER equations for other life-stage groups can be found on the Estimated Energy Requirement (EER) Calculations and Physical Activity (PA) Values card at the back of the book.

$$\text{Adult man: EER} = 662 - (9.53 \times \text{age}) + \text{PA} \times (15.91 \times \text{weight} + 539.6 \times \text{height})$$

$$\text{Adult woman: EER} = 354 - (6.91 \times \text{age}) + \text{PA} \times (9.36 \times \text{weight} + 726 \times \text{height})$$

To solve these equations, you must insert your age in years, your weight in kilograms (kg), and your height in meters (m). The abbreviation *PA* indicates a physical activity value that is based on whether a person is sedentary, low active, active, or

> **Estimated Energy Requirement (EER)** The average energy intake needed for a healthy person to maintain weight.

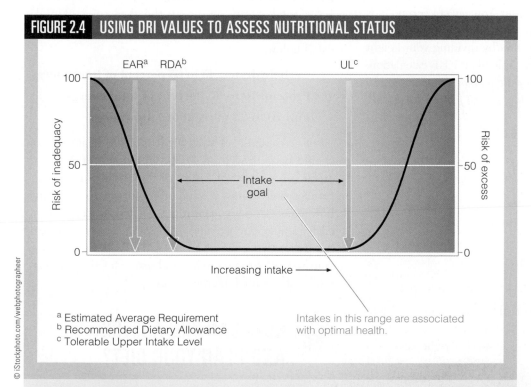

FIGURE 2.4 USING DRI VALUES TO ASSESS NUTRITIONAL STATUS

EAR[a]   RDA[b]   UL[c]

Risk of inadequacy

Risk of excess

Intake goal

Increasing intake →

Intakes in this range are associated with optimal health.

[a] Estimated Average Requirement
[b] Recommended Dietary Allowance
[c] Tolerable Upper Intake Level

> Calculating your Estimated Energy Requirement (EER) is a simple way for you to approximate how many calories to consume daily to maintain your current weight.

By determining your nutrient intake values and comparing them to the appropriate DRI values, you can easily determine whether you are likely getting too little, too much, or just the right amount of a given nutrient.

© iStockphoto.com/webphotographeer

TABLE 2.2 PHYSICAL ACTIVITY (PA) CATEGORIES AND VALUES[a]

Activity Level	Physical Activity (PA) Value — Men	Physical Activity (PA) Value — Women	Description
Sedentary	1.00	1.00	No physical activity aside from that needed for independent living.
Low Active	1.11	1.12	1.5 to 3 miles/day at 2 to 4 miles/hour in addition to the light activity associated with typical day-to-day life.
Active	1.25	1.27	3 to 10 miles/day at 2 to 4 miles/hour in addition to the light activity associated with typical day-to-day life.
Very Active	1.48	1.45	10 or more miles/day at 2 to 4 miles/hour in addition to the light activity associated with typical day-to-day life.

[a]These values only apply to healthy-weight, nonpregnant, nonlactating adults. Values for children, pregnant or lactating women, and overweight or obese individuals are different.

Source: Institute of Medicine. Dietary Reference Intakes for energy, carbohydrate, fiber, fat, fatty acids, cholesterol, protein, and amino acids. Washington, DC: National Academies Press; 2005.

very active. The lower your PA value, the less active you are, and consequently the lower your EER is. Table 2.2 provides examples of these activity categories and their corresponding PA values. Note that at every age, active individuals need more energy than do their sedentary counterparts.

Calculating your EER is not difficult, as it requires only that you insert your age, PA value, weight, and height into the correct equation. You can convert pounds to kilograms by dividing your weight in pounds by 2.2. You can convert feet to meters by dividing your height in feet by 3.3. Here is an example of an EER calculation.

Kyung-Soon is a 38-year-old woman who weighs 115 pounds (52.3 kg), is 5 feet 4 inches (5.3 feet, or 1.6 m) tall, and has a low activity level.

Kyung-Soon's EER is calculated as follows.

$$EER = 354 - (6.91 \times age) + PA \times$$
$$(9.36 \times weight + 726 \times height)$$
$$= 354 - (6.91 \times 38) + 1.12 \times$$
$$(9.36 \times 52.3 + 726 \times 1.6)$$
$$= 354 - 262.6 + 1.12 \times (489.5 + 1161.6)$$
$$= 91.4 + 1.12 \times 1651.1$$
$$= 1,941 \ kcal/day$$

Take the time right now to determine your own EER. Knowing how much energy you require on a daily basis can serve as a guide to help establish the right balance of energy in your diet.

Acceptable Macronutrient Distribution Range (AMDR)
The recommended range of intake for a given energy-yielding nutrient, expressed as a percentage of total daily caloric intake.

Acceptable Macronutrient Distribution Ranges After you establish whether you are consuming the right amount of calories, you must determine whether your distribution of energy sources—carbohydrates, proteins, and fats—is healthy. To answer this question, the Acceptable Macronutrient Distribution Ranges were developed. An **Acceptable Macronutrient Distribution Range (AMDR)**, expressed as a percentage of total energy, represents the ideal range of intake for a given class of energy-yielding nutrient, and are listed below.

Carbohydrates: 45 to 65 percent of total energy

Protein: 10 to 35 percent of total energy

Fat: 20 to 35 percent of total energy

For example, if your EER is 2,400 kcal/day, you should be getting from 1,080 (45 percent of 2,400) to 1,560 (65 percent of 2,400) of those daily kilocalories from carbohydrates. It is important for you to determine how many calories you currently get and should get from each group of energy-yielding macronutrients. Maintaining a diet that adheres to the AMDRs both decreases your risk of chronic disease and promotes the adequate intake of essential micronutrients.

2-4 HOW CAN YOU ASSESS AND PLAN YOUR DIET?

If you complete a dietary assessment using a food record and investigate your DRI, EER, and AMDR values, you can determine whether your nutrient and energy intakes are likely adequate. However, because this process can be rather cumbersome, it is usually only done in a research environment or when nutrient inadequacies are suspected. Luckily, several additional process-simplifying tools are available. A number of reputable organizations have streamlined dietary assessment by summarizing what a healthy diet generally "looks like." In other words, they have attempted to answer the question "What kinds of foods should I consume, and how much of them should I eat?" by formulating general nutritional guidelines such as the Dietary Guidelines for Americans and the MyPlate food guidance system.

2-4a Food Guidance Systems

Several government agencies are involved in assuring the United States' health. In fact, some of these agencies have been collecting and providing information about nutrition and health for more than a century. One such agency is the United States Department of Agriculture (USDA), which published its first set of nutritional recommendations for Americans in 1894. The USDA has continued to publish recommendations for the American public since that time. In fact, over the past 100 years, a succession of federally supported recommendations—all designed to provide guidelines for dietary planning—have been published.[4] Two of these publications are shown in Figure 2.5. Each such **USDA Food Pattern** publication (formerly called USDA Food Guide) categorizes nutritionally similar foods into *food groups* and makes recommendations regarding the number of servings of each food group that should be consumed daily.

You may already be aware of some of the USDA's food groups. For example, poultry, eggs, meat, and seafood have typically been grouped together because they are all excellent sources of high-quality protein. However, the composition and number of food groups has changed over the years as science has evolved and socioeconomic times have changed. For instance, during the Great Depression, food was scarce and malnutrition was common. In response, the USDA shifted the focus of the Food Guide to recommend relatively inexpensive, high-fat foods such as milk, peanuts, and cheese. These foods provide maximal energy and nutrients at minimal cost. By contrast, when the *Basic 4* food groups were unveiled in 1956, the USDA Food Guide highlighted a bread/cereal group and a vegetable/fruit group, reflecting both a positive economic climate and growing evidence that these types of foods provided important vitamins and minerals to the diet. As you will soon learn, the current Food Patterns report has five food groups, each contributing an important set of nutrients.

In 1980, the U.S. Department of Health and Human Services (DHHS) and the USDA published an expanded dietary recommendation called the **Dietary Guidelines for Americans**. This new series of guidelines

USDA Food Pattern A USDA publication that categorizes nutritionally similar foods into food groups and makes recommendations regarding the number of servings of each food group that should be consumed daily.

Dietary Guidelines for Americans A series of recommendations that provide specific nutritional guidance and advice about physical activity, alcohol intake, and food safety.

FIGURE 2.5 DIETARY GUIDANCE HAS EVOLVED TREMENDOUSLY OVER THE YEARS

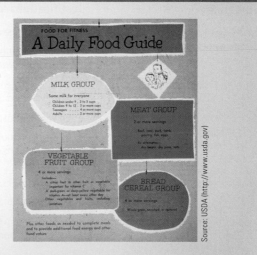

The U.S. Department of Agriculture has published numerous food-related recommendations over the past century. Here, Depression Era (left) and "Basic 4" (right) recommendations are expressed visually.

provided specific nutritional guidance and advice about physical activity, alcohol intake, and food safety. Since its introduction, the Dietary Guidelines for Americans have been revised every five years to reflect new scientific information. In addition to overarching advice concerning healthy eating, the Dietary Guidelines for Americans now incorporate the USDA Food Patterns. Indeed, such guidelines are designed to evolve as the understanding of nutrition and health does. You can follow these changes on the USDA website for years to come (http://www.usda.gov).

2-4b 2015 Dietary Guidelines for Americans

The 2015 Dietary Guidelines for Americans, the nation's current nutritional recommendations targeted primarily toward policymakers as well as nutrition and health professionals (not the general public), were developed by a highly esteemed committee of nutrition scientists. These experts systematically evaluated published scientific literature to formulate dietary recommendations to optimize the health of children and adults living in the United States. It is noteworthy that, like previous editions of the Dietary Guidelines, this version applies only to children 2 years and older. However, as recently mandated by federal law, beginning with the 2020 edition the Dietary Guidelines will expand to include infants and toddlers (from birth to 2 years of age), as well as additional guidance for women who are pregnant.

eating patterns (food patterns) A combination of foods and beverages that constitute a person's overall dietary intake.

Whereas previous editions of the Dietary Guidelines focused primarily on individual dietary components such as food groups and nutrients, the current version focuses more on **eating patterns** (also referred to as **food patterns**), the combinations of foods and beverages that constitute a person's complete dietary intake over time. This is because people do not compartmentalize their diets in terms of food groups and nutrients. Rather, people eat foods that collectively make up their eating patterns, which potentially have cumulative effects on health. As such, the committee stressed the importance of tailoring overall food patterns to an individual's personal preferences, enabling Americans to choose the diet that is right for them. The 2015 Dietary Guidelines document also delineates specific nutrients that are generally under-consumed in the U.S. population (nutrients of concern) and others that are too-often consumed in excess.

As summarized in Table 2.3, the 2015 edition of Dietary Guidelines provides five overarching guidelines that encourage healthy eating patterns and associated shifts in choices and behaviors to meet this goal. The fact that all segments of our society have a role to play in supporting healthy choices is also stressed. The guidelines also embody the idea that "a healthy eating pattern is not a rigid prescription, but rather, an adaptable framework in which individuals can enjoy foods that meet their personal, cultural, and traditional preferences

TABLE 2.3	2015 DIETARY GUIDELINES FOR AMERICANS
Guideline	**Justifications and Recommendations**
Follow a healthy eating pattern across the lifespan.	• All food and beverage choices matter. • Choose a healthy eating pattern at an appropriate calorie level to help achieve and maintain a healthy body weight, support nutrient adequacy, and reduce the risk of chronic disease.
Focus on variety, nutrient density, and amount.	• To meet nutrient needs within calorie limits, choose a variety of nutrient-dense foods across and within all food groups in recommended amounts.
Limit calories from added sugars and saturated fats and reduce sodium intake.	• Consume an eating pattern low in added sugars, saturated fats, and sodium. • Cut back on foods and beverages higher in these components to amounts that fit within healthy eating patterns.
Shift to healthier food and beverage choices.	• Choose nutrient-dense foods and beverages across and within all food groups in place of less healthy choices. • Consider cultural and personal preferences to make these shifts easier to accomplish and maintain.
Support healthy eating patterns for all.	• Everyone has a role in helping to create and support healthy eating patterns in multiple settings nationwide, from home to school to work to communities.

Source: Adapted from U.S. Department of Agriculture and U.S. Department of Health and Human Services. 2015–2020 Dietary Guidelines for Americans, 2015–2020. 8th edition. Washington, DC Government Printing Office. December 2015.

and that fit within their budgets."[5] Furthermore, they emphasize an underlying premise that nutritional needs should be met primarily from foods, and beverages (not supplements) and that all forms of foods, including fresh, canned, dried, and frozen, can be included in healthy eating patterns. The five guidelines and examples of healthy eating patterns that translate and integrate the recommendations into overall healthy ways to eat are summarized in the following sections.

Guideline 1: Follow a Healthy Eating Pattern Across the Lifespan

The 2015 Dietary Guidelines recommend following a healthy eating pattern throughout life. This includes taking into account all food and beverage choices, because both provide energy and nutrients. Choosing a healthy eating pattern at an appropriate calorie level is emphasized to help achieve and maintain healthy body weight, support nutrient adequacy, and reduce the risk of chronic disease. But, it is equally important to recognize that there is no "one-size-fits-all" construct when it comes to what is a healthy eating pattern. Indeed, we all have different tastes when it comes to foods—many of which are based in our cultural and family traditions. Recognizing this, the 2015 Dietary Guidelines provide three examples of such food patterns: the Healthy U.S.-Style Eating Pattern, the Healthy Mediterranean-Style Eating Pattern, and the Healthy Vegetarian Eating Pattern. Details about these eating patterns are provided later in this chapter.

In addition to consuming a healthy eating pattern, Americans are encouraged to meet the Physical Activity Guidelines for Americans which are described in detail in Chapter 11 of this book. Briefly, these guidelines urge adults to participate in at least 150 minutes of moderate-intensity physical activity weekly and perform muscle-strengthening exercises on two or more days each week. Following these guidelines is important because regular physical activity helps people maintain and/or achieve a healthy weight.

Guideline 2: Focus on Variety, Nutrient Density, and Amount

Although the focus of the 2015 Dietary Guidelines is on dietary patterns, not individual foods, you may be wondering what types of foods should be emphasized in a healthy eating pattern. To answer this question, it is recommended that you strive to meet the following goals related to what types of foods to include:

- A variety of vegetables from all of the subgroups: dark green, red and orange, legumes (beans and peas), and starchy

- A variety of fruits, especially whole fruits
- A variety of grains, at least half of which should be whole grains
- Fat-free or low-fat dairy, including milk, yogurt, cheese, and/or fortified soy beverages
- A variety of protein foods, including seafood, lean meats and poultry, eggs, legumes, nuts, seeds, and soy products
- Healthy oils

Important to these recommendations are the concepts of **moderation**, consuming foods in reasonable amounts so as to not consume an excess of calories; **variety**, consuming different foods within a particular food group (e.g., vegetables) and across food groups, and **nutrient density**, the relative ratio of a particular nutrient in a food to its total calories. Nutrient-dense foods and beverages contain many beneficial compounds, such as vitamins and minerals, and fewer unhealthy ones, such as saturated fats and added sugars, relative to their caloric values. Ideally, nutrient-dense foods are those that retain their naturally occurring components, such as dietary fiber. Unless they are altered during preparation, all vegetables, fruits, whole grains, seafood, eggs, beans, peas, unsalted nuts, seeds, lean meats, fat-free and low-fat milk and milk products, and poultry are nutrient-dense foods. Faithfully choosing a food pattern that consists of mostly nutrient-dense foods means that an occasional indulgence such as ice cream will not dramatically impact a person's health or body weight. Focusing on nutrient-dense food patterns also helps to ensure that a person is not consuming an excess of calories, and that essential nutrients that are important to health are abundant.

Guideline 3: Limit Calories from Added Sugars, Saturated Fats, and Alcohol; Reduce Sodium

In addition to choosing moderate amounts of healthy foods, there are other foods and food components that the 2015 Dietary Guidelines recommend limiting (Figure 2.6). These include saturated fatty acids, *trans* fatty acids, added sugars, sodium, and alcohol. This is because excess consumption of these food components and beverages is associated with increased risks for cardiovascular disease, type 2 diabetes, some

moderation A nutrition basic related to choosing overall serving sizes that fit within caloric needs.

variety A nutrition basic related to consuming different types of foods and beverages within a single food group and across food groups.

nutrient density The relative ratio of a food's amount of nutrients to its total calories.

FIGURE 2.6   KEY POINTS FROM THE 2015 DIETARY GUIDELINES FOR AMERICANS

A healthy eating pattern includes:

Fruits

Vegetables

Protein

Dairy

Grains

Oils

A healthy eating pattern limits:

Limit

The 2015 Dietary Guidelines for Americans stress the importance of consuming moderate portions from each of the food groups and limiting consumption of saturated and *trans* fats, added sugars, and sodium.

Source: U.S. Department of Agriculture and U.S. Department of Health and Human Services. 2015–2020 Dietary Guidelines for Americans, 2015–2020. 8th edition. Washington, DC Government Printing Office. December 2015.

forms of cancer, and obesity. The following specific recommendations should be adhered to:

- Saturated fats: Strive to have less than 10 percent of calories come from saturated fats.

- *Trans* fats: Consume as little industrially produced *trans* fats (from partially hydrogenated oils) as possible; naturally occurring *trans* fats (from beef and dairy) do not need to be limited.

- Sodium: Consume less than 2,300 milligrams per day.

- Added sugars: Should constitute less than 10 percent of calories consumed.

- Alcohol: If consumed, it should be in moderation, which is defined as up to one drink per day for women and up to two drinks per day for men, and only by adults of legal drinking age.

It is noteworthy that, unlike the 2010 Dietary Guidelines, which recommended limiting consumption of dietary cholesterol to 300 milligrams/day, the 2015 edition does not make this suggestion. This change is because the weight of scientific evidence no longer supports a causal relationship between cholesterol consumption, per se, and health. Nonetheless, because many foods (e.g., high-fat dairy and beef) that contain cholesterol also tend to contain high levels of saturated fats, current recommendations for a healthy dietary pattern continue to be somewhat low in cholesterol. A few foods—notably egg yolks and some shellfish—are rich in cholesterol but not saturated fats, however. As such, these foods can be consumed along with a variety of other protein-rich food choices without concern about excessive cholesterol intake.

Special Recommendations about Alcohol

Alcoholic beverages may contain calories from both alcohol and other ingredients. So, if consumed, their caloric contributions to overall energy intake should be within the various limits of a healthy eating pattern. Otherwise, alcohol consumption can easily lead to weight gain and, eventually, obesity. The 2015 Dietary Guidelines strongly recommend that women who are capable of becoming or trying to become pregnant or are breastfeeding should completely avoid consuming alcohol. And, those who choose to drink alcohol should be cautious about mixing caffeine and alcohol together or consuming them at the same time. In any case, it is not recommended that individuals begin drinking alcohol, or drink more, for any reason, and if alcohol is consumed, it should be used in moderation. Moderate alcohol consumption is defined as up to one drink per day for women and up to two drinks per day for men, and only by adults of legal drinking age. One alcoholic drink is defined as containing 14 grams (0.6 fluid ounces) of pure alcohol: 12 fluid ounces of regular beer, 5 fluid ounces of wine, or 1.5 fluid ounces of 80-proof spirits.

Guideline 4: Shift to Healthier Food and Beverage Choices

The third guideline put forth in the 2015 Dietary Guidelines describes shifts that are needed to align current dietary intakes common in the United States to recommendations. This is because most Americans would benefit from shifting food choices both within and across food groups and from current food choices to more nutrient-dense choices. For instance,

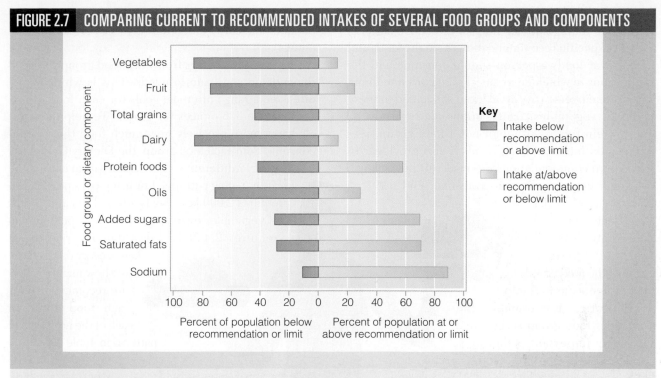

Source: U.S. Department of Agriculture and U.S. Department of Health and Human Services. 2015–2020 Dietary Guidelines for Americans, 2015–2020. 8th edition. Washington, DC Government Printing Office. December 2015.

about three-fourths of the population consumes too few vegetables, fruits, dairy, and healthy oils. Conversely, most Americans exceed the recommendations for added sugars, saturated fats, and sodium. These differences between current and recommended intakes are shown in Figure 2.7. The key to making successful shifts in this regard is recognizing the many cultural and personal preferences that should be taken into account. Additionally, making changes to eating patterns can be overwhelming. But it is important to remember that every food choice is an opportunity to move toward a healthy eating pattern, and small shifts in these choices—over a course of a week, day, or even a meal—can make a big difference. For instance, choosing low-sodium pinto beans over those containing substantial amounts of added salt could be one change. Other examples would be choosing low-fat over full-fat milk, or lean ground beef over fattier versions. Another small change might include increasing the variety of protein food choices, which in turn adds to a person's nutrient-dense food profile.

Guideline 5: Support Healthy Eating Patterns for All

The final guideline in the 2015 edition of the Dietary Guidelines recognizes that "everyone has a role in helping to create and support healthy eating patterns." For instance, school policies designed to enhance the school food setting can lead to improvements in what children purchase and eat while in school. Indeed, almost all aspects of our society, including federal and local governments, educational institutions, and private business, can impact the ease at which we can meet our dietary and physical activity goals. Among the many strategies and suggestions made in this regard are the following:

- Foster partnerships among food producers, suppliers, and retailers to increase access to foods that align with healthy food patterns.
- Promote development and availability of foods that align with healthy food patterns.
- Identify and support policies and programs that promote healthy eating and physical activity patterns.
- Encourage participation in physical activity programs offered in various settings.
- Provide nutrition assistance programs that support education and promotional activities tailored to the needs of the community.

2-4c Healthy Eating Patterns, and Food Groups

As mentioned previously, the Dietary Guidelines provide three examples of healthy eating patterns: the Healthy U.S.-Style Eating Pattern, the Healthy

Mediterranean-Style Eating Pattern, and the Healthy Vegetarian Eating Pattern. Each of these food patterns provides specific recommendations as to the kinds and amounts of foods a person should consume and the proportions in which he or she should consume them. There are currently five food groups included in these patterns: vegetables, fruits, grains, dairy products, and protein foods. Although not a food group, recommended intake of healthy oils is also provided. In addition, recommendations are made for how many extra calories a person can consume (and not gain weight) if he or she selects nutrient-dense options in all the food groups. Some of the food groups are further separated into subcategories. For example, the vegetables food group is separated into five important subgroups: dark green (e.g., broccoli and spinach), red and orange (e.g., tomatoes and pumpkins), beans and peas (e.g., kidney beans and lentils), starchy (e.g., white potatoes and corn), and other (e.g., iceberg lettuce and onions).

How many servings from each food group you need depends on your caloric needs. This is why you will often see a range of intake goals for each food group; for instance 3–6 ounces for grains. To help individuals know more precisely how much food they should consume from each food group, the Dietary Guidelines support 12 different sets of values within each of the healthy eating patterns. For instance, a person who requires 2,000 kcal/day is advised to eat 2½ cups of vegetables every day, whereas a person who requires 2,400 kcal/day should strive to eat 3 cups every day. You can see how many servings are recommended for each food group in each of the healthy food patterns in Table 2.4.

MyPlate A visual food guide that illustrates the most important food intake pattern recommendations of the 2015 Dietary Guidelines for Americans.

The U.S. Department of Agriculture defines a food as being "natural" if it contains no artificial ingredients and is minimally processed.

© Joan Hall/Shutterstock.com

TABLE 2.4 AMOUNTS OF EACH FOOD GROUP RECOMMENDED (PER DAY) FOR EACH OF THE 2015 DIETARY GUIDELINES HEALTHY DIETARY PATTERNS AND MYPLATE FOOD GUIDANCE SYSTEM

Food Group	Healthy Food Pattern[a]			Dietary Significance
	U.S.-Style	Mediterranean-Style	Vegetarian	
Vegetables	1–2½ cups	1–2½ cups	1–2½ cups	Vegetables are rich sources of potassium, vitamin C, folate, dietary fiber, and vitamins A and E.
Fruits	1–2 cups	1–2½ cups	1–2 cups	Fruits are good sources of folate, vitamin C, vitamin A, and fiber.
Grains	3–6 ounces	3–6 ounces	3–6½ ounces	Grains are a major source of B vitamins, iron, magnesium, selenium, energy, and dietary fiber.
Dairy	2–3 cups	2 cups	2–3 cups	Dairy products are major sources of calcium, potassium, vitamin D, and protein.
Protein foods	2–5½ ounces	2–6½ ounces	1–3½ ounces	Protein foods are rich sources of protein, magnesium, iron, zinc, B vitamins, vitamin D, energy, and potassium.
Oils	15–27 grams	15–27 grams	15–27 grams	Polyunsaturated oils, like those found in vegetable oils and oily fish, are recommended.
Limit on calories for other uses (% of calories)[b]	150–270 kcal (8–15%)	100–260 kcal (8–15%)	170–290 kcal (11–19%)	

[a]Recommended amounts depend on age, sex, and physical activity level. Personalized recommendations can be generated at the MyPlate website (http://www.choosemyplate.gov).
[b]Assumes food choices to meet food group recommendations are in nutrient-dense forms. Calories from added sugars, added refined starches, solid fats, alcohol, and/or to eat more than the recommended amount of nutrient-dense foods are accounted for under this category.

Are legumes protein foods or vegetables?

The term "legume" refers to several types of dried beans (sometimes called pulses) such as kidney beans, pinto beans, white beans, black beans, garbanzo beans (chickpeas), lima beans, split peas, lentils, and edamame (green soybeans). Legumes, which are technically vegetables and therefore rich in fiber and other plant-source nutrients, are also excellent sources of protein as well as other nutrients, such as iron, that are more typically found in seafood, meats, and poultry. As such, legumes can be counted as either a vegetable or protein food when it comes to keeping track of how many servings you are getting from these categories. Because fresh green peas and green beans are much lower in protein, however, they should be counted only as vegetables.

2-4d MyPlate Illustrates How to Put Recommendations into Practice

Because many people do not know how many calories they need on a daily basis and therefore do not know which version of the many dietary patterns to use, the USDA and U.S. Department of Health and Human Services have developed **MyPlate**, a visual food guide that illustrates the most important food pattern intake recommendations put forth in the 2015 Dietary Guidelines. The MyPlate graphic reminds Americans to eat healthful amounts of the five food groups, using a familiar mealtime visual: a place setting. As you can see in Figure 2.8, recommended daily intakes of four food groups (fruits, vegetables, grains, and protein foods) are illustrated proportionately on the plate, whereas the fifth food group, dairy, is represented by the round glass (or bowl) on the upper-right periphery of the plate. Note that the MyPlate graphic does not specify the numbers of servings recommended by the healthy eating patterns. The recommended amount of each food group depends on a person's age, sex, and physical activity level. The creators of MyPlate wanted to encourage each person to determine how much food is needed on an individual basis.

Choosemyplate.gov Is a Powerful Tool To determine your own food needs, you must log on to the MyPlate website (www.choosemyplate.gov) and provide personal information (age, sex, and physical activity level) under the SuperTracker section, which can be found within the Online Tools tab. Once you have provided your personal information, the MyPlate website generates

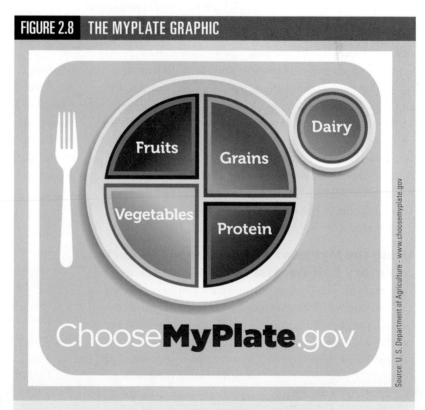

FIGURE 2.8   THE MYPLATE GRAPHIC

Source: U.S. Department of Agriculture - www.choosemyplate.gov

MyPlate illustrates the Dietary Guidelines for Americans recommended food consumption pattern as a consumer-friendly graphic.

individualized recommendations regarding food-intake patterns, serving sizes, and menu selection. In all, there are 12 different sets of recommended dietary patterns for the general population. This corresponds to the 12 different energy-intake levels listed in the USDA Food Patterns, on which MyPlate was based. An example of a personalized food plan, created for a 20-year-old woman with low physical activity, is illustrated in Figure 2.9. Food plans can also be generated for pregnant and breast-feeding women by clicking the For Moms/Moms-to-Be section. In addition to personalized food plans, the MyPlate website provides in-depth information about the types and amounts of foods that fit into each food group, assistance with meal planning, and material for special subgroups, such as vegetarians. For example, using "What's Cooking" you can plan meals and see how they contribute to your overall MyPlate recommendations. Nutrient compositions and serving sizes of individual foods can be found in *MyFoodapedia*, and tips about eating well on a budget are also available. You can even set goals for making small shifts in your diet using the MyPlate, MyWins feature of the website. At the time this textbook was revised, the USDA had not fully updated the MyPlate website. Consequently, there may be slight differences between what is written in this book and what is available on the MyPlate website.

Using the MyPlate Food Tracker to Conduct a Dietary Self-Assessment

The MyPlate website offers an excellent opportunity for you to conduct a self-assessment of your own dietary intake using the free **Food Tracker**. (To access this tool, click SuperTracker on the MyPlate homepage.) Food Tracker allows you to assess whether your diet meets your recommended healthy eating pattern (e.g., whether you are consuming enough servings of milk), as well as recommendations put forth in the DRIs (e.g., whether you are consuming your RDA for calcium). To use Food Tracker, you must keep track of everything you eat and drink for at least one day and then enter the information into the website. Ideally, you should keep a food record for three days, however, and one day should be on the weekend. Other tips for keeping an accurate food record include the following.

- *Detail is important.* Include as much detail as possible for all foods and beverages consumed. Brand names and preparation methods will help improve accuracy. For mixed dishes (or composite foods), estimate the amount of each ingredient. For example, "salad" might include lettuce, tomatoes, eggs, and so on. The more detail included in the diet record, the more accurate the analysis can be.

- *Estimate or measure serving sizes accurately.* Ideally, weigh or measure foods using standard household devices such as measuring cups or a kitchen scale. If this is not possible, estimate serving sizes as carefully as possible. Some restaurants provide detailed information about the amounts of food served. Such detail can be very helpful.

Food Tracker A component of the MyPlate website that allows individuals to conduct dietary self-assessments.

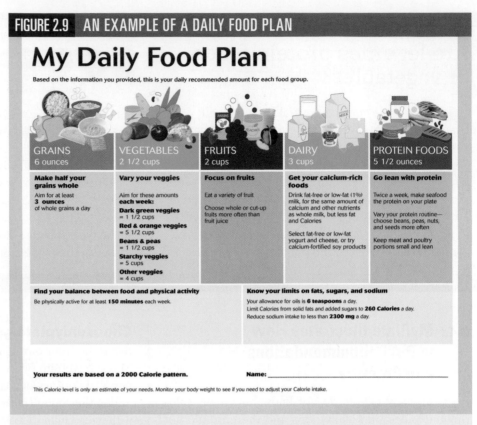

FIGURE 2.9 AN EXAMPLE OF A DAILY FOOD PLAN

There are 12 different daily food plans in the MyPlate food guidance system. Each food plan is based on a particular energy requirement. The food plan shown here is for a 2,000 kcal/day diet.

Source: U. S. Department of Agriculture—http://www.choosemyplate.gov/myplate/index.aspx

- *Choose representative, normal days.* Choose days that are representative of your typical eating patterns, and avoid special days such as holidays and birthdays. Because being sick or under unusual stress can influence food preferences and overall intake, it is best to avoid these circumstances as well.
- *Do not change your normal eating patterns.* When keeping diet records, people often alter their eating patterns to be more convenient or healthful. Resist this temptation, as it is especially important that diet records reflect normal intake.

When these guidelines are followed, the MyPlate Food Tracker can be a powerful tool in helping you determine whether the foods and beverages you typically choose are likely to encourage or deter you from being well.

2-5 HOW CAN YOU USE FOOD LABELS TO PLAN A HEALTHY DIET?

Understanding the recommendations put forth by the Dietary Guidelines and MyPlate is an excellent starting place for healthful eating. However, other resources are available to help you choose the appropriate amounts of nutritious foods and make the right food-related decisions

every day. One of the most important resources is the food label, which contains a vast amount of information that can help you optimize your health.

2-5a Food Labels

Although paying attention to food labels is an excellent way to obtain nutrition-related information, this was not always the case. In fact, consistent nutrition labeling was not established for foods until 1973, when the U.S. Food and Drug Administration (FDA) implemented a series of rules to help consumers become aware of food's nutrient content. The FDA now requires that most packaged foods that contain more than one ingredient have the following information printed on their labels.

- Product name and place of business
- Product net weight
- Product ingredient content (from most abundant to least abundant ingredient)
- Company name and address
- Country of origin (for some, but not all, products)
- Product code (UPC bar code)
- Product dating, if applicable
- Religious symbols, if applicable (such as kosher)
- Safe-handling instructions, if applicable (such as for raw meats)

The Buy Fresh Buy Local Campaign

Today, many people are interested in purchasing and consuming foods grown and manufactured in their local areas. This trend has developed in response to a variety of factors, including economic downturn, concerns about energy costs and the safety of foods produced in other countries, and a belief that locally grown foods might be more nutritious or better tasting than those that have been transported across the country. Although there are no federal guidelines defining what is meant by *local*, many cities and individual neighborhoods now host food cooperatives and other venues, such as farmers' markets, at which local produce is labeled and sold. Even if you do not see the appeal of buying local, you might be interested in a food's country of origin. You can usually find this information by reading its food label.

The Buy Fresh Buy Local campaign, as advertised in this logo from footroutes.org, encourages consumers to purchase locally produced and manufactured goods.

- Special warning instructions, if applicable (such as for aspartame and peanuts)
- Nutrition Facts panel outlining specified nutrient information

Of special interest to many people is the **Nutrition Facts panel**, a required component of most food labels that provides information about the nutrient content of the food. The FDA mandates that several critical elements be listed on every Nutrition Facts panel. For example, the food's serving size must be noted. In most foods, standard serving sizes have been established so that a person can easily compare the nutrient contents of similar foods. For instance, one can compare the amount of iron in a single serving of two similar breakfast cereals by looking at the amounts listed on their Nutrition Facts panels. Because the serving sizes of the two cereals are the same, their iron contents can be easily compared. In addition to serving size, the label must list the total energy (Calories), total carbohydrates (including dietary fiber), sugar, and protein per serving. An example of a food label that includes a Nutrition Facts panel is shown in Figure 2.10.

Nutrition Facts panels must also provide information concerning specific nutrients that individuals should try to limit. These include total fat, saturated fat, *trans* fat, cholesterol, and sodium. Conversely, because

Nutrition Facts panel A required component of most food labels that provides information about the nutrient content of the food.

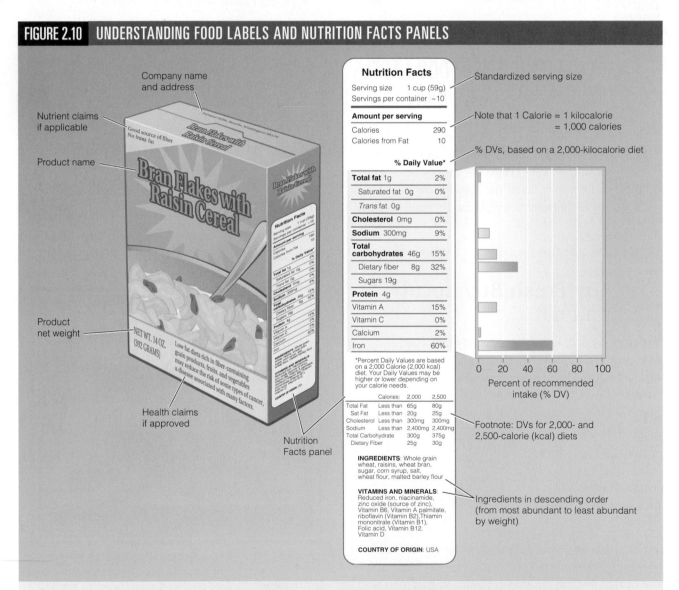

FIGURE 2.10 UNDERSTANDING FOOD LABELS AND NUTRITION FACTS PANELS

The FDA mandates that food labels provide specific information. A Nutrition Facts panel, required on most food labels, helps consumers choose their foods wisely.

Why Don't All Foods Have Food Labels?

Although most purchased foods have food labels, you may have noticed that some do not. For instance, fresh fruits and vegetables, and fish are currently not required to carry Nutrition Facts panels. Nonetheless, many grocers provide their customers with nutrition-related information voluntarily, often in the form of a poster or pamphlet. Other products such as gums and candies that come in very small packages are not required to carry nutrition labels either. In addition, foods produced in limited quantities by small business entities do not need to carry nutrition labels. To obtain nutrition information for these products, you frequently must contact the manufacturer or check the product's website.

individuals are encouraged to consume more dietary fiber, vitamin C, vitamin A, calcium, and iron, the FDA requires that information concerning these "shortfall" nutrients be included on Nutrition Facts panels as well. In this way, food labels can help consumers choose foods that specifically meet current dietary intake goals. The federal government continually evaluates and updates its food labeling requirements. Exemplifying this ever-evolving process is a relatively new requirement that nutrient information is included on some raw meat packaging. Another example of food labeling regulatory changes is the FDA's decision in 2015 to no longer require country-of-origin labeling on beef and pork, unless they are imported products sold in consumer-ready packages.[6] You can learn more about food labels by visiting the FDA's website.

Daily Values Although Nutrition Facts panels provide an impressive amount of nutrition information, actually using this information to plan a diet can be somewhat challenging. For instance, a cereal box's Nutrition Facts panel may tell you how much vitamin C is in a serving of the cereal, but how do you know if that amount is a little or a lot? To help answer this question, Nutrition Facts panels include a standard called the **Daily Value (DV)**, which gives consumers a benchmark as to whether a food is a good source of a particular nutrient. Most importantly, Daily Values allow consumers to compare one food to another quickly and easily.

There are two basic types of DVs. The first type, used for select vitamins and minerals, represents a nutrient's recommended daily intake for a person who requires approximately 2,000 kcal/day. For example, vitamin C's DV is 60 milligrams/day. Thus, a cereal that provides 30 milligrams of vitamin C per serving fulfills half of the DV for vitamin C. Another example of this type of DV is evident in Figure 2.10. By reading the Nutrition Facts label, you can see that one serving of this cereal provides 15 percent of an average person's DV for vitamin A.

The second type of DV represents a nutrient's upper limit: it presents a daily amount that you should try *not to exceed*. This type of DV is used for total fat, saturated fat, cholesterol, sodium, and total carbohydrates. For example, one serving of the cereal illustrated in Figure 2.10 provides 2 percent of an average person's daily upper limit of total fat. If a food's package is big enough, actual upper limit DV amounts are provided in addition to percentages. As you can see in the lower portion of the food label in Figure 2.10, the saturated fat DV of a person who requires 2,000 kcal/day is 20 grams. This amount increases to 25 grams for a person who requires 2,500 kcal/day.

In addition to understanding the daily requirements and upper limits of nutrients in a food, you can use DVs to determine whether a food is a good source of a particular nutrient. A food that provides less than 5 percent of a nutrient's DV is considered low in that nutrient. Conversely, a food that provides at least 20 percent of a nutrient's DV is high in that nutrient. Thus, the cereal illustrated in Figure 2.10 is considered high in vitamin A and low in saturated fat.

2-5b Nutrient Content Claims and Health Claims

Food packaging often provides additional nutrition-related information that can help you plan a healthy diet. For example, a **nutrient content claim** describes in a very consumer-friendly way how much of a nutrient (or its *content*) is in a food. Nutrient content claims include phrases like "sugar free," "low sodium," and "good source of." Thus, if a person wants to increase his or her fiber intake, he or she can consistently choose foods labeled "good source of fiber." The use of these claims is regulated by the FDA; a selection of approved definitions is provided in Table 2.5.

Daily Value (DV) A benchmark as to whether a food is a good source of a particular nutrient. May represent a nutrient's recommended daily intake or upper limit.

nutrient content claim An FDA-regulated word or phrase that describes how much of a nutrient (or its content) is in a food.

TABLE 2.5 SELECTED FDA-APPROVED NUTRIENT CONTENT CLAIMS

Wording	Description
"Light" or "Lite"	If 50 percent or more of the regular product's calories are from fat, fat must be reduced by at least 50 percent, as compared to the regular product. If less than 50 percent of the regular product's calories are from fat, fat must be reduced by at least 50 percent, or calories must be reduced by at least one-third, as compared to the regular product.
"Reduced Calories"	Product contains at least 25 percent fewer calories per serving than a regular product.
"Calorie Free"	Product contains less than 5 kcal (Calories) per serving.
"Fat Free"	Product contains less than 0.5 g fat per serving.
"Low Fat"	Product contains 3 g or less fat per serving.
"Saturated Fat Free"	Product contains less than 0.5 g saturated fat and less than 0.5 g *trans* fatty acids per serving.
"Low in Saturated Fat"	Product contains 1 g saturated fat per serving and derives 15 percent or fewer calories from saturated fat.
"Cholesterol Free"	Product contains less than 2 mg cholesterol per serving. Cholesterol claims are only allowed if a food contains 2 g or less saturated fat per serving.
"Low in Cholesterol"	Product contains 20 mg or less cholesterol per serving.
"Sodium Free"	Product contains less than 5 mg sodium per serving.
"Low in Sodium"	Product contains 140 mg or less sodium per serving.
"Sugar Free"	Product contains less than 0.5 g sugar per serving. This does not include alcohol sugars.
"High," "Rich in," or "Excellent Source of"	Product contains 20 percent or more of the Daily Value per serving. This wording may describe protein, vitamins, minerals, dietary fiber, or potassium.
"Good Source of," "Contains," or "Provides"	Product contains 10 to 19 percent of the Daily Value per serving.
"More," "Added," "Extra," or "Plus"	Product contains 10 percent or more of the Daily Value per serving. This wording may describe protein, vitamins, minerals, dietary fiber, or potassium.
"Fresh"	A raw food that has not been frozen, heat-processed, or otherwise preserved.
"Fresh Frozen"	Food was quickly frozen while still fresh.

Source: Adapted from U.S. Food and Drug Administration. Food Labeling Guide. Available from: http://www.fda.gov/Food/GuidanceRegulation/GuidanceDocumentsRegulatoryInformation/LabelingNutrition/ucm2006828.htm.

Some food manufacturers also include information about potential health benefits of foods or food components on their products' packaging. This type of information is called a **health claim**. Health claims are quite different from nutrient content claims, which simply state that a food contains or does not contain a particular nutrient. Health claims must be supported by sufficient scientific evidence that increased consumption of a nutrient or food component impacts health positively. For example, in order to print "eating oatmeal significantly decreases your risk of heart disease"on a box of oatmeal, the manufacturer must first confirm that this health claim is supported by a series of conclusive experimental studies. Like other parts of a food's package, all health claims must be approved by the FDA.

Manufacturers can make two kinds of health claims: regular health claims and qualified health claims. Both kinds of claims concern the relationship between a specific food (or food component) and a health-related condition. However, while a **regular health claim** is supported by considerable research, a **qualified health claim** has less scientific backing and must be accompanied by a disclaimer (or qualifier) statement. You can usually tell that a health claim is qualified because it contains a statement such as, "However, the FDA has determined that this evidence is limited and not conclusive." As new research emerges, some new health claims are approved while others are disapproved. Some examples of approved health claims can be found in Table 2.6.

health claim An FDA-approved statement that describes a specific health benefit of a food or food component.

regular health claim A health claim that is supported by considerable scientific research.

qualified health claim A health claim that has less scientific backing and must be accompanied by a disclaimer (or qualifier) statement.

TABLE 2.6 SELECTED, APPROVED HEALTH CLAIMS THAT CAN BE INCLUDED ON FOOD LABELS

Approved Claims	Food Eequirement	Required or Suggested Wording
Whole-grain foods and risk of heart disease and certain cancers	Product must contain at least 51% whole-grain ingredients by weight	"Diets rich in whole-grain foods and other plant foods and low in total fat, saturated fat, and cholesterol may reduce the risk of heart disease and some cancers."
Potassium and the risk of high blood pressure and stroke	Must be good source of potassium and low in sodium, total fat, saturated fat, and cholesterol	"Diets containing foods that are a good source of potassium and that are low in sodium may reduce the risk of high blood pressure and stroke."
Fluoridated water and reduced risk of dental carries	Bottled water must meet federal standards for fluoride content	"Drinking fluoridated water may reduce the risk of [dental caries or tooth decay]".
Substitution of saturated fat in the diet with unsaturated fatty acids to reduce risk of heart disease	Food must be low fat and low cholesterol	"Replacing saturated fat with similar amounts of unsaturated fats may reduce the risk of heart disease. To achieve this benefit, total daily calories should not increase."
Calcium and vitamin D and osteoporosis	Food must be high in calcium and/or vitamin D, and phosphorus content cannot exceed calcium content	"Adequate calcium throughout life, as part of a well-balanced diet, may reduce the risk of osteoporosis." "Adequate calcium and vitamin D, as part of a well-balanced diet, along with physical activity, may reduce the risk of osteoporosis."
Sodium and hypertension	Product must be low in sodium; must also include physician statement urging individuals with high blood pressure to consult their physicians if claim defines high or normal blood pressure	"Diets low in sodium may reduce the risk of high blood pressure, a disease associated with many factors."
Fruits and vegetables and cancer	Food must be low fat, and naturally a good source of vitamin A, vitamin C, or fiber	"Low fat diets rich in fruits and vegetables (foods that are low in fat and may contain dietary fiber, vitamin A, or vitamin C) may reduce the risk of some types of cancer, a disease associated with many factors."
Folate and neural tube defects	Product must be a good source of folate while not providing excess levels of vitamin A and vitamin D	"Healthful diets with adequate folate may reduce a woman's risk of having a child with a brain or spinal cord defect."
Soluble fiber and risk of coronary heart disease	Food must be low in saturated fat, cholesterol, and total fat and provide one or more of the following: oat bran, rolled oats, whole oat flour, whole-grain barley; and must contain at least 0.75 g of soluble fiber in each serving	"Soluble fiber from foods such as this, as part of a diet low in saturated fat and cholesterol, may reduce the risk of heart disease."

Source: Adapted from U.S. Food and Drug Administration. Guidance for industry: A food labeling guide (11. Appendix C: Health Claims). Available at: http://www.fda.gov/Food/GuidanceRegulation/GuidanceDocumentsRegulatoryInformation/LabelingNutrition/ucm064919.htm.

2-6 CAN YOU PUT THESE CONCEPTS INTO ACTION?

One benefit of studying nutrition is that much of what you learn can be applied directly to your own life. Remember that though dietary assessment is only one component of a complete nutritional assessment, it is an important first step toward a lifetime of health and nutritional awareness. You now have the information needed to assess your own diet and begin to choose the right foods to improve it. Consider the example of Linda, a 21-year-old college student. In the following vignettes, Linda uses the same basic knowledge you just learned to assess her diet and make changes that affect her health for the better.

2-6a Step 1: Set the Stage and Set Your Goals

After reading about dietary assessment, Linda decides to make sure her own diet is adequate. Linda is generally in good health, but has lately felt overly tired, even when she gets plenty of sleep. After doing some research, Linda learns that iron and vitamin deficiencies might be responsible for her symptoms. As such, she wants to know whether her diet lacks adequate amounts of these or other nutrients.

2-6b Step 2: Assess Your Nutritional Status

First, Linda examines her anthropometric measurements to determine whether there is any evidence of overall malnutrition. Linda is 5 feet, 3 inches (1.6 meters) tall and weighs 135 pounds (61.4 kilograms). Because she is likely at a healthy body weight, Linda concludes that she is probably not consuming too few or too many calories. However, because weight and height are not good indicators of overall nutritional adequacy, Linda decides to conduct a dietary self-assessment using a three-day diet record and the MyPlate Food Tracker. She records everything she eats and drinks for three days, paying close attention to portion sizes. Linda carefully notes every component of more complex foods such as the lasagna served in the cafeteria.

Next, Linda logs on to the MyPlate website and enters her information into its database. Using this free software, she is able to compare her dietary intake to the DRI values of all the required vitamins, minerals, and macronutrients, as well as to her EER. In addition, Linda compares her dietary intakes of certain food groups to those recommended by the 2015 Dietary Guidelines for Americans, USDA Food Patterns, and MyPlate.

The results of Linda's dietary analysis indicate that she consumes almost all of the necessary nutrients at levels above their AIs or between their RDAs and ULs. Further, Linda's total energy intake is acceptable. However, her percentage of calories coming from fat is 40 percent—higher than recommended—and her daily fiber intake is below the recommended 20 to 25 grams. Linda's intake of iron is only 13 milligrams/day (72 percent of her RDA), and her vitamin B_{12} intake is only 1.9 µg/day (79 percent of her RDA). This suggests that Linda may have inadequate intakes of fiber, iron, and vitamin B_{12}. Linda does some research on why the body requires iron and vitamin B_{12} and is surprised to learn that both are needed for energy production—this might explain why she has been so tired. With this new knowledge, Linda decides to increase her intakes of iron and vitamin B_{12}, decrease her fat intake, and increase her fiber consumption.

Reading all the information found on food labels can be useful when deciding which option best fits your dietary goals.

© Ditty_about_summer/Shutterstock.com

2-6c Step 3: Set the Table to Meet Your Goals

Using USDA nutrient composition databases, information on the MyPlate website, and a food composition table, Linda learns that lean meats and fortified breakfast cereals are good sources of iron that do not supply high amounts of fat. She also learns that meat and dairy products contain vitamin B_{12}, and that fiber is found in whole-grain products, peas and lentils, and in some fruits and vegetables. Linda begins to eat more of these foods and limit her total fat intake. She also begins to read Nutrition Facts panels on packaged foods and, when possible, to choose foods with lower total fat contents. In the cafeteria, Linda looks at cereal box labels and begins to eat high-fiber, fortified breakfast cereals instead of low-fiber, unfortified ones. (Fortified foods are those to which nutrients have been added during manufacturing.) When cooking for herself, Linda chooses foods containing at least 20 percent of the DV for iron, looks for products that are rich sources of whole grains, and makes sure she eats enough lean meat and low-fat dairy products.

2-6d Step 4: Compare Your Plan and Your Assessment: Did You Succeed?

After a few weeks, Linda wants to know whether the changes she made to her diet have improved her nutrient intake. She does another dietary self-assessment and

2-6e There Is No Time Like the Present

As evidenced by this example, dietary self-assessment and planning can be applied easily to anyone's life—including your own. Now that you know more about measuring your nutritional status and completing a dietary assessment, you can use the many reference standards (such as DRI values), dietary recommendations (e.g., the 2015 Dietary Guidelines), and nutrition tools (e.g., Nutrition Facts panels) to conduct your own assessment and choose a diet that fits your personal nutritional needs. Some professors require students to conduct a dietary self-assessment as a class assignment. If this is the case for you, then you will garner firsthand experience with and understanding of these procedures. If not, you are enthusiastically advised to do so on your own.

Whether required or optional, a dietary self-assessment will make the rest of this course more meaningful because you will be able to apply the insights you gain to your own diet and overall health. Experts agree that the food habits you establish now will not only affect your success in college but also influence your eating patterns and health for years to come. Take the time to set yourself up for success by eating right and being as healthy as possible.

Doing what you can right now to improve your health will help you live a longer and healthier life.

learns that these simple dietary changes have resulted in adequate intakes of iron, vitamin B_{12}, and fiber and a reduced total fat intake. Linda also notes that she has felt less tired and has been able to concentrate on her studies for longer periods of time. It appears that Linda has succeeded, but just to make sure that she has not overlooked anything and that her overall health is good, she makes an appointment with her campus health care provider.

STUDY TOOLS 2

READY TO STUDY? IN THE BOOK, YOU CAN:

☐ Rip out the Chapter Review Card, which includes key terms and chapter summaries.

ONLINE AT WWW.CENGAGEBRAIN.COM, YOU CAN:

☐ Learn about superfoods in a short video.

☐ Explore the various parts of a food label.

☐ Interact with figures from the text to check your understanding.

☐ Prepare for tests with quizzes

☐ Review the key terms with Flash Cards.

3 | Body Basics

LEARNING OUTCOMES

3-1 Understand the connection between chemistry and nutrition.

3-2 Compare and contrast cells, tissues, organs, and organ systems.

3-3 Describe the process of digestion.

3-4 Describe the process of nutrient absorption and circulation.

3-5 Understand the process of nutrient metabolism and excretion.

Ikon Images/Getty Images

After finishing this chapter go to **PAGE 71** for **STUDY TOOLS.**

3-1 WHY LEARN ABOUT CHEMISTRY WHEN STUDYING NUTRITION?

Your body requires over 40 different nutrients to satisfy its nutritioinal needs. These nutrients come directly from a variety of complex foods that you ingest daily. The first step in this process takes place in the gastrointestinal tract, where food is systematically broken down into its most basic components. Only once they are broken down can nutrients leave the gastrointestinal tract and circulate through the extensive network of blood and lymph vessels that make up your circulatory system. Nutrients are taken up by the cells in your body where they undergo amazing chemical transformations that ultimately sustain your life. To ensure that these activities take place under optimal conditions, nonstop communication networks—your endocrine and nervous systems—orchestrate these events. In this chapter, you will learn about the chemical and physiological functions that take place every time you eat, and you will gain an appreciation for the intricate and varied tasks required to nourish your body.

Chemistry is fundamental to the study of nutrition. Not only are nutrients chemicals themselves, but the utilization of nutrients in the body involves a vast number of chemical reactions. The organization of atoms into molecules, molecules into macromolecules, macromolecules into cells, cells into tissues, tissues into organs, and organs into organ systems is indeed remarkable (see Figure 3.1). Your entire body is made of and fueled by the nutrients contained in food. In order to appreciate these life-sustaining functions, it is important to first have a basic understanding of chemistry—the science of matter.

3-1a Atoms Make Up the World around You

It is difficult to imagine the existence of something that you cannot see, taste, touch, or hear. Yet the **atom**, though invisible to the human eye, is the basic unit that makes up the world around you (matter). Some atoms have an equal number of positively and negatively charged particles, and therefore are neutral. This is not the case for

atom The basic unit of matter that makes up the world around us.

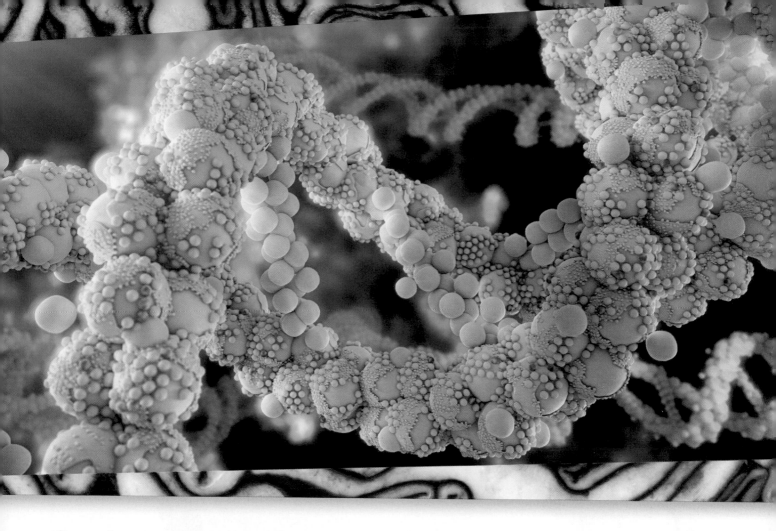

all atoms, however. Depending on the type and number of particles it contains, an atom can have a positive or a negative charge. A charged atom is called an **ion**. Ions serve many vital functions in the body. For example, ions such as calcium (Ca^{2+}) and magnesium (Mg^{2+}) are required for muscle contraction and nerve function. Note the plus sign (+) denotes positively charged ions. Other important ions found in the human body include sodium (Na^+), potassium (K^+), chloride (Cl^-), iodide (I^-), and fluoride (F^-). The minus sign (–) indicates that these are negatively charged ions.

Ions with opposite charges are often attracted to one another, a phenomenon that can result in the formation of a chemical bond. For example, because the ions sodium (Na^+) and chloride (Cl^-) have opposite charges, they form a chemical bond, creating sodium chloride (NaCl), which you know as table salt. You may have also heard the term *electrolyte* before. Although the terms *ion* and *electrolyte* are often used interchangeably, they do not actually mean the same thing. An **electrolyte** is a molecule that, when submerged in water, separates into individual ions. When this happens, the molecule is *ionized*. Because excessive sweating can

cause a depletion of water and electrolytes from the body, athletes are often encouraged to drink beverages that replenish both of these important components. In fact, the sports drink Gatorade® was developed by researchers at the University of Florida to help the school football players (the Gators) sustain healthy fluid–electrolyte balances while exercising.

3-1b Elements and Molecules

An **element** is a pure substance made up of only one type of atom. There are 118 known elements, and approximately 98 of these are naturally occurring. Of the 98 naturally occurring elements, only 20 are essential to human health. Perhaps surprisingly, just six elements—carbon, oxygen, hydrogen, nitrogen, calcium, and phosphorus—account for

ion An atom that has a positive or negative electrical charge.

electrolyte A molecule that, when submerged in water, separates into individual ions.

element A pure substance made up of only one type of atom.

AP Photo/Jon Elswick

FIGURE 3.1 LEVELS OF ORGANIZATION IN THE BODY

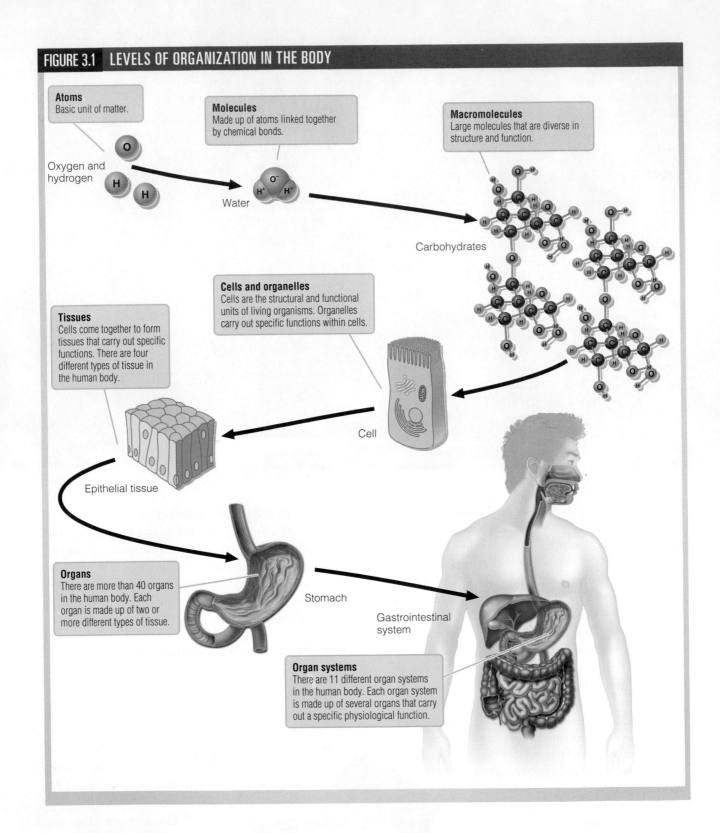

Atoms
Basic unit of matter.

Oxygen and hydrogen

Molecules
Made up of atoms linked together by chemical bonds.

Water

Macromolecules
Large molecules that are diverse in structure and function.

Carbohydrates

Cells and organelles
Cells are the structural and functional units of living organisms. Organelles carry out specific functions within cells.

Cell

Tissues
Cells come together to form tissues that carry out specific functions. There are four different types of tissue in the human body.

Epithelial tissue

Organs
There are more than 40 organs in the human body. Each organ is made up of two or more different types of tissue.

Stomach

Gastrointestinal system

Organ systems
There are 11 different organ systems in the human body. Each organ system is made up of several organs that carry out a specific physiological function.

99 percent of your body mass. These important elements provide the raw materials needed to form the molecules essential to life.

A **molecule** is formed when two or more atoms join together by one or more chemical bonds. Amazingly, such bonds enable a relatively small number of atoms to form millions of different molecules. Without chemical bonds, the molecular world would fall apart. In fact, you can think of chemical bonds as the glue that holds atoms together within a molecule. Some molecules are small

molecule A unit of two or more atoms joined together by chemical bonds.

and simple, whereas others can be very large and complex, consisting of thousands of atoms bonded together. Carbohydrates, lipids, proteins, and nucleic acids (DNA and RNA) are examples of large, complex molecules (or *macromolecules*) that are vital to the functions of cells. The elements that comprise these large molecules come from the nutrients in foods that you eat.

Understanding a Molecular Formula A **molecular formula** is a representation of the number and type of atoms present in a molecule. For example, glucose, an important source of energy in your body, has a molecular formula of $C_6H_{12}O_6$. This formula tells you that one molecule of glucose consists of six carbon (C_6), 12 hydrogen (H_{12}), and six oxygen (O_6) atoms. You may recognize the molecular formula for water, H_2O. Water is comprised of two hydrogen atoms (H_2) and one oxygen (O) atom. Note that, when an element is not followed by a number, it means that there is only one atom of that type of element present. When more than one molecule of a substance is present, the total number of molecules is placed before the molecular formula. For example, the molecular formula for three molecules of water is written $3H_2O$ (see Figure 3.2).

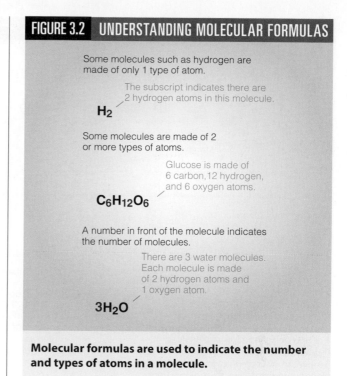

FIGURE 3.2   UNDERSTANDING MOLECULAR FORMULAS

Some molecules such as hydrogen are made of only 1 type of atom.

The subscript indicates there are 2 hydrogen atoms in this molecule.

H_2

Some molecules are made of 2 or more types of atoms.

Glucose is made of 6 carbon, 12 hydrogen, and 6 oxygen atoms.

$C_6H_{12}O_6$

A number in front of the molecule indicates the number of molecules.

There are 3 water molecules. Each molecule is made of 2 hydrogen atoms and 1 oxygen atom.

$3H_2O$

Molecular formulas are used to indicate the number and types of atoms in a molecule.

3-2 HOW ARE CELLS, TISSUES, ORGANS, AND ORGAN SYSTEMS RELATED?

Molecules such as carbohydrates, proteins, lipids, water, and nucleic acids are basic to life as we know it. However, these molecules in and of themselves do not always function independently. Rather, some molecules function as building blocks, forming structural and functional units called **cells**. Cells make up tissues, which in turn function as building blocks for organs. Finally, organs work together as components of organ systems. The human body has 11 organ systems, all of which are pertinent to the study of nutrition.

3-2a Passive and Active Transport Mechanisms

Cells are like microscopic cities—full of activity. Cell structures called **organelles** are the components within body cells that carry out very specific functions. Organelles subsist in a gel-like fluid called **cytoplasm**, which fills every living cell in your body. Cells are surrounded by a protective membrane that regulates the movement of substances into and out of the cell. The movement of nutrients and other substances across a cell membrane occurs through a variety of processes, which are referred to collectively as *transport mechanisms*. A transport mechanism that does not require energy (ATP) is called a **passive transport mechanism**, whereas one that does require energy is called an **active transport mechanism**.

Passive Transport Mechanisms As illustrated in Figure 3.3, there are three types of passive transport mechanisms: osmosis, simple diffusion, and facilitated diffusion. **Osmosis** is the diffusion of water molecules across a cell membrane. Fluids in the body contain different types of dissolved substances, referred to as **solutes**. The concentrations of dissolved substances (solutes) inside and outside the cell determines the movement (direction)

molecular formula A representation of the number and type of atoms present in a molecule.

cell A structural and functional unit that makes up body tissues.

organelles A structure that is responsible for a specific function within a cell.

cytoplasm A gel-like fluid that fills every living cell.

passive transport mechanism A transport mechanism that does not require energy (ATP) to move substances across cell membranes.

active transport mechanism A transport mechanism that requires energy (ATP) to move substances across cell membranes.

osmosis The passive movement of water molecules across a cell membrane from a solution with low solute concentration to a solution with a higher solute concentration.

solutes A dissolved substance; a component of a solution.

FIGURE 3.3 PASSIVE TRANSPORT MECHANISMS

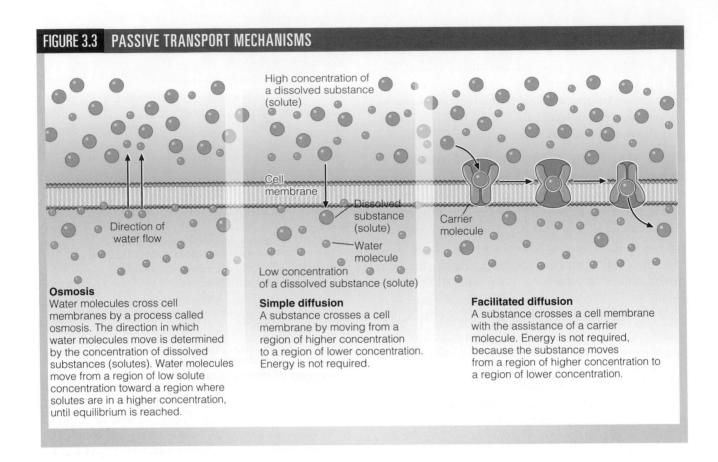

Osmosis
Water molecules cross cell membranes by a process called osmosis. The direction in which water molecules move is determined by the concentration of dissolved substances (solutes). Water molecules move from a region of low solute concentration toward a region where solutes are in a higher concentration, until equilibrium is reached.

Simple diffusion
A substance crosses a cell membrane by moving from a region of higher concentration to a region of lower concentration. Energy is not required.

Facilitated diffusion
A substance crosses a cell membrane with the assistance of a carrier molecule. Energy is not required, because the substance moves from a region of higher concentration to a region of lower concentration.

of water molecules. Specifically, water molecules move towards a higher concentration of dissolved substances, until the concentration of dissolved substances is equalized on both sides of the cell membrane.

Simple diffusion enables substances to cross cell membranes from a region of higher concentration to a region of lower concentration without using energy in the process. In simple diffusion, a substance moves without assistance, like a raft floating downstream. In **facilitated diffusion**, a substance moves passively from a region of higher concentration to a region of lower concentration, but only with the assistance of a carrier molecule. Similar to the example of a raft moving downstream with the water current (simple diffusion),

you can think of carrier molecules in facilitated diffusion as oars that help guide a boat in a specific direction.

Active Transport Mechanisms Some substances must cross cell membranes against the prevailing concentration gradient, moving from a region of lower concentration to one of a higher concentration. Because energy is needed to pump the molecule across a cell membrane, this type of transport mechanism is considered active. **Carrier-mediated active transport** is a transport mechanism whereby a substance uses energy (ATP) to move from a region of lower concentration to a region of higher concentration with the assistance of a carrier molecule. The process of carrier-mediated active transport is similar to that of a motorboat because it requires energy to move upstream (see Figure 3.4).

3-2b Cell Organelles

As previously mentioned, every cell contains small structures called organelles that carry out specialized functions that are critical to life. Some organelles produce substances necessary for cellular activity, while others function as waste-disposal systems, assisting the degrading and recycling of worn-out cellular components. Organelles called *mitochondria* serve as power

simple diffusion A passive transport mechanism whereby substances move from a region of higher concentration to a region of lower concentration without using energy or the assistance of a carrier molecule.

facilitated diffusion A passive transport mechanism whereby a substance moves from a region of higher concentration to a region of lower concentration with the assistance of a carrier molecule.

carrier-mediated active transport An active transport mechanism whereby a substance moves from a region of lower concentration to a region of higher concentration with the assistance of a carrier molecule and energy.

FIGURE 3.4 CARRIER-MEDIATED ACTIVE TRANSPORT

Low concentration of dissolved substances (solutes)

Dissolved substance (solute)

Carrier molecule

Water

Energy (ATP)

High concentration of a dissolved substance

A substance crosses a cell membrane with the assistance of a carrier molecule. Energy (ATP) is required, because the substance moves from a lower concentration to a higher concentration.

stations, converting the chemical energy of energy-yielding nutrients (carbohydrates, lipids, and proteins) into energy that is usable by cells (ATP). Another organelle, the *nucleus*, houses the genetic material (DNA) that provides the blueprint for protein synthesis. Figure 3.5 provides an overview of a cell and its components.

3-2c The Four Types of Tissue

In multicellular organisms, the next level of complexity is the tissue. A **tissue** is formed when cells of similar structure and function group together to accomplish a common task. The human body contains four different types of tissue:

- Epithelial
- Connective
- Muscle
- Neural

Your skin and the inner lining of your organs are made of **epithelial tissue**, which helps to protect the body. **Connective tissue** supports, connects, and anchors body structures—it is the glue that holds the body together. Tendons, cartilage, and some parts of bones are examples of connective tissue. The body contains several types of **muscle tissue**, all of which are used for movement. *Skeletal muscle*, a type of muscle tissue under voluntary control, allows you to move the various parts of your body. There are approximately 640 different skeletal muscles in the human body that are under your active control. *Smooth muscle* tissue, which you cannot control voluntarily, is embedded within the epithelial lining of organs such as the esophagus, stomach, and small intestine. Smooth muscle is even found in the interior lining of blood vessels; it is this layer of smooth muscle that allows vessels to constrict (become narrower) and relax (become wider).

tissue An aggregation of similarly structured and functioning cells that have grouped together to accomplish a common task.

epithelial tissue Tissue that helps protect the body.

connective tissue Tissue that supports, connects, and anchors structures in the body.

muscle tissue Tissue that is used for movement.

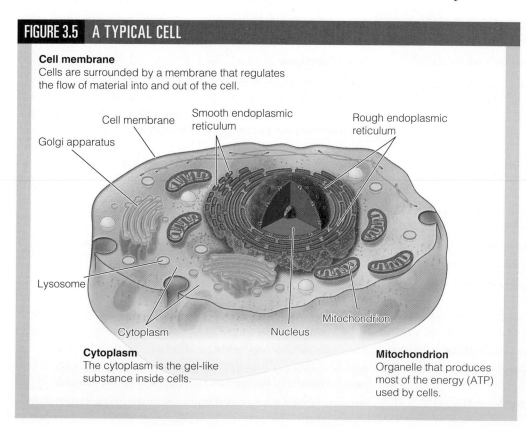

FIGURE 3.5 A TYPICAL CELL

Cell membrane
Cells are surrounded by a membrane that regulates the flow of material into and out of the cell.

Cell membrane

Golgi apparatus

Smooth endoplasmic reticulum

Rough endoplasmic reticulum

Lysosome

Cytoplasm

Nucleus

Mitochondrion

Cytoplasm
The cytoplasm is the gel-like substance inside cells.

Mitochondrion
Organelle that produces most of the energy (ATP) used by cells.

Because smooth muscles are not under our conscious control, they are referred to as involuntary muscles—their movement is regulated in part by the nervous system. The final type of tissue is **neural tissue**, which makes up the brain, spinal cord, and nerves. Neural tissue plays an important communicative role in the body. The four types of tissue are illustrated in Figure 3.6.

3-2d Organs and Organ Systems

In the human body, the four types of tissue comprise more than 40 organs, which in turn make up 11 unique organ systems. An **organ** consists of two or more different types of tissue functioning together to perform a variety of related tasks. An *organ system* is formed when several organs work together. Each organ in the organ system carries out its own important physiological function, but also contributes to the overall function of the system. For example, the digestive system is composed of several organs that work together to break down food. The major organ systems and their basic functions are summarized in Table 3.1.

neural tissue Tissue that facilitates communication throughout the body.

organ Two or more different types of tissue functioning together to perform a variety of related tasks.

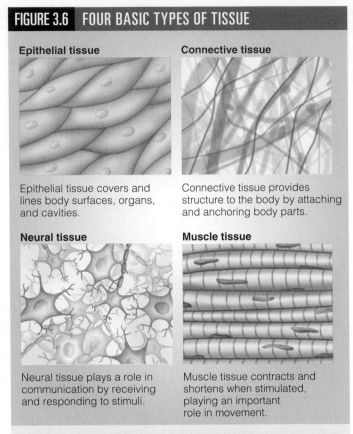

FIGURE 3.6 FOUR BASIC TYPES OF TISSUE

Epithelial tissue

Epithelial tissue covers and lines body surfaces, organs, and cavities.

Connective tissue

Connective tissue provides structure to the body by attaching and anchoring body parts.

Neural tissue

Neural tissue plays a role in communication by receiving and responding to stimuli.

Muscle tissue

Muscle tissue contracts and shortens when stimulated, playing an important role in movement.

Epithelial, connective, neural, and muscle tissue make up all the organs in the human body.

TABLE 3.1 ORGAN SYSTEMS AND RELATED MAJOR FUNCTIONS

Organ System	Major Organs and Structures	Major Functions
Circulatory	Heart, blood vessels, blood, lymph vessels, lymph nodes, and lymph	Transports nutrients, waste products, gases, and hormones.
Digestive	Mouth, esophagus, stomach, small intestine, large intestine, liver, gallbladder, pancreas, and salivary glands	Governs the physical and chemical breakdown of food into chemicals that can be absorbed into the blood. Eliminates solid waste.
Endocrine	Endocrine glands such as pituitary gland, thyroid gland, and adrenal glands	Produces and releases hormones that control physiological functions such as reproduction, hunger, satiety, blood glucose regulation, metabolism, and stress response.
Immune	Lymph vessels, bone marrow, and lymphatic tissue	Protects against foreign bodies, such as bacteria and viruses, and helps regulate cell division and cell death.
Integumentary	Skin, hair, nails, and sweat glands	Protects against pathogens and helps regulate body temperature.
Muscular	Smooth, cardiac, and skeletal muscle	Assists in voluntary and involuntary body movements.
Nervous	Brain, spinal cord, nerves, and sensory receptors	Interprets and responds to information. Controls the basic senses, movement, and intellectual functions.
Reproductive	Gonads, genitals, and mammary glands (breasts)	Carries out reproductive functions and is associated with sexual characteristics, sexual function, sexual behaviors, and nourishing the newborn.
Respiratory	Lungs, nose, mouth, throat, and trachea	Governs gas exchange between the blood and air.
Skeletal	Bones, cartilage, and joints	Provides support and structure to the body. The marrow of some bones produces blood cells. Also provides a storage site for certain minerals.
Urinary	Kidneys, bladder, and ureters	Removes metabolic waste products from the blood and regulates water balance.

Organ Systems Work Together to Maintain Balance The ability of organs and organ systems to work together requires constant communication within the body. In other words, the right hand must know what the left hand is doing. The body has two well-developed communication systems that coordinate physiologic functions: the nervous system and the endocrine system. The nervous system communicates information via nerve cells (neurons), whereas the endocrine system communicates by releasing **hormones** into the blood. Hormones are chemical messengers that coordinate and regulate the activity of specific cells in the body. Together, the nervous and endocrine systems continuously monitor our internal environment. When a change or imbalance is detected, the body initiates adaptive physiological responses that help to counter these changes and restore balance. These adaptive responses allow us to maintain **homeostasis**, which is a state of balance or equilibrium. Homeostasis is an important concept in nutrition. Even everyday events such as eating can disrupt the body's delicate, internal balance. For example, when a carbohydrate-rich meal causes a rise in blood glucose, the pancreas releases the hormone insulin. This response lowers blood glucose to normal levels and restores homeostasis.

© glayan/Shutterstock.com

3-3 WHAT HAPPENS DURING DIGESTION?

In order to nourish your body, your digestive system methodically disassembles the complex food molecules into simpler basic nutrient components. Although this arduous task requires many organ systems, it is primarily the digestive system that gets the job done. Once food is consumed, it travels through various regions of the digestive tract, undergoing physical and chemical transformations—a journey that takes 24 to 72 hours. By the time it reaches the small intestine, food barely resembles the substance you ingested. Each region of the gastrointestinal (GI) tract makes its own unique contribution to the digestive process, which comes to completion when nutrients are able to move out of the digestive tract and into the blood or lymph. In addition, the digestive system is also responsible for eliminating any undigested food residue remaining in the GI tract. These complex processes are explained in detail throughout the following sections.

3-3a The Digestive System

Your digestive system, comprised of the digestive tract and accessory organs, is uniquely designed to physically and chemically transform food into its basic components—nutrients. The digestive or GI tract can be thought of as a hollow tube that runs from the mouth to the anus (see Figure 3.7). The inner cavity that spans the entire length of the GI tract is called the **lumen**. Organs that make up the GI tract include the mouth, esophagus, stomach, small intestine, and large intestine. The amount of time it takes for food to travel the entire length of the GI tract is called **transit time**. Although there are different methods used to measure transit time in humans, on average, 40 to 50 hours passes between the time food enters the mouth and the time its residue exits the body.

Organs such as the salivary glands, liver, gallbladder, and pancreas are not part of the digestive tract. Rather, these organs are referred to as **accessory organs**, which play an important role in the process of digestion. Together, the GI tract and accessory organs carry out three important functions:

1. **Digestion**: the physical and chemical breakdown of food into a form that allows nutrients to be absorbed.

2. **Absorption**: the movement of nutrients out of the GI tract into the blood or lymph.

3. **Elimination**: the process whereby solid waste is removed from the body.

hormones Chemical messengers released into the blood by the endocrine system that coordinate the activities of specific cells in the body.

homeostasis A state of balance or equilibrium.

lumen The cavity that spans the entire length of the GI tract.

transit time The amount of time it takes for food to travel the entire length of the GI tract.

accessory organs Organs that are part of the digestive system, and assist with the process of digestion.

digestion The physical and chemical breakdown of food into a form that allows nutrients to be absorbed.

absorption The movement of nutrients out of the GI tract and into the blood or lymph.

elimination The process whereby solid waste is removed from the body.

FIGURE 3.7 ORGANS OF THE DIGESTIVE SYSTEM

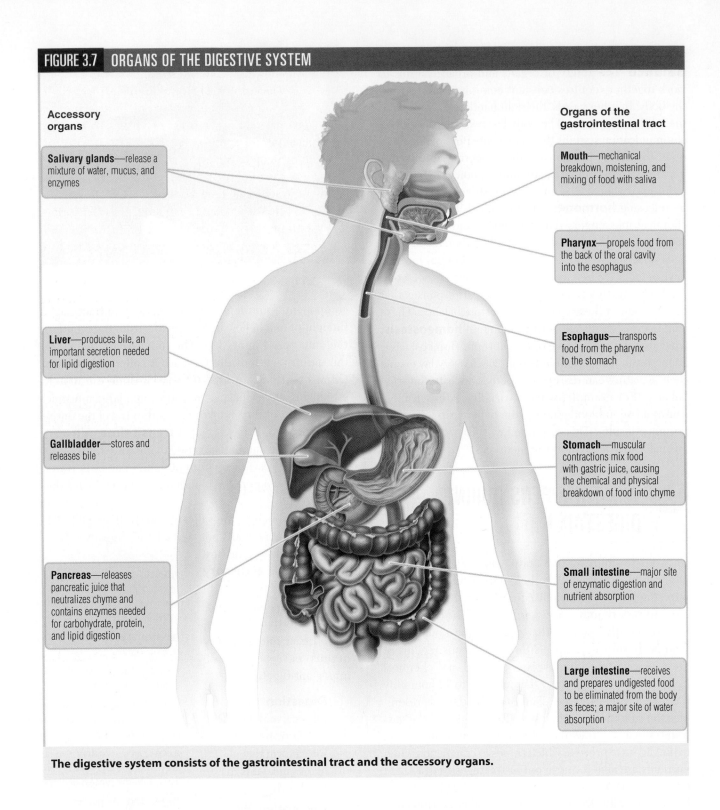

Accessory organs

Salivary glands—release a mixture of water, mucus, and enzymes

Liver—produces bile, an important secretion needed for lipid digestion

Gallbladder—stores and releases bile

Pancreas—releases pancreatic juice that neutralizes chyme and contains enzymes needed for carbohydrate, protein, and lipid digestion

Organs of the gastrointestinal tract

Mouth—mechanical breakdown, moistening, and mixing of food with saliva

Pharynx—propels food from the back of the oral cavity into the esophagus

Esophagus—transports food from the pharynx to the stomach

Stomach—muscular contractions mix food with gastric juice, causing the chemical and physical breakdown of food into chyme

Small intestine—major site of enzymatic digestion and nutrient absorption

Large intestine—receives and prepares undigested food to be eliminated from the body as feces; a major site of water absorption

The digestive system consists of the gastrointestinal tract and the accessory organs.

Movement and Secretions Aid the Process of Digestion

The process of digestion takes place in a highly regulated and organized manner. Like the conductor of an orchestra, neural and hormonal signals coordinate the movement of food through the GI tract. A vigorous, wave-like muscular contraction called **peristalsis** propels the food from one region of the GI tract to the next (see Figure 3.8). The rate of peristalsis increases and decreases to ensure that the food mass moves along the GI tract at the appropriate rate. At certain points throughout the GI tract, a circular band of muscle called a

peristalsis A vigorous, wave-like muscular contraction that propels food from one region of the GI tract to the next.

FIGURE 3.8 PERISTALSIS

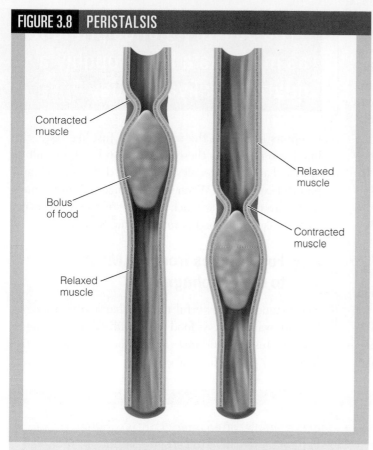

Contracted muscle

Bolus of food

Relaxed muscle

Relaxed muscle

Contracted muscle

Peristalsis consists of a series of wavelike, rhythmic muscular contractions. This action propels food forward through the GI tract.

sphincter regulates the flow of food. Sphincters act like one-way valves, opening and closing in response to neural and hormonal signals. Relaxation of the sphincter allows the food to flow forward. Once the food reaches the next organ, the sphincter closes to prevent the food from flowing backward (see Figure 3.9).

The innermost lining of the digestive tract, the **mucosa**, consists mainly of epithelial tissue that produces and releases a variety of substances that facilitate the process of digestion. An **enzyme**, for example, is a biological catalyst that accelerates chemical reactions. Enzymes released into the GI tract facilitate reactions that aid in the chemical breakdown of nutrients. In addition to various enzymes, some epithelial cells in the mucosal lining release hormones into the blood that regulate the rate at which food moves through the GI tract. Another important substance that aids digestion is mucus, which keeps the inner lining of the GI tract moist and provides protection from irritating (and sometimes harmful) substances.

3-3b Digestion Begins in the Mouth

In truth, the process of digestion begins even before food enters your mouth—the mere thought, smell, and sight of food serve as a wake-up call to the digestive system. Sometimes referred to as the *cephalic phase*, this response optimizes digestion by stimulating the muscles that move food through the digestive tract and by triggering the release of substances that help break down food chemically. The process of

sphincter A circular band of muscle that regulates the flow of food through the GI tract.

mucosa The innermost lining of the gastrointestinal tract.

enzyme A biological catalyst that accelerates a chemical reaction.

FIGURE 3.9 SPHINCTERS REGULATE THE FLOW OF FOOD

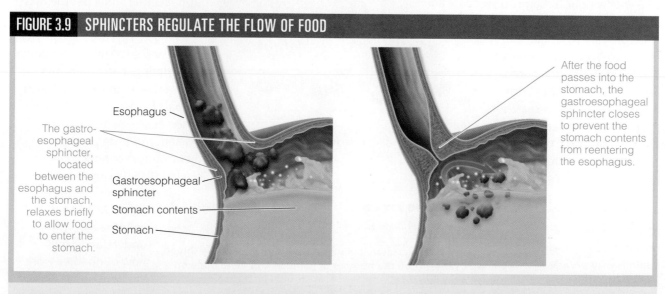

Esophagus

The gastroesophageal sphincter, located between the esophagus and the stomach, relaxes briefly to allow food to enter the stomach.

Gastroesophageal sphincter

Stomach contents

Stomach

After the food passes into the stomach, the gastroesophageal sphincter closes to prevent the stomach contents from reentering the esophagus.

Sphincters are circular bands of muscles located throughout the gastrointestinal tract that regulate the flow of food.

digestion begins when the forceful grinding action of the teeth physically tears food apart into smaller, more manageable pieces. The presence of food in the mouth also stimulates the release of **saliva** from the salivary glands. As food is broken apart, it mixes with saliva, making the food moist and easier to swallow. This also allows the digestive enzymes in saliva to begin the process of chemical digestion of carbohydrates and lipids while they are still in the mouth.

Saliva serves another useful purpose—it helps you to taste your food. Taste is a complex process that occurs when food dissolves in saliva. When this happens, taste molecules react with receptors found in taste buds that reside on the surface of your tongue. Once stimulated, taste receptors send neural signals to the brain, which allows you to experience basic taste sensations such as sweet, sour, salty, bitter, and savory. However, the ability to experience the thousands of different flavors that you enjoy in your food is also the result of aroma molecules. Aroma molecules stimulate

saliva A secretion released into the mouth by the salivary glands that moistens food and starts the chemical process of digestion.

> "The salivary glands produce as much as a liter (roughly a quart) of saliva per day."

receptors located in the nasal cavity. Just like taste, the brain also interprets these signals as well. As a result of taste and aroma molecules, your food delivers a rich and flavorful experience. When you are congested, aroma molecules are not able to reach receptors in the nasal cavity, which makes your food taste bland and flavorless.

3-3c Food Moves from the Mouth to the Esophagus

The tongue is a powerful muscle that assists in chewing and swallowing. As food mixes with saliva, the tongue manipulates the food mass and pushes it up against the hard, bony palate that makes up the roof of your mouth.

The Science of Taste

We all have food likes and dislikes, but some people are pickier about their food than others. Whereas some people enjoy the bitter, acidic taste of a morning cup of coffee, others may find it offensive. There is a long-held belief that food likes and dislikes are determined by the number and type of structures people have on their tongues. In reality, the small bumps that give your tongue a rough appearance are not actually taste buds. Rather, these structures are called papillae. Taste buds (cells) are nestled within papilla, each housing approximately 50 to 100 taste buds. According to some researchers, the more papillae and taste buds people have, the more sensitive they are to bitter-tasting vegetables such as broccoli, onions, and kale. However, this long-held belief may not accurately explain why people dislike certain foods.

Some people, often referred to as supertasters, are extremely sensitive to certain compounds found in bitter-tasting vegetables. Two compounds in particular, phenylcarbamide and propylthiouracil, may help explain why certain people are particularly sensitive to bitter-tasting foods. According to some researchers, the ability to detect these two bitter-tasting compounds is partially due to genetics. It appears that some people may have a variation of a gene that results in increased sensitivity to bitterness. Therefore, similar to other genetic traits, the acute sense of taste may also, in part be determined by your genes. Learning more about the science of taste provides another example of how advances in scientific investigation allow us to challenge long-held beliefs. As a result, we have a better understanding of how our bodies work and function.[1]

The ability to detect bitter-tasting compounds is partially due to genetics rather than the number of papillae and taste buds on a person's tongue.

As illustrated in Figure 3.10, swallowing takes place in two phases. First, as you prepare to swallow, your tongue directs the soft, moist mass of chewed food, now referred to as a **bolus**, to the region at the back of your mouth, an area known as the **pharynx**. The pharynx is the shared space between the oral and nasal cavities. This first phase of swallowing occurs under voluntary control, but once the bolus reaches the pharynx, the second (involuntary) phase of swallowing begins. At this point, the bolus is ready to enter the **esophagus**, a narrow muscular tube that ends at the stomach.

3-3d The Esophagus Delivers Food to the Stomach

During the involuntary phase of swallowing, the upper-back portion of the mouth (called the soft palate) lifts upward, blocking the entrance to the nasal cavity. This helps guide the bolus into the correct passageway—the esophagus. This movement also causes the *epiglottis*, a cartilage flap, to cover the *trachea*, the airway leading to the lungs. If it were not for the epiglottis, food would readily lodge in the trachea causing us to choke. Once the bolus moves past this dangerous intersection, the voluntary and involuntary phases of swallowing are ready for the next bite of food.

The esophagus is lubricated and protected by a thin layer of mucus, which facilitates the passage of food. Peristalsis propels the food through the esophagus toward the stomach, where the bolus encounters the first of several sphincters in the GI tract, the **gastroesophageal sphincter**. As the bolus approaches the end of the esophagus, the gastroesophageal sphincter relaxes long enough for it to pass into the stomach. Once this occurs, the sphincter closes, preventing the contents of the stomach from reentering the esophagus. The entire trip from the pharynx to the stomach takes only 6 to 10 seconds.

bolus A soft, moist mass of chewed food.

pharynx A region at the back of the mouth that serves as the shared space between the oral and nasal cavities.

esophagus A narrow muscular tube that begins at the pharynx and ends at the stomach.

gastroesophageal sphincter A circular muscle that regulates the flow of food from the esophagus to the stomach.

FIGURE 3.10 VOLUNTARY AND INVOLUNTARY PHASES OF SWALLOWING

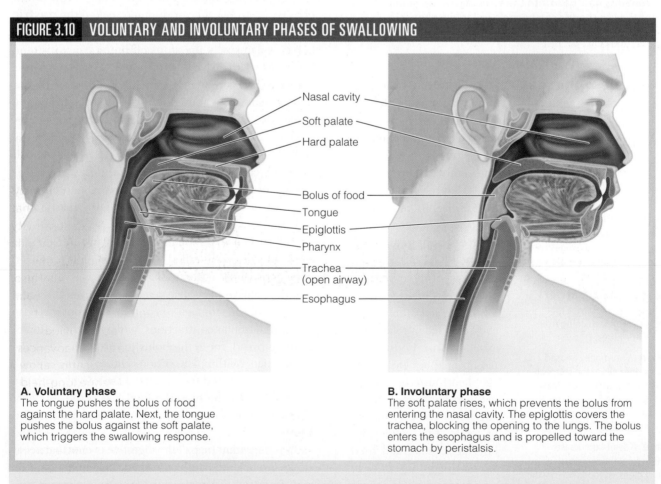

Nasal cavity
Soft palate
Hard palate
Bolus of food
Tongue
Epiglottis
Pharynx
Trachea (open airway)
Esophagus

A. Voluntary phase
The tongue pushes the bolus of food against the hard palate. Next, the tongue pushes the bolus against the soft palate, which triggers the swallowing response.

B. Involuntary phase
The soft palate rises, which prevents the bolus from entering the nasal cavity. The epiglottis covers the trachea, blocking the opening to the lungs. The bolus enters the esophagus and is propelled toward the stomach by peristalsis.

The first phase of swallowing is under our control (voluntary) whereas the second phase is not (involuntary).

FIGURE 3.11 GASTROESOPHAGEAL REFLUX DISEASE (GERD)

The mucosal lining of the esophagus can become inflamed when repeatedly exposed to the acidic stomach contents.

Gastro-esophageal sphincter

Gastroesophageal reflux disease (GERD) is caused by dysfunction of the gastroesophageal sphincter. If the gastroesophageal sphincter weakens, the stomach contents flow back into the esophagus. The lining of the esophagus can become inflamed when it is exposed to the contents of the stomach.

gastroesophageal reflux disease (GERD) A condition caused by chronic reflux of the stomach contents into the esophagus, irritating its lining.

gastric juice Digestive secretions that consist mainly of water, hydrochloric acid, digestive enzymes, mucus, and intrinsic factor.

gastric mucosal barrier A thick, gel-like substance that protects the stomach lining from acidic gastric juices.

gastrin A hormone released by endocrine cells in the stomach lining that stimulates the release of gastric juice and causes the muscular wall of the stomach to contract vigorously.

chyme A semi-liquid paste resulting from the mixing of partially digested food with gastric juice in the stomach.

Gastroesophageal Reflux Disease The gastroesophageal sphincter is sometimes unable to prevent the contents of the stomach from reentering the esophagus, which is referred to as *gastric reflux* (see Figure 3.11). Over time, chronic reflux can result in a condition called **gastroesophageal reflux disease (GERD)**, whereby the lining of the esophagus becomes irritated, causing a burning sensation in the chest (commonly referred to as heartburn). Fortunately, most people are able to manage heartburn by changing their lifestyles. Avoiding large meals and certain types of foods (e.g., caffeinated beverages, mint, and fried foods) are often effective prevention and management strategies. However,

if left untreated, GERD can lead to serious conditions such as inflammation of the esophagus. For this reason, it is important for anyone experiencing GERD to seek medical attention and treatment.[2]

3-3e The Stomach Mixes and Stores Food

The stomach is uniquely equipped to carry out two important functions: (1) mixing food with the gastric secretions that aid in chemical digestion and (2) temporarily storing food. Specialized cells in the lining of the stomach produce a variety of substances that aid in the process of digestion. These digestive secretions, collectively referred to as **gastric juice**, consist mainly of water, hydrochloric acid, digestive enzymes, mucus, and intrinsic factor, a protein that is essential to vitamin B_{12} absorption.

Your stomach produces more than 2 quarts (roughly 2 liters) of gastric juice daily. Although the hydrochloric acid and digestive enzymes in gastric juice would normally damage most tissues, the lining of the stomach is protected by a layer of mucus called the **gastric mucosal barrier**. This thick, tenacious gel-like substance explains why the stomach lining can withstand this hostile, corrosive environment. Damage to this protective layer of mucus can result in inflammation or *gastritis*, which can subsequently lead to the formation of a gastric ulcer. Contrary to popular belief, the majority of ulcers are not caused by stress or by eating spicy foods. Most ulcers are caused by a small, spiral-shaped bacterium called *Helicobacter pylori* (*H. pylori*). However, some medications (such as aspirin) and chronic, excessive consumption of alcohol can cause ulcers as well.[3]

The presence of food in the stomach stimulates the release of a hormone called **gastrin**, which is produced by endocrine cells in the stomach's lining. Gastrin stimulates the release of gastric juice and causes the muscular wall of the stomach to contract vigorously. These powerful muscular contractions, much like the action of kneading bread, force the bolus to mix with the acidic gastric juice. Within 3 to 5 hours after eating a meal, the partially digested food is mixed thoroughly with the gastric juice. By the time the food leaves the stomach, it has been transformed into a semi-liquid paste called **chyme**.

It is important that all the digestive events that occur in the stomach are completed before chyme enters the small intestine. To make sure that this happens, the stomach serves as a temporary storage facility. As your

The majority of ulcers are caused by this small spiral-shaped bacterium called *Helicobacter pylori* (*H. pylori*).

3-3f The Small Intestine Completes the Digestion Process

The small intestine, a narrow, 20-foot-long tube with a diameter of about 1 inch, is the primary site of chemical digestion and nutrient absorption. After leaving the stomach, chyme passes into the first segment of the small intestine, the **duodenum**. The duodenum also receives secretions from the accessory organs—the gallbladder and the pancreas (see Figure 3.12).

The lining of the small intestine has a large surface area, which is aptly suited for the processes of digestion and nutrient absorption. As illustrated in Figure 3.13, the inner lining of the small intestine is arranged in large rounded folds that face inward (toward the lumen of the small intestine). These rounded folds are covered with millions of small finger-like projections called **villi**. Each villus contains blood capillaries and a lymphatic vessel (lacteal), both of which circulate nutrients once absorbed.

Another way to think about the inner lining of the small intestine is to imagine a bathroom rug folded like an accordion. The folds in the rug represent the large rounded folds, whereas each tiny loop that covers the surface of the rug represents a villus. Villi (the plural form of villus) consist of specialized epithelial cells called **enterocytes**. Lined up closely, side-by-side,

stomach fills with food, its walls expand (much like an accordion). This stretching triggers a neural response, signaling the brain that the stomach is becoming full. In turn, the brain sends a signal that hunger has been satisfied, causing a person to stop eating. Feeling the sensation of fullness is an important aspect of regulating food intake.

The **pyloric sphincter**, located at the base of the stomach, regulates the flow of chyme into the small intestine. With each peristaltic wave, a few teaspoons of chyme squeeze through the sphincter as it briefly relaxes. The remaining chyme tumbles back and forth in the stomach, allowing for even more mixing. This slow release of chyme allows the small intestine to prepare for its role in the digestive process.

A number of factors influence the rate of **gastric emptying**—the process by which food leaves the stomach and enters the small intestine. The consistency of a food, for example, is an important factor. Because solid foods take more time to liquefy than liquids, they must remain in the stomach longer. Even the nutrient composition of a food impacts the rate of gastric emptying. In general, high-fat foods make a person feel full for longer periods of time.[4] This is because high-fat foods take more time to digest and therefore remain in the stomach longer than foods that contain little fat. Beyond consistency and nutrient composition, volume, or how much a person eats, can also influence the rate of gastric emptying. Large meals increase the strength of peristaltic contractions, which in turn increase the rate of gastric emptying.

pyloric sphincter A circular muscle that regulates the flow of chyme from the stomach into the small intestine.

gastric emptying The process by which food leaves the stomach and enters the small intestine.

duodenum The first segment of the small intestine.

villi Small finger-like projections that cover the inner lining of the small intestine (note that villus is the singular form of villi).

enterocytes Epithelial cells that make up villus.

> "The combined surface area of the villi and microvilli that make up the small intestine is approximately the size of a standard tennis court."

FIGURE 3.12 OVERVIEW OF THE SMALL INTESTINE AND ACCESSORY ORGANS

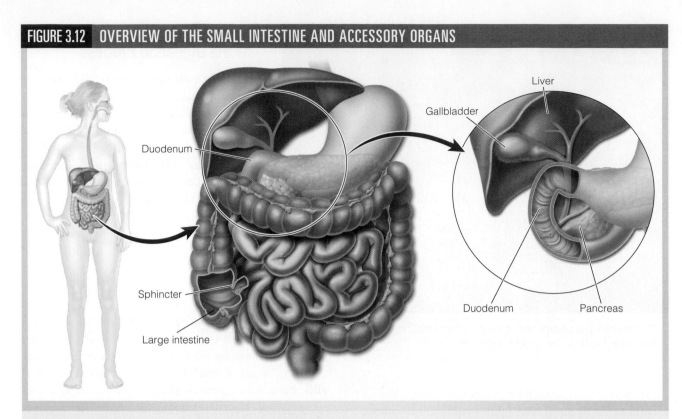

Duodenum

Sphincter

Large intestine

Gallbladder

Liver

Duodenum

Pancreas

The liver, gallbladder, and pancreas (accessory organs) work closely with the small intestine to facilitate the process of nutrient digestion.

enterocytes are held together by interlocking proteins, referred to as **tight junctions**. Sometimes tight junctions develop gaps, allowing partially digested food particles or other substances to leak through. This can trigger an immunological response, which is known as a food allergy.

Enterocytes are covered on their lumenal (outward) surfaces with thousands of minute projections, each of which is called a **microvillus**. It is here, on the microvilli (the plural form of microvillus), where the final stages of nutrient digestion take place.

The enterocytes also produce digestive enzymes. These enzymes facilitate the chemical reactions that break food particles down into the smallest components yet. The small intestine also releases hormones that help regulate digestion by coordinating the release of secretions from the pancreas and gallbladder.

These actions ensure that nutrient digestion and absorption in the small intestine are rapid and efficient. Indeed, shortly after the arrival of chyme in the small intestine, the final stages of digestion are usually complete.

3-3g The Pancreas and Gallbladder Play Important Roles in Digestion

The pancreas plays many important roles in digestion. It protects the small intestine from the acidity of chyme and it also supplies various enzymes necessary for digestion (see Figure 3.14). The arrival of chyme in the small intestine stimulates the pancreas to release **pancreatic juice**, a mixture of water, bicarbonate, and digestive enzymes. Pancreatic juice is released into a duct that empties directly into the upper region (duodenum) of the small intestine. Being a particularly strong base, the bicarbonate quickly neutralizes the acidic chyme as it enters the duodenum. Another fluid that plays an important role in digestion—especially when fatty foods are consumed is **bile**. Although bile is produced in the liver, it is stored in the gallbladder for quick release into the small intestine. When high-fat foods are consumed, the gallbladder releases bile, which acts like a detergent. Bile causes large globules of fat to disperse into

tight junctions Interlocking proteins between adjacent enterocytes that create an impermeable barrier.

microvillus A tiny finger-like projection. Microvilli cover the lumenal surfaces of the enterocytes that line lumen of the small intestine.

pancreatic juice A mixture of water, bicarbonate, and various enzymes released by the pancreas.

bile A fluid produced in the liver and released by the gallbladder that disperses large globules of fat into smaller droplets that are easier to digest.

FIGURE 3.13 ABSORPTIVE SURFACE OF THE SMALL INTESTINE

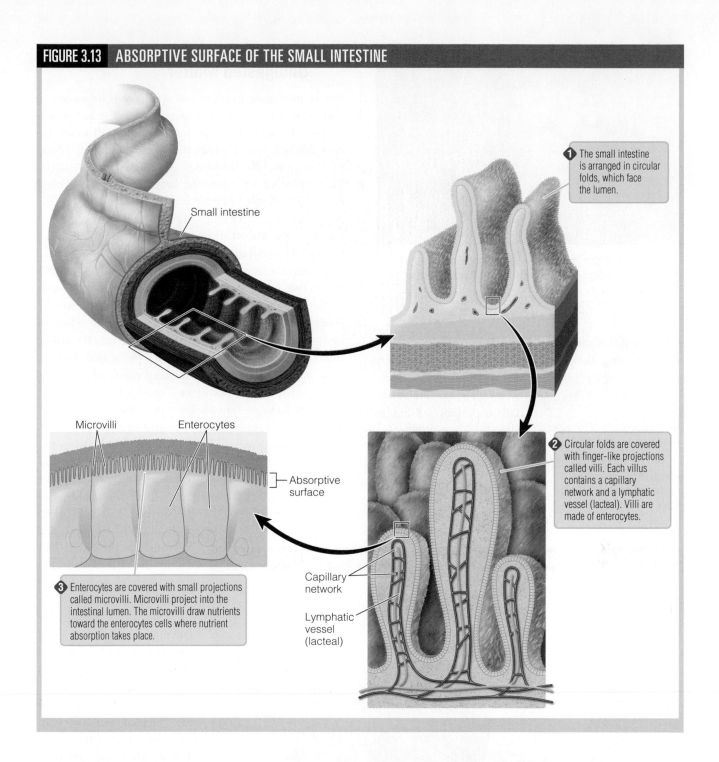

Small intestine

❶ The small intestine is arranged in circular folds, which face the lumen.

❷ Circular folds are covered with finger-like projections called villi. Each villus contains a capillary network and a lymphatic vessel (lacteal). Villi are made of enterocytes.

Microvilli Enterocytes

Absorptive surface

❸ Enterocytes are covered with small projections called microvilli. Microvilli project into the intestinal lumen. The microvilli draw nutrients toward the enterocytes cells where nutrient absorption takes place.

Capillary network

Lymphatic vessel (lacteal)

smaller fat droplets that are easier to digest. This process, called **emulsification**, enables fats and water to form a smooth, uniform mixture.

Gallstones The formation and accumulation of hard, pebble-like deposits inside the gallbladder, referred to as *gallstones*, can interfere with the normal flow of bile. Some people with gallstones have no symptoms, while others experience extreme pain. When gallstones block the passage of bile on its way from the gallbladder to the small intestine, the gallbladder can become enlarged and inflamed, causing pain. Although the exact cause of gallstone formation is not clear, it is more common in women than men, and the risk of developing gallstones increases with age. Other risk factors associated with gallstone formation include obesity, rapid weight loss, and pregnancy. Often, surgical removal of the gallbladder is the only

emulsification The breakdown of large fat globules into smaller droplets that aids in the overall process of digestion.

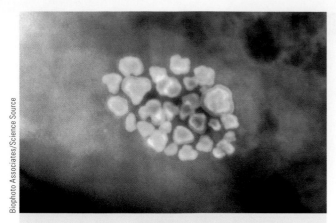

This x-ray of the gallbladder shows gallstones, which are composed of calcium, cellular debris, and cholesterol.

3-3h The Large Intestine Eliminates Undigested Matter

Before discussing nutrient absorption, it is important to understand the events that take place in the last region of the GI tract—the large intestine. Not all foods are completely digested, which means that undigested food residue (mostly the remains of plant-based foods) exit the small intestine and enter the large intestine. The large intestine is shaped like an inverted letter U (∩) and is approximately 5 feet long. The first portion of the large intestine, the **colon**, receives approximately 2 liters of undigested material from the small intestine every day. A sphincter separating the small from the large intestine serves two main functions (1) to prevent the contents of the large intestine from flowing backward into the small intestine and (2) to prevent the premature flow of undigested food from the small intestine into the colon. Following the colon is the **rectum**, the segment of the large intestine that leads to the anal canal, which opens to the outside of the body (see Figure 3.15).

Materials entering the large intestine consist mostly of undigested remains from plant-based foods, water, bile, and electrolytes. Muscles embedded within the wall of the colon squeeze the undigested food residue, slowly propelling the material forward. As material moves through the various regions of the colon, massive amounts of water and electrolytes are absorbed and returned to the blood for reuse by the body. This exemplifies the body's ability to reclaim its important resources.

way to treat this painful condition. Because bile is necessary for fat digestion, you may wonder how people get by without their gallbladders. Although a person may initially experience difficulty with fat digestion after gallbladder removal, the liver continues to produce bile, which subsequently can be released into the small intestine.

colon The first portion of the large intestine.

rectum The segment of the large intestine that leads to the anal canal.

GI microbiota The natural microbial population that resides in the GI tract.

The slow, propulsive movement of the large intestine also provides an ideal environment for bacteria to grow and flourish. Although bacteria reside throughout the entire GI tract, their density is greatest in the colon. The number and variety of bacteria residing in the large intestine is astronomical—more than 400 different species call your large intestine home. This natural microbial population, also referred to as the **GI microbiota**, helps maintain a healthy environment throughout the GI tract. Some bacteria residing

FIGURE 3.14 THE PANCREAS HAS AN IMPORTANT ROLE IN NUTRIENT DIGESTION

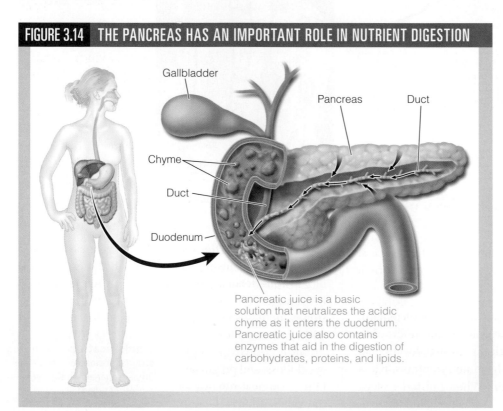

Gallbladder

Pancreas

Duct

Chyme

Duct

Duodenum

Pancreatic juice is a basic solution that neutralizes the acidic chyme as it enters the duodenum. Pancreatic juice also contains enzymes that aid in the digestion of carbohydrates, proteins, and lipids.

FIGURE 3.15 OVERVIEW OF THE LARGE INTESTINE

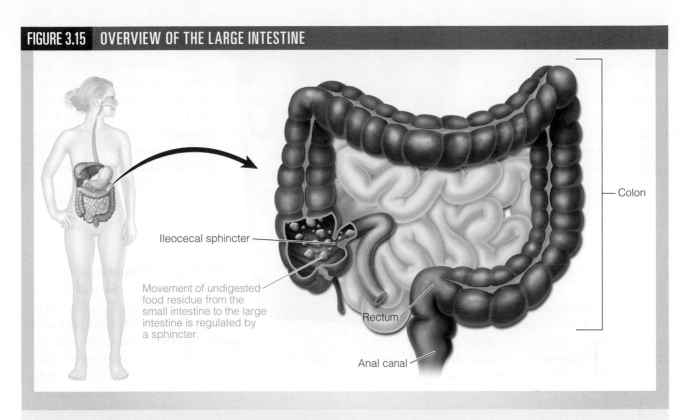

Colon

Ileocecal sphincter

Movement of undigested food residue from the small intestine to the large intestine is regulated by a sphincter.

Rectum

Anal canal

The large intestine has three general regions: the colon, the rectum, and the anal canal.

Power and Syred / Science Source

Millions of bacteria thrive throughout the gastrointestinal tract.

in your colon break down undigested food residue; others produce nutrients such as vitamin K, certain B vitamins, and some lipids. A healthy and diverse GI microbiota can help inhibit the growth of disease-causing bacteria. There is also evidence that the microbial population residing in your colon may reduce risk to certain diseases such as colon cancer.

To help establish and maintain a healthy intestinal microbiota, it is important to consume both probiotic and prebiotic foods. A **probiotic food** is one that contains live microorganisms that are beneficial in the body. Fermented foods such as yogurt, sauerkraut, sour pickles, and certain soy products are examples of probiotic foods. Some dietary supplements also supply probiotic bacteria. A **prebiotic food** is typically fiber-rich and may stimulate the proliferation of the microbial population in the large intestine by providing a source of nourishment. Some health care professionals believe that consuming both probiotic and prebiotic foods provides a powerful defense against disease-causing bacteria.[5]

Once the remaining material completes its journey through the large intestine, it is ready to be eliminated from the body. This solid waste, now called **feces**, consists mainly of undigested and unabsorbed matter, dead cells, secretions from the GI tract,

probiotic food A food that contains live bacteria, some of which thrive in the colon.

prebiotic food A typically fiber-rich food that may stimulate the growth of the microbial population in the large intestine.

feces Solid waste consisting mainly of undigested and unabsorbed matter, dead cells, secretions from the GI tract, water, and bacteria.

water, and bacteria. As the feces approaches the end of the colon, it passes into the rectum, which serves as a holding chamber. An accumulation of feces causes the walls of the rectum to stretch, signaling the need to defecate. Unlike other sphincters in the GI tract, the sphincter located between the rectum and the anal canal is under voluntary control. This enables a person to determine whether the time is right for waste elimination. When the sphincter relaxes, the feces move into the anal canal and are expelled from the body.

© Christopher Gardiner/Shutterstock.com

The consistency of feces depends mainly on its water content. If undigested food residue moves too quickly through the colon, a sufficient amount of water cannot be extracted from it. This results in loose, watery feces referred to as *diarrhea*. Prolonged diarrhea can cause excessive water loss from the body, which can lead to dehydration. Conversely, when the contents in the large intestine move too slowly, too much water may be absorbed. This can cause the fecal matter to become hard and dry, resulting in a condition called *constipation*. Constipation can make elimination difficult and can put excessive strain on the muscles in the colon wall.

Intestinal Disorders Diarrhea and constipation are not the only conditions that can cause intestinal discomfort. **Irritable bowel syndrome (IBS)** is a disorder that typically affects the lower GI tract, causing bouts of cramping, bloating, diarrhea, and constipation. Because IBS is not associated with any known structural abnormalities, it is considered a functional (as opposed to anatomical) disorder. Although the underlying cause of IBS is

irritable bowel syndrome (IBS) A disorder that typically affects the lower GI tract, causing bouts of cramping, bloating, diarrhea, and constipation.

Nutrient + Pharmaceutical = Nutraceutical

The term *nutraceutical* is used to describe foods that are consumed to prevent and/or treat certain diseases. Similarly, the term *functional foods* refers to foods that provide health benefits beyond basic nutrition. Though these terms are recent inventions, the perception that certain foods may have medicinal value is not new: many ancient civilizations used foods to treat a variety of ailments. Today, the use of foods to treat certain disorders is viewed by some as a type of alternative medicine. However, this practice has gained considerable recognition in the medical community as a viable way to manage and treat some health problems. The consumption of a probiotic food to promote intestinal health is a good example of how nutraceuticals might contribute to overall well-being. Researchers are currently exploring whether probiotics can relieve conditions such as infectious diarrhea, irritable bowel syndrome, inflammatory bowel disease, tooth decay, and vaginal infections.

© Elena Elisseeva/Shutterstock.com

Other foods believed to have therapeutic purposes include cranberry juice (to help prevent urinary tract infections), soy (for the management of menopausal symptoms), garlic (to lower blood cholesterol levels), and certain types of fish (to prevent heart disease). Although consumption of neutraceuticals is a growing trend, it is important to recognize that some of the health benefits attributed to medicinal foods may be due to the placebo effect.[6]

Phytochemicals found in cranberries may help protect against the recurrence of urinary tract infections. This is an example of a nutraceutical food.

yet to be determined, some believe that it is caused by elevated levels of a neurohormone released by nerves residing in the GI tract wall. This neurohormone, called serotonin, is also found in the brain. Whereas low levels of serotonin in the brain are associated with depression, elevated levels in the lower GI tract are thought to increase a person's sensitivity to pain caused by stretching of the colon wall. Because emotional stress can adversely affect GI function, people with IBS are encouraged to have healthy outlets such as yoga, exercise, or other forms of physical activity. Diet is also important in managing IBS. Some foods can trigger muscle spasms in the colon. This is why it is important for people with IBS to identify which foods help manage IBS, and which foods make it worse.

While the terms *irritable bowel syndrome* and *inflammatory bowel disease* may sound similar, they refer to very different conditions. **Inflammatory bowel disease (IBD)** is a group of inflammatory conditions that includes ulcerative colitis and Crohn's disease. Both are autoimmune diseases, and are characterized by inflammation of the lining of the lower GI tract. Many researchers believe that IBD develops when exposure to something in the environment causes an immunologic response. Factors believed to trigger this response include certain proteins, viruses, and bacteria. When exposure triggers the production of antibodies, the intestinal lining becomes inflamed. IBD's signs and symptoms include diarrhea, fatigue, weight loss, abdominal pain, diminished appetite, and on occasion, rectal bleeding.[7] Prolonged periods of inflammation can permanently damage the delicate lining of the large intestine.

3-4 WHAT HAPPENS AFTER DIGESTION? NUTRIENT ABSORPTION AND CIRCULATION

In the previous section you learned how food moves through the GI tract and is systematically broken down. When the process of digestion is complete, nutrients are ready to be absorbed. Nutrient absorption is the movement of nutrients from the lumen of the GI tract into either the blood or the lymph (Figure 3.16). Although some nutrient absorption takes place through the lining of the stomach, the vast majority of nutrients are absorbed in the small intestine.

3-4a Nutrient Absorption

In the small intestine, microvilli move in a sweeping action to trap and pull nutrients toward the enterocytes that line each villus. During absorption, nutrients move into and out of the enterocytes by both passive and active transport mechanisms. Once absorbed, nutrients enter either the blood or lymphatic circulatory system.

Nobody absorbs 100 percent of all the nutrients consumed in food. This is because a variety of factors can affect nutrient absorption. The extent to which a nutrient is absorbed into the enterocyte is called its **bioavailability**. The bioavailability of a particular nutrient can be influenced by physiological conditions, other dietary components, and certain medications. For example, the physiological needs of the body can influence the amount of iron absorbed, a function that protects you from iron toxicity. Some nutrients can markedly affect the bioavailability of other nutrients. For example, the presence of vitamin C can enhance the absorption of certain forms of iron, and too much dietary fiber can decrease the bioavailability of certain minerals such as calcium, iron, and zinc. It is important to be aware of the impact that certain nutrients and drugs have on nutrient bioavailability. Certain diseases such as celiac disease can also impact nutrient bioavailability. When a person with celiac disease consumes foods that contain gluten, the body produces antibodies that damage the absorptive surface of the small intestine. As a result, nutrient absorption (bioavailability) is impaired and a person can develop nutrient deficiencies.

3-4b Nutrient Circulation

Once food has been digested and nutrients have been absorbed, the body's next task is to transport the nutrients throughout the body. This is accomplished by veins and arteries that comprise the blood circulatory system and by lymphatic

inflammatory bowel disease (IBD) Inflammatory conditions such as ulcerative colitis and Crohn's disease that affect the lining of the lower GI tract.

bioavailability The extent to which a nutrient is absorbed into the enterocyte.

FIGURE 3.16 NUTRIENT ABSORPTION AND CIRCULATION

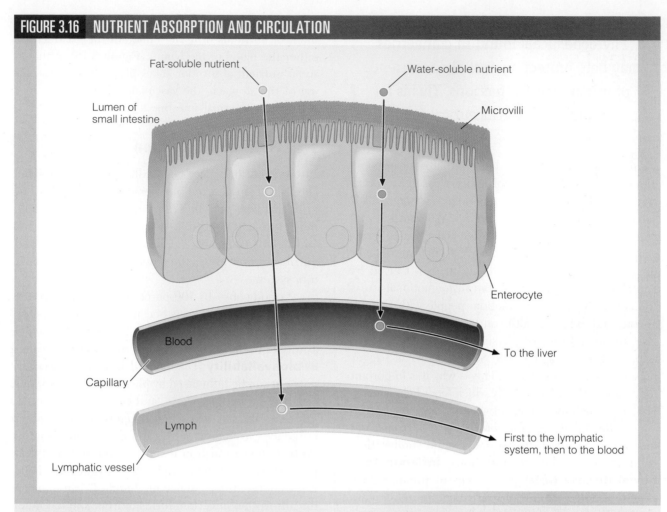

Fat-soluble nutrient

Water-soluble nutrient

Lumen of small intestine

Microvilli

Enterocyte

Blood

To the liver

Capillary

Lymph

First to the lymphatic system, then to the blood

Lymphatic vessel

Nutrient absorption includes both entry into and exit out of enterocytes. Nutrients then circulate away from the intestine in either blood or lymph. Water-soluble nutrients are circulated in blood, whereas fat-soluble nutrients are circulated in lymph. Eventually, the lymph releases the nutrients into the blood.

vessels that comprise the lymphatic circulatory system. These extensive systems both deliver the nutrients to where they are needed in the body, and aid in the elimination of waste products generated by cells.

The Cardiovascular and Lymphatic Systems Circulate Nutrients to other parts of the body

Upon absorption, water-soluble nutrients enter the bloodstream through the capillaries contained within each villus. Once water-soluble nutrients enter the bloodstream, they circulate directly to the liver. Thus, the liver is the recipient of nutrient-rich blood that circulates directly from the small intestine. The liver then regulates the use of the nutrients to suit the body's needs. Small amounts of some nutrients

lacteal A lymphatic vessel found in an intestinal villus into which nutrients circulate after they are absorbed.

are stored in the liver, but most either undergo chemical modification or are released directly into blood, and circulated to where they are needed in the body.

The lymphatic system also plays an important role in the circulation of fat-soluble nutrients (mostly lipids and some fat-soluble vitamins) away from the small intestine. Each villus contains a lymphatic vessel—a **lacteal**—through which the nutrients are absorbed. Each lacteal connects to a larger network of lymphatic vessels that circulate a translucent liquid called lymph. Though the circulatory route of the lymphatic system initially bypasses the liver, it eventually delivers the nutrients to the blood. Once in the blood, nutrients can be taken up and used by cells.

Other examples of the excretory process include the exhalation of air through the nose and mouth (to rid the body of carbon dioxide), and the active process of sweating (to rid the body of water and salt).

Celiac Disease

Celiac disease is an autoimmune disease that causes an inflammatory response to gluten, a specific protein found in a variety of cereal grains such as wheat, rye, barley, and possibly oats. When a person with celiac disease consumes gluten-containing foods, the lining of the small intestine becomes damaged. Researchers only came to understand celiac disease and how to treat it within the last 50 years. They now know that people with celiac disease (or gluten-sensitive enteropathy) experience an immunological response to gluten. More specifically, consumption of gluten triggers the production of antibodies that attack the intestinal microvilli, causing them to flatten. As a result, nutrient absorption is impaired. Because of the progressive damage done to the inner lining of the small intestine, people with celiac disease often experience diarrhea, weight loss, and malnutrition. Because its signs and symptoms are similar to other common GI disorders, celiac disease is often misdiagnosed. When it is suspected, a blood test may be performed to confirm whether a person actually has celiac disease. It is not clear if the occurrence of celiac disease is on the rise or if diagnostic tests have improved, and therefore, it is more readily detected. Nonetheless, once diagnosed, a person with celiac disease can live symptom-free by eliminating gluten from the diet.[8]

For people with celiac disease, it is important to buy foods that are gluten free. A variety of products are available in most grocery stores.

© Keith Homan/Shutterstock.com

3-5 HOW ARE NUTRIENTS METABOLIZED AND HOW ARE METABOLIC WASTE PRODUCTS EXCRETED FROM THE BODY?

After nutrients are circulated away from the GI tract, they are delivered to and taken up by cells. The work of the digestive system is complete—at least until the next meal. Once a nutrient enters a cell, the complex process of nutrient metabolism begins. **Metabolism** is the sum of chemical processes that occur within a living cell to maintain life. In short, when a cell uses a nutrient, it metabolizes it. The process of metabolism generates waste products such as ammonia and carbon dioxide that must be removed from cells through the excretory process. Some metabolic processes utilize nutrients to build and support the body's structures, such as bones and muscle. Others transfer the energy stored in energy-yielding nutrients to the powerful chemical bonds of ATP, the major energy currency of cells. ATP is a perfectly designed, intricate molecule that fuels thousands of chemical reactions required to sustain life.

The processes by which cells use and store energy involve a series of chemical reactions referred to collectively as **energy metabolism**. Energy metabolism enables cells to transform the chemical energy stored in certain nutrients ultimately providing cells with adequate ATP.

3-5a Metabolic Pathways

Metabolic processes are generally complex, requiring many steps. Recall that cells contain mitochondria, which are organelles that play an important role in converting the chemical energy found in certain nutrients into ATP. This transformation occurs through a series of interrelated chemical reactions collectively called a **metabolic pathway**. Each chemical reaction in a metabolic pathway requires the help of at least one enzyme. To function properly, enzymes often require the assistance of certain vitamins and minerals. Thus, although they

> **metabolism** The sum of chemical processes that occur within a living cell to maintain life.
>
> **energy metabolism** Chemical reactions that enable cells to use and store energy.
>
> **metabolic pathway** A series of interrelated chemical reactions that require enzymes.

themselves do not provide cells with a source of energy, vitamins and minerals play an important role in energy metabolism.

Some metabolic pathways break down complex molecules into simpler ones, releasing energy and heat in the process. This process is referred to as **catabolism**. Conversely, a series of metabolic reactions that uses energy to construct a complex molecule from simpler ones is referred to as **anabolism**. Thus, catabolism and anabolism can be thought of, respectively, as series of "break it" and "make it" chemical reactions (see Figure 3.17).

Catabolism and the Release of Energy

Energy-yielding nutrients (carbohydrates, proteins, and lipids) store energy in their chemical bonds. However, energy-yielding nutrients must first undergo a series of chemical reactions in order for mitochondria to synthesize ATP. Often, the chemical reactions that make up a catabolic pathway require oxygen to function. These are referred to as **aerobic**, or oxygen requiring. Catabolic pathways that can function under conditions of low oxygen availability are referred to as **anaerobic**. Both aerobic and anaerobic pathways play important roles in making energy available to cells. For example, when you perform a moderate-intensity physical exercise, such as jogging, for an extended period of time, your lungs and heart work strenuously to pump oxygen-rich blood throughout your body so that ATP can be generated through aerobic pathways. During more intense exercise, such as strength training, it can be difficult for your lungs and heart to deliver adequate amounts of oxygen to active muscle cells. When this occurs, anaerobic catabolic pathways help sustain you by generating enough ATP to prevent muscle fatigue. There is a limit to how much energy can be generated under conditions of limited oxygen availability, however. This is why muscles tire more quickly during high-intensity workouts.

catabolism A series of metabolic reactions that breaks down a complex molecule into simpler ones, releasing energy in the process.

anabolism A series of metabolic reactions that uses energy to construct a complex molecule from simpler ones.

aerobic A metabolic pathway that requires oxygen to function.

anaerobic A metabolic pathway that can function under conditions of low oxygen availability.

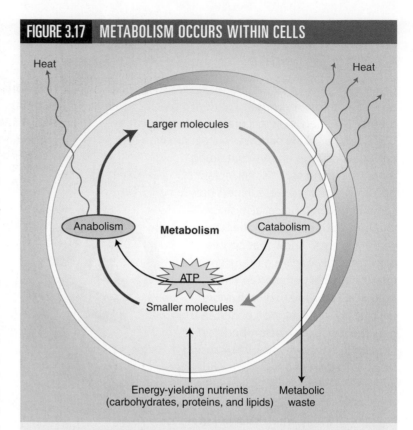

FIGURE 3.17 METABOLISM OCCURS WITHIN CELLS

Catabolism breaks down large molecules into small molecules, releasing energy (ATP). Anabolism assembles small molecules into large molecules, which requires energy (ATP).

When there is a need to generate ATP rapidly, muscles rely primarily on anaerobic metabolic pathways.

Courtesy of Kathy Beerman

Anabolism and Making New Molecules

Whereas catabolic pathways usually result in the release of energy, anabolic pathways serve a different function—the synthesis of new molecules and/or the storage of energy. An example of an anabolic process at work is the construction of the many proteins needed for tissue growth, maintenance, and repair. When a person has more energy than needed cells use anabolic pathways to store energy-yielding nutrients for later use—sort of like depositing money in the bank. For example, glucose molecules not immediately needed for energy can be converted to the glucose-storage molecule glycogen. When glycogen stores are full, cells turn the remaining glucose into fatty acids, which are subsequently stored as body fat. These anabolic reactions provide cells with an energy reserve, a critical component for survival when food is limited and/or energy requirements are high.

3-5b Excretion of Metabolic Waste Products

Not only does blood deliver nutrients and oxygen to cells, but equally important, blood also transports metabolic waste products away from cells. The thousands of metabolic reactions taking place in your cells generate more than just energy; metabolic waste products such as water, carbon dioxide, and nitrogen-containing compounds are produced as well. Although the term elimination refers to the removal of solid waste (feces) from the body, the term **excretion** refers to the removal of metabolic waste products. The removal of solid and metabolic waste products are both vital to your health.

The main excretory organs involved in the removal of metabolic waste products include the liver, kidneys, lungs, and skin. Collectively, these organs help prevent the accumulation of toxic waste products by aiding in their removal. For example, the liver converts ammonia— a nitrogen-containing by-product of protein breakdown—into a less toxic substance called urea, which is released into the blood. The kidneys then filter the urea out of the blood so it can be excreted from the body in urine. This is why people with impaired kidney function often undergo a treatment called dialysis. Dialysis involves a special machine that performs similar functions of healthy kidneys. Without dialysis, toxic metabolic waste products would accumulate in the blood, eventually resulting in death.

© zulufoto/Shutterstock.com

excretion The removal of metabolic waste products produced in cells.

4 | Carbohydrates

LEARNING OUTCOMES

4-1 Define and distinguish simple carbohydrates.

4-2 Define and distinguish complex carbohydrates.

4-3 Explain how the body digests, absorbs, and circulates carbohydrates.

4-4 Describe how glucose is utilized and stored in the body, and how hormones regulate levels of glucose in the blood.

4-5 Differentiate between Type 1 and Type 2 diabetes.

4-6 Summarize carbohydrate intake recommendations.

After finishing this chapter go to **PAGE 96** for **STUDY TOOLS.**

4-1 WHAT ARE SIMPLE CARBOHYDRATES?

Carbohydrates comprise a diverse group of compounds produced primarily by plants. Some carbohydrates contain only one or two "sugar" molecules, whereas others comprise hundreds of them. As the number of sugar molecules increases, so does the size and complexity of the carbohydrate. Carbohydrates, which are plentiful in a variety of foods, serve many essential functions within the body. As such, they are an important part of a healthy, well-balanced diet. In this chapter, carbohydrate chemistry, digestion, dietary sources, functions, and dietary requirements will be examined.

carbohydrates An organic compound made up of one or more sugar molecules.

monosaccharide A carbohydrate consisting of a single sugar molecule.

disaccharide A carbohydrate consisting of two monosaccharides bonded together.

simple carbohydrates (or **simple sugars**) A category of carbohydrates comprised of monosaccharides and disaccharides.

Carbohydrates are organic molecules made up of carbon, hydrogen, and oxygen atoms (Figure 4.1). The simplest type of carbohydrate is often referred to as a *sugar*—a carbohydrate molecule that cannot be broken down further to produce other sugars. Most people think of sugar as a substance used to sweeten their foods. Although this is true, the uses and functions of sugars extend far beyond sweetening. For example, cells use a special type of sugar called glucose as an important source of energy. A carbohydrate consisting of a single sugar molecule is called a **monosaccharide**, and a carbohydrate consisting of two sugar molecules bonded together is called a **disaccharide**. Because of the small sizes of these molecules, monosaccharides and disaccharides are referred to as **simple carbohydrates** (or **simple sugars**).

4-1a Monosaccharides

There are hundreds of different naturally occurring monosaccharides, but the three that are most plentiful in food are glucose, fructose, and galactose. Although the structures of these sugars differ, they all have one thing in common: each contains six carbon atoms (see Figure 4.2).

Photosynthesis literally means "putting together from light."

Glucose Glucose, the most abundant monosaccharide in the human body, is produced when chlorophyll-containing plants use energy from sunlight to combine carbon dioxide (CO_2) and water (H_2O). This process, called **photosynthesis**, provides plants with an important source of energy (see Figure 4.3). Plants can store glucose by converting it into a larger, more complex carbohydrate called starch and fiber. When you consume plant-based foods, your body breaks down (digests) these large starch molecules into glucose, providing you with an important source of energy. In this way, plants and animals are interconnected in the delicate balance that exists in nature.

Fructose Fructose is a naturally occurring monosaccharide found primarily in honey, fruits, and vegetables. Although it is the most abundant sugar in fruits and vegetables, the majority of fructose consumed in the Western diet comes from foods and beverages made with high-fructose corn syrup. **High-fructose corn syrup (HFCS)** is a widely used sweetener consisting of glucose and fructose that is manufactured from cornstarch. It is found in soft drinks, cereals, fruit juice beverages, soups, and a multitude of other foods. In the early 1970s, food manufacturers began using

glucose The most abundant monosaccharide in the human body; used extensively for energy.

photosynthesis A process whereby chlorophyll-containing plants produce glucose by combining carbon dioxide (CO_2) and water (H_2O) using energy harvested from sunlight.

fructose A naturally occurring monosaccharide found primarily in honey, fruits, and vegetables.

high-fructose corn syrup (HFCS) A widely used sweetener consisting of glucose and fructose that is manufactured from cornstarch.

FIGURE 4.1 CLASSIFICATION OF CARBOHYDRATES

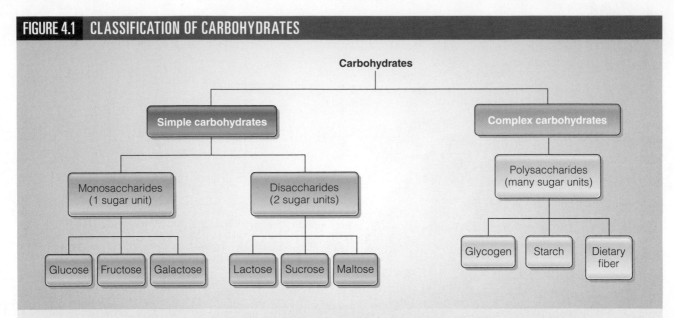

Carbohydrates are classified as either simple or complex. Simple carbohydrates include monosaccharides and disaccharides, whereas complex carbohydrates include polysaccharides such as glycogen, starch, and fiber.

FIGURE 4.2 THE STRUCTURES OF MONOSACCHARIDES

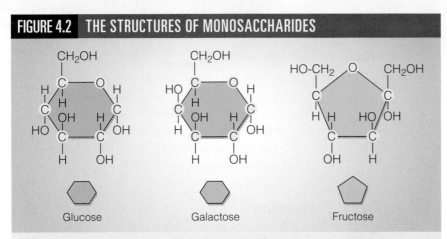

Although glucose, galactose, and fructose are all monosaccharides that contain six carbon atoms, each has a slightly different chemical structure.

HFCS as an inexpensive replacement for sucrose. As illustrated in Figure 4.4, the consumption of HFCS (per capita) steadily increased, reaching an all-time high in 1999.[1] Perhaps due to health concerns associated with HFCS, the average per capita consumption of HFCS has declined in recent years. The U.S. Department of Agriculture estimates that, on average, Americans now consume approximately 127 kcal from HFCS each day, a 30 percent decrease since 2000.[2]

The liver converts the majority of fructose to glucose, which is then used as a source of energy. However, an abundance of fructose in the diet, regardless if it comes from sucrose or HFCS, may be potentially harmful to your health. Of particular concern is a condition called nonalcoholic fatty liver disease. This can occur when the liver converts excess fructose into fat rather than glucose. As a result, fat accumulates in the liver. Researchers have a long way to go before they fully understand the relationship between excess fructose consumption and nonalcoholic fatty liver disease.[3] Nonetheless, most health experts would agree that excessive amounts of added sugar in your diet can be detrimental to your health.

Galactose Glucose and **galactose** look rather similar. However, slight differences in their chemical structures result in important differences in their physiological functions. Few foods contain galactose in its free state. Rather, the majority of dietary galactose comes from a naturally occurring disaccharide in dairy products. Like fructose, the majority of galactose in the body is converted to glucose and used for energy.

galactose A monosaccharide that exists primarily as part of a naturally occurring disaccharide found in dairy products.

FIGURE 4.3 PHOTOSYNTHESIS

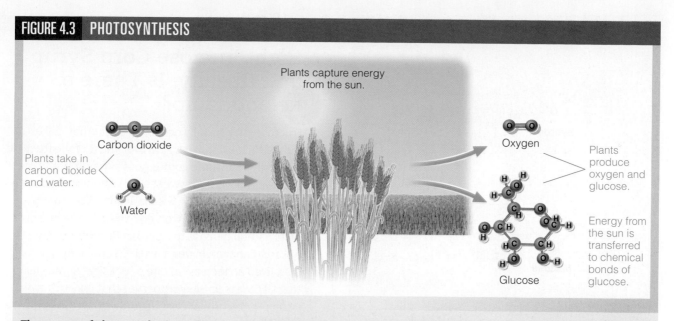

Plants capture energy from the sun.

Carbon dioxide

Plants take in carbon dioxide and water.

Water

Oxygen

Plants produce oxygen and glucose.

Glucose

Energy from the sun is transferred to chemical bonds of glucose.

The process of photosynthesis combines carbon dioxide and water in the presence of sunlight to produce glucose and oxygen.

FIGURE 4.4 TRENDS IN CONSUMPTION OF FOODS AND BEVERAGES SWEETENED WITH HIGH-FRUCTOSE CORN SYRUP AND TABLE SUGAR: 1970–2014

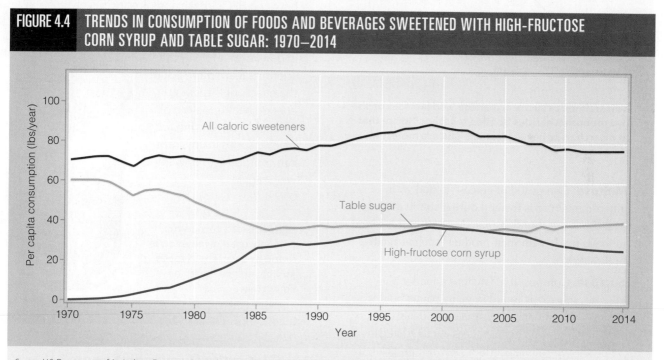

Source: U.S. Department of Agriculture, Economic Research Service. 2012. Table 51-Refined cane and beet sugar: estimated number of per capita calories consumed daily, by calendar year. Table 52-High fructose corn syrup: estimated number of per capita calories consumed daily, by calendar year. Table 53-Other sweeteners: estimated number of per capita calories consumed daily, by calendar year. Sugar and Sweeteners Yearbook 2012. Available from: http://www.ers.usda.gov/data-products/sugar-and-sweeteners-yearbook-tables.aspx.

4-1b Disaccharides

Disaccharides consist of two monosaccharides bonded together. Present in a wide variety of foods, the most common disaccharides are lactose, maltose, and sucrose. Note that, in all of these disaccharides, at least one of the monosaccharides in the pair is glucose (see Figure 4.5).

• **Lactose**: Composed of galactose joined with glucose, lactose is the most abundant carbohydrate in milk and many other dairy products (e.g., yogurt, cheese, and ice cream).

lactose A disaccharide comprised of galactose joined with glucose. It is the most abundant carbohydrate in milk and many other dairy products.

FIGURE 4.5 DISACCHARIDES

Glucose — O — Fructose
Sucrose

- Sucrose consists of glucose bonded to fructose.
- Another name for sucrose is table sugar.
- Sucrose is often added to foods to make them sweet.

Glucose — O — Glucose
Maltose

- Maltose consists of glucose bonded to glucose.
- Very few foods contain maltose.

Galactose — O — Glucose
Lactose

- Lactose consists of glucose bonded to galactose.
- Another name for lactose is milk sugar.
- Lactose is found in milk and most products made from milk.

The disaccharides sucrose, maltose, and lactose consist of two monosaccharides bonded together. Notice that each disaccharide has at least one glucose molecule.

- **Maltose**: Composed of glucose joined with glucose, maltose is formed during starch digestion. Maltose is not found in many foods, but it is found in foods made with malt products such as malted milkshakes.

- **Sucrose**: Composed of fructose joined with glucose, sucrose is found in many plants (and is especially abundant in sugar cane and sugar beets). Crushing these plants produces a juice that can be processed to make a thick brown liquid called molasses. Further treatment and purification of molasses forms pure crystallized sucrose, otherwise known as refined table sugar. Because most people enjoy the intense sweetness of sucrose, it is often added to foods.

maltose A disaccharide composed of glucose joined with glucose. It is not found in many foods.

sucrose A disaccharide composed of fructose joined with glucose. It is naturally occurring and is most abundant in sugar cane and sugar beets.

High-Fructose Corn Syrup and Obesity: Is There a Relationship?

Fructose is a naturally occurring sugar that makes fruits deliciously sweet. However, most of the fructose Americans consume comes not from fruit but from foods and beverages sweetened with high-fructose corn syrup (HFCS). HFCS is made from cornstarch that has been treated with enzymes to convert some of its natural glucose to fructose. Recently, scientists observed that an increase in HFCS consumption has paralleled an increasing rate of obesity.[4] Some claim that HFCS has unique properties that may be responsible for this correlation, but studies have not shown a definite causal relationship between the consumption of HFCS and weight gain. The association between HFCS intake and obesity is likely a product of excess consumption of energy-dense foods and beverages in general, not HFCS *per se*. Until scientists learn more about the effects of this sweetener on health, we can all agree that consuming too many calories in any form can contribute to unwanted weight gain.

High-fructose corn syrup, a widely used sweetener, is found in a variety of foods and beverages—such as soft drinks.

© Warongdech/Shutterstock.com

4-1c Naturally Occurring Sugars and Added Sugars

Foods such as fruits, vegetables, and milk contain naturally occurring sugars, while many processed foods contain added sugars or syrups. On food labels, the term "sugar" usually accounts for both added and naturally occurring sugar. However, the FDA is in the process of changing the information and appearance of food labels. One of the proposed changes pertains to added sugars, and will require food manufacturers to provide information regarding the amount (grams) of added sugars per

TABLE 4.1   ADDED SWEETENERS COMMONLY LISTED ON FOOD LABELS

Brown rice syrup	Invert sugar
Brown sugar	Lactose
Concentrated fruit juice sweetener	Levulose
Confectioner's sugar	Maltose
Corn syrup	Maple sugar
Dextrose	Molasses
Fructose	Natural sweeteners
Glucose	Raw sugar
Granulated sugar	Sucrose
High-fructose corn syrup	Turbinado sugar
Honey	White sugar

serving.[5] As you can see in Table 4.1, many types of sugars and syrups can be used to sweeten foods and beverages. White sugar, brown sugar, corn syrup, honey, and molasses are common added sugars. Although added sugars are chemically identical to naturally occurring sugars, foods with naturally occurring sugars usually contain other important nutrients in addition to their carbohydrates. For example, besides sucrose, fruits are naturally rich in vitamins, minerals, and fiber. By contrast, foods with large amounts of added sugars (e.g., soft drinks, cakes, and candy) often have little nutritional value beyond the calories they contain.

In recent years, consumption of added sugars has decreased slightly throughout the United States. Today, the average American consumes about 82 grams (20 teaspoons) of added sugars—a total of 328 kcal—every day.[6] Although this may not sound like much, the calories can add up quickly. Nearly one-half of the added sugars in the American diet come from soft drinks.[7] Substantial evidence suggests that a greater consumption of sugar-sweetened beverages is associated with increased body weight in adults.[8] This is not surprising when you consider that the average 12-ounce can of sweetened soft drink contains eight teaspoons of added sugar—the equivalent of 130 kcal. Consistent with the World Health Organization's recommendation that adults limit their daily intake of calories from added sugars to 5–10 percent, the American Heart Association also recommends that added sugars be limited to 25 grams (100 kcal) per day for women and 38 grams (152 kcal) for men.[9]

Alternative Sweeteners Some people believe that natural sweeteners, such as honey, present a healthier alternative to processed sweeteners, such as refined white sugar. However, like refined sugars, natural sweeteners possess limited nutritional value beyond the energy they contain. To add sweetness without increasing the caloric contents of foods, food manufacturers often use artificial low-calorie sweeteners such as saccharin, aspartame, and acesulfame K. In addition to these artificial sweeteners, food manufacturers use natural substitutes such as sugar alcohols and stevia, a sweet-tasting herbal dietary product. None of these substances are actual sugars, but each provides a sweet taste nonetheless.

For individuals who want to reduce their caloric intake, choosing a specific artificial sweetener may not be as simple as you might think. Some artificial sweeteners lose their sweetness when heated and therefore cannot be used for baking or cooking. Others are not chemically stable and become bitter over time. In order to use alternative sweeteners safely and effectively, it is important to understand some of their properties.

- **Saccharin:** There was once concern that saccharin might cause cancer, but recent studies support its safety for human consumption. Saccharin is extremely sweet, very stable, and inexpensive to produce. Commercial products with saccharin include Sweet 'N Low® and Sugar Twin®.

- **Aspartame:** Although it has the same energy content as sucrose (4 kcal per gram), aspartame is almost 200 times sweeter. The food industry uses aspartame in many sugar-free beverages, but because it is not heat stable, aspartame cannot be used in products that require cooking. Although the U.S. Food and Drug Administration (FDA) has ruled aspartame safe for most people, some individuals claim that it causes adverse effects such as headaches, dizziness, nausea, and seizures. While several studies have reported that aspartame increases the occurrence of cancer in laboratory animals, hundreds of studies have failed to find similar effects in humans.[10] People with the genetic disorder phenylketonuria (PKU)

The average American consumes 20 teaspoons (328 kcal) of added sugars every day.

©istockphoto.com/ NoDerog

should not consume products sweetened with aspartame because they are not able to metabolize it properly. Commercial products with aspartame include NutraSweet® and Equal®.

- **Acesulfame K:** Used extensively throughout Europe, acesulfame K was approved in 1998 for use in the United States, where it is sold under the trade name Sweet One®. Unlike aspartame, this artificial sweetener is heat stable and is used in a wide variety of prebaked commercial products.

- **Sucralose:** Even though sucralose is roughly 600 times sweeter than sucrose, it provides minimal calories because it is difficult for the body to digest and absorb. Since it is water soluble and stable, sucralose is used in a broad range of foods and beverages. Commercial products with sucralose include Splenda® and Altern®.

- **Sugar alcohols:** Sugar alcohols occur naturally in plants (particularly fruits) and have half the sweetness and calories of sucrose (roughly 2.5 kcal per gram). Sorbitol, mannitol, and xylitol (the most common sugar alcohols) are often found in sugar-free products such as chewing gums, breath mints, candies, toothpastes, mouthwashes, and cough syrups. One advantage of sugar alcohols is that, unlike sucrose, they do not promote tooth decay. When consumed in excessive amounts however, sugar alcohols can cause diarrhea.

- **Stevia:** The term *stevia* is used to describe a group of related plants that grow in some regions of Central and South America, as well as in the southwestern region of the United States. The leaves of stevia plants are particularly sweet, a fact long recognized by the native peoples of these semi-arid regions. Stevia is essentially free of calories and is considerably sweeter than table sugar. Commercial products with stevia include Truvia® and NuStevia™.

Stevia, an herb that grows in semi-arid climates, is essentially calorie free and is considerably sweeter than table sugar.

- **Agave:** The Agave plant grows in warm, dry regions of the United States as well as in parts of South America. It is the same plant that is used to make tequila. Although there may be some health benefits associated with Agave in its natural, native state, these benefits are lost after the sugars are extracted to produce a thick, sweet syrup. In fact, you may be surprised to learn that agave has more calories per tablespoon than table sugar (60 vs. 48 kcal, respectively). Because Agave is derived from plants, some people believe it is a "natural" healthier alternative compared to other sweeteners. However, there is not much research to support this claim, and like all other sweeteners, it is best to limit your intake.[11]

WHAT ARE COMPLEX CARBOHYDRATES?

4-2

In contrast to the simple carbohydrates that contain one or two monosaccharides, **complex carbohydrates** (or **polysaccharides**) are comprised of many monosaccharides bonded together. The types and arrangements of sugar molecules determine the shape and form of the polysaccharide. For example, some polysaccharides have an orderly linear appearance, whereas others are highly branched like branches of a tree. Three of the most common polysaccharides—starch, glycogen, and fiber—are discussed next.

4-2a Starch

Recall that plants generate glucose through the process of photosynthesis. To store this important source of energy, plants convert the glucose to starch. Thus, starch is made entirely of glucose molecules bonded together. The glucose molecules in starch are arranged in either an orderly unbranched linear chain (**amylose**) or a highly branched configuration called **amylopectin**. (see Figure 4.6). Plants typically contain a mixture of these two types of starch. Examples of starchy foods include grains (corn, rice, and wheat), products made from them (pasta and bread), and legumes (lentils and split peas). Potatoes and hard winter squashes are also good sources of starch.

complex carbohydrate (or **polysaccharide**) A category of carbohydrates comprised of many monosaccharides bonded together.

amylose A type of starch that consists solely of glucose molecules arranged in a linear, unbranched chain

amylopectin A type of starch that consists solely of glucose molecules in a highly branched configuration.

FIGURE 4.6   PLANTS STORE GLUCOSE AS STARCH

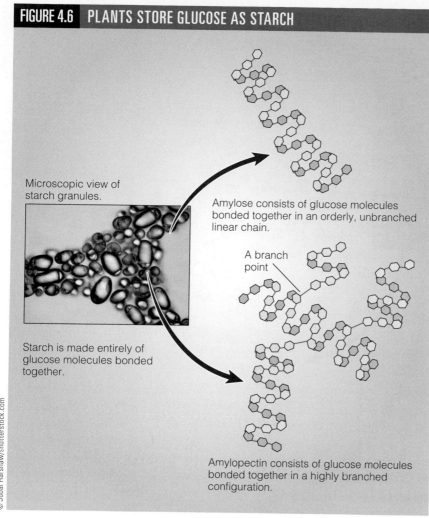

Microscopic view of starch granules.

Amylose consists of glucose molecules bonded together in an orderly, unbranched linear chain.

A branch point

Starch is made entirely of glucose molecules bonded together.

Amylopectin consists of glucose molecules bonded together in a highly branched configuration.

© Jubal Harshaw/shutterstock.com

from that found in starch. Because humans lack the enzymes needed to break these chemical bonds, fiber is virtually indigestible in the human small intestine (see Figure 4.8). As a result, undigested fiber passes from the small to the large intestine relatively intact. Once in the large intestine, fiber is broken down by intestinal bacteria. There is evidence that dietary fiber helps promote the growth of beneficial bacteria in the large intestine, which in turn helps inhibit the growth of disease-causing (pathogenic) bacteria.[12]

Dietary fiber is commonly classified on the basis of its solubility in water. **Soluble fiber** attracts water and turns to gel as it passes through the GI tract, slowing the process of digestion. Soluble fiber is found in oat bran, barley, nuts, seeds, beans, lentils, peas, and some fruits and vegetables. It is also found in psyllium, a common fiber supplement. Some types of soluble fiber may help lower risk of heart disease. **Insoluble fiber** is found in foods such as wheat bran, vegetables, and whole grains. It adds bulk to the stool, which aids in the process of elimination.[13]

4-2b Glycogen

Whereas plants store extra glucose as starch, the human body stores small amounts of excess glucose in the form of **glycogen**. Glycogen is a polysaccharide found primarily in liver and skeletal muscle. Like amylopectin the glucose molecules in glycogen are arranged in a highly branched configuration (see Figure 4.7). The numerous branches of glucose molecules found in glycogen can be broken down quickly, providing cells with an immediate source of energy when needed. However, unlike glycogen stored in skeletal muscle, only liver glycogen can release its glucose into the blood.

4-2c Fiber

The term **fiber** (or **dietary fiber**) refers to a diverse group of polysaccharides found in a variety of plant-based foods such as whole grains, legumes, vegetables, and fruits. Different foods contain different types of dietary fiber. The type of chemical bond found in fiber differs

Possible Health Benefits Although fiber is not an essential dietary component, it is clearly an important part of a healthy diet. Some studies suggest that increased fiber consumption is associated with lower risk of certain diseases. The evidence supporting the health benefits of fiber-rich foods is so substantive that the FDA has approved several fiber-related health claims (see Table 4.2).[14] Though fiber is not a magic bullet, its credentials are impressive. Consumption of soluble fiber can help lower blood cholesterol levels, promote satiety, and lower blood glucose levels.[15] In addition to these health-promoting benefits, soluble fiber may also function as a prebiotic by promoting the growth of beneficial bacteria in the GI tract.[16]

glycogen A polysaccharide found primarily in liver and skeletal muscle that is comprised of glucose molecules.

fiber (or **dietary fiber**) A diverse group of plant polysaccharides that are not digestible by human enzymes.

soluble fiber Dietary fiber that tends to dissolve or swell in water.

insoluble fiber Dietary fiber that remains relatively unchanged in water.

FIGURE 4.7 ANIMALS STORE GLUCOSE AS GLYCOGEN

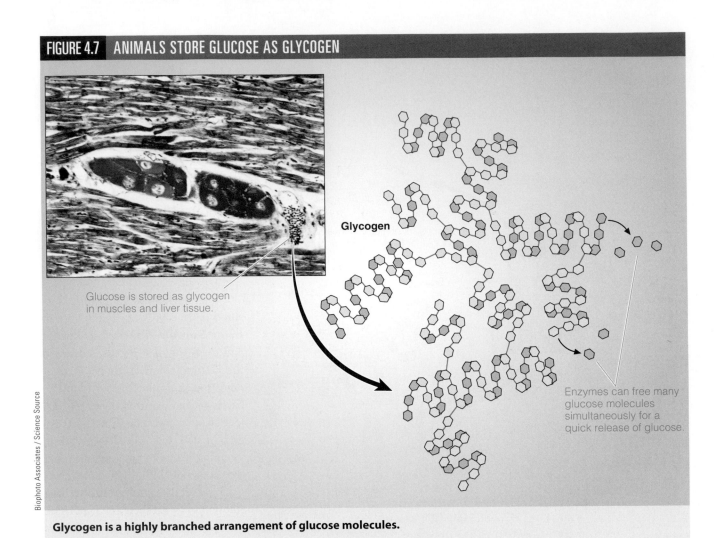

Glucose is stored as glycogen in muscles and liver tissue.

Glycogen

Enzymes can free many glucose molecules simultaneously for a quick release of glucose.

Biophoto Associates / Science Source

Glycogen is a highly branched arrangement of glucose molecules.

FIGURE 4.8 HUMANS ARE UNABLE TO DIGEST DIETARY FIBER

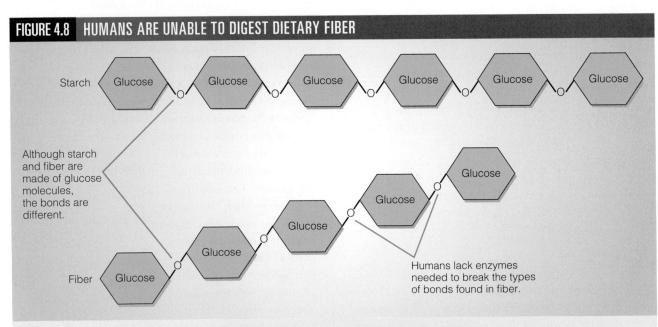

Starch

Although starch and fiber are made of glucose molecules, the bonds are different.

Fiber

Humans lack enzymes needed to break the types of bonds found in fiber.

When humans consume dietary fiber, it remains intact in the gastrointestinal tract. This is because humans lack digestive enzymes that are needed to break the bonds in fiber.

Because humans do not have the enzymes needed to digest fiber, it passes through the GI tract relatively intact, increasing fecal mass. As a result, feces move through the colon more quickly, which may help prevent or alleviate constipation. Not only can hard, dry feces make elimination more difficult, but it can also contribute to **diverticular disease** (or **diverticulosis**), a condition associated with persistent constipation. Constipation can cause undue straining during a bowel movement, which in turn can lead to the formation of pouches called *diverticula* that protrude along the colon wall (see Figure 4.9). **Diverticulitis** occurs when the diverticula become infected or inflamed (note that the suffix–*itis* refers to inflammation). Symptoms of diverticulitis include cramping, diarrhea, fever, and, occasionally, bleeding from the anus. Preventing conditions such as diverticular disease is another reason why a diet high in fiber may be beneficial to one's health.

Some manufacturers add fiber to their products, making them functional foods. This extra fiber may provide important health benefits.

Getty Images Publicity/Getty Images

Whole-Grain Foods To ensure adequate fiber intake, it is important to eat a variety of fruits and vegetables every day. It is likewise important to eat **whole-grain foods**. The nutritional value of grains is greatest when all three components of the grain—bran, germ, and endosperm are present (see Figure 4.10). Whereas the **bran** contains most of the grain's fiber, the **germ** contains much of its vitamins and minerals. The **endosperm** contains mostly starch. Sometimes the bran layer of the wheat kernel is removed during the milling process to produce a product known as refined flour. As a result, foods made with refined flour have very little fiber. When reading food labels, it is important to look for the phrases "whole-grain cereals" and "whole-wheat flour" because foods made with simply "wheat flour" are not necessarily good sources of fiber.

4-3 HOW ARE CARBOHYDRATES DIGESTED, ABSORBED, AND CIRCULATED?

The ultimate goal of carbohydrate digestion is to break down large, complex molecules such as starch into small, absorbable monosaccharides. Carbohydrates undergo extensive chemical transformations as they move through the GI tract. The bonds that hold disaccharides and starches together are broken with the help of digestive enzymes, and the resulting monosaccharides are absorbed into the bloodstream.

4-3a Starch Digestion

The long, arduous process of breaking starch into individual glucose molecules begins in the mouth. Chemical digestion starts when the salivary glands release saliva, which contains the enzyme **salivary amylase**. Salivary amylase breaks some of the chemical bonds that join glucose molecules. This results in the formation of

diverticular disease (or **diverticulosis**) A condition whereby pouches called *diverticula* form along the colon wall.

diverticulitis A condition whereby the diverticula become infected or inflamed.

whole-grain foods Cereal grains that contain bran, endosperm, and the germ in the same relative proportions as exist naturally.

bran The outer portion of a grain that contains most of the fiber.

germ The portion of a grain that contains most of the vitamins and minerals.

endosperm The portion of a grain that contains mostly starch.

salivary amylase An enzyme released from the salivary glands that breaks the chemical bonds in starch.

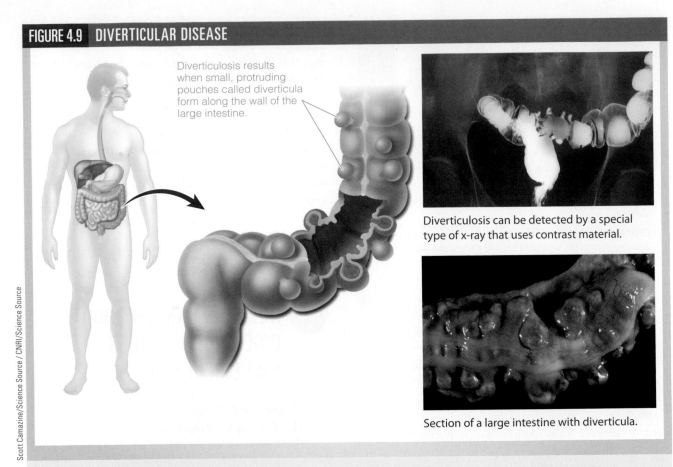

FIGURE 4.9 DIVERTICULAR DISEASE

Diverticulosis results when small, protruding pouches called diverticula form along the wall of the large intestine.

Diverticulosis can be detected by a special type of x-ray that uses contrast material.

Section of a large intestine with diverticula.

Diverticular disease is most common in older adults who do not consume adequate amounts of fiber-rich foods. Diverticulitis is the inflammation of diverticula.

multiple, shorter chains of glucose molecules. However, because food stays in the mouth only a short time, very little starch digestion actually takes place there.

Once the partially digested starch enters the stomach, the acidic environment stops the enzymatic activity of salivary amylase. The short chains of glucose enter the small intestine, where they encounter **pancreatic amylase**, an enzyme made in the pancreas and delivered to the small intestine as part of the pancreatic juice. Pancreatic amylase picks up where salivary amylase left off, breaking the chemical bonds that hold the glucose molecules together. The glucose chains get shorter and shorter, eventually forming the disaccharide maltose. The final enzyme in the sequence of starch digestion is **maltase**, which is made in the cells (enterocytes) that line the small intestine. Once maltase is released into the intestinal lumen,

pancreatic amylase An enzyme released from the pancreas that splits chemical bonds that hold glucose molecules together, eventually forming maltose.

maltase An intestinal enzyme that digests maltose, releasing two glucose molecules.

FIGURE 4.10 ANATOMY OF A WHEAT KERNEL

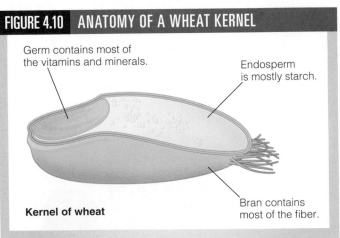

Germ contains most of the vitamins and minerals.

Endosperm is mostly starch.

Kernel of wheat

Bran contains most of the fiber.

Foods made with whole-wheat flour are more nutritious than foods made with refined wheat flour because the bran and the germ have not been removed. Each component contributes important nutrients needed for good health.

the last remaining chemical bond in maltose is split. This results in two free (unbound) glucose molecules that are readily absorbed into the blood (see Figure 4.11).

FIGURE 4.11   THE DIGESTION OF STARCH

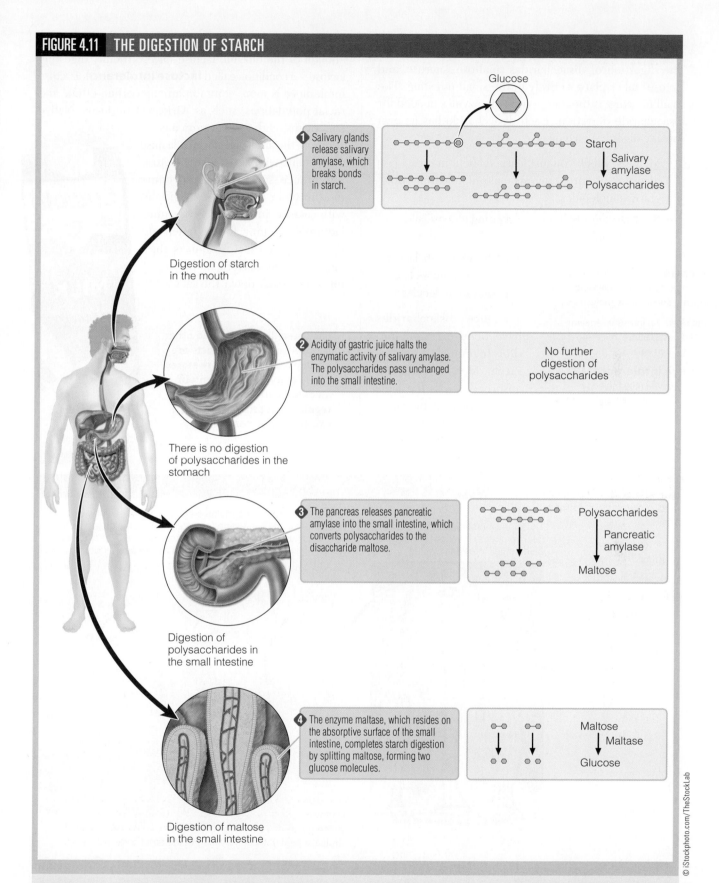

Glucose

1 Salivary glands release salivary amylase, which breaks bonds in starch.

Starch

Salivary amylase

Polysaccharides

Digestion of starch in the mouth

2 Acidity of gastric juice halts the enzymatic activity of salivary amylase. The polysaccharides pass unchanged into the small intestine.

No further digestion of polysaccharides

There is no digestion of polysaccharides in the stomach

3 The pancreas releases pancreatic amylase into the small intestine, which converts polysaccharides to the disaccharide maltose.

Polysaccharides

Pancreatic amylase

Maltose

Digestion of polysaccharides in the small intestine

4 The enzyme maltase, which resides on the absorptive surface of the small intestine, completes starch digestion by splitting maltose, forming two glucose molecules.

Maltose

Maltase

Glucose

Digestion of maltose in the small intestine

© iStockphoto.com/TheStockLab

Through the process of digestion, starch is broken down into molecules of glucose. The majority of starch digestion takes place in the small intestine.

4-3b Disaccharide Digestion

The digestion of disaccharides (maltose, sucrose, and lactose) takes place entirely in the small intestine. The small intestine is the lone source of enzymes needed for disaccharide digestion. Each disaccharide has its own specific digestive enzyme:

- Maltase (which you just learned about in the digestion of starch) digests maltose, releasing two glucose molecules.

- **Sucrase** digests sucrose, releasing glucose and fructose molecules.

- **Lactase** digests lactose, releasing glucose and galactose molecules.

Once disaccharides have been digested into their component monosaccharides, they can be absorbed across the cells (enterocytes) of the small intestine and circulated in the blood (see Figure 4.12).

sucrase An intestinal enzyme that digests sucrose, releasing glucose and fructose molecules.

lactase An intestinal enzyme that digests lactose, releasing glucose and galactose molecules.

lactose intolerance A condition whereby the body does not produce enough of the enzyme lactase, making it difficult to digest lactose.

Lactose Intolerance People who do not produce enough of the enzyme lactase have difficulty digesting lactose—a condition called **lactose intolerance**. Lactose intolerance is more common among certain ethnic and racial populations, such as African Americans, Native Americans, and Asian Americans. In addition, the ability to produce lactase can decline with age, making lactose intolerance somewhat more common in older adults. When people with lactose intolerance consume lactose-containing foods, much of the undigested lactose enters the large intestine. Bacteria in the large intestine break down the lactose,

Lactose-reduced milk and other lactose-free dairy products are available at most grocery stores. These dairy products contain all of the nutrients found in regular milk products with the exception of lactose.

© LunaseeStudios/Shutterstock.com

FIGURE 4.12 DISACCHARIDE DIGESTION

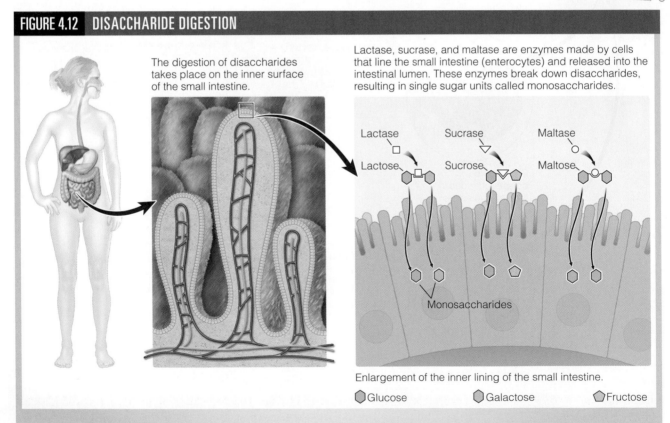

The digestion of disaccharides takes place on the inner surface of the small intestine.

Lactase, sucrase, and maltase are enzymes made by cells that line the small intestine (enterocytes) and released into the intestinal lumen. These enzymes break down disaccharides, resulting in single sugar units called monosaccharides.

Lactase Sucrase Maltase

Lactose Sucrose Maltose

Monosaccharides

Enlargement of the inner lining of the small intestine.

◯ Glucose ◯ Galactose ⬠ Fructose

Enzymes made by the small intestine are needed for disaccharide digestion.

The Evolution of Lactose Tolerance

All human infants are born with the ability to produce the enzyme lactase and therefore are able to digest lactose; the most abundant carbohydrate in human milk. However, in many cultures around the world, the "genetic switch" that regulates lactase production turns off after childhood. These individuals, referred to as being lactose intolerant, experience severe GI discomfort when consuming foods/beverages that contain lactose. Today, however, millions of people around the world can digest lactose and continue to consume dairy products throughout their lives. In terms of the evolutionary clock, the transition from lactose intolerance to lactose tolerance persistence appears to have transpired over a relatively short period of time. Evolutionary biologists theorize that a mutation in the gene that regulates lactase production may have occurred when humans started to domesticate milk-producing livestock such as cows, goats, and sheep. This evolutionary phenomenon, referred to as lactase persistence, enabled people in certain parts of the world to evolve into dairy-consuming societies. It is possible that lactose intolerance is the genetic "default setting," and that lactose tolerance is an acquired trait. During times of food scarcity, the ability to consume milk and other dairy products may have been advantageous to survival, allowing the trait to be passed on over many generations. Yet, two-thirds of the world's adult population today remains lactose intolerant, meaning that the spread of this mutation was not global. As scientists learn more about the genes that regulate lactase production, we will better understand our ancestral past, and how dietary adaptations helped ensure survival in varying global environments.[17]

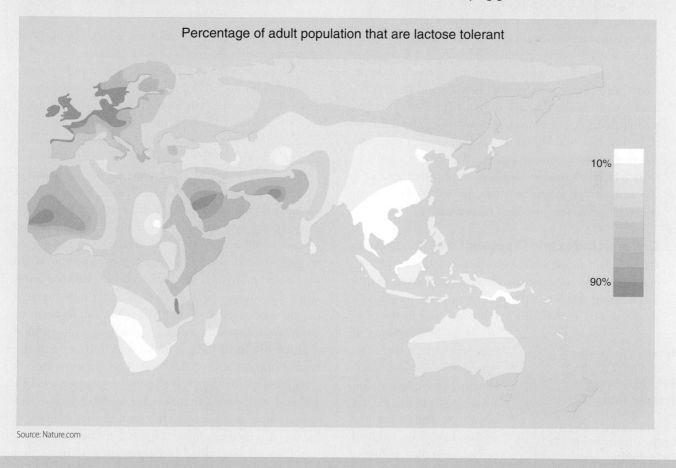

Percentage of adult population that are lactose tolerant

10%

90%

Source: Nature.com

which produces gas and sometimes causes other symptoms such as abdominal cramping and bloating. Most people with lactose intolerance can consume small amounts of low-lactose dairy products such as yogurt and cheese. The availability of lactose-free products and lactase-containing preparations makes it relatively easy for those with lactose intolerance to enjoy dairy products and obtain important nutrients such as calcium.

4-3c Monosaccharide Absorption

Once disaccharide and starch digestion is complete, the resulting monosaccharides (glucose, galactose, and fructose) are taken up by the cells (enterocytes) lining the small intestine and subsequently released into the blood. The blood then carries the monosaccharides directly to the liver. Because monosaccharides enter the bloodstream relatively quickly after consumption, a rise in blood glucose levels can be detected shortly after you eat most carbohydrate-rich foods. However, not all carbohydrates have the same effect on blood glucose levels. Some foods cause blood glucose levels to rise quickly and remain elevated, while others elicit a more subdued or gradual increase.

The rise in blood glucose following the ingestion of a food is called the **glycemic response** (see Figure 4.13). Scientists have long believed that simple carbohydrates cause a greater glycemic response than complex carbohydrates. However, this is not always the case: starchy foods containing complex carbohydrates, such as potatoes, refined cereal products, white bread, some whole-grain breads, and white rice, tend to elicit similar glycemic responses than many simple carbohydrate-rich such as soft drinks. Scientists now recognize that in addition to a carbohydrate's chemical complexity, other factors can also influence a food's ability to raise blood glucose levels. For example, whether a food is raw or cooked, ground or whole, or processed or unprocessed are all factors that must be considered.

Glycemic Index and Glycemic Load The **glycemic index (GI)**, a rating system based on a scale of 0 to 100, can be used to compare the glycemic responses elicited by different foods. In this system, the blood glucose response of a particular food is compared to the consumption of 50 grams of pure glucose. In this case, pure glucose serves as a reference for comparison, and is assigned the highest GI value: 100. The GI value of a given food is determined experimentally: subjects consume the amount of food that contains 50 grams of carbohydrate, and their blood glucose is measured. Using this standardized experiment, researchers can compare the glycemic response elicited by any food to that of eating 50 grams of pure glucose. Foods that elicit glycemic responses similar to that of pure glucose are considered high-GI foods (GI $\geq$ 70), whereas those that cause a lower and/or more gradual rise in blood glucose are considered low-GI foods (GI $\leq$ 55).

One limitation of using GI values to compare the effects of foods on glycemic response is that the amount of carbohydrate found in a typical serving of a food is not taken into account. Rather, GI values are based on a standard amount of carbohydrate (50 grams), which may or may not represent the average amount that a person would normally eat. For example, to consume 50 grams of carbohydrates from carrots, a person would need to eat more than a pound of carrots, which is unrealistic. To counter this problem, another rating system is sometimes used to evaluate the glycemic response of foods. In this system, foods are assigned a value called a **glycemic load (GL)**, which takes into account the typical portion size of the food. The glycemic load is calculated by dividing the glycemic index of a food by 100 and multiplying that number by the grams carbohydrate in a single serving the food. As you can see in Table 4.3, a food can have a high GI, yet have a low GL.

There is considerable interest in whether a diet emphasizing low-GL foods may be beneficial to health.[18] Although some health advocates believe that a diet based solely on foods with low GI and GL values can greatly improve long-term health, the truth is that diets are significantly more complex than that. It is important to recognize that the GI and GL values associated with

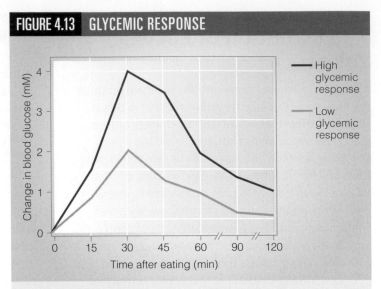

FIGURE 4.13   GLYCEMIC RESPONSE

Some foods cause blood glucose levels to rise quickly and remain elevated (high glycemic response), while others elicit a more gradual increase (low glycemic response).

glycemic response The change in blood glucose following the ingestion of a food.

glycemic index (GI) A rating system based on a scale of 0 to 100 used to compare the glycemic responses elicited by different foods.

glycemic load (GL) A rating system used to compare the glycemic responses associated with different foods that takes into account the typical portion of food consumed.

TABLE 4.3 COMPARING GLYCEMIC INDEX AND GLYCEMIC LOAD VALUES OF SELECTED FOODS

	Low Glycemic Index* (< 55)	Medium Glycemic Index (58–69)	High Glycemic Index (≥70)
Low Glycemic Load** (≤10)	All-bran cereal Apples Carrots Chocolate Peanuts Plums Strawberries Sweet corn	Beets Cantaloupe Pineapple Table sugar	Popcorn Watermelon Waffles Whole wheat flour Bread
Medium Glycemic Load (11–19)	Apple juice Bananas Fettuccine Orange juice Sourdough wheat Bread	Cola soft drink Life Cereal® New potatoes Wild rice	Cheerios® Gatorade® Shredded wheat
High Glycemic Load (≥20)	Lentils Linguine Macaroni Spaghetti	Couscous Raisins White rice	Bagels Baked Russet potato Cornflakes French fries

Foods listed with a yellow background are foods with similar GI and GL classifications. Foods listed with a blue background are foods with differing GI and GL classifications.

*Based on 50 g of glucose as a reference food

**Glycemic load is the glycemic index divided by 100 and multiplied by the number of grams of carbohydrate per food or beverage serving

Source: Foster-Powell K., Holt SHA., Brand-Miller JC. International table of glycemic index and glycemic load values. American Journal of Clinical Nutrition. 62:2002;5-56. Available from: http://www.ajcn.org/cgi/content/full/76/1/5

The concentration of glucose in your blood fluctuates throughout the day, depending on when you eat, what you eat, and how much you eat. After several hours without eating, blood glucose decreases. A low level of glucose in the blood is referred to as **hypoglycemia**. Conversely, blood glucose increases after one eats a carbohydrate-rich meal. Elevated levels of glucose in the blood is referred to as **hyperglycemia**. Blood glucose levels are the lowest in the morning after an overnight fast, returning to normal shortly after eating. Because your cells need energy throughout the day, the hormones insulin and glucagon work vigilantly to keep blood glucose levels within an acceptable range at all times (see Figure 4.14). **Insulin** and **glucagon** are secreted by the pancreas in response to fluctuations (increases and decreases, respectively) in blood glucose levels.

foods are determined experimentally and do not consider other factors, such as how a food is prepared, its combination with other foods, and individual physiological differences. Until nutritional scientists know more, it is important to make healthy carbohydrate choices by increasing your consumption of whole grains, fruits, and vegetables, and decreasing your consumption of foods made with refined flour and added sugar.

4-4 HOW DOES YOUR BODY REGULATE AND USE GLUCOSE?

Once absorbed, monosaccharides circulate directly to the liver, where the majority of galactose and fructose molecules are converted into other compounds—most notably glucose. Some glucose molecules are converted to ribose, a constituent of many vital compounds, including ATP, RNA, and DNA. Although monosaccharides serve numerous functions within the body, the ability to convert the energy contained in glucose into ATP is probably the most noteworthy role.

4-4a Insulin

After a person eats carbohydrate-rich foods, blood glucose levels quickly rise, leading to hyperglycemia. This in turn prompts the pancreas to increase its release of insulin. Insulin has several important effects, all of which help to lower the level of glucose in the blood.

Insulin is capable of communicating only with cells (e.g., those in adipose tissue and skeletal muscle) that have built-in "receivers" located on the outer surface of their cell membranes. These receivers, referred to as **insulin receptors**, bind insulin, in turn signaling the cell to take up glucose from the blood. Once glucose is inside the cell, it can be used as a source of energy (ATP). Insulin has other far-reaching effects in the body.

hypoglycemia A condition characterized by low blood glucose.

hyperglycemia A condition characterized by high blood glucose.

insulin A hormone secreted by the pancreas in response to elevated levels of blood glucose.

glucagon A hormone secreted by the pancreas in response to low levels of blood glucose.

insulin receptors Specialized proteins located on the outer membranes of certain types of cells that bind insulin and signal the cell to take up glucose.

FIGURE 4.14 RELEASE OF INSULIN AND GLUCAGON FROM THE PANCREAS

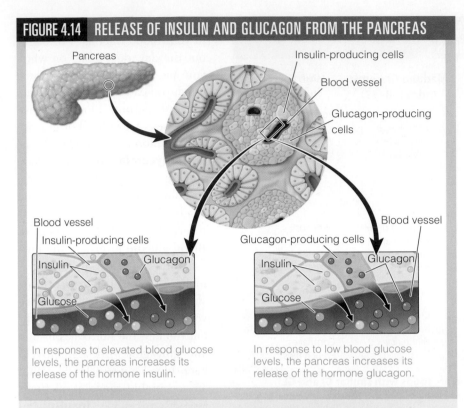

In response to elevated blood glucose levels, the pancreas increases its release of the hormone insulin.

In response to low blood glucose levels, the pancreas increases its release of the hormone glucagon.

The hormones insulin and glucagon are made and released by the pancreas. Both hormones play an important role in blood glucose regulation.

For example, insulin promotes the storage of excess glucose as glycogen in liver and skeletal muscle cells. This is often important after a meal when there may be more glucose available than what is needed by the body. However, only small amounts of glucose can be stored this way. Once this limit is reached, glucose is redirected to metabolic pathways that convert it to fat. Unlike glycogen storage, the body has a seemingly endless capacity to store body fat. In addition to stimulating the storage of excess glucose as glycogen and fat, insulin also plays an important role in the preservation of skeletal muscle. All of these actions work together to help bring blood glucose levels back down to normal—an excellent example of how the body maintains glucose homeostasis. The hormonal regulation of blood glucose by insulin is illustrated in Figure 4.15.

4-4b Glucagon

Within several hours after eating, blood glucose levels begin to decline. This shifts the hormonal balance away from insulin and toward glucagon. The primary function of glucagon is to increase the level of glucose in the blood. The brain is particularly sensitive to low blood glucose levels, and even a relatively small drop in blood glucose (hypoglycemia) can make a person feel nauseous, dizzy, anxious, lethargic, and irritable. This is one reason why it is hard to concentrate when you have not eaten for a long period of time.

To increase glucose availability during these times, glucagon stimulates the breakdown of energy-storing liver glycogen into glucose. These glucose molecules are released into the blood, increasing glucose availability to the rest of the body. The term for this metabolic process—**glycogenolysis**—literally means the breakdown ("lysis") of glycogen. Liver glycogen can supply glucose for approximately 24 hours before being depleted.

The breakdown of liver glycogen is an effective short-term solution for providing cells with glucose. However, because this reserve can be quickly depleted, the body must soon find an alternative glucose source. As glycogen stores dwindle, glucagon stimulates another metabolic process called gluconeogenesis. **Gluconeogenesis** is the synthesis of glucose from noncarbohydrate sources. Taking place mostly in the liver, gluconeogenesis synthesizes glucose mainly from amino acids derived from protein associated with skeletal muscle. You may be wondering why cells do not convert fat into glucose. Although fat can provide cells with plentiful amounts of energy, the carbon atoms contained in fat cannot be utilized to make glucose.

Ketones Spare the Use Of Amino Acids for Glucose Production

Gluconeogenesis increases glucose availability, but too much can have negative consequences; the body cannot rely on the use of amino acids from the breakdown of muscle protein for very long. To minimize loss of muscle, the body reduces its dependence on glucose by using an alternative energy source called **ketones**. These organic compounds are made from fatty acids under conditions of limited glucose availability. Ketones produced from the metabolism of body fat are released into the blood and used primarily by the brain for an alternate energy source. This glucose-sparing response helps minimize the loss of muscle protein by lessening the cellular demand for glucose.

glycogenolysis The breakdown of glycogen into glucose.

gluconeogenesis The synthesis of glucose from noncarbohydrate sources.

ketones An organic compound used as an alternative energy source under conditions of limited glucose availability.

FIGURE 4.15 HORMONAL REGULATION OF BLOOD GLUCOSE

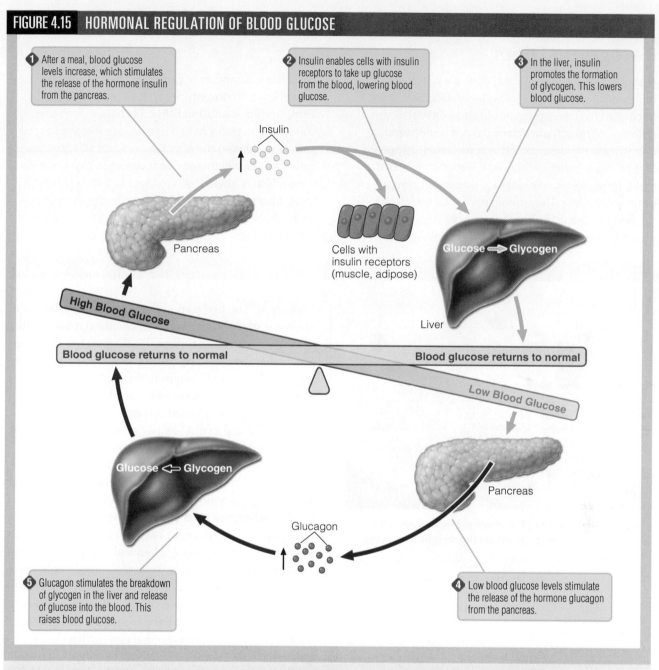

① After a meal, blood glucose levels increase, which stimulates the release of the hormone insulin from the pancreas.

② Insulin enables cells with insulin receptors to take up glucose from the blood, lowering blood glucose.

③ In the liver, insulin promotes the formation of glycogen. This lowers blood glucose.

Insulin

Pancreas

Cells with insulin receptors (muscle, adipose)

Glucose ⇌ Glycogen

Liver

High Blood Glucose

Blood glucose returns to normal

Blood glucose returns to normal

Low Blood Glucose

Glucose ⇐ Glycogen

Pancreas

Glucagon

⑤ Glucagon stimulates the breakdown of glycogen in the liver and release of glucose into the blood. This raises blood glucose.

④ Low blood glucose levels stimulate the release of the hormone glucagon from the pancreas.

Insulin lowers blood glucose and promotes energy storage. Glucagon increases blood glucose by promoting the breakdown of liver glycogen.

Ketone synthesis is not without its own consequences, however. When excessive ketones accumulate in the blood, a condition called **ketosis** can occur, causing a variety of complications, including loss of appetite. In fact, this is one reason why many popular low-carbohydrate diets help some people lose weight.

Epinephrine Stimulates Quick Glucose Release

Whereas glucagon is involved in the day-to-day regulation of blood glucose, another hormone, **epinephrine**, can also stimulate glycogenolysis. Under stressful conditions, the adrenal glands release epinephrine, which acts on both the liver and skeletal muscles. As a result, glycogen is quickly broken down, and glucose is released. Sometimes called the fight-or-flight reaction, this response helps ensure that glucose is available during extreme

ketosis A condition characterized by excessive ketone accumulation in the blood.

epinephrine A hormone released from the adrenal glands that stimulates glycogenolysis in emergency situations.

Ketogenic Diets

Ketogenic (low-carbohydrate) diets are not new to the medical world; therapeutic ketogenic diets were once used to prevent seizures in children with severe epilepsy. Although physicians did not understand exactly why the diet worked in some cases, it appeared that ketones somehow altered the metabolism in the brain. Today, ketogenic diets are employed primarily for weight loss. They mimic starvation by forcing the body to use fat rather than carbohydrates as its primary energy source. Some experts believe that chronic consumption of carbohydrate-rich foods cause insulin levels to rise, which can lead to weight gain. Thus, limiting starch and refined sugars in one's diet should theoretically help a person lose weight. Several studies evaluating the effectiveness and safety of low-carbohydrate weight-loss diets reported that these diets were effective alternatives to low-fat diets.[19] Although low-carbohydrate diets appear safe in the short term, some experts have raised doubts about their long-term effectiveness.

© Liv friis-larsen/Shutterstock.com

Some popular weight loss diets promote ketosis as the key to burning body fat, losing weight, and living a healthy life.

glycolysis An anaerobic metabolic pathway made up of a series of chemical reactions that splits glucose into two three-carbon molecules.

pyruvate The end product of glycolysis; formed by the breakdown of glucose.

tricarboxylic acid (TCA) cycle (also called the Krebs cycle) An oxygen-requiring metabolic pathway consisting of a series of chemical reactions that ultimately generate energy (ATP).

diabetes mellitus A group of metabolic disorders characterized by elevated levels of glucose in the blood.

circumstances. However, recall that only the liver can release glucose into the blood, whereas glucose resulting from glycogen breakdown in skeletal muscle is used strictly within the muscle cells.

4-4c Glucose as an Energy Source

A rich source of energy, glucose can be used by every cell in the body to make ATP. This conversion is accomplished by means of a catabolic pathway (see Chapter 3). The first step in the metabolic breakdown of glucose is **glycolysis**, a series of chemical reactions that splits glucose, a six-carbon molecule, into two three-carbon molecules called **pyruvate**. Because oxygen is not required for any of the steps in this pathway, glycolysis is considered to be an anaerobic metabolic pathway. The amount of energy released through glycolysis is small—enough to make two ATPs per glucose molecule—but there is much more energy yet to be harvested.

When conditions are right, the pyruvate molecules can be broken down further to yield additional energy. This requires another metabolic pathway called the **tricarboxylic acid (TCA) cycle**, or also the Krebs cycle, that functions under oxygen-rich conditions. This series of oxygen-requiring chemical reactions, referred to as aerobic metabolism, releases considerably more energy. Thus, the total net yield from one molecule of glucose can range between 36–38 ATPs. In addition to ATP, the metabolic breakdown of glucose also generates waste products such as water and carbon dioxide are also generated (see Figure 4.16). Recall that metabolic waste products such as these are eliminated from the body via expired air, urine, and sweat.

4-5 WHAT IS DIABETES?

Diabetes mellitus is a group of metabolic disorders characterized by elevated levels of glucose in the blood—otherwise known as hyperglycemia. There are different types of diabetes, which have been classified in a variety of ways since their discovery. Diabetes was once categorized according to the typical age of onset, but was later reclassified based on whether insulin injections

FIGURE 4.16 GLUCOSE METABOLISM

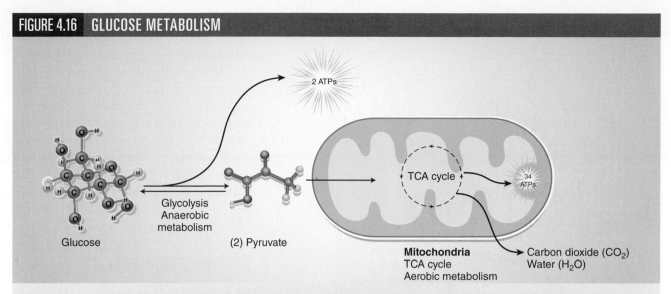

Glycolysis is an anaerobic pathway that splits glucose, a six-carbon molecule, into two three-carbon molecules called pyruvate. The amount of energy released from glycolysis is small—enough to make just two ATPs. Under oxygen-rich conditions, pyruvate can be broken down further through an aerobic metabolic pathway called the tricarboxylic acid (TCA) cycle. This results in more energy being released—enough to make about 34 ATPs. The total net yield from one molecule of glucose can range between 36–38 ATPs. Metabolic waste products resulting from glucose metabolism include water and carbon dioxide.

were required as a treatment. Because there were many exceptions associated with both these classifications, the American Diabetes Association developed a new system of diabetes classification based on etiology, or underlying cause. Today, the two main types of diabetes are referred to as type 1 diabetes and type 2 diabetes. A third type, gestational diabetes, is discussed in Chapter 10.

4-5a Type 1 Diabetes

Type 1 diabetes occurs when the pancreas is no longer able to produce insulin. Without insulin, certain cells (e.g., skeletal muscle and adipose cells) cannot take up glucose, causing blood glucose levels to become dangerously high. Approximately 5 to 10 percent of all people with diabetes have type 1. Although type 1 diabetes can develop at any age, it typically develops during childhood and early adolescence.

Type 1 diabetes develops when a person's immune system produces antibodies that mistakenly attack and destroy the insulin-producing cells of the pancreas (see Figure 4.17). This inappropriate immunologic response is thought to be triggered by an environmental factor such as exposure to a virus. For this reason, type 1 diabetes is classified as an autoimmune disease, an illness that occurs when an abnormal immunological response results in the destruction of body tissues. Although the majority of people who develop type 1 diabetes have no family history of the disease, most experts agree that a genetic tendency does increase a person's risk.

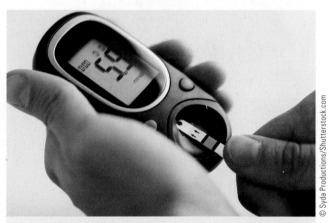

Glucometers are an important part of diabetes management. It is important for people who have diabetes to monitor blood glucose levels daily.

Destruction of the insulin-producing pancreatic cells and the subsequent inability to produce insulin cause blood glucose levels to become dangerously elevated (severe hyperglycemia). Symptoms tend to develop rapidly, and because of their severity, are not easily ignored. Signs and symptoms associated with type 1 diabetes include rapid weight loss, extreme thirst, and frequent urination. Type 1 diabetes also

type 1 diabetes A form of diabetes whereby the pancreas is no longer able to produce insulin, causing blood glucose levels to become dangerously high.

FIGURE 4.17 TYPE 1 DIABETES

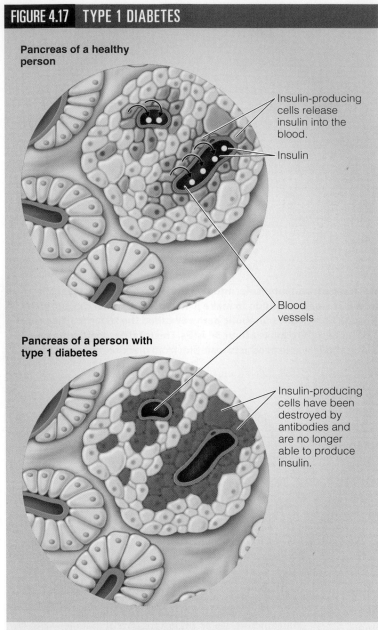

Pancreas of a healthy person

Insulin-producing cells release insulin into the blood.

Insulin

Blood vessels

Pancreas of a person with type 1 diabetes

Insulin-producing cells have been destroyed by antibodies and are no longer able to produce insulin.

Type 1 diabetes is caused by an autoimmune disorder in which antibodies attack and destroy the insulin-producing cells of the pancreas.

glucometer A medical device used to monitor the concentration of glucose in the blood.

type 2 diabetes A form of diabetes whereby insulin resistance impairs the ability of certain cells to take up glucose from the blood.

insulin resistance A condition whereby insulin receptors throughout the body are less responsive to insulin.

causes a person to feel hungry and weak because cells are "starved" for energy. In fact, diabetes is often described as "starvation in the midst of plenty."

For people with type 1 diabetes, an external source of insulin is necessary to control blood glucose levels. Insulin can be administered through a regimen of multiple daily injections or by an insulin pump. Without an exogenous source of insulin, type 1 diabetes is fatal. When diagnosed with type 1 diabetes, a person must learn to balance insulin treatments with a healthy diet and physical activity. A medical device called a **glucometer** helps those with type 1 diabetes monitor the concentration of glucose in the blood so they can know how much insulin to administer. Using a small sample of blood, these user-friendly devices provide immediate feedback regarding an individual's blood glucose levels.

4-5b Type 2 Diabetes

Type 2 diabetes is by far the most common form of diabetes - 90 to 95 percent of people with diabetes fall into this category. In the United States, type 2 diabetes has become so widespread that an estimated 29.1 million people or 9.3 percent of the population have diabetes; about 1 out of every 11 people.[20] Although type 2 diabetes can occur at any age, it most frequently develops in adults, middle-aged, and older. However, the rising prevalence of obesity in children and teens throughout the United States has contributed to a new and alarming trend: an escalation in the number of children and teens diagnosed with type 2 diabetes.[21] Given this trend, type 2 diabetes can no longer be thought of as a condition that only affects adults.

Unlike people with type 1 diabetes, most people with type 2 diabetes have normal or even elevated levels of insulin in their blood. Type 2 diabetes is caused by **insulin resistance**, meaning that insulin receptors are less responsive to insulin. Because insulin-requiring cells do not respond appropriately to insulin's signal, the amount of glucose taken up from the bloodstream is notably diminished. When blood glucose levels rise beyond the normal range, a person begins to experience symptoms associated with type 2 diabetes. Symptoms tend to develop gradually and are often ignored—this is why type 2 diabetes can go undiagnosed for many years. Some of the early symptoms associated with type 2 diabetes include fatigue, frequent urination, and excessive thirst.

There are numerous risk factors associated with the development of type 2 diabetes (see Table 4.4). Although a person has little control over some of these risk factors such as genetics, one can make informed lifestyle choices to help prevent the disease. Studies show that even individuals who are genetically susceptible to the development of type 2 diabetes can significantly reduce their risk by eating a variety of healthy foods, staying physically fit, and maintaining a recommended body weight.[22] The last point is particularly important because obesity is a profoundly significant risk factor associated with type 2 diabetes. Approximately 80 percent of people diagnosed with type 2 diabetes are overweight or obese. Furthermore, a particular distribution of body fat can pose additional risk for type 2 diabetes. More specifically, body fat stored in the abdominal region of the body presents a greater risk than does that stored in the lower regions of the body.[23] The link between excess adiposity and type 2 diabetes is not entirely clear, but scientists believe that certain hormones or other substances made and released by adipose tissue may trigger insulin resistance. Thus, when fat accumulates in the body, the production of these insulin-resistant hormones increase.

© Tyler Olson/Shutterstock.com

People with diagnosed diabetes, on average, have medical expenditures approximately 2.3 times higher than what expenditures would be in the absence of diabetes. Although recent technological advances have engineered more effective, faster acting forms of insulin, this has contributed to rising costs of insulin, which has tripled since 2002.

This also may explain why weight loss, and subsequent loss of body fat can help manage and, in some cases, reverse type 2 diabetes.

TABLE 4.4 RISK FACTORS ASSOCIATED WITH TYPE 2 DIABETES

Having a close family member (mother, father, brother, sister) with type 2 diabetes	Being over 45 years of age
For women, a history of gestational diabetes	Being of African, Hispanic, Native American, or Pacific Island descent
Physical inactivity (physically active less than three times per week)	Having high blood pressure (greater than or equal to 140/90 mmHg)
Being overweight, especially if body fat stores are primarily in the abdomen (central adiposity)	Having low HDL cholesterol (less than 35 milligrams per deciliter)
Prediabetes where blood glucose levels are higher than normal, but not high enough to be classified as diabetes	Having high triglycerides in the blood (greater than 250 milligrams per deciliter)

Adapted from National Diabetes Education Program, Department of Health and Human Services, National Institute of Diabetes and Digestive and Kidney Diseases, and U.S. Centers for Disease Control and Prevention. Diabetes Risk Factors. Available from: http://ndep.nih.gov/am-i-at-risk/DiabetesRiskFactors.aspx.

4-5c Preventing Complications Associated with Diabetes

As the prevalence of obesity in the United States continues to climb, so does the prevalence of type 2 diabetes. If these trends continue, researchers estimate that 18 million Americans will have type 2 diabetes by the year 2020.[24] The greatest concern raised by these statistics is that diabetes can lead to other serious health problems. For example, diabetes increases one's risk of having a heart attack or stroke. It can also lead to a loss of feeling in the feet and hands, blindness, limb amputation, and impaired kidney function. These long-term complications are largely attributable to the harmful effects of hyperglycemia on blood vessels and nerves.

The good news is that maintaining a near-normal level of blood glucose can often prevent

What the Pima Indians Have Taught Us about Type 2 Diabetes

Because of their unusually high rates of obesity and type 2 diabetes, researchers have studied the Pima Indians living in the American Southwest for over 30 years.[25] During this time, researchers have also studied a group of genetically related Pima Indians still living in a remote area in Mexico. By contrast, this second group of Pima experiences very low rates of type 2 diabetes and obesity. Scientists have concluded that, although both groups of Pima have similar genetics, only those living in an environment of abundant food and reduced physical activity tend to develop type 2 diabetes. Those who maintain traditional diets and

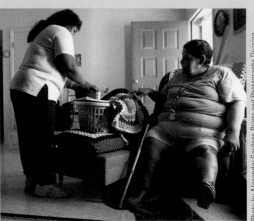

Photo by Alexander Gardner, Prints and Photographs Division, Library of Congress, LC-USZ62-126404 / AP Images/J. Pat Carter

Genetics and environment both play a role in the development of type 2 diabetes. Research conducted on the Pima Indians has provided evidence that a healthy lifestyle can help prevent type 2 diabetes, even in a genetically susceptible population.

physically active lifestyles are likely to remain free of diabetes. The Pima Indians' story exemplifies the fact that alterations in diet and physical activity can either lead to or help prevent type 2 diabetes—even in a genetically susceptible population.

---

these complications. In the case of type 1 diabetes, one must take care to balance carbohydrate intake and exercise with insulin injections. In the case of type 2 diabetes, one can often control blood glucose through a combination of weight management, regular exercise, and a healthy diet. In fact, most of the nutritional guidelines developed to manage type 2 diabetes can also be applied by anyone who wants to maintain good health. This means adhering to a balanced diet that emphasizes fruits, vegetables, whole grains, legumes, low-fat dairy products, low-fat meat, and foods prepared with minimum amounts of added sugar.

4-6 WHAT ARE THE RECOMMENDATIONS FOR CARBOHYDRATE INTAKE?

Although carbohydrates are not technically essential nutrients, they are nonetheless an important part of a healthy diet because they provide energy and dietary fiber. Some tissues, such as the brain, rely extensively on glucose for energy. Furthermore, there is evidence that consumption of certain carbohydrates may help prevent chronic diseases. Thus, to ensure that individuals make healthy carbohydrates a part of their eating patterns, the 2015 Dietary Guidelines for Americans emphasizes the importance of eating a variety of brightly colored vegetables (dark-green, red, and orange), whole grains, and legumes such as dried peas and beans. To establish a healthy eating pattern, the number of servings of these important, nutrient-rich foods coincides with a person's appropriate calorie level. To address changing needs throughout the various stages of life, recommended daily servings from each food group (vegetables, fruits, whole grains, and legumes) have been established for 12 different calorie levels ranging from 1,000 kcal/day to 2,200 kcal/day. Thus, the recommended number of servings of vegetables, fruits, whole grains, and legumes depends on a person's energy requirements. The emphasis on whole grains is particularly important because refined grain-based foods often contain high levels of solid fats, added sugars, and sodium. By making this dietary change, you can improve your overall health in many ways.

4-6a Dietary Reference Intakes for Carbohydrates

The Institute of Medicine's Dietary Reference Intake (DRI) values for carbohydrates were developed primarily to ensure that the brain has adequate glucose for its energy needs. The Recommended Dietary Allowance (RDA) for carbohydrates—the minimum amount of glucose needed by the brain each day—is 130 grams for children and adults. For most adults, a well-balanced diet that includes two servings of fruit, three servings of vegetables, and six servings of grains every day easily provides the recommended amount of carbohydrates. In terms of overall energy distribution, the Acceptable Macronutrient Distribution Range (AMDR) suggests that 45 to 65 percent of your total caloric intake should come from carbohydrates.

In addition to ensuring that daily glucose needs are met, the Institute of Medicine's recommendations aim to minimize one's risk for chronic disease and to promote optimal health. To this end, they suggest that, in addition to the amount of carbohydrates you consume, it is important to pay attention to the types and sources of carbohydrates you choose. Although a candy bar and a serving of breakfast cereal contain roughly the same amount of carbohydrate (24 grams), a 2-ounce candy bar has approximately 275 kcal, whereas a 1-cup serving of breakfast cereal has approximately 150 kcal. Because some carbohydrate-rich foods are more nutrient-dense (and thus nutritional) than others, one must weigh a number of factors before deciding which carbohydrate-containing foods to consume.

4-6b Making the Right Choices

When it comes to carbohydrate-rich foods, there are many from which to choose. Although the choice can be overwhelming, health experts agree that the best strategy is to maximize your intake of foods that are nutrient-dense and provide plenty of fiber, vitamins, and minerals. Conversely, it is important to minimize your intake of foods high in fat and added sugar. Following the 2015 Dietary Guidelines for Americans can help eliminate much of the guesswork when it comes to determining which carbohydrate-rich foods to choose. One general guideline is to consume more fruits, vegetables, and whole-grain foods;

TABLE 4.5 ADDED SUGARS IN SELECTED FOOD ITEMS

Food	Serving Size	Added Sugar (g)[a]
Soft drink	12 oz	43
Milkshake	10 oz	36
Fruit punch	8 oz	38
Chocolate candy	1.5 oz	24
Sweetened breakfast cereal	1 cup	15
Yogurt with fruit	1 cup	33
Ice cream	1 cup	28
Cake with frosting	1 slice	28
Cookies	2 (medium)	14
Jam or jelly	1 tbsp	2

[a] 1 teaspoon equals 4.75 grams.
Source: U.S. Department of Agriculture database for the added sugars content of selected foods, Release 1 (2006). Available from: http://www.ars.usda.gov/Main/docs.htm?docid=12107

these foods provide naturally occurring sugars, fiber, and plenty of vitamins and minerals. The Choose MyPlate food guidance system, which emphasizes whole grains, fruit, and vegetables as the foundation of a healthy diet, is another useful tool for evaluating your carbohydrate food choices.[26]

One more key recommendation of the 2015 Dietary Guidelines is to consume less than 10 percent of calories per day from added sugars. To achieve this goal, it is best to consume foods with high amounts of added sugar (e.g., cookies, soda, sugary cereals, and heavy syrups) sparingly. Table 4.5 lists the average amounts of added sugar in selected foods and beverages.

Although a candy bar and a serving of whole-grain breakfast cereal contain roughly the same amount of carbohydrate (24 grams), the breakfast cereal is more nutrient-dense and is high in dietary fiber.

A Balanced Diet Necessitates a Healthy Intake of Dietary Fiber As you have learned, there are many health benefits associated with dietary fiber. One way to add more fiber to your diet is to read nutrition labels and choose those foods labeled "a good source of fiber." It is equally important when considering grain-based foods to look for terms such as *whole grain* and *whole wheat* on food packaging. The current recommendation for adults is to make at least one-half of total grain consumption whole-grain foods, and to minimize refined grain foods such as white bread, rolls, bagels, and grain-based desserts. Refined grain-rich, ready-to-eat cereals often offer little nutritional value (aside from calories and added micronutrients) and very little dietary fiber. Examples of whole-grain foods are brown rice, pasta made with whole wheat, and whole-grain bread. Outside of grain-based products, fruits and vegetables should also contribute to one's daily intake of dietary fiber. The Institute of Medicine recommends that adults consume at least 14 grams of dietary fiber per 1,000 kcal. The average daily intake of fiber in the United States is about one-half this amount (approximately 16 grams of dietary fiber total per day).[27]

Most experts agree that a diet high in fiber is beneficial to one's health. Although some fibers may have a tendency to bind minerals such as calcium, zinc, iron, and magnesium, it is doubtful that the consumption of dietary fiber in recommended amounts affects mineral status in healthy adults. However, a few words of caution must be stated regarding the overconsumption of fiber: more is not always better. A sudden and/or large increase in fiber intake may cause GI problems such as diarrhea or constipation. As such, a person should increase their fiber intake gradually and give the body time to adjust. It is also wise to increase fluid intake as one increases fiber—this can help alleviate common problems such as constipation.

STUDY TOOLS 4

READY TO STUDY? IN THE BOOK, YOU CAN:

☐ Rip out the Chapter Review Card, which includes key terms and chapter summaries.

ONLINE AT WWW.CENGAGEBRAIN.COM, YOU CAN:

☐ Learn about diabetes in a short video.

☐ Explore carbohydrate digestion in the GI tract with an animation.

☐ Interact with figures from the text to check your understanding.

☐ Prepare for tests with quizzes

☐ Review the key terms with Flash Cards.

NUTR ONLINE

ACCESS TEXTBOOK CONTENT ONLINE— INCLUDING ON SMARTPHONES!

Includes Videos & Other Interactive Resources!

MANAGE MY COURSE ∨ STUDENT

NUTR2

CHAPTER 1

Why Does Nutrition Matter?

CHAPTER 2

Choosing Foods Wisely

4LTR PRESS

5 | Protein

LEARNING OUTCOMES

5-1 Understand what makes up the basic building blocks of a protein and describe what makes each protein unique.

5-2 Describe the process by which cells make proteins.

5-3 Explain the significance of a protein's shape.

5-4 Define and understand *genetics* and *epigenetics* and how they relate to protein synthesis.

5-5 Explain protein digestion, absorption, and circulation.

5-6 Describe the major functions of proteins in the body.

5-7 Explain how the body recycles and reuses amino acids.

5-8 Calculate the amount of protein you need on a daily basis.

5-9 Describe vegetarianism and variations of vegetarian diets.

5-10 Know the consequences of protein deficiency and excess.

© monofaction/Shutterstock.com

After finishing this chapter go to **PAGE 122** for **STUDY TOOLS.**

5-1 WHAT ARE PROTEINS?

The term *protein* was derived more than 170 years ago from the Greek *prota*, meaning "of first importance."[1] Indeed, proteins are the most abundant organic substances in the body, making up at least 50 percent of its dry weight.[2] But, why is protein important, and what is the relationship between protein in the foods you eat and the protein found in your body?

In this chapter, you will learn what proteins are, what foods are good sources of them, how protein is digested, and how the products of protein digestion—amino acids—are used by the body to make the proteins it needs to function. You will also learn about eating patterns, such as vegetarianism, that can influence protein intake; how certain proteins in foods cause allergic reactions in some people; and how your dietary choices, genetics, and environment can interact to influence the proteins you make and your risk of developing various diseases. With this information, you can begin to make informed decisions about which dietary protein sources to choose and how to optimize long-term health and well-being.

Perhaps the first question you should ask when beginning your study of protein is, "What are proteins, makes each protein unique and what makes proteins different from other nutrients?" **Proteins** are large molecules composed of one or more **polypeptide** chains. A polypeptide chain is made up of building blocks called **amino acids**, which are joined together by chemical bonds referred to as **peptide bonds**. Chemically distinct from the other energy- yielding macronutrients (carbohydrates and lipids), all proteins contain not only carbon and hydrogen but also nitrogen. Proteins come in a variety of sizes: some are very simple, comprised of only a few amino acids, while others contain thousands of amino-acid building blocks. Most proteins are of intermediate size, containing 250 to 300 amino acids. Because they vary so greatly in size, proteins can

protein A nitrogen-containing macronutrient made from amino acids.

polypeptide A chain of amino acid subunits joined together through peptide bonds.

amino acid A nitrogen-containing subunit that combines with other amino acids to form proteins.

peptide bond A chemical bond that joins amino acids.

be classified based on their number of amino acids: *dipeptides* have two amino acids, *tripeptides* have three, and so forth.

5-1a Amino Acid Structure

The numerous proteins in the body are amazingly diverse. The key to this variety lies not only in the number and types of amino acids each protein contains, but also in the order in which the amino acids are linked together. To understand protein diversity, you must first understand the basic components of an amino acid and what makes each one unique.

Amino acids have three *common* components:

- A central carbon atom bonded to a hydrogen atom,
- A nitrogen-containing **amino group**, and
- A carboxylic acid group.

Collectively, these three components of amino acids are referred to as the *common structure*. In the body, the amino and carboxylic acid groups almost always exist in weakly charged states. These charges cause proteins to twist and bend, ultimately giving rise to the protein's shape. As you will learn later in the chapter, it is the shape of a protein that imparts its function. In addition to the

common structure, each amino acid contains a unique side-chain group called an **R-group**; it is the structure of the R-group that distinguishes each amino acid from the others, and the subtle differences in the R-groups give each amino acid its distinctive chemical and physical nature. For example, some of the R-groups are negatively charged, some are positively charged, and some have no charge at all. Figure 5.1 illustrates a generic amino acid and some examples of R-groups.

5-1b Essential, Nonessential, and Conditionally Essential Amino Acids

Just 20 different amino acids combine to form all the proteins required by the human body. Amino acids can be categorized as essential (or indispensable), nonessential (or dispensable), or conditionally essential (or conditionally indispensable).[3] You can see the names of all the necessary amino acids and their classifications as essential, nonessential, or conditionally essential in Table 5.1. The nine essential amino acids are those

amino group The nitrogen-containing component of an amino acid.

R-group The side-chain component of an amino acid that distinguishes it from other amino acids.

you must consume in your diet because your body cannot make them at all, or cannot make them in the required amounts. For most people, the remaining 11 amino acids are nutritionally nonessential because they can be synthesized from other compounds. To do this, the body transfers an amino group from an essential amino acid to an **α-keto acid**, which is, in essence, an amino acid without an amino group. This process, called **transamination**, results in the synthesis of the nonessential amino acids.

Under some conditions, however, the body is unable to synthesize one or more of the nonessential amino acids. Some infants (especially those born prematurely) cannot synthesize several of the traditionally nonessential amino acids, for example. These amino acids are therefore considered conditionally essential because they must be obtained from the diet until the baby matures.[4] Fortunately, sufficient amounts of the conditionally essential amino acids can typically be obtained from human milk or regular infant formula. Because the milk produced by women who deliver prematurely does not contain some essential amino acids in amounts sufficient to meet the needs of their infants, premature infants' diets are usually supplemented with the proper amino acids.[5]

Certain diseases such as **phenylketonuria (PKU)** can also cause generally nonessential amino acids to become conditionally essential. People with PKU do not produce one of the enzymes required

α-keto acid A compound similar to an amino acid that does not have an amino group; used to synthesize nonessential amino acids.

transamination The process by which nonessential amino acids are synthesized.

phenylketonuria (PKU) A disorder caused by deficiency of an enzyme needed to convert the essential amino acid phenylalanine to the normally nonessential amino acid tyrosine; as a result, tyrosine cannot be made and becomes conditionally essential.

Animal-derived foods such as meat, milk, fish, and eggs are considered complete protein sources because they contain all of the essential amino acids.

FIGURE 5.1   COMPONENTS OF AN AMINO ACID

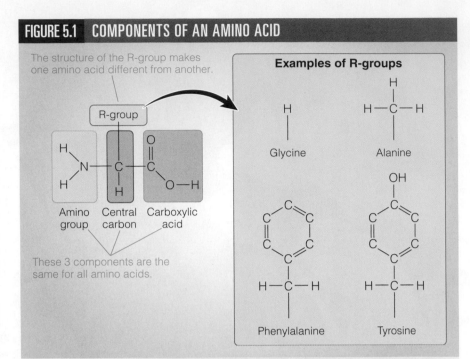

The structure of the R-group makes one amino acid different from another.

R-group

Amino group · Central carbon · Carboxylic acid

These 3 components are the same for all amino acids.

Examples of R-groups

Glycine · Alanine · Phenylalanine · Tyrosine

Amino acids have four parts: an amino group, a central carbon, a carboxylic acid group, and an R-group.

TABLE 5.1   ESSENTIAL, NONESSENTIAL, AND CONDITIONALLY ESSENTIAL AMINO ACIDS

Essential	Nonessential
Histidine	Alanine
Isoleucine	Arginine[a]
Leucine	Asparagine
Lysine	Aspartic acid
Methionine	Cysteine[a]
Phenylalanine	Glutamic acid
Threonine	Glutamine[a]
Tryptophan	Glycine[a]
Valine	Proline[a]
	Serine
	Tyrosine[a]

[a]These amino acids are also classified as conditionally essential amino acids; for some people, these amino acids must be consumed in the diet.

Source: Institute of Medicine. Dietary Reference Intakes for energy, carbohydrate, fiber, fat, fatty acids, cholesterol, protein, and amino acids. Washington, DC: National Academies Press; 2005.

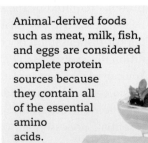

©iStock.com/krivicm

to convert the essential amino acid phenylalanine to the normally nonessential amino acid tyrosine. Thus, tyrosine becomes conditionally essential. In all, six amino acids are considered conditionally essential.

5-1c Not All Proteins in Food Are Created Equal

As with carbohydrates, some foods generally contain more protein than do others. Meat, poultry, fish, eggs, dairy products, and nuts contain more protein (per gram) than grains, fruits, and vegetables. Leguminous plants such as soybeans, dried beans, lentils, peas, and peanuts are unique in that they are associated with bacteria that can take nitrogen from the air and incorporate it into amino acids, which legumes use to make their own proteins. This is why legumes tend to contain more protein than most other plants and why the 2015 Dietary Guidelines for Americans states that legumes can "count" as either a vegetable or protein food.

Even foods with the same amounts of total protein can contain different combinations of amino acids. For example, both 1 cup of cottage cheese and 1 cup of cooked lima beans provide about 15 g of protein. The specific amino acids present in these two types of foods are quite different, however. Proteins of animal origin (like those found in cottage cheese) provide amino acids

© Kerdkanno/Shutterstock.com

that are in "higher demand" by the body than those found in plant-derived foods (like lima beans). This is because the former proteins generally have greater amounts of essential amino acids than do the latter.

A food that supplies an adequate and balanced amount of all essential amino acids is considered a **complete protein source**, whereas one that lacks or supplies low amounts of one or more of the essential amino acids is considered an **incomplete protein source**. An essential amino acid that is insufficient or absent in an incomplete protein source is called its **limiting amino acid**. When a limiting amino acid is not available, your body is unable to make any of the proteins that contain that particular amino acid—even if you have all of the other essential amino acids in full supply. As you might expect, meat, poultry, fish, eggs, and dairy products are complete protein sources, and plant-based products are incomplete protein sources.

Protein Complementation You may be wondering how people who only eat plant-based foods (vegetarians) get all of their essential amino acids if their diets are limited to incomplete proteins. The answer to this question is that diverse foods with different incomplete proteins can be combined to provide adequate amounts of all the essential amino acids. This dietary practice, called

---

Living with PKU

People with phenylketonuria (PKU) are unable to make the amino acid tyrosine from the amino acid phenylalanine. This is a problem not only because tyrosine is a necessary component of many proteins, but also because high levels of the phenylalanine in the blood can damage the brain. For this reason, almost all babies born in the United States are screened for PKU soon after birth. Children with PKU must follow a diet that restricts their intake of many protein-rich foods (e.g., dairy, nuts, meats, and eggs) because they contain high amounts of phenylalanine. They also need to avoid foods and medications made with aspartame, including many diet sodas, because this artificial sweetener contains phenylalanine. Adults with PKU must also be careful to not consume too much phenylalanine, although they do not usually need to be as vigilant as children.

---

complete protein source A food that supplies an adequate and balanced amount of all the essential amino acids.

incomplete protein source A food that lacks or supplies low amounts of one or more of the essential amino acids.

limiting amino acid An essential amino acid that is insufficient or absent in an incomplete protein source.

Complete and incomplete proteins are sometimes said to carry high and low biological value, respectively.

protein complementation, is customary around the world, especially in regions that traditionally rely heavily on plant-based foods for protein.[6] Examples of commonly consumed foods whose proteins complement each other are rice and beans, or corn and beans. Both rice and corn have several limiting amino acids (for example, lysine) but provide adequate amounts of others (e.g., methionine). By contrast, dried beans and other legumes tend to be limiting in methionine but provide adequate amounts of lysine. In general, protein complementation allows diets containing a variety of plant-based protein sources to provide all of the necessary essential amino acids.

5-2 HOW DO CELLS MAKE PROTEINS?

protein complementation The combining of diverse foods with different incomplete proteins to provide adequate amounts of all the essential amino acids.

cell signaling The process by which a cell is notified that it should make a particular protein.

transcription The process by which mRNA is constructed using DNA as a template; occurs in the cell's nucleus.

messenger RNA (mRNA) A chemical that carries the instructions contained in DNA outside of the nucleus.

You now know the fundamental concepts related to what makes up a protein, how proteins differ from each other, and why some foods are considered better sources of protein than others. But to really understand why proteins are an essential part of your diet, you must also understand how your body converts the proteins you eat into proteins that function in your body. The following sections describe the process of protein synthesis, which involves three basic steps:

1. Cell signaling,
2. Transcription, and
3. Translation.

Figure 5.2 illustrates the three steps that comprise the process of protein synthesis.

5-2a Step 1: Cell Signaling

Virtually every cell in your body needs and makes proteins. However, different cells need different proteins, and the amounts needed at different times can be highly variable. As a result, protein synthesis within a particular cell is neither consistent nor random; it is tightly regulated by the amounts and types of proteins needed by the cell and the entire body at a given time. For example, synthesis of the proteins needed for calcium absorption in your small intestine is turned on or off depending on your body's need for calcium, ensuring that calcium availability is maintained at optimal levels. Because protein synthesis is not an ongoing process, a cell must be notified when to make a particular protein. This communicative process, **cell signaling**, conveys physiological conditions or cellular needs to the nucleus of the cell, just like an indoor/outdoor thermometer conveys the outside temperature to the inside of a house.

5-2b Step 2: Transcription

Cell signaling initiates the second step of protein synthesis, **transcription**, whereby a specific type of ribonucleic acid (RNA), called **messenger RNA (mRNA)**, is constructed using DNA as a template. One way to think about this process is that it is like reading a cookbook.

Cell signaling functions somewhat like an indoor/outdoor thermometer. Cues from outside the cell can "turn on" the protein-synthetic machinery located in the cell's nucleus.

A chemical called **deoxyribonucleic acid (DNA)** provides the fundamental instructions for protein synthesis. Found in a cell's nucleus, coiled strands of DNA combine with special proteins to form **chromosomes**, which serve as the complete, organized cookbook. There are 23 different chromosomes in human cells. However, in most cells each chromosome has a matched pair. In total, there are 46 chromosomes in each cell except for egg and sperm cells. Each chromosome is subdivided into thousands of units, each of which is like an individual recipe. Each recipe, or **gene**, provides the instructions and list of ingredients needed to make a protein. In other words, a gene (the recipe) tells a cell (the cook) which amino acids are needed and in what order they must be arranged to synthesize a protein (the food).

In order for protein synthesis to take place, the information contained in the DNA's genetic code must be communicated from inside the nucleus to the cytoplasm of the cell where proteins are made. To accomplish this process, the instructions contained in the targeted portion of the DNA strand are converted to a chemical called messenger ribonucleic acid (mRNA). A series of mRNA subunits bind to the targeted gene coding for the protein that needs to be synthesized. These mRNA subunits then join, forming a strand of mRNA that is essentially a mirror image of the part of the DNA molecule that carries the instructions for protein synthesis. The newly formed strand of mRNA separates from the DNA, exits the nucleus, and enters the cytoplasm where it participates in the next step of protein synthesis: translation.

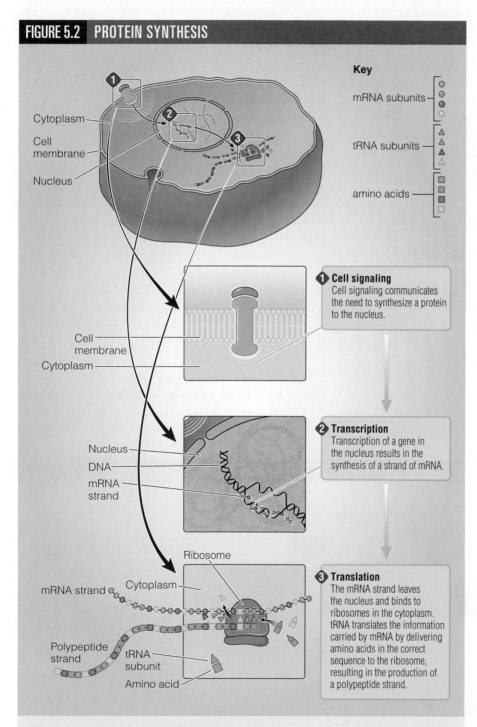

FIGURE 5.2 PROTEIN SYNTHESIS

Key

mRNA subunits

tRNA subunits

amino acids

Cytoplasm
Cell membrane
Nucleus

Cell membrane
Cytoplasm

1 Cell signaling
Cell signaling communicates the need to synthesize a protein to the nucleus.

Nucleus
DNA
mRNA strand

2 Transcription
Transcription of a gene in the nucleus results in the synthesis of a strand of mRNA.

Ribosome
Cytoplasm
mRNA strand

Polypeptide strand
tRNA subunit
Amino acid

3 Translation
The mRNA strand leaves the nucleus and binds to ribosomes in the cytoplasm. tRNA translates the information carried by mRNA by delivering amino acids in the correct sequence to the ribosome, resulting in the production of a polypeptide strand.

The steps involved in protein synthesis are three-fold: cell signaling, transcription, and translation.

deoxyribonucleic acid (DNA)
A chemical in the nucleus of a cell that provides the instructions for protein synthesis.

chromosomes A substance comprised of coiled strands of DNA and special proteins; found in a cell's nucleus.

gene A subunit of a chromosome that tells a cell which amino acids are needed and in what order they must be arranged to synthesize a specific protein.

5-2c Step 3: Translation

Once the mRNA strand is outside the nucleus, the strand binds to a cytoplasmic organelle called a **ribosome**, on which the third step of protein synthesis occurs. **Translation**, the process whereby amino acids are joined via peptide bonds, requires another form of RNA called **transfer ribonucleic acid (tRNA)**. tRNA units carry amino acids to the ribosome to be assembled into a peptide chain. For translation to proceed, the ribosome moves along the mRNA strand, reading its sequence. The sequence of the mRNA, in turn, instructs specific tRNAs to transfer the amino acids they are carrying to the ribosome. One by one, amino acids join together via peptide bonds to form a growing peptide chain. When translation is complete, the newly formed protein separates from the ribosome. This is not the final step in the formation of a new protein, however; the final structure and shape of the protein are yet to be determined.

5-3 WHY IS A PROTEIN'S SHAPE CRITICAL TO ITS FUNCTION?

After a polypeptide chain is released from the ribosome, it must fold (and sometimes combine with other polypeptide chains and substances) to form the complex shape and structure of the final protein. Because a protein's final shape is critical to its ultimate function within the body, it is important to understand the many levels of protein structure—and what can happen when something goes wrong.

ribosome A cellular component primarily involved in the assembly of proteins by involving mRNA and tRNA and the process of translation.

translation The process by which amino acids are joined via peptide bonds; occurs in the cell's cytoplasm.

transfer ribonucleic acid (tRNA) A chemical that carries amino acids to a ribosome to be assembled into a protein.

primary structure (or **primary sequence**) The most basic level of protein structure; determined by the number and sequence of amino acids in a single peptide chain.

sickle cell anemia (or **sickle cell disease**) A disease whereby a small alteration in the DNA results in the production of defective, misshapen molecules of the protein hemoglobin within red blood cells.

5-3a Primary Structure

The most basic level of protein structure, the **primary structure** (or **primary sequence**), is determined by the number and sequence of amino acids in a single peptide chain (see Figure 5.3). A protein's primary structure is determined by the genetic code embedded in the strands of DNA. Each peptide chain has a unique primary structure. Consequently, each protein is a unique molecule. Although understanding the concept of primary structure is relatively simple, take a moment to contemplate the enormous number of primary structures that can be made from just 20 amino acids. Consider, as an analogy, the English alphabet, which has a similar number of letters. As you know, the number and variety of words that can be constructed from just 26 letters is astounding. Some words are short; some are long. Some words contain just a few different letters; others contain many different letters. The same holds true for proteins: some are short; some are long; some contain a handful of different amino acids, whereas others contain all 20 amino acids. The possibilities are seemingly endless.

The primary structure of a protein is critical to its function because it determines the protein's most basic chemical and physical characteristics. Thus, a change in a protein's primary structure can profoundly affect the ability of the protein to do its job. Sometimes, alterations to a protein's primary structure can be caused by inherited genetic variations in the DNA code. An example of a disease caused by an inherited genetic variation is **sickle cell anemia** (or **sickle cell disease**). Sickle cell anemia is caused by a small alteration in the DNA code that ultimately results in the production of

FIGURE 5.3 PRIMARY STRUCTURE OF A PROTEIN

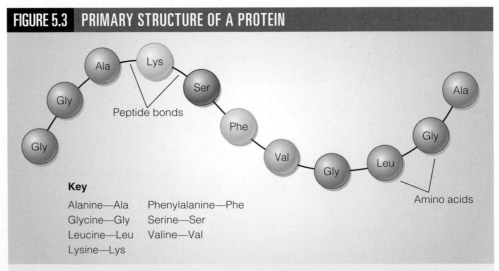

Key

Alanine—Ala	Phenylalanine—Phe
Glycine—Gly	Serine—Ser
Leucine—Leu	Valine—Val
Lysine—Lys	

The number and sequence of amino acids in a single polypeptide chain is referred to as a protein's primary structure.

The number and variety of proteins that the body can make are as astonishing as the number of words that can be formed with the letters of the English alphabet.

Why Are Some People Especially Prone to Sickle Cell Anemia?

Scientists hypothesize that, hundreds of years ago, the genetic alteration responsible for sickle cell anemia somehow protected people from the serious and sometimes deadly disease malaria.[8] As a result, individuals with the sickle cell anemia gene (or genes) were able to survive malaria epidemics, passing on their genetic code to their offspring. Today, millions of people have one or two genes for sickle cell anemia. Those with two genes have full-blown signs and symptoms of the disease, whereas those with one sickle cell gene are less affected, if at all. This genetic variation exists all over the world, especially in those with African, Mediterranean, Middle Eastern, or Indian ancestry. In the United States, sickle cell anemia is most common in people of black African heritage: 1 in 500 African Americans has the disease.

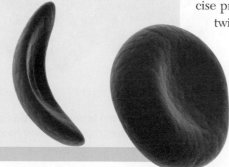

Normal red blood cells are disc-shaped (like the ones on the right), whereas those in people with sickle cell anemia are crescent-shaped (like the one on the left).

defective, misshapen molecules of the protein hemoglobin within red blood cells.[7] Because hemoglobin is responsible for carrying oxygen and carbon dioxide in the blood, complications such as fatigue and increased risk of infections sometimes occur and can be serious. When sickle-shaped red blood cells accumulate, they can damage the delicate lining of blood vessels and cause pain. Often, a person with sickle cell anemia will need to have a blood transfusion in order to have sufficient ability to deliver oxygen to their cells.

5-3b Secondary and Tertiary Structures

Polypeptide chains are relatively linear molecules. However, functional proteins are anything but linear—most have three-dimensional shapes. Because the backbone of the polypeptide chain is made of a series of weakly charged amino and carboxylic acid groups, these charges attract and repel each other like magnets. This causes portions of the polypeptide to fold into an organized and predictable pattern—the **secondary structure** of the protein. The two most common folding patterns are the **α-helix**, similar in shape to a spiral staircase, and the **ß-folded sheet**, similar in shape to a folded paper fan. Both folding patterns are illustrated in Figure 5.4.

The next level of protein complexity, **tertiary structure**, involves additional folding caused by interactions between the amino acids' R-groups. This folding transforms the entire protein into an even more complex, three-dimensional structure. Imagine, for example, what would happen to a folded paper fan if you were to crumple it gently in your hand. The folds (analogous to a protein's secondary structure) would remain, but they might be further bent and twisted in some regions (analogous to a protein's tertiary structure). This is similar to what happens to a protein when its tertiary structure is formed. It is this exact, precise process of folding and twisting that gives each protein its unique shape. And, it is this shape that allow proteins to recognize and interact with other proteins.

secondary structure Organized and predictable folds that develop in portions of a peptide chain because charged portions of the amino acid backbone attract and repel each other.

α-helix A common secondary structure folding pattern that resembles the shape of a spiral staircase.

ß-folded sheet A common secondary structure folding pattern that resembles the shape of a folded paper fan.

tertiary structure Additional folding of a peptide chain that develops because of interactions between the amino acids' R-groups.

FIGURE 5.4 SECONDARY STRUCTURE OF A PROTEIN

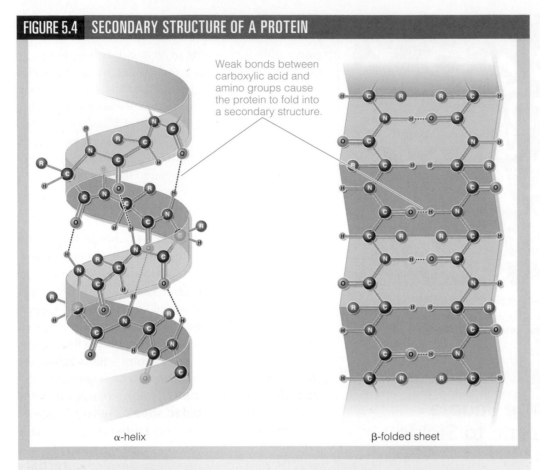

Weak bonds between carboxylic acid and amino groups cause the protein to fold into a secondary structure.

α-helix

β-folded sheet

Weak chemical bonds fold proteins into three-dimensional spiral- and fan-like shapes, resulting in secondary structures.

the quaternary structure, a nonprotein component called a **prosthetic group** must sometimes be positioned precisely within a protein for it to function. Prosthetic groups often contain minerals that are needed for the protein to carry out its purpose. Hemoglobin is an example of a protein with quaternary structure and prosthetic groups, because it is made from four separate polypeptide chains, each of which contains an iron-containing prosthetic group called heme. Heme is the portion of hemoglobin that actually transports the oxygen and carbon dioxide gases in the blood.

5-3c Quaternary Structure and Prosthetic Groups

The fourth level of protein structure is called **quaternary structure**. Quaternary structure occurs when two or more polypeptide chains join together, as shown in Figure 5.5. This level of complexity is somewhat like putting two or three crumpled paper fans together. Not all proteins have a quaternary structure—only those made from more than one polypeptide chain. In addition to

quaternary structure The most complex level of protein structure; occurs when two or more peptide chains join together.

prosthetic group A nonprotein component of a protein that often contains minerals needed for the protein to carry out its purpose.

FIGURE 5.5 QUATERNARY STRUCTURE AND PROSTHETIC GROUPS OF HEMOGLOBIN

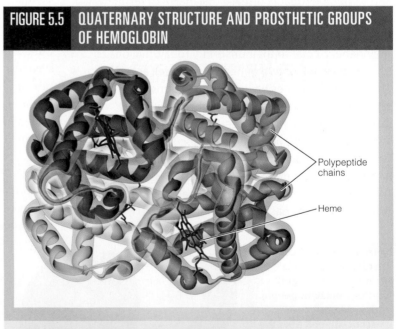

Polypeptide chains

Heme

Hemoglobin is made from four polypeptide chains and four iron-containing prosthetic groups called heme.

5-3d A Protein's Shape Determines Its Function

You now know how a protein's primary, secondary, tertiary, and quaternary structures come to be, and how they determine the protein's final shape. Shape is critical to a protein's ability to carry out its function. When a protein's shape is disrupted, the protein can no longer complete the task for which it was designed.

Denaturation

One way a protein's three-dimensional shape can be altered is by **denaturation**. Denaturation is akin to flattening out one of the pieces of paper that makes up a folded paper fan protein. Flattening out the paper results in a fan that probably does not work, just as denaturation can cause a protein to lose its function. Compounds and conditions that denature proteins are called denaturing agents. These include physical agitation (e.g., shaking), heat, detergents, acids, alkaline (basic) solutions, salts, alcohol, and heavy metals (e.g., lead and mercury).

Heat denatures the proteins in egg white, causing it to transform from a clear gel-like substance into a soft opaque solid.

©istock.com/Vladimir Glazkov

5-4 WHAT IS MEANT BY GENETICS AND EPIGENETICS?

As you now know, the DNA found in a cell's nucleus contains the instructions for all the proteins synthesized in the body. Each chromosome, consisting of double-stranded DNA, contains hundreds of genes. Each gene, in turn, contains the genetic code (information) to guide protein synthesis. Most chromosomes can code for many thousands of proteins. When a sperm, cell, containing 23 chromosomes, fertilizes an ovum (egg cell) which also has just 23 chromosomes, the offspring receives half its DNA from the father and the other half from the mother. Thus, it is at the moment of conception that an individual's **genotype** (genetic makeup) is determined. Except for identical twins, no two individuals have the same genotype (which in part explains the vast diversity among

humans). Some genetic differences impart individual physical characteristics, such as eye and hair color, which do not really affect your health. Such physical characteristics in part determine one's **phenotype**, the observable physical or biochemical characteristics of an organism. Although alterations in genes affecting phenotype do not usually affect one's health, alterations in other genes can have more serious consequences.

5-4a Mutations

An alteration in a gene sometimes results in a protein with an altered amino acid sequence—in other words, its primary structure is changed. When such an alteration occurs due to a genetic modification, it is called a **mutation**. Although some mutations have no measurable effects, others do. Sickle cell anemia, for example, is a disease caused by a mutation that results in decreased oxygen transport. Other mutations, such as the one involved in PKU, can influence metabolism. Still others alter protein synthesis so that cells experience uncontrollable growth—a condition that can result in cancer. Mutations present in the DNA of egg and sperm cells can be passed on to offspring and are therefore considered to be *inherited*. PKU and sickle cell anemia are examples of inherited DNA mutations. Still, there are other types of mutations that are random occurrences such as the one that causes Down Syndrome.

5-4b Epigenetics

A person's genotype is coded deep within the DNA that makes up his or her chromosomes. However, scientists recently learned that the connection between genes and physiology is actually much more complex than that. For example, DNA can be modified in ways that regulate whether a particular gene is expressed.

denaturation The process by which a protein's three-dimensional structure is altered.

genotype The particular DNA inherited from one's parents.

phenotype The observable physical or biochemical characteristics of an organism.

mutation An alteration in a gene that occurs due to a chance genetic modification.

Personalized Diet Prescriptions Based on Genetics

Nutrition scientists can now directly test what they have long thought: nutrition interacts with genetics to influence health. For example, modern technology has made it much easier to study how dietary factors influence whether a gene is turned on or turned off. Additionally, emerging studies suggest that our dietary choices also likely determine some epigenetic alterations. Many scientists and health professionals hope that someday everyone will be able to inexpensively and noninvasively find out what type of diet is best for their personal genetic makeup and health.

It is possible that someday you will be able to have a personal diet prescription that matches your distinct genetic and epigenetic makeup. For instance, increased consumption of some micronutrients might help decrease risk of heart disease in some individuals more so than in others.

© Rawpixel.com/Shutterstock.com

The term **epigenetics** refers to inheritable changes in gene expression that do not involve changes in the DNA sequence itself. A person's **epigenome**, therefore, refers to all of the chemical compounds that attach to the strands of DNA or proteins associated with DNA that collectively influencing gene expression by turning genes on or off. Even if two people have exactly the same genetic sequences in their DNA, as is the case with identical twins, they may differ in terms of how those genes are expressed due to epigenetic variation. Interestingly, like mutations in DNA, some epigenetic differences can be passed on to the next generation. Scientists now think that epigenetic modifications may play important roles in the development of many chronic degenerative diseases such as cancer, type 2 diabetes, and cardiovascular disease.

Nutrition and Epigenetics Growing evidence that nutritional status may affect long-term epigenetic modifications is of great importance to the field of nutrition.

epigenetics Alterations in protein synthesis that do not involve changes in the DNA sequence.

epigenome A network of chemical variations around DNA that together determine which genes are active in a particular cell.

For example, there is strong evidence that babies who are malnourished during fetal life, but then experience accelerated growth in childhood, may be at increased risk for cardiovascular disease and type 2 diabetes as adults, partly due to epigenetic modifications in gene expression.[9] Whether later alterations in a person's environment, such as better nutrition, can reverse this effect remains to be discovered. Clearly, epigenetics is an exciting new area of nutrition research.

5-5 HOW ARE PROTEINS DIGESTED, ABSORBED, AND CIRCULATED?

Protein digestion requires the splitting of peptide bonds that hold the amino acids together. Once this occurs, the released amino acids are absorbed and circulated in the blood to all the body's cells, where the amino acids are used for protein synthesis. This is somewhat like disassembling someone else's house and then using the materials to build another house that perfectly fits your own personal needs. The body also efficiently and systematically breaks down and recycles its own proteins when they become old and nonfunctional. In fact, you can think of your body as having its own protein-recycling center. The stages of protein digestion, absorption, and circulation are shown in Figure 5.6 and described next.

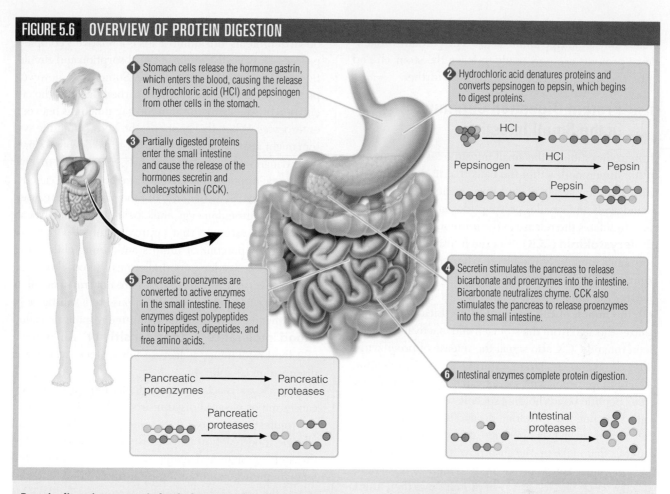

FIGURE 5.6 OVERVIEW OF PROTEIN DIGESTION

1 Stomach cells release the hormone gastrin, which enters the blood, causing the release of hydrochloric acid (HCl) and pepsinogen from other cells in the stomach.

2 Hydrochloric acid denatures proteins and converts pepsinogen to pepsin, which begins to digest proteins.

3 Partially digested proteins enter the small intestine and cause the release of the hormones secretin and cholecystokinin (CCK).

HCl

Pepsinogen $\xrightarrow{\text{HCl}}$ Pepsin

$\xrightarrow{\text{Pepsin}}$

5 Pancreatic proenzymes are converted to active enzymes in the small intestine. These enzymes digest polypeptides into tripeptides, dipeptides, and free amino acids.

4 Secretin stimulates the pancreas to release bicarbonate and proenzymes into the intestine. Bicarbonate neutralizes chyme. CCK also stimulates the pancreas to release proenzymes into the small intestine.

Pancreatic proenzymes $\longrightarrow$ Pancreatic proteases

Pancreatic proteases

6 Intestinal enzymes complete protein digestion.

Intestinal proteases

Protein digestion occurs in both the stomach and small intestine via pepsin, pancreatic proteases, and intestinal proteases.

5-5a Protein Digestion Begins in the Stomach

Before you can use the proteins in the foods you eat, they must be broken down into their component amino acids. Although a small amount of mechanical digestion of protein occurs as you chew your food, chemical digestion of protein does not begin until the food comes in contact with specialized cells in the stomach. The presence of food causes these stomach cells to release the hormone gastrin, which, in turn, triggers other cells in the stomach lining to release hydrochloric acid (HCl), mucus, and pepsinogen. **Pepsinogen** is an inactive form of **pepsin**, an enzyme needed for protein digestion. In general, an inactive precursor of an enzyme is called a **proenzyme** (sometimes called a **zymogen**). Once the components of the gastric juices are released, the chemical digestion of proteins can begin via the two-step process described next.

First, hydrochloric acid disrupts the weak chemical bonds responsible for the protein's secondary, tertiary, and quaternary structures. This process of denaturation straightens out the complex protein structure, helping to expose the peptide bonds to the digestive enzymes present in the stomach and small intestine. It is important to note that the peptide bonds present in the primary structure of the protein are still intact. Second, hydrochloric acid converts the proenzyme pepsinogen into its active form, pepsin. Pepsin is an example of a **protease**, an enzyme that breaks peptide bonds between amino acids. Note that the stomach does not produce the active protease enzyme pepsin. Instead, it produces and stores the inactive (or safe) proenzyme pepsinogen. This protects the stomach from the active enzyme's protein-digesting function until it is needed. As a result of the actions of stomach acid and enzymes,

pepsinogen The inactive form of the enzyme pepsin.

pepsin An enzyme needed for protein digestion.

proenzyme (also called a **zymogen**) An inactive precursor of an enzyme.

protease A type of enzyme that breaks peptide bonds between amino acids.

proteins are partially digested to shorter polypeptide chains and some free amino acids. The partially broken-down proteins are now ready to leave the stomach and enter the small intestine to be digested further.

5-5b Protein Digestion Continues in the Small Intestine

Protein digestion in the small intestine takes place both in the lumen and within the cells that line it. Initiating this cascading series of digestive events, the arrival of amino acids and smaller polypeptides in the small intestine stimulates the release of the hormones **secretin** and **cholecystokinin (CCK)** from the intestinal cells. These hormones then enter the blood, where they travel to the pancreas. Secretin signals the pancreas to release bicarbonate (the same substance that makes up baking soda) into the lumen of the small intestine. Bicarbonate neutralizes the acid from the stomach and inactivates pepsin. Secretin and CCK also signal the release of proenzymes from the pancreas. Upon entering the small intestine, each of these proenzymes is activated and is then able to break peptide bonds holding specific sequences of amino acids together. The resulting di- and tripeptides are further broken down by a multitude of proteases produced in the cells that make up the absorptive surface of the small intestine. This usually results in the complete breakdown of proteins into their amino acid constituents.

5-5c Amino Acid Absorption and Circulation

When protein digestion is complete, amino acids can enter the cells (enterocytes) of the small intestine. Most amino acids are absorbed in the duodenum, where they enter the blood and circulate to the liver for further processing.

secretin A hormone, secreted by intestinal cells, that signals the release of sodium bicarbonate and proteases from the pancreas.

cholecystokinin (CCK) A hormone, secreted by intestinal cells, that signals the release of bile from the gallbladder and proenzyme proteases from the pancreas.

food allergy A condition whereby the body's immune system responds to a food-derived peptide as if it were dangerous.

food intolerance (or **food sensitivity**) A condition whereby the body reacts negatively to a food or food component, but does not mount an immune response.

anaphylaxis A rapid immune response that causes a sudden drop in blood pressure, rapid pulse, dizziness, and a narrowing of the airways.

Food Allergies and Intolerances The breakdown of proteins into amino acids is a relatively complete process—it typically results in the absorption and circulation of amino acids (not proteins). Sometimes, however, partially digested proteins are absorbed or enter the circulation by squeezing through the gap junctions between enterocytes. When this happens, the body's immune system might respond as if these partial-breakdown products were dangerous. In such cases, the person is said to have an *allergic response*, or what is more commonly called a **food allergy**.[10] The majority of food allergies are caused by proteins present in eggs, milk, peanuts, soy, and wheat. Researchers estimate that approximately 2 percent of adults and 5 percent of infants and young children in the United States have food allergies.[11] Note, however, that all adverse reactions to foods are not true food allergies. A negative physiological response to a substance in a food that does not trigger an immune response is called a **food intolerance** (or **food sensitivity**). An example of a food intolerance is lactose intolerance, which was discussed in Chapter 4.

An allergic reaction to a particular food protein often causes minor physical discomforts such as skin rashes or gastrointestinal distress. For some people, however, an allergic food reaction can be frightening and even life threatening. Signs and symptoms usually develop within a few minutes to an hour after eating the food, depending on the type of food allergy a person has and how his immune system reacts to it. The most common signs and symptoms of food allergies are listed here:

- Tingling in the mouth;
- Hives, itching, or eczema;
- Swelling of the lips, face, tongue, throat, or other parts of the body;
- Wheezing, nasal congestion, or trouble breathing;
- Abdominal pain, diarrhea, nausea, or vomiting;
- Dizziness, lightheadedness, or fainting.

In a severe case, a person may experience more extreme symptoms, such as **anaphylaxis**. Anaphylaxis is a rapid immune response that causes a sudden drop in blood pressure, rapid pulse, dizziness, and a narrowing of the airways. This can, almost immediately,

More than 3 million people are allergic to peanuts or tree nuts in the United States.

impair the person's ability to breath. When this occurs, emergency treatment is critical. Anaphylaxis due to food allergies is responsible for thousands of emergency room visits and as many as 200 deaths in the United States each year.[12]

What to Do If You Have a Food Allergy If you have a food allergy, it is important to know which foods to avoid. It is especially important to read food labels carefully; the U.S. Food and Drug Administration (FDA) requires that all foods containing the most common allergens be labeled as such. Wearing a medical alert bracelet or necklace may be advantageous in case an allergenic food is accidentally ingested. In the case of children with food allergies, parents should notify friends, family members, childcare providers, and school personnel so that they can help avoid exposure to offending foods. It is also important to make sure that the child knows which foods to avoid and to ask for help if needed. Because of the severity of some food allergies, it is important for parents and schools to work together to create a food allergy management plan for emergency care.

5-6 WHY DO YOU NEED PROTEINS AND AMINO ACIDS?

Once amino acids are circulated away from the GI tract, your body uses them to make the thousands of proteins it needs via protein synthesis. Using the previous analogy, this is the stage at which you would use all of the disassembled materials from someone else's house to build one of your own. The deconstructed materials could be used to build walls, cabinets, stairways, or furniture—each of these household items has its own function. Similarly, the proteins that your body makes can be classified into general categories based on their functions (see Table 5.2). Some proteins, such as those in your muscles, are used for movement. Others, such as the hormone insulin, are used to regulate blood glucose. Some proteins can be broken down and used for energy, and some amino acids can be converted to glucose. In the following sections, you will learn more about the various types of proteins and amino acids your body needs, as well as other ways that they are used to promote functionality and health.

TABLE 5.2	MAJOR FUNCTIONS OF PROTEINS IN THE BODY	
Function	**Description**	**Selected Examples**
Structure	Proteins that make up the basic structure of tissues such as bones, teeth, and skin	• Bone matrix proteins (such as hydroxyapetite) • Collagen in skin, teeth, ligaments, and tendons • Keratin in hair and fingernails
Catalysis	Proteins (enzymes) that facilitate chemical reactions	• Lingual lipase • Pancreatic amylase • Pepsin
Movement	Proteins found in muscles, ligaments, and tendons	• Actin and myosin in muscle • Elastin in ligaments
Transport	Proteins involved in the movement of substances across cell membranes and within the circulatory system	• Glucose and sodium transporters in cell membranes • Vitamin A-binding protein, which transports vitamin A in blood
Communication	Protein hormones, cell-signaling proteins, and neurotransmitters	• Insulin and glucagon • Cholecystokinin (CCK)
Protection	Proteins that constitute the skin and immune system	• Collagen in skin • Fibrinogen in blood clots • Antibodies
Regulation of fluid balance	Proteins that regulate the distribution of fluid in the body's various compartments via the process of osmosis	• Albumin
Regulation of pH	Proteins that take up and release hydrogen ions (H^+) to maintain appropriate pH of body fluids and tissues	• Hemoglobin

5-6a Proteins Provide Structure

Being constituents of muscles, skin, bones, hair, and fingernails, proteins comprise most of the structural materials in your body. Collagen, for instance, is a structural protein that forms a supporting matrix in bones, teeth, ligaments, and tendons. Proteins are also important structural components of cell membranes and organelles. The synthesis of structural proteins is especially important during periods of active growth and development, such as infancy and adolescence.

5-6b Enzymes Catalyze Chemical Reactions

The myriad chemical reactions in your body are driven by a class of protein molecules called enzymes. These biological catalysts speed up chemical reactions without being consumed or altered in the process. Without the catalytic functions of enzymes, the thousands of chemical reactions needed by your body to function would simply not occur or, at best, would occur at very slow rates. Examples of enzymes you have already learned about are amylase and pepsin, which catalyze reactions needed to digest carbohydrates and proteins, respectively.

5-6c Muscle Proteins Facilitate Movement

Protein is also necessary for movement, which results from the contraction and relaxation of the many muscle fibers in the body. Muscles are involved in both voluntary and involuntary movements such as those needed for cardiovascular function and physical activity, respectively. Nearly half of the body's protein is present in skeletal muscle, and adequate protein intake is required to form and maintain muscle mass and function throughout life. Although there are many proteins related to movement, perhaps the most important are actin and myosin, which make up much of the machinery needed for muscles to contract and relax. This is why protein deficiency can cause muscle wasting and weakness.

© Markgraf/Shutterstock.com

antibody (or **immunoglobulin**) A protein, produced by the immune system, that helps fight infection.

Proteins are essential to movement.

5-6d Some Proteins Serve as Transporters

Amino acids are also used to make transport proteins, which are responsible for escorting substances into and around the body, as well as across cell membranes. For example, absorption of many nutrients (such as calcium) requires one or more transport proteins to help the nutrients cross the lumen-facing portion of membranes surrounding enterocytes. Protein deficiency can decrease the body's production of intestinal transport proteins, resulting in secondary malnutrition. In addition to facilitating the transport of substances across cell membranes, many proteins are critical for the transport of nutrients and other substances in the blood. Examples of circulating transport proteins include hemoglobin, which transports gases (oxygen and carbon dioxide), and a variety of binding proteins that transport hormones and fat-soluble vitamins in the blood.

5-6e Hormones and Cell-Signaling Proteins Are Critical Communicators

Tissues and organs have a variety of ways to communicate with each other, and most of these methods involve proteins. Although not all hormones are proteins, most are, including secretin, gastrin, insulin, and glucagon are proteins. Beyond hormonal communicators, there are also specialized proteins embedded in cell membranes that communicate information about the extracellular environment to the intracellular space. Some of these proteins are involved in the cell-signaling process that initiates protein synthesis itself. Others regulate cellular metabolism. Together, hormones and cell-signaling proteins make up part of the body's critical communication network. Thus, protein deficiency can have profound effects on your body's ability to coordinate all of its functions.

5-6f Proteins Protect the Body

One of the most vital and basic functions of the proteins in your body is protecting it from physical danger and infection. For instance, skin is mainly made of proteins that form a barrier between the outside world and your internal environment. If your skin gets cut, blood clots, which are produced by a series of clotting proteins, close off the possible entry point to infection. And, if a bacterium or other foreign substance does enter the body, your immune system responds by producing **antibody** (or **immunoglobulin**) proteins to help fight the infection. Antibodies bind to foreign substances so they can be destroyed. Protein deficiency can make it difficult for the body to prevent and fight certain diseases because its natural defense systems become weakened. This is why infection and illness often accompany protein deficiency.

5-6g Fluid Balance Is Regulated in Part by Proteins

Another function of proteins is regulating how fluids are distributed in the body. As you might know, most of your body is made of water. This important fluid is found both inside and outside of cells. The fluid outside of cells can be subdivided into that found in blood and lymph vessels (intravascular fluid) and that found between cells (extravascular fluid). The amount of fluid in these spaces is tightly regulated by a variety of means, some of which involve proteins. **Albumin**, a protein present in relatively high quantities in the blood plays such a role. As the heart beats, blood is pumped out of the heart and into blood vessels that become increasingly narrower. As the pressure builds, the fluid portion of the blood is squeezed out of the tiny capillaries. Albumin, which remains in the blood vessels, becomes more concentrated as more fluid is lost. The high concentration of albumin draws the fluid that was once squeezed out of the blood vessel back into it (see Figure 5.7). Severe protein deficiency can impair albumin synthesis, resulting in low levels of albumin in the blood and causing fluid to accumulate in the space surrounding tissues. This condition, called **edema**, can sometimes be

albumin A protein present in the blood that plays an important role in regulating fluid balance.

edema A condition whereby low levels of albumin in the blood cause fluid to accumulate in body tissues or cavities.

FIGURE 5.7 REGULATION OF FLUID BALANCE BY ALBUMIN

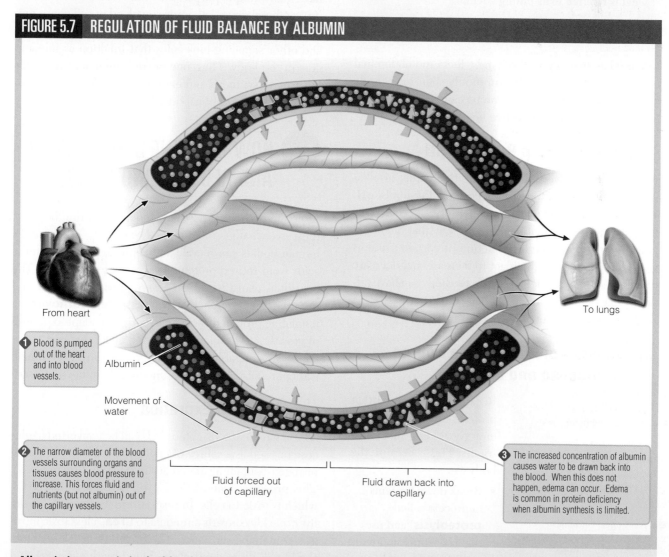

From heart

To lungs

1 Blood is pumped out of the heart and into blood vessels.

Albumin

Movement of water

2 The narrow diameter of the blood vessels surrounding organs and tissues causes blood pressure to increase. This forces fluid and nutrients (but not albumin) out of the capillary vessels.

Fluid forced out of capillary

Fluid drawn back into capillary

3 The increased concentration of albumin causes water to be drawn back into the blood. When this does not happen, edema can occur. Edema is common in protein deficiency when albumin synthesis is limited.

Albumin is a protein in the blood that helps regulate fluid balance within and around tissues.

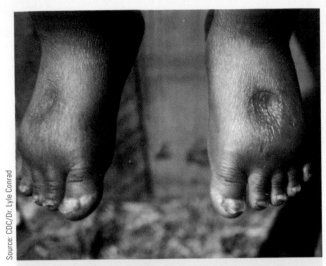

Because protein is needed for fluid balance, severe protein deficiency can cause edema. Pressure has been applied to the tops of these feet, demonstrating what is referred to as pitting edema.

observed as swelling in the hands, feet, and abdominal cavity. Although edema is commonly seen in severely malnourished individuals, it can be caused by other factors as well, such as congestive heart failure.

5-6h Proteins Help Regulate pH

Proteins are involved in regulating the acidity or alkalinity of your body fluids, referred to as the fluids' pH. Your body must maintain certain pH levels in each of its fluids to ensure optimal health. One way that blood pH is maintained is through the action of certain proteins such as hemoglobin, which act to increase and decrease the blood's pH as needed. As such, the body can have difficulty maintaining optimal pH balance during periods of severe protein deficiency.

5-6i Amino Acids Provide a Source of Glucose and Energy

proteolysis The breakdown of proteins.

protein turnover The continual coordinated process of protein breakdown and synthesis.

urea A nitrogen-containing substance produced by the body when it breaks down amino acids; expelled in the urine.

Recall from Chapter 1 that proteins are categorized as energy-yielding macronutrients. This is because the body can:

• Break down proteins (a process called **proteolysis**) and use their constituent amino acids for energy (ATP) production;

• Chemically transform some amino acids into glucose (a process called gluconeogenesis);

• Convert excess amino acids into fat (a process called lipogenesis).

Together, these processes help the body:

1. generate ATP to power chemical reactions even when glucose and fat availability is limited,

2. maintain blood glucose at appropriate levels, and

3. store excess energy when dietary protein intake is more than adequate.

5-6j Amino Acids Serve Many Additional Purposes

In addition to serving as the building blocks for proteins and providing a source of energy, amino acids themselves have many unique purposes in your body. Some regulate protein breakdown, others are involved in cell communication, and still others are converted to neurotransmitters and other signaling molecules that function as messengers in the body. Thus, one needs amino acids not only for protein synthesis and energy production but also for a multitude of additional functions.

HOW DOES THE BODY RECYCLE AND REUSE AMINO ACIDS?

Although proteins serve many functions within the body, they eventually—inevitably—wear out. Fortunately, human bodies can recycle and reuse most of the amino acids from retired proteins to synthesize new ones. The continual coordinated process of breaking down and resynthesizing protein is known as **protein turnover**. By regulating protein turnover, the body can adapt to periods of growth and development during childhood and maintain relatively stable amounts of protein during adulthood without requiring enormous amounts of protein from food.

5-7a Nitrogen Excretion

As you have learned, amino acids can be converted to glucose or used as a source of energy. For this to occur, the nitrogen-containing amino group must first be removed. This process (deamination) produces ammonia (NH_3), which is toxic to cells. In response to its production, the liver quickly converts ammonia to **urea**, a less toxic nitrogen-containing substance. The urea is then released into the blood, filtered out of the blood by the kidneys, and excreted in the urine.

5-7b Nitrogen Balance and Protein Status

Protein turnover results in a somewhat complex flux (or remodeling) of amino acids in your body every day. Measuring protein turnover can provide health professionals with important information about overall protein status. One's protein status can be assessed by comparing protein *intake* to nitrogen *loss* in body secretions such as urine, sweat, and feces.[13] When nitrogen loss equals nitrogen intake, the body is in **neutral nitrogen balance**. When nitrogen loss exceeds intake, as can occur during starvation, illness, or stress, a person is in **negative nitrogen balance**. When nitrogen intake exceeds loss, as can occur during childhood or recovery from an illness, a person is in **positive nitrogen balance**. Knowing whether a person is in neutral, positive, or negative nitrogen balance can help clinicians diagnose and treat certain disease states and physiologic conditions. For example, people who are on dialysis because of kidney failure often experience negative nitrogen balance and therefore require specialized nutritional support. Conversely, growing children should be in a state of positive nitrogen balance. If this is not the case, protein intake may need to be increased. Because the need for protein shifts throughout the course of one's life, dietary recommendations for proteins and amino acids have been developed to take into account both nitrogen balance and the need for specific amino acids during different life-stage periods.

5-8 HOW MUCH PROTEIN DO YOU NEED?

You need to consume dietary protein for two major reasons:

1. to supply adequate amounts of the essential amino acids, and
2. for the additional nitrogen needed to make the nonessential amino acids and other nonprotein, nitrogen-containing compounds such as DNA. As such, recommendations for dietary amino acid and overall protein consumption reflect these needs.

5-8a Dietary Reference Intakes (DRIs) for Amino Acids

To begin with, consider how much of each essential amino acid you need to eat every day. Currently, the best estimates can be obtained from the Institute of Medicine's Recommended Dietary Allowances (RDAs), which are shown in Figure 5.8.[14] Note that these values are presented in units of milligrams per kilogram per day (mg/kg/day) because they represent requirements of the essential amino acids *relative to body size*: the larger you are, the more essential amino acids you need. For example, because the RDA for the essential amino acid valine in adults is 4 mg/kg/day, a woman weighing roughly 140 lb (64 kg) would require 4 mg × 64 kg (256 mg/day) of this amino acid in her diet. A larger person weighing roughly 200 lb (91 kg) would require 4 mg × 91 kg (364 mg) of valine each day.

Because the DRI committee concluded that there is no compelling evidence that high intake of any of the essential amino acids poses known health risks, the Institute of Medicine did not establish Tolerable Upper Intake Levels (ULs) for them.

Growing children are in a state of positive nitrogen balance because they consume more nitrogen than they excrete.

© Yuganov Konstantin/Shutterstock.com

---

neutral nitrogen balance A bodily state whereby nitrogen intake equals nitrogen loss.

negative nitrogen balance A bodily state whereby nitrogen intake is less than nitrogen loss.

positive nitrogen balance A bodily state whereby nitrogen intake is greater than nitrogen loss.

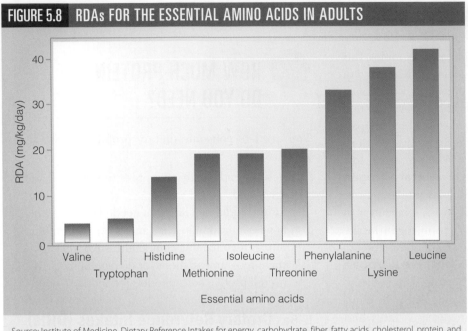

FIGURE 5.8 RDAs FOR THE ESSENTIAL AMINO ACIDS IN ADULTS

Source: Institute of Medicine. Dietary Reference Intakes for energy, carbohydrate, fiber, fatty acids, cholesterol, protein, and amino acids. Washington, DC: National Academies Press; 2005.

5-8b Dietary Reference Intakes (DRIs) for Protein

Perhaps of greater interest are recommendations for overall protein consumption. Indeed, RDAs for total protein intake have also been published and are found in the DRI card located at the back of this book. Although not all protein sources are created equal, researchers generally agree that most diets in affluent countries, such as the United States, provide a balanced mix of all the essential amino acids. Therefore, the dietary recommendations for protein intake do not distinguish between people who consume high-quality proteins and those who do not.

The RDA values for protein are expressed in two ways. The first, grams per day (g/day), reflects requirements for a typical person in a particular population group. These recommended protein intakes increase with age and are somewhat higher for men than women because, in general, men are larger than women and have more muscle mass. Using this set of values, a typical college-age man needs 56 g/day of protein, whereas a comparable woman needs 46 g/day, which is enough protein for adults to maintain neutral protein balance.

The second way that an RDA value for protein can be expressed is as grams per kilogram body weight per day (g/kg/day). Like those for the essential amino acids, these recommendations adjust for body weight. The Institute of Medicine recommends that healthy adults consume 0.8 g/kg/day of protein. For example, an adult weighing roughly 140 lb (64 kg) requires about 51 g of protein (0.8 g × 64 kg) every day, regardless of whether the adult is a woman or a man. A person could easily get this much protein by eating a bowl of wheat flake cereal and low-fat milk (12 g protein) for breakfast, a hamburger (24 g protein) for lunch, and a bean burrito (15 g protein) for supper. Again, this amount of protein in a person's diet is adequate to maintain neutral

It is important for athletes to get adequate amounts of protein by eating sufficient amounts of high-quality protein foods.

protein balance. However, there are times when adults may require additional protein. For instance, they might need more protein to support growth and/or repair when working out or recovering from an illness or injuries.

During infancy, the most rapid phase of growth in the life cycle, protein requirements (when adjusted for body weight) are relatively high. Protein requirements also increase during pregnancy and lactation, because additional protein is needed to support fetal growth and milk production.[15] Because healthy people show little evidence of harmful effects of high protein intake, no UL values are set for this macronutrient.

5-8c Experts Debate Whether Athletes Need More Protein

Although many people believe that athletes have higher protein requirements than nonathletes, this is a topic of active debate. The DRI committee that established recommendations for amino acid and protein intake considered this question carefully. The committee concluded that physically active people likely require similar amounts of protein on a body-weight basis, and that adult athletes can generally estimate their protein requirements as would other adults by using the same mathematical formula of 0.8 g/kg/day. On the other hand, in a position statement published in 2009, the American College of Sports Medicine concluded that protein intakes of 1.2 to 1.7 g/kg/day for endurance and strength-trained athletes may be beneficial.[16] Similarly, the International Society of Sports Nutrition recommends protein intakes of 1.4 to 2.0 g/kg/day for physically active individuals.[17] Thus, scientists continue to grapple with this issue.

5-8d Additional Recommendations for Protein Intake

Aside from the RDA values, several other sets of recommendations for protein intake are also available. For example, the Institute of Medicine's Acceptable Macronutrient Distribution Ranges (AMDRs) recommend that you consume 10 to 35 percent of your energy as protein. Using this advice, consider a moderately active college student with an energy requirement of 2,000 kcal/day. How much protein should this student consume? To answer this question, you must first determine that $0.10 \times 2,000$ kcal and $0.35 \times 2,000$ kcal (200 to 700 kcal) should come from protein. This translates to 200 kcal ÷ 4 kcal/gram, and 700 kcal ÷ 4 kcal/gram of protein (50 to 175 g). Remember that 1 g of protein supplies 4 kcal of energy. One medium hamburger patty and one cup of skim milk provide approximately 25 and 10 grams of

Are Protein Supplements Helpful for Athletes?

Although there is very little evidence that protein supplementation is beneficial in terms of athletic performance, a handful of studies suggest that supplementation with certain amino acids—especially branched-chain amino acids (BCAA) such as valine and leucine—may help slow muscle breakdown during intensive training.[18] In response, the International Society of Sports Nutrition published a position paper on protein and exercise stating that, under certain circumstances, supplementation with branched-chain amino acids may improve exercise performance and recovery from exercise.[17] Nonetheless, the bottom line recommendations for athletes remain:

1. eat a well-balanced diet that provides sufficient energy and an appropriate mix of carbohydrates, fats, and protein, and

2. train long and hard.

protein, respectively, making the recommended amount of protein quite easy to obtain, especially at the lower end of the range.

The 2015 Dietary Guidelines for Americans and the accompanying MyPlate food guidance system provide additional recommendations concerning intake of high-protein foods. Aside from supporting the AMDR for protein (10 to 35 percent of calories from protein), the Guidelines specifically encourage a range of intakes from fat-free or low-fat milk, lean meats, eggs, fish, nuts/seeds, and legumes to support optimal health. These food groups represent nutrient-dense, high-protein foods. More specifically, the current Dietary Guidelines recommend 1 to 6 ounces of lean meat and 2 to 3 cups of fat-free or low-fat dairy products daily, depending on caloric needs and whether you tend to choose foods that align most closely to a US-style, Mediterranean-style, or vegetarian eating pattern. Periodic consumption of legumes, such as dried beans and peas, is also encouraged. Note that 3 ounces of lean meat is generally equivalent to a small steak, lean hamburger patty, chicken breast, or a piece of fish. Portions of meat served in restaurants and cafeterias are often much larger than this. To determine precisely how many servings are recommended for you, visit the MyPlate website (http://www.choosemyplate.gov).

It is important to understand basic recommendations concerning how much of each essential amino acid you should consume. You can calculate your total protein requirement using the MyPlate food guidance system and can otherwise refer to MyPlate for suggestions that will help you meet your goals. But what if you were to decide not to eat meat or other high-quality protein sources? How would this affect your health, and what might you do to make sure that your diet was adequate?

© stepmorem/Shutterstock.com

Su vegetarianism, practiced by many Taiwanese Buddhists, prohibits the consumption of fetid (meaning smelly) vegetables such as garlic, shallot, onion, and coriander, in addition to animal-based products.

5-9 CAN VEGETARIAN DIETS BE HEALTHY?

People have many different reasons for deciding which foods they will and will not eat. This seems especially true for meat and other animal-derived products. For example, some religious groups avoid some or all types of animal-based foods. Economic considerations and personal preference can also determine whether people eat meat and/or which types of meat they choose. In fact, it is very likely that you or someone you know avoids eating some or all animal-based products. Because these types of foods tend to provide high-quality protein, as well as a multitude of other essential nutrients, it is important to consider the effect of animal-based food consumption (or the lack thereof) on issues related to nutritional status.

5-9a There Are Several Forms of Vegetarianism

Most people have heard the term *vegetarian* (from the Latin *vegetus*, meaning "whole," "sound," "fresh," or "lively"). What does this word actually mean? The term was first used in 1847 by the Vegetarian Society of the United Kingdom to refer to a person who does not eat any meat, poultry, fish, or their related products such as milk and eggs. Today, the term **vegetarian** usually indicates a person who does not consume any, or consumes only some, foods and beverages made from animal-based products. Most vegetarians consume dairy products and eggs. Such a practitioner is called a **lacto-ovo-vegetarian**.[19] Alternatively, a

vegetarian A person who does not consume or consumes only some foods and beverages made from animal products.

lacto-ovo-vegetarian A vegetarian who consumes dairy products and eggs in an otherwise plant-based diet.

Vegetarian diets can provide all the essential nutrients, but care must be taken to make sure sufficient protein, iron, calcium, zinc, and vitamin B$_{12}$ are consumed.

© jreika/Shutterstock.com

lactovegetarian includes dairy products—but not eggs—in his diet. A vegetarian who avoids all animal-based products is referred to as a **vegan**. Thus, when people tell you they are vegetarians, you might want to ask what type they are.

5-9b Vegetarian Diets Sometimes Require Thoughtful Choices

Do vegetarians necessarily have any special nutritional risks? The answer to this question depends on what kind of vegetarian a person is. In general, a well-balanced lacto-ovo- or lactovegetarian diet can easily provide adequate protein, energy, and micronutrients. Dairy products and eggs are convenient sources of high-quality protein and many vitamins and minerals. However, because meat is often the primary source of bioavailable iron, eliminating it can make it difficult to meet your iron requirements. Furthermore, vegans may be at increased risk of being deficient in several micronutrients, including calcium, zinc, iron, and vitamin B_{12}.[20] This risk is increased further during pregnancy, lactation, and periods of growth and development such as infancy and adolescence.[21] It is especially important that vegetarians consume sufficient amounts of plant-based foods rich in these micronutrients.

5-9c Special Dietary Recommendations for Vegetarians

Because some types of vegetarian diets pose certain nutritional risks, it is important to follow special dietary strategies if you decide to make this choice.

The MyPlate food guidance system specifically recognizes protein, iron, calcium, zinc, and vitamin B_{12} as nutrients that vegetarians should focus on and makes specific recommendations as to how to get adequate amounts of these substances. Special food patterns for vegetarians are also included in the 2015 Dietary Guidelines for Americans.

In addition, it is noteworthy that some meat replacements, such as cheese, can be very high in calories, saturated fat, and cholesterol. Lower-fat versions should be chosen, and they should be consumed in moderation. The following comments and suggestions can help ensure optimal health in individuals who choose to become vegetarians. Note that these recommendations are especially pertinent to vegans.

- Select protein sources that are naturally low in fat, such as skim milk and legumes.

- Minimize the use of high-fat cheese as a meat replacement.
- If you do not consume dairy products, consider drinking calcium-fortified, soy-based beverages. These can provide calcium in amounts similar to milk, are usually low in fat, and do not contain cholesterol.
- Add vegetarian meat substitutes, such as tofu, to soups and stews to boost protein without adding saturated fat and cholesterol.
- Recognize that most restaurants can accommodate vegetarian modifications to menu items by substituting meatless sauces, omitting meat from stir-fry recipes, and adding vegetables or pastas in place of meat.
- Consider eating out at Asian or Indian restaurants, as they often offer a varied selection of high-protein, nutrient-dense, vegetarian dishes.
- Be mindful of getting enough vitamin B_{12}. Because this vitamin is naturally found only in foods that come from animals, vitamin B_{12}-fortified foods or dietary supplements may be necessary for vegans.

The key to a healthy vegetarian diet, as with any diet, is to enjoy a wide assortment of foods and to consume them in moderation. Because no single food provides all the nutrients the body needs, eating a variety of foods can help ensure that vegetarians get the necessary nutrients and substances that promote good health.

WHAT ARE THE CONSEQUENCES OF PROTEIN DEFICIENCY AND EXCESS?

Although generally not a concern in industrialized countries, even for those who consume a vegetarian diet, protein deficiency is more commonly seen in regions where the amount and variety of foods are limited. Protein deficiency is also seen in adults with some debilitating conditions such as acquired immune deficiency syndrome (AIDS) or cancer. Because proteins and amino acids are so important to optimal health, protein deficiency can have significant health implications—some of which are described next.

lactovegetarian A vegetarian who consumes dairy products (but not eggs) in an otherwise plant-based diet.

vegan A vegetarian who consumes no animal products.

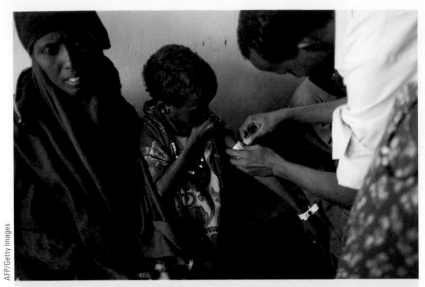
Marasmus results from severe, chronic, overall malnutrition.

protein is needed to make several components of your immune system, as well as the skin and membranes that keep pathogens from entering the body. The World Health Organization estimates that PEM plays a role in at least 2.5 million child deaths each year, many of which are complicated by infection.[21]

Severe PEM actually encompasses a spectrum of malnutrition. At the extremes are two distinct types of severe PEM, and between them are conditions that combine features of both.[22] At one end of the spectrum is a condition called **marasmus**, which results from severe, chronic, overall malnutrition. When a child develops marasmus, his fat and muscle tissue are depleted, and the skin hangs in loose folds, with the bones clearly visible beneath the skin. Children with marasmus tend, at first, to be alert and ravenously hungry, but, with increasing severity, they become apathetic and lose their appetites. Clinicians often say that marasmus represents the body's survival response to long-term, chronic dietary insufficiency.

The other extreme type of PEM, called **kwashiorkor**, is distinguished from marasmus by the presence of severe edema (swelling) in the extremities. Edema

5-10a Protein Deficiency in Early Life

Protein deficiency is rare during the first months of life when infants consume most of their energy from human milk or infant formula. However, once weaned from these high-quality protein sources to foods that lack adequate protein, infants become at greater risk for protein deficiency. Because protein-deficient diets generally also lack energy, protein deficiency is often referred to as **protein-energy malnutrition (PEM)**. Children with this condition are typically deficient in one or more micronutrients as well; so PEM is considered a condition of overall malnutrition. PEM has many implications for childhood health. For example, children with PEM are at great risk for infection and illness. Recall that

protein-energy malnutrition (PEM) A condition whereby protein deficiency is accompanied by a deficiency in energy, and usually, one or more micronutrients.

marasmus A form of PEM characterized by extreme wasting of muscle and loss of adipose tissue.

kwashiorkor A form of PEM characterized by severe edema in the extremities and sometimes the abdomen.

The word marasmus comes from Greek, meaning "starvation," whereas kwashiorkor is believed to come from the Ga language of coastal Ghana, meaning "the sickness the baby gets when the new baby comes."

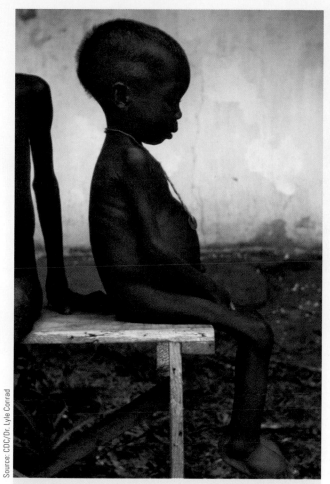

Children with kwashiorkor often have distended abdomens (ascites), edema in their hands and feet, cracked and peeling skin, and an apathetic nature. These children are at especially increased risk for infections.

case.[23] Researchers now believe that many of the signs and symptoms of kwashiorkor are the result of micronutrient deficiencies, such as vitamin A deficiency, in combination with infection or other environmental stressors. In addition, emerging research suggests that alterations in a child's gastrointestinal microbes might play a role.[24]

5-10b Protein Deficiency in Adults

PEM can also occur in adults. Unlike children, however, adults with PEM rarely experience kwashiorkor. Instead, they generally develop marasmus. There are many causes of PEM in adulthood, including inadequate dietary intake, such as sometimes occurs in alcoholics and those with eating disorders; protein malabsorption, such as occurs with some gastrointestinal disorders such as celiac disease; excessive and chronic blood loss; cancer; infection; and injury (especially burns).[25]

Adults with PEM sometimes experience extreme muscle loss because the body's muscles are broken down to provide glucose and energy. Fat accumulation in the liver and edema are common, and adults with severe PEM experience decreased function of many vital physiological systems, including the cardiovascular, renal (kidneys), digestive, endocrine, and immune systems. Treatment of PEM in adults is often long and difficult. For example, if the cause is infection, treatment may involve both dietary intervention and the use of antibiotics. By contrast, if protein deficiency is a result of an eating disorder, psychological counseling becomes a key component of the health care plan. Regardless of its cause, effective treatment of adult PEM presents a special challenge to any medical team.

5-10c Protein Excess

Protein deficiency can result in serious health concerns, but what about protein excess? Contrary to popular belief, high-protein diets do not appear to cause adverse health outcomes (e.g., osteoporosis, kidney problems, heart disease, obesity, and cancer) in most people. This conclusion was confirmed by the DRI committee, which carefully considered the peer-reviewed literature related to the potential health consequences of high-protein diets. In fact, the upper limit of the AMDR for protein intake (35 percent of energy from protein) was developed not because there was evidence that additional protein might

is sometimes present in children with marasmus, but those with kwashiorkor usually have more extensive edema, which typically starts in the legs but often occurs throughout the entire body. Remember that one of the bodily functions of protein is regulation of fluid balance. Children with kwashiorkor sometimes have large, distended abdomens due to fluid accumulation in the abdominal cavity. This condition is referred to as **ascites**. Because malnourished children often have intestinal parasites, worms sometimes contribute to this abdominal distension as well. Children with kwashiorkor are often apathetic and sometimes have cracked and peeling skin, enlarged fatty livers, and sparse unnaturally blond or red hair. Although many characteristics of kwashiorkor were once thought simply to be caused by protein deficiency, this does not appear to be the

ascites A condition characterized by fluid accumulation in the abdominal cavity.

pose a health risk, but solely to complement the recommendations for carbohydrate and fat intakes. Nonetheless, high intakes of protein are often accompanied by high intakes of fat, saturated fat, and cholesterol. Because these dietary components are risk factors for heart disease, it is important to choose a variety of lean and low-fat protein foods, such as those recommended by the MyPlate food guidance system.

5-10d High Red Meat Consumption

Growing evidence suggests that chronically elevated intake of red meat (beef, lamb, and pork) or processed meats (bacon, sausage, hot dogs, ham, and cold cuts) is associated with increased risk for colorectal cancer.[26] As a result, the World Cancer Research Fund and the American Institute for Cancer Research recommended in 2007 that individuals limit their intakes of red meat to no more than 18 oz (500 g) each week and eat very little processed meat.[27] On average, this would be about 2.6 ounces (70 g) of meat each day—an amount less than that recommended by

© HLPhoto/Shutterstock.com

the 2015 Dietary Guidelines and MyPlate, if one were to eat only red meat to fulfill the protein foods requirement. Importantly, the panel of experts who made this recommendation emphasized that they do not suggest avoiding all meat or foods of animal origin. Clearly, these foods can be a valuable source of many essential nutrients and should be considered part of a healthy diet. Like several other issues related to protein nutrition, this topic continues to be one of active debate.

STUDY TOOLS 5

READY TO STUDY? IN THE BOOK, YOU CAN:

☐ Rip out the Chapter Review Card, which includes key terms and chapter summaries.

ONLINE AT WWW.CENGAGEBRAIN.COM, YOU CAN:

☐ Explore protein digestions in the GI tract with an animation.

☐ Interact with figures from the text to check your understanding.

☐ Prepare for tests with quizzes

☐ Review the key terms with Flash Cards.

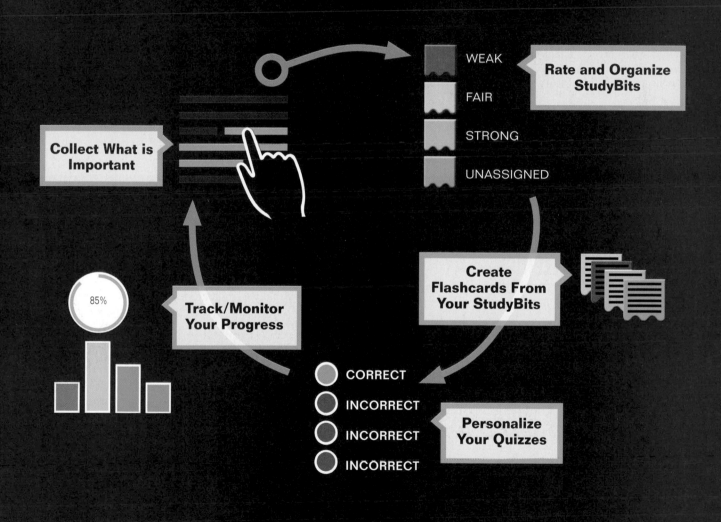

NUTR
ONLINE
STUDY YOUR WAY
WITH STUDYBITS!

Rate and Organize StudyBits

WEAK

FAIR

STRONG

UNASSIGNED

Collect What is Important

Create Flashcards From Your StudyBits

Track/Monitor Your Progress

85%

CORRECT

INCORRECT

INCORRECT

INCORRECT

Personalize Your Quizzes

4LTR PRESS

6 | Lipids

LEARNING OUTCOMES

6-1 List and describe the major types, functions, and structures of lipids.

6-2 Differentiate among essential, conditionally essential, and nonessential fatty acids.

6-3 Describe the differences among mono-, di-, and triglycerides.

6-4 Discuss the functions of phospholipids and sterols.

6-5 Explain the processes of triglyceride digestion, absorption, and circulation.

6-6 List and describe the types and functions of the various lipoproteins.

6-7 Explain the importance of lipids to your health.

6-8 Understand and be able to implement dietary recommendations for lipids.

After finishing this chapter go to **PAGE 150** for **STUDY TOOLS.**

6-1 WHAT ARE LIPIDS?

The term **lipid** refers to a diverse and important category of organic macronutrients that are relatively insoluble in water and tend to be soluble in organic solvents such as fingernail polish remover (acetone) and paint thinner (turpentine).

Lipids (often referred to as fats and oils) are required for hundreds if not thousands of physiological functions in the body. Body fat (which consists mainly of lipids) protects vital organs, stores excess calories, and is a source of some hormones. Lipids also serve as structural components of cell membranes. Lipids also make your favorite foods flavorful—butter, cream, olive oil, and well-marbled beef all taste delicious because of lipids. However, the thought of fatty foods conjures up only images of unhealthy living for many people. And some people shop for "fat-free"

lipid An organic macronutrient that is relatively insoluble in water and relatively soluble in organic solvents.

hydrophobic A substance that does not easily mix with water.

oil A type of lipid that is liquid at room temperature.

fat A type of lipid that is solid at room temperature.

foods and try to avoid fats altogether. Food manufacturers have even developed fat substitutes to replace some of the fats normally found in food. Although diets high in fat and its inherent calories can lead to health complications such as obesity and heart disease, getting enough of the right types of fat is just as essential to optimal health as avoiding excess fat and the wrong kinds of fat. In this chapter, you will learn about the variety of fats and oils in foods and why, although they should be consumed in moderation, they are vital to your health.

There are many types of lipids, and in order to understand their functions in foods and impact on health, it is important to appreciate how lipids differ, both chemically and physiologically.

6-1a Fats and Oils

The first step to understanding lipids is learning some basic terminology. Most lipids are **hydrophobic**, or water "fearing." This means that they do not easily mix with water. A lipid that is liquid at room temperature is called an **oil**, and one that is solid at room temperature is called a **fat**. There are many different types of lipids, but the most common ones in your body and the foods you

eat include the fatty acids, tri-glycerides, phospholipids, ste-rols, and fat-soluble vitamins. Although most of these sub-stances are discussed in detail throughout this chapter, fat-soluble vitamins are addressed in Chapter 7.

6-1b Fatty Acids

Fatty acids are made entirely of carbon, hydrogen, and oxy-gen atoms, and comprise the most abundant type of lipid in your body and the foods you

FIGURE 6.1 FATTY ACID STRUCTURE

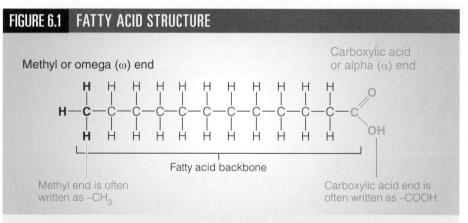

Methyl or omega (ω) end

Carboxylic acid or alpha (α) end

Methyl end is often written as –CH₃.

Fatty acid backbone

Carboxylic acid end is often written as –COOH.

All fatty acids have three components: a methyl or ω end (–CH₃), a fatty acid backbone made from carbon and hydrogen atoms and a carboxylic acid or α end (–COOH).

eat (see Figure 6.1). A chain of carbon atoms forms the backbone of each fatty acid. One end of this car-bon chain, called the **alpha (α) end**, contains a *car-boxylic acid* group (–COOH). The other end, called the **omega (ω) end**, contains a *methyl* group (–CH₃). Both in the body and in foods, most fatty acids do not exist in their free (unbound) form. Instead, they are com-ponents of larger molecules, such as triglycerides and phospholipids.

fatty acids The most abundant type of lipid in the body and foods; comprised of a chain of carbons with a methyl (–CH₃) group on one end and a carboxylic acid (–COOH) group on the other.

alpha (α) end The end of a fatty acid that contains the carboxylic acid (–COOH) group.

omega (ω) end The end of a fatty acid that contains the methyl (–CH₃) group.

There are hundreds of types of fatty acids which are distinguished from one another both by the number of carbons they contain and by the types and locations of chemical bonds holding the carbon atoms together. These variations influence the physical properties of the fatty acids and the roles they play in your body and the foods you eat.

6-1c Number of Carbons (Chain Length)

The number of carbon atoms in a fatty acid determines its **chain length**, as illustrated in Figure 6.2. In fact, it may be helpful to think of a fatty acid as if it were a chain, with each link representing a carbon atom. Most naturally occurring fatty acids have an even number of carbon atoms: chains usually span from 12 to 22 carbons, although some may be as short as four or as long as 26 carbons. A fatty acid with fewer than eight carbon atoms is called a **short-chain fatty acid**; one with 8 to 12 carbon atoms is called a **medium-chain fatty acid**; and one with more than 12 carbon atoms is called a **long-chain fatty acid**.

The chain length of a fatty acid affects its chemical properties and physiological functions. For example, chain length influences the temperature at which a fatty acid melts (its *melting point*), and lipids constructed predominantly of short-chain fatty acids are generally oils or even gases. Chain length also affects solubility in water: short-chain fatty acids are generally more water soluble than long-chain fatty acids. Because the human body is mostly water, it is relatively easy to absorb and transport water-soluble substances such as short-chain fatty acids. Conversely, the body needs more complex processes to absorb, transport, and use dietary lipids that are more water insoluble, such as the long-chain fatty acids.

6-1d Number and Positions of Double Bonds

Aside from the number of carbon atoms they contain, fatty acids also differ in the types of chemical bonds between their carbon atoms (see Figure 6.3). A fatty acid's carbon–carbon bonds can be either single bonds or double bonds. If a fatty acid contains only carbon–carbon single bonds in its backbone, it is a **saturated fatty acid (SFA)**; if it contains at least one carbon–carbon double bond, it is an **unsaturated fatty acid**. A fatty acid with just one double bond in its backbone is a **monounsaturated fatty acid (MUFA)**, and one with more than one double bond is a **polyunsaturated fatty acid (PUFA)**.

As with chain length, the number of double bonds can influence the physical nature of a fatty acid. As you can see in Figure 6.3, each carbon atom in an SFA is surrounded (or saturated) by hydrogen atoms. Saturation prevents the fatty acid from bending. Because of this rigidity, SFAs form straight chains. Foods with a high proportion of saturated fatty acids tend to be solid at room temperature because the straight-chain SFAs are highly organized and stack together in a tight

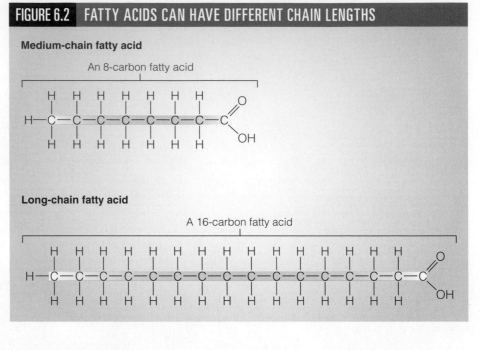

FIGURE 6.2 **FATTY ACIDS CAN HAVE DIFFERENT CHAIN LENGTHS**

Medium-chain fatty acid

An 8-carbon fatty acid

Long-chain fatty acid

A 16-carbon fatty acid

chain length The number of carbon atoms in a fatty acid.

short-chain fatty acid A fatty acid with fewer than eight carbon atoms.

medium-chain fatty acid A fatty acid with 8 to 12 carbon atoms.

long-chain fatty acid A fatty acid with more than 12 carbon atoms.

saturated fatty acid (SFA) A fatty acid that contains only carbon–carbon single bonds in its backbone.

unsaturated fatty acid A fatty acid that contains at least one carbon–carbon double bond in its backbone.

monounsaturated fatty acid (MUFA) A fatty acid that contains one carbon–carbon double bond in its backbone.

polyunsaturated fatty acid (PUFA) A fatty acid that contains more than one carbon–carbon double bond in its backbone.

array. Foods containing large amounts of SFAs (such as butter) tend likewise to be solids (fats) at room temperature.

Compared with SFAs, unsaturated fatty acids (those with double bonds) have fewer hydrogen atoms associated with the carbon backbones—this allows them to bend. In fact, whenever there is a carbon–carbon double bond, there is a kink or bend in the fatty acid backbone. These bends cause unsaturated fatty acids to become disorganized, preventing them from being densely packed. In general, organized molecules such as SFAs are solid at room temperature, while disorganized molecules like PUFAs are liquid. MUFAs have chemical characteristics that lie between those of SFAs and PUFAs, making them thick liquids or semisolids at room temperature. You may have noticed that olive oil, which is high in MUFAs, is a thick oil at room temperature.

6-1e Understanding *Cis* versus *Trans* Fatty Acids

Unsaturated fatty acids differ in the ways that their hydrogen atoms are arranged around the carbon–carbon double bonds. In most naturally occurring fatty acids, the hydrogen atoms are positioned on the same side of the double bond. This is called a **cis double bond** (see Figure 6.4). When the hydrogen atoms are positioned on opposite sides of the double bond, it is called a **trans double bond**. Unlike *cis* double bonds, *trans* double bonds do not cause bending. A fatty acid containing at least one *trans* double bond is called a **trans fatty acid**. *Trans* fatty acids have fewer bends in their backbones than do their *cis* counterparts. For this reason, *trans* fatty acids are also more likely to be solid (fats) at room temperature.

Trans Fatty Acids in Food *Trans* fatty acids are found naturally in some foods, such as dairy and beef products. However, most dietary *trans* fatty acids are produced commercially via a

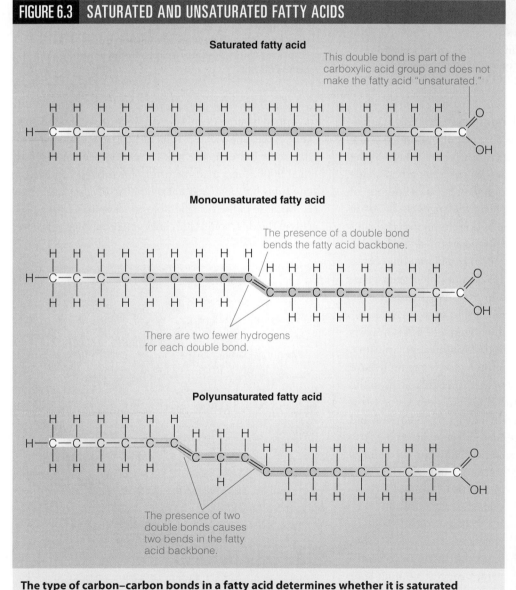

FIGURE 6.3 SATURATED AND UNSATURATED FATTY ACIDS

Saturated fatty acid

This double bond is part of the carboxylic acid group and does not make the fatty acid "unsaturated."

Monounsaturated fatty acid

The presence of a double bond bends the fatty acid backbone.

There are two fewer hydrogens for each double bond.

Polyunsaturated fatty acid

The presence of two double bonds causes two bends in the fatty acid backbone.

The type of carbon–carbon bonds in a fatty acid determines whether it is saturated or unsaturated.

cis double bond A carbon–carbon double bond in which the hydrogen atoms are positioned on the same side of the double bond.

trans double bond A carbon–carbon double bond in which the hydrogen atoms are positioned on opposite sides of the double bond.

trans fatty acid A fatty acid containing at least one *trans* double bond.

Tropical oils, such as palm and coconut oils, contain a relatively high amount of saturated fatty acids.

FIGURE 6.4 CIS VERSUS TRANS FATTY ACIDS

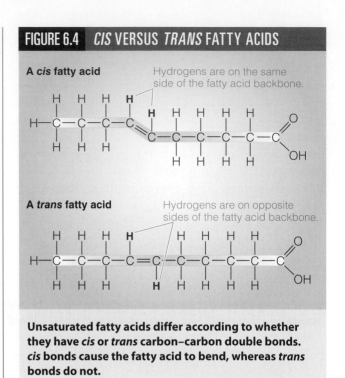

A *cis* fatty acid — Hydrogens are on the same side of the fatty acid backbone.

A *trans* fatty acid — Hydrogens are on opposite sides of the fatty acid backbone.

Unsaturated fatty acids differ according to whether they have *cis* or *trans* carbon–carbon double bonds. *cis* bonds cause the fatty acid to bend, whereas *trans* bonds do not.

process called **partial hydrogenation**. Partial hydrogenation converts oils, such as corn oil, into solid fats, such as margarine or shortening, by converting many of the carbon–carbon double bonds into carbon–carbon single bonds. This is achieved through the chemical addition of hydrogen atoms; hence the term *hydrogenation*. Aside from decreasing the number of double bonds (and thus increasing the number of carbon–carbon single bonds), the process of partial hydrogenation converts some of the remaining *cis* double bonds to *trans* double bonds, causing the lipid to become high in *trans* fatty acids. Partial hydrogenation has long been used in food manufacturing because adding partially hydrogenated lipids imparts desirable food texture and reduces spoilage. Crackers, pastries, bakery products, shortening, and margarine have long been the main sources of the *trans* fatty acids in the diet.[1] However, a recent trend toward decreasing *trans* fatty acid intake has resulted in new food preparation and processing methods that reduce or eliminate *trans* fatty acids in many foods. For example, many fast-food chains have switched from frying their foods in high-*trans* fatty acid shortening to *trans* fat–free vegetable oils.

Many public health agencies (such as the USDA and the Institute of Medicine) suggest limiting your intake of *trans* fatty acids. These recommendations (described in more detail later on) were developed because some studies show that certain industrially produced *trans* fatty acids increase the risk for cardiovascular disease.[2] Since 2006, food manufacturers have been required to state *trans* fatty acid content on their Nutrition Facts panels. Because of the considerable concern about *trans* fatty acids in the diet, some cities are even declaring themselves

Olive oil is a thick liquid because it contains a relatively high amount of MUFAs.

partial hydrogenation
A process whereby liquid oil is converted into solid fat by changing many of the carbon–carbon double bonds into carbon–carbon single bonds; some *trans* fatty acids are also produced.

You can tell how much *trans* fatty acid a food has by looking at its Nutrition Facts panel.

Alpha (α) Naming System The *alpha* (α) naming system for fatty acids is based on the positions and types of double bonds relative to the carboxylic acid (α) end of the fatty acid. The basic formula for constructing an alpha (α) name is [*cis* or *trans*]*a*, [*cis* or *trans*]*b*–*x*:*y*, where *x* signifies the total number of carbon atoms and *y* signifies the total number of double bonds. In this formula, there are two double bonds. The type (*cis* or *trans*) and number of carbons from the carboxylic acid (α) end of each double bond (signified as *a* and *b*) are separated by a comma.

As an example, consider a fatty acid with the following characteristics: 18 carbons and two *cis* double bonds—one between the ninth and tenth carbons from the carboxylic acid (α) end and the other between the twelfth and thirteenth carbons from the carboxylic acid (α) end. The fatty acid's name is constructed beginning with an 18, signifying that there are 18 carbons. The number 2 is added to form 18:2, signifying that there are two double bonds. Next, the two double bonds' locations are added to the front of the name: 9,12–18:2. Remember, the locations are determined by counting from the carboxylic acid (α) end. Finally, because both double bonds are *cis* double bonds, the name is modified to *cis*9,*cis*12–18:2. This fatty acid is illustrated in Figure 6.5.

Omega (ω) Naming System An alternate system for naming a fatty acid is the omega (ω) system. In this system, the number of carbons and double bonds are again distinguished as *x*:*y* (for example, 18:2). However, in the omega naming system, fatty acids are characterized and grouped on the basis of the first double bond's distance from the methyl (ω; omega) end of the molecule. If the first double bond is located between the third and fourth carbons from the omega end, the fatty

"*trans* fat–free zones." In 2015, the U.S. Food and Drug Administration (FDA) announced that it was no longer going to allow the inclusion of partially hydrogenated oils, which are the major source of industrially produced *trans* fatty acids, in foods.[3] This decision was based on science-driven evidence that these compounds increase risk for cardiovascular disease. However, this shift in the *trans* fatty acid content of the U.S. food supply will not happen overnight. Indeed, the FDA is allowing the food industry three years to gradually phase out its use of partially hydrogenated vegetable oils. Consequently, this new regulation will not really take effect until sometime in 2018. It is noteworthy that this ban on *trans* fatty acids only applies to those produced industrially. *Trans* fatty acids from natural sources (primarily beef and dairy) will still be allowed in foods because there is no evidence that these types of *trans* fats are harmful to health.

6-1f Fatty Acids Are Named for Their Structures

There are several methods used to name or describe fatty acids. In addition to common names that often reflect food sources, fatty acids are also named according to structural properties such as the number of carbons atoms, the number and types of carbon–carbon double bonds, and the position of the double bonds.

FIGURE 6.5 THE ALPHA (α) NAMING SYSTEM

The first double bond (which is *cis*) from the alpha end is on the ninth carbon atom.

omega (methyl) end

alpha (carboxylic acid) end

The second double bond (which is *cis*) from the alpha end is on the twelfth carbon atom.

The alpha naming system for fatty acids requires that you count how far the double bonds are from the alpha (carboxylic acid) end of the carbon backbone. Using this system, this fatty acid's name is *cis*9, *cis*12–18:2.

acid is an **omega-3 (ω-3) fatty acid**. If the first double bond is located between the sixth and seventh carbons from the omega end, it is an **omega-6 (ω-6) fatty acid**. There are also ω-7 and ω-9 fatty acids. It is noteworthy that only unsaturated fatty acids can be named using the omega nomenclature system, because it requires that there be at least one carbon–carbon double bond in the fatty acid.

Common Names In addition to the alpha and omega naming systems, fatty acids are also referred to by their common (non-numerical) names, which usually reflect a prominent food source of the fatty acid. For example, *palmitic acid* (16:0) is found in palm oil, and *arachidonic acid* (*cis*5,*cis*8,*cis*11,*cis*14–20:4; from *arachis*, meaning legume or peanut) is found in peanut butter.

omega-3 (ω-3) fatty acid A fatty acid in which the first double bond is located between the third and fourth carbons from the omega (ω) end.

omega-6 (ω-6) fatty acid A fatty acid in which the first double bond is located between the sixth and seventh carbons from the omega (ω) end.

linoleic acid An essential ω-6 fatty acid with 18 carbons and two double bonds.

linolenic acid (or **α-linolenic acid**) An essential ω-3 fatty acid with 18 carbons and three double bonds.

6-2 WHAT ARE ESSENTIAL, CONDITIONALLY ESSENTIAL, AND NONESSENTIAL FATTY ACIDS?

There are potentially hundreds of different fatty acids, each with its own distinct structure, physiologic properties, and name. Although foods provide a variety of different fatty acids, many of which are needed for optimal health, only two are considered essential. Because cells are unable to make these fatty acids, it is important that you get them from food.

6-2a The Essential Fatty Acids: Linoleic Acid and Linolenic Acid

The two essential fatty acids are **linoleic acid** and **linolenic acid** (also called **α-linolenic acid**). Linoleic acid has 18 carbons and two double bonds, and it is an ω-6 fatty acid. Linolenic acid has 18 carbons and three double bonds, and it is an ω-3 fatty acid. Linoleic acid and linolenic acid are essential nutrients because the body cannot insert double bonds in the ω-3 and ω-6 positions of fatty acids. As such, you need linolenic acid and linoleic acid to make all of the other ω-3 and ω-6 fatty acids, respectively, and to make other related substances. For example, as illustrated in Figure 6.6, linoleic acid is used to make *arachidonic acid* (a 20-carbon, ω-6 fatty acid).

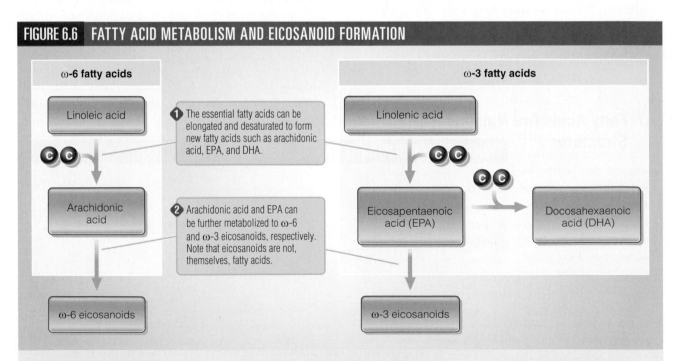

FIGURE 6.6   FATTY ACID METABOLISM AND EICOSANOID FORMATION

Linoleic acid and linolenic acid can be converted, via elongation and desaturation, to longer-chain polyunsaturated fatty acids, some of which are used to make hormone-like eicosanoids.

Not All *Trans* Fatty Acids Are Created Equal

There are literally hundreds of fatty acids found in foods, and as you know, they differ from one another in terms of their chain length and, for unsaturated varieties, the number, placement, and configuration of their double bonds. Similarly, *trans* fatty acids can differ as well. This is important because various forms of *trans* fatty acids influence health in different ways. Those produced during the partial hydrogenation of vegetable oils can have negative effects on heart health, but other naturally occurring forms demonstrate potent anticarcinogenic effects. The most well-studied naturally occurring *trans* fatty acid is commonly referred to as conjugated linoleic acid (CLA; *cis*9, *trans*11–18:2) because, like linoleic acid (*cis*9, *cis*12–18:2), it has 18 carbons and 2 double bonds. However, unlike linoleic acid, which has two single bonds between its double bonds, the double bonds in CLA are separated by only one single bond. Furthermore, one of CLA's double bonds has a *trans* configuration. The most common dietary CLA is primarily found in dairy and beef fat, and its common name is rumenic acid. Because naturally occurring *trans* fatty acids, such as rumenic acid, are thought to promote health (or at least not pose health risks), the FDA does not regulate their presence in foods.

© Eric Isselee/Shutterstock.com

Linoleic acid (*cis*9, *cis*12–18:2)

Conjugated linoleic acid (*cis*9, *trans*11–18:2)

Similarly, linolenic acid can be converted to *eicosapentaenoic acid* (EPA), a 20-carbon, ω-3 fatty acid, which can subsequently be converted to *docosahexaenoic acid* (DHA), a 22-carbon, ω-3 fatty acid. This is accomplished by increasing the number of carbon atoms in the chain, a process called **elongation**, and by increasing the number of doubled bonds in the fatty acid chain, or through **desaturation**. These important long-chain PUFAs have many important functions in the body. For example, they are necessary for normal cell function, forming structural components of cell membranes, and the regulation of gene expression. They also serve as precursors to hormone-like compounds called eicosanoids.

Conversion of Essential Fatty Acids to Eicosanoids

The essential fatty acids and some other long-chain fatty acids can be used to make other important compounds that are not themselves fatty acids, but are nonetheless vital for health. One example are the **eicosanoids**, a collection of substances with diverse, hormone-like effects. Eicosanoids are involved in regulating the immune and cardiovascular systems and act as chemical messengers that influence a variety of physiologic functions.[4] Your body produces both ω-3 and ω-6 eicosanoids, which have somewhat opposing actions. The ω-6 eicosanoids (produced from linoleic acid) tend to cause inflammation, stimulate blood clot formation, and induce the constriction of blood vessels, whereas the ω-3 eicosanoids (produced from linolenic acid) tend to reduce inflammation, stimulate dilation (or relaxation) of the blood vessel

elongation The process whereby fatty acid chain length is increased by the addition of carbon atoms.

desaturation The process whereby a fatty acid's carbon–carbon single bonds are converted to double bonds.

eicosanoids A diverse group of hormone-like compounds that help regulate the immune and cardiovascular systems and act as chemical messengers.

People who consume foods high in ω-3 fatty acids (such as many cold-water fish) may have lower risk for conditions related to inflammation, such as heart disease.

6-2b Essential Fatty Acid Deficiency

Because of the almost endless supply of linoleic and linolenic acids stored in adipose tissue, essential fatty acid deficiencies do not generally occur in otherwise healthy people. Primary essential fatty acid deficiency is rare, generally occurring only in those with very poor health. Secondary fatty acid deficiencies can occur in conjunction with diseases that disrupt lipid absorption or utilization, such as cystic fibrosis. When an essential fatty acid deficiency is present, infections are common and wound healing may be slow. Children with essential fatty acid deficiencies also exhibit slow growth.

6-2c Conditionally Essential Fatty Acids in Infancy

In addition to linoleic and linolenic acids, other fatty acids such as arachidonic acid and DHA may be conditionally essential during infancy. Although adults can synthesize these fatty acids from linoleic acid and linolenic acid, respectively, babies may not be able to make them because they cannot produce the needed enzymes in sufficient amounts. As might be expected, human milk contains ample amounts of these conditionally essential fatty acids, and many infant formulas are now fortified with arachidonic acid and DHA, making them more like human milk in this regard.

6-2d Dietary Sources of Fatty Acids

Although most foods contain a mixture of fatty acids, some are especially high in a particular fatty acid or group of fatty acids. For example, some nuts (such as walnuts),

walls, and inhibit blood clotting. Both ω-3 and ω-6 eicosanoids are important to health, and the body can shift its relative production in response to its needs and relative availability of the parent compounds (linoleic and linolenic fatty acids) in the body.

Dietary choices can influence the amount and types of eicosanoids that you make. For example, Alaska natives who consume high amounts of ω-3 fatty acids from fish and marine mammals (such as whales and seals) have enhanced physiologic responses that are stimulated by ω-3 eicosanoids. Consequently, Alaska natives who consume traditional diets tend to take more time to form blood clots than people who consume fewer ω-3 fatty acids.[5] Research suggests that alterations in the balance of ω-3 to ω-6 eicosanoids may influence a person's risk of conditions related to inflammation, such as heart disease and cancer.[6] This is why experts often recommend that people regularly consume fish. These recommendations will be described in more detail later in this chapter.

Omega-3 fatty acids are sometimes added to foods (like peanut butter and margarine) or fed to farm animals (like chickens) so that their food products (like eggs) are biofortified with these fatty acids.

seeds, and oils (such as those made from soybean, saf-flower, and corn) are abundant in linoleic acid. Linolenic acid is especially plentiful in canola (or rapeseed), soybean, and flaxseed oils, as well as in some nuts, such as walnuts. Note that some foods, such as soybean oil and walnuts, are good sources of both essential fatty acids. Because many of these foods and oils are common to the U.S. diet, getting adequate amounts of the essential amino acids is not a problem for most people.

Although long-chain ω-3 fatty acids such as EPA and DHA are not considered essential, some experts believe that it is important to consume adequate amounts of them because higher intake of these substances is associated with lower risk for heart disease and stroke. This is likely due to the potent anti-inflammatory effects of these compounds in the body. EPA and DHA are especially abundant in fatty fish and other seafood, while lesser amounts are found in meats and eggs. Longer-chain ω-6

fatty acids are also found in meat, poultry, and eggs. In addition, many food manufacturers fortify their products with these important fatty acids to provide a greater variety of foods rich in ω-3 fatty acids.

6-2e Dietary Sources of Nonessential Saturated and Unsaturated Fatty Acids

Saturated fatty acids and most unsaturated fatty acids are not essential to your diet because your body can synthesize them when needed. Nevertheless, a healthy diet typically includes a variety of nonessential fatty acids, which (like essential fatty acids) serve as a source of energy, insulation, and protection. In general, animal-based foods supply the majority of dietary SFAs, whereas plant-derived foods supply the majority PUFAs. There are exceptions, however: some tropical oils, such as coconut and palm oils, contain relatively high amounts of SFAs, and many oily fish have high levels of PUFAs. Monounsaturated fatty acids are supplied by both plant- and animal-based foods. The relative amounts of SFAs, MUFAs, and PUFAs in commonly consumed fats and oils are summarized in Figure 6.7.

Essential Fatty Acids versus Essential Oils

An essential fatty acid is one that the body needs but must get from the diet because we cannot make it ourselves. This is very different from an *essential oil*, which is an oil derived from a natural substance, usually either for its purported healing properties or fragrance. Examples of essential oils include camphor and eucalyptus, which are believed to help relieve congestion. Thus, whereas the aroma or topical application of essential oils may provide valuable psychological or physical benefits, these oils are not for consumption.

Essential oils contain compounds that are added to medicinal preparations for their scents or putative health benefits. These essential oils are not the same as essential fatty acids, which are an important part of a healthy diet.

© fotohunter/Shutterstock.com

6-3 WHAT IS THE DIFFERENCE BETWEEN MONO-, DI-, AND TRIGLYCERIDES?

As previously mentioned, most fatty acids do not exist in their free (unbound) form. Instead, they comprise various parts of larger, more complex molecules called monoglycerides, diglycerides, and triglycerides. As implied by their names, these substances are defined by the number of fatty acids present in their chemical structures. Whereas a **monoglyceride** molecule has only one fatty acid attached to a molecule of glycerol, a **diglyceride** molecule consists of one glycerol molecule with two fatty acids attached, and a **triglyceride** is made up of one glycerol molecule with three fatty acids attached. The fatty acids can be saturated, monounsaturated, polyunsaturated, or a mixture thereof. In each case, the fatty acid molecule (or molecules) are attached to

monoglyceride A lipid comprised of one fatty acid attached to a glycerol backbone.

diglyceride A lipid comprised of two fatty acids attached to a glycerol backbone.

triglyceride A lipid comprised of three fatty acids attached to a glycerol backbone.

FIGURE 6.7 DISTRIBUTION OF FATTY ACID TYPES IN COMMONLY CONSUMED LIPIDS

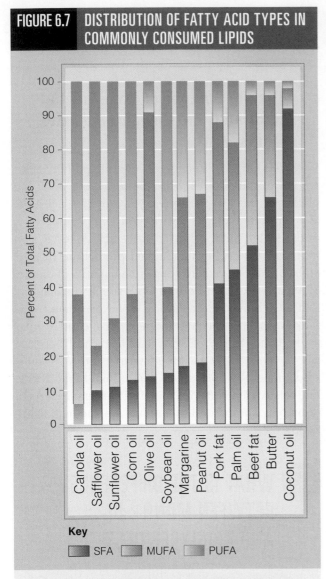

Dietary lipids contain different relative amounts of saturated fatty acids (SFAs), monounsaturated fatty acids (MUFAs), and polyunsaturated fatty acids (PUFAs).

Source: USDA National Nutrition Database for Standard Reference, Release 18, 2005.

a glycerol backbone, which by itself is made up of three carbon atoms (see Figure 6.8). The metabolic process by which fatty acids combine with glycerol to form triglycerides is called **lipogenesis**.

---

lipogenesis The metabolic process by which fatty acids combine with glycerol to form triglycerides.

lipolysis The metabolic process by which a triglyceride's fatty acids are removed from the glycerol backbone.

ß-oxidation The metabolic breakdown of fatty acids into 2-carbon units that are used to produce ATP.

FIGURE 6.8 TRIGLYCERIDE STRUCTURE

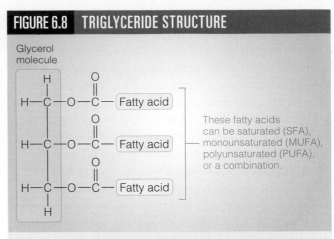

These fatty acids can be saturated (SFA), monounsaturated (MUFA), polyunsaturated (PUFA), or a combination.

A triglyceride consists of a glycerol molecule and three fatty acids.

6-3a Triglycerides Are Rich Sources of Energy

Compared to other energy-yielding macronutrients, triglycerides represent the body's richest source of energy. As you may recall from Chapter 1, the complete breakdown of 1 gram of fatty acids yields approximately 9 kilocalories which is more than twice the yield from 1 gram of carbohydrate or protein (4 kilocalories). Therefore, gram for gram, high-fat foods yield more calories than do other foods. Note that, before a triglyceride can be used for energy, its fatty acids must be removed from the glycerol backbone. This metabolic process is called **lipolysis**. The subsequent breakdown of fatty acids to produce ATP is called **ß-oxidation**, which splits the carbon–carbon bonds in the fatty acid chain into several 2-carbon units. It is these 2-carbon molecules that are metabolized further to yield numerous molecules of ATP.

In addition to using fatty acids as an immediate energy source, the body can convert them to other energy-yielding organic compounds called ketones. Recall from Chapter 4 that the production of ketones from fatty acids, called *ketogenesis*, occurs when glucose availability is low. Ketogenesis is important because some tissues (such as brain, heart, skeletal muscle, and kidney tissue) can metabolize ketones to synthesize ATP, thus slowing down the mobilization and metabolism of muscle protein for energy during periods of low calorie or low carbohydrate intake.

6-3b Storage of Excess Triglycerides in Adipose Tissue

What happens when the energy available to your body exceeds its energy needs? When you have consumed more calories than are needed, fatty acids not required for energy or other functions are stored as triglycerides in adipose tissue and, to a lesser extent, skeletal muscle. Adipose tissue consists primarily of a specialized type of cell called an **adipocyte**. Adipocytes can accumulate large amounts of triglycerides, thus serving as a reservoir of energy. Adipose tissue is located in many parts of your body, including directly under your skin (**subcutaneous adipose tissue**) and around the vital organs in your abdomen (**visceral adipose tissue**). Many of your body's organs and tissues (such as the kidneys and breasts) also have adipose tissue associated with them. Not only does this help protect these organs, it also provides them with a readily available energy source.

Compared to glycogen, the storage of excess energy as triglycerides has two major advantages. First, because triglycerides are not stored with water which takes up space, a large amount can fit into a small space. Second, as previously discussed, a gram of pure triglyceride stores over twice the energy as does a gram of carbohydrate. Consequently, your body can store much more energy in 1 pound of adipose tissue than in 1 pound of liver glycogen. Indeed, the body has a seemingly infinite ability to store excess energy in adipose tissue, whereas its capacity to store glycogen is limited.

Insulin Stimulates Lipogenesis and Inhibits Lipolysis The hormone insulin stimulates the storage of triglycerides during times of energy excess, such as after a meal, by causing adipocytes (and to a lesser extent skeletal muscle) to take up glucose and fatty acids from the blood. Insulin also stimulates the conversion of excess glucose to fatty acids (another form of lipogenesis), which are in turn incorporated

into triglycerides. Finally, insulin inhibits lipolysis. The increased lipogenesis and decreased lipolysis experienced after a meal help direct excess glucose and fatty acids to adipose tissue, where they are stored as triglycerides for later use.

6-3c Triglycerides Needed for Insulation

In addition to providing energy and protecting internal organs from injury, adipose tissue also insulates the body. Humans rely on subcutaneous adipose tissue to keep warm, and people with very little body fat often have difficulty regulating body temperature. In fact, a physiological response to insufficient body fat is the growth of very fine hair on the body. This hair, called **lanugo**, makes up for the absence of subcutaneous adipose tissue to some extent by providing a layer of external insulation. The presence of lanugo is common in very malnourished and underweight people, such as those with the eating disorder anorexia nervosa.[7] For this reason, the appearance of lanugo can serve as an important sign of malnutrition during a clinical assessment.

6-4 WHAT ARE PHOSPHOLIPIDS AND STEROLS?

Along with triglycerides, phospholipids and sterols are major types of lipids found throughout the body. These substances are essential components of cell membranes and are involved in the transport of lipids in your bloodstream. However, because the body can synthesize all the phospholipids and sterols it needs, there are no dietary requirements for either of them. Still, because they are commonly found in the foods you eat, an awareness of these two substances is important to complete your understanding of lipid nutrition.

Adipocytes are specialized cells that store excess energy as triglycerides.

adipocyte A specialized cell, found in adipose tissue, that can accumulate large amounts of triglycerides.

subcutaneous adipose tissue Adipose tissue located directly under the skin.

visceral adipose tissue Adipose tissue located around the vital organs in the abdomen.

lanugo Very fine hair that grows as a physiological response to insufficient body fat.

6-4a Phospholipids

A **phospholipid** is similar to a triglyceride in that both contain a glycerol molecule bonded to fatty acids (see Figure 6.9). However, instead of having three fatty acids (like a triglyceride), a phospholipid has only two. Replacing the third fatty acid is a phosphate-containing **hydrophilic** (or water loving, meaning that it mixes easily with water) **head group**. There are many different types of head groups, but the most common are choline, ethanolamine, inositol, and serine.

Phospholipids are **amphipathic**, meaning they contain both hydrophilic and hydrophobic regions. Having both properties is very advantageous in the body. While the head group (the hydrophilic region) of each phospholipid molecule is attracted to water, the fatty acids (the hydrophobic region) repel water. This structure allows phospholipids to both act as major components of cell membranes and play important roles in the digestion, absorption, and transport of lipids throughout your body. Their amphipathic structure is also helpful in terms of how phospholipids can be used in the food industry. For example, the phospholipid *lecithin* is often used in foods such as mayonnaise and salad dressings as a stabilizer. Lecithin prevents the lipid- and water-soluble components of these foods from separating from each other.

In the body, phospholipids are the main structural component of cell membranes (see Figure 6.10). Cell membranes consist of two layers (a bilayer) of phospholipids with proteins distributed throughout this bilayer. The hydrophilic head groups of these back-to-back layers point to the inside and outside of the cell, both of which are predominantly water. The hydrophobic fatty acids face toward each other, forming a water-free, lipid-rich zone. The incorporation of two layers of amphipathic phospholipids allows cell membranes to carry out their functions effectively, while remaining stable in a predominantly watery environment. Phospholipids serve other functions as well: some supply fatty acids for cellular metabolism; others activate enzymes important to blood clotting and cell replication; and still others donate their fatty acids for eicosanoid production or act as carriers of fatty substances throughout the body. Phospholipids also emulsify fats and oils so that they can mix with water and other hydrophilic fluids. You will learn more about emulsification later in this chapter.

FIGURE 6.9 PHOSPHOLIPID STRUCTURE

Phospholipids consist of a glycerol, a hydrophilic head group, and two hydrophobic fatty acids.

phospholipid A lipid composed of a glycerol molecule bonded to two hydrophobic fatty acids and a hydrophilic head group.

hydrophilic Water loving; mixes easily with water.

head group A phosphate-containing, hydrophilic chemical structure that serves as a component of a phospholipid.

amphipathic A characteristic of a substance that contains both hydrophilic and hydrophobic portions.

Phospholipids such as lecithin are added to some salad dressings to keep their hydrophilic and hydrophobic components mixed.

© Patty Orly/Shutterstock.com

FIGURE 6.10   CELL MEMBRANE MADE FROM PHOSPHOLIPID BILAYER

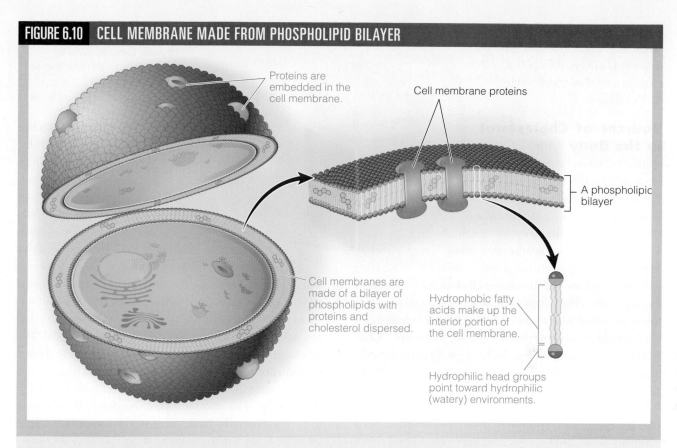

Proteins are embedded in the cell membrane.

Cell membrane proteins

A phospholipid bilayer

Cell membranes are made of a bilayer of phospholipids with proteins and cholesterol dispersed.

Hydrophobic fatty acids make up the interior portion of the cell membrane.

Hydrophilic head groups point toward hydrophilic (watery) environments.

The amphipathic nature of phospholipids allows cell membranes to carry out their functions.

6-4b Sterols

A **sterol** is different from other lipids in that it consists of a distinct multi-ring structure. There are many types of sterols, but the most abundant and widely discussed is **cholesterol** (see Figure 6.11). You may have heard of the potentially unhealthy relationship between elevated blood cholesterol and heart disease. However, you may not be as familiar with the many important functions that cholesterol serves in your body. For instance, cholesterol is used to synthesize bile acids. A **bile acid**, which plays a critical role in digestion and absorption of lipids, consists of a cholesterol molecule attached to a very hydrophilic subunit, making a bile acid molecule amphipathic (much like a phospholipid). This structure allows bile acids to help break up dietary lipids in the intestine, preparing them for chemical digestion and subsequent absorption. Cholesterol is also a component of cell membranes, where it helps maintain flexibility. In addition to these functions, cholesterol is integral to the synthesis

FIGURE 6.11   STRUCTURE OF CHOLESTEROL

sterol A type of lipid with a multiple-ring chemical structure.

cholesterol An abundant and widely discussed sterol that is used to synthesize bile acids and a variety of hormones such as testosterone.

bile acid An amphipathic substance that plays a critical role in digestion and absorption of lipids; consists of a cholesterol molecule attached to a very hydrophilic subunit.

CHAPTER 6: Lipids   137

of many of the reproductive hormones (such as testosterone and estrogen), vitamin D energy metabolism, calcium homeostasis, and electrolyte (salt) balance.

Sources of Cholesterol in the Body

Almost every tissue in your body can make cholesterol—this is especially true for liver tissue. As such, cholesterol is not an essential nutrient. Many dietary factors influence how much cholesterol a person synthesizes, however. For example, eating a low-calorie or low-carbohydrate diet decreases cholesterol synthesis in some people.[8] This is not the case for everyone, however, because carbohydrate intake may only affect cholesterol synthesis in those with certain genetic make-ups.[9] Cholesterol production can also be lowered by certain medications (commonly referred to as *statins*), such as those taken by people at elevated risk for heart disease.

In addition to the cholesterol made by cells, the body obtains cholesterol from animal-derived foods such as shellfish, meat, butter, eggs, and liver (see Figure 6.12). Plants do not produce substantial amounts of cholesterol. So, exclusively plant-based diets are relatively very low in this substance. Because cholesterol is synthesized by the body, vegans who do not eat animal-based products are not at risk of cholesterol deficiency.

Phytosterols Are Sterol-Like Plant Compounds

Although plants make very small amounts of cholesterol, some contain relatively large amounts of sterol-like compounds called **phytosterols**. An interesting group of phytosterols occurs naturally in corn, wheat, rye, and other plants. Similar sterols are produced commercially and marketed under various brand names, such as Benecol®. These products are often found in butter substitutes, yogurt drinks, salad dressings, and even dietary supplements. Some studies suggest that consuming roughly 500 to 2,000 milligrams of phytosterols a day may decrease blood cholesterol, lowering one's risk for cardiovascular disease.[10] Because a typical serving of a plant sterol–fortified table spread contains about 1,000 milligrams of the sterol, you would need to consume one or two servings daily to reach this goal. The mechanisms by which plant sterols decrease blood cholesterol are not well understood. However, most plant sterols are not easily absorbed and appear to bind cholesterol in the intestine. This may increase cholesterol elimination in the feces. Similarly, some plant sterols can cause loss of fat in the feces, leading to diarrhea and possibly decreased absorption of fat-soluble vitamins.

> Cholesterol was first identified in solid form in 1769 by François Poulletier de la Salle.

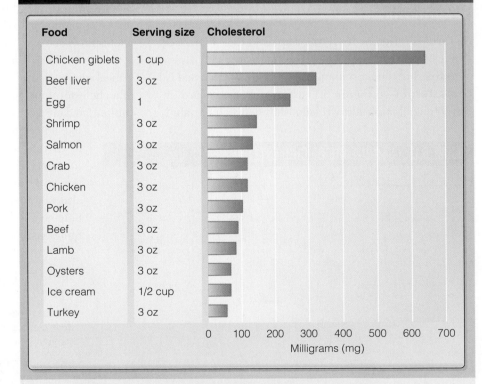

FIGURE 6.12 CHOLESTEROL CONTENT OF SELECTED FOODS

Food	Serving size	Cholesterol
Chicken giblets	1 cup	
Beef liver	3 oz	
Egg	1	
Shrimp	3 oz	
Salmon	3 oz	
Crab	3 oz	
Chicken	3 oz	
Pork	3 oz	
Beef	3 oz	
Lamb	3 oz	
Oysters	3 oz	
Ice cream	1/2 cup	
Turkey	3 oz	

Milligrams (mg): 0 100 200 300 400 500 600 700

Source: USDA National Nutrition Database for Standard Reference, Release 18, 2005.

phytosterol A sterol-like compound made by plants.

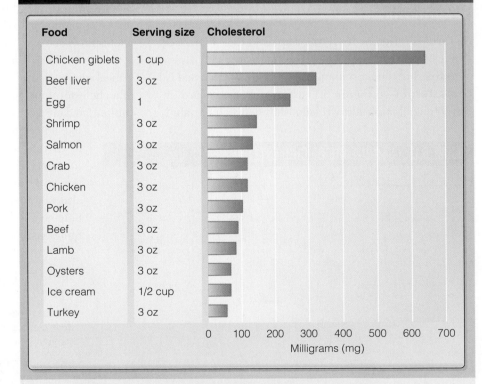

6-5 HOW ARE TRIGLYCERIDES DIGESTED, ABSORBED, AND CIRCULATED?

Once ingested, lipids must be digested, absorbed, and circulated away from the small intestine. The basic goal of triglyceride digestion is to separate (or cleave) most of the fatty acids from their glycerol backbones—in other words, lipolysis. This process, which involves several enzymes and other secretions produced by the gastro-intestinal tract and accessory organs, is relatively more complicated than the digestion of the other macronutrients (carbohydrates and proteins). This is because it is more difficult for your body, which is made principally of water-soluble compounds, to digest water-insoluble molecules such as lipids than water-soluble molecules such as sugars and proteins (see Figure 6.13). Still, as you will soon understand, the digestion of dietary lipids occurs in your mouth, stomach, and small intestine every time you eat a meal containing fats or oils.

6-5a Triglyceride Digestion Begins in Your Mouth

A small portion of triglyceride digestion occurs in your mouth. As chewing breaks food apart, **lingual lipase**, an enzyme produced by your salivary glands, begins to remove fatty acids from the glycerol molecules. After the food is swallowed, lingual lipase accompanies the bolus into your stomach, where the enzyme continues to cleave additional fatty acids from glycerol.

> **lingual lipase** An enzyme, produced by the salivary glands, that cleaves fatty acids from glycerol molecules.

FIGURE 6.13 OVERVIEW OF TRIGLYCERIDE DIGESTION

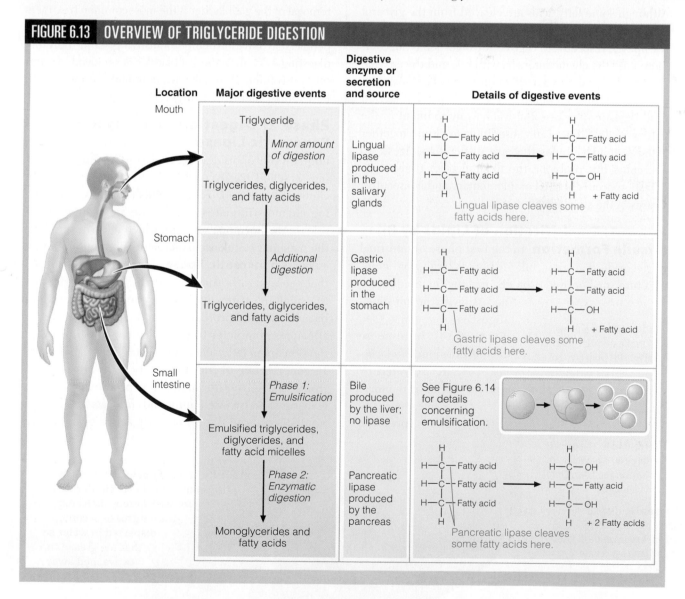

6-5b Triglyceride Digestion Continues in Your Stomach

The second stage of triglyceride digestion begins when food enters your stomach, stimulating the release of gastrin from specialized stomach cells. Recall from Chapter 3 that gastrin is a hormone that stimulates the release of gastric juices in the stomach. Gastrin also causes the muscular wall of the stomach to vigorously contract. Circulating in the blood, gastrin quickly stimulates the release of **gastric lipase**, an enzyme produced in the stomach. Gastric lipase, a component of the gastric juices, picks up where lingual lipase left off by further cleaving fatty acids from the glycerol molecules.

6-5c Triglyceride Digestion Is Completed in Your Small Intestine

Although some fatty acids are cleaved from the glycerol backbones in the mouth and stomach by lingual and gastric lipases, respectively, triglyceride digestion is not complete until the chyme interacts with bile and the enzyme pancreatic lipase in your small intestine. The mixing of chyme with bile, an emulsifier, is an important step in lipid digestion because the watery environment of the gastrointestinal tract can cause lipids to clump together in large lipid globules that are difficult to digest. To overcome this problem, the final stage of triglyceride digestion occurs in two complementary and consecutive phases in the small intestine.

Phase 1: Emulsification of Lipids by Bile–Micelle Formation In the first phase of intestinal triglyceride digestion, bile disperses large lipid globules into smaller lipid droplets, making the lipids more accessible to the intestine's digestive enzymes (see Figure 6.14). In response to the hormone *cholecystokin* (CCK) secreted by enterocytes lining the small intestine in the presence

gastric lipase An enzyme, produced in the stomach, that cleaves fatty acids from glycerol molecules.

micelle A small droplet of fat formed, via emulsification, in the small intestine.

emulsification The process whereby large lipid globules are broken down and stabilized into smaller lipid droplets.

pancreatic lipase An enzyme, produced in the pancreas, that completes triglyceride digestion by cleaving fatty acids from glycerol molecules.

of chyme, the gallbladder contracts and releases bile into the duodenum. Recall from Chapter 3 that bile is made by the liver and stored in the gallbladder until needed. Bile is comprised of a mixture of bile acids, cholesterol, and phospholipids. Both bile acids and phospholipids are amphipathic. Because the hydrophilic and hydrophobic components of bile are attracted to water and lipids, respectively, the large lipid globules are pulled apart into smaller droplets when they mix with the bile and phospholipids. Bile then surrounds each newly formed droplet, or **micelle**, stabilizing it in the intestine. The process whereby large lipid globules are broken down and stabilized into micelles is called **emulsification**.

When a person's bile contains an excess amount of cholesterol in relation to its other components, gallbladder disease may develop. Recall from Chapter 3 that the accumulation of cholesterol, calcium, and cellular debris in bile can lead to the formation of gallstones. Surgical removal of the gallbladder is the most common treatment for persistent gallstone-related problems. Because some people may have difficulty emulsifying—and therefore digesting—fat after the gallbladder is removed, doctors often recommend initially avoiding high-fat meals after undergoing gallbladder surgery.

Phase 2: Digestion of Triglycerides by Pancreatic Lipase Emulsification by itself does not complete lipid digestion; fatty acids remaining attached to glycerol molecules still need to be chemically cleaved. To accomplish this, lipid-containing chyme stimulates cells in the small intestine to release the hormone *secretin*, which in turn stimulates the pancreas to release pancreatic juice containing the enzyme **pancreatic lipase**. Recall from Chapter 5 that secretin, a hormone made by enterocytes lining the small intestine, also signals the release of sodium bicarbonate and proteases from the pancreas. Pancreatic lipase completes triglyceride digestion by cleaving the remaining fatty acids from their glycerol molecules. In general, two of the three fatty acids are removed from the triglyceride molecules, producing a monoglyceride and two free (unbound) fatty acids. The final products of lipid digestion (fatty acids, glycerol, and

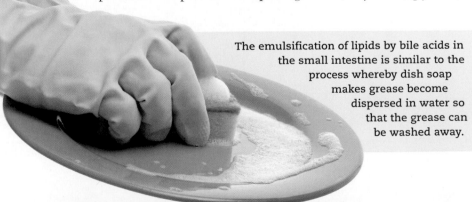

The emulsification of lipids by bile acids in the small intestine is similar to the process whereby dish soap makes grease become dispersed in water so that the grease can be washed away.

©istock.com/flubydust

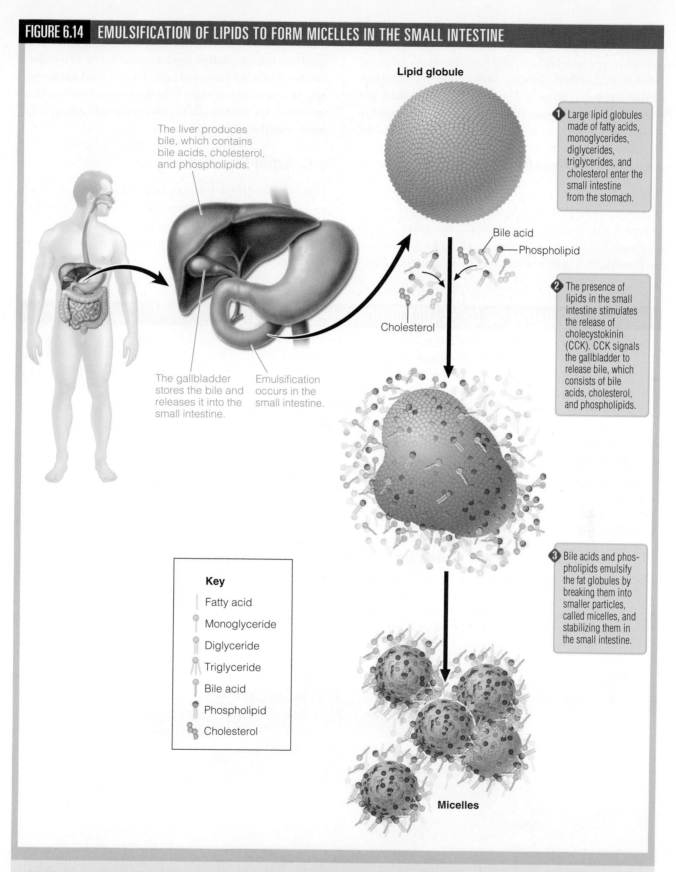

Lipid globule

The liver produces bile, which contains bile acids, cholesterol, and phospholipids.

The gallbladder stores the bile and releases it into the small intestine.

Emulsification occurs in the small intestine.

Bile acid

Phospholipid

Cholesterol

1 Large lipid globules made of fatty acids, monoglycerides, diglycerides, triglycerides, and cholesterol enter the small intestine from the stomach.

2 The presence of lipids in the small intestine stimulates the release of cholecystokinin (CCK). CCK signals the gallbladder to release bile, which consists of bile acids, cholesterol, and phospholipids.

3 Bile acids and phospholipids emulsify the fat globules by breaking them into smaller particles, called micelles, and stabilizing them in the small intestine.

Key

| Fatty acid |
| Monoglyceride |
| Diglyceride |
| Triglyceride |
| Bile acid |
| Phospholipid |
| Cholesterol |

Micelles

Intestinal emulsification of lipids requires bile acid- and phospholipid-containing bile, which is produced in the liver and stored in the gallbladder.

monoglycerides) are then taken up into the intestinal cells and circulated to the rest of the body. This process requires special handling because many of these substances are hydrophobic, while both the interiors of the intestinal cells and the circulatory system are hydrophilic. Once again, it is the amphipathic property of phospholipids that makes lipid absorption possible.

6-5d Lipid Absorption

Lipid absorption is accomplished in one of two ways, depending on how hydrophilic the lipid is (see Figure 6.15). Because they are relatively water soluble (hydrophilic), short- and medium-chain fatty acids can be transported into intestinal cells unassisted. More

hydrophobic compounds (such as long-chain fatty acids, monoglycerides, and cholesterol) must first be repackaged into micelles within the intestinal lumen before they can move into the intestinal cells. Once a lipid-containing micelle comes into contact with the lumenal surface of an intestinal cell, the micelle's contents are released into the interior of the enterocyte.

6-5e Lipid Circulation

Upon crossing the intestinal wall, dietary lipids are ready to be circulated away from the gastrointestinal tract for delivery to the body's tissues. Like absorption, the method by which a lipid is circulated depends on how hydrophilic it is.

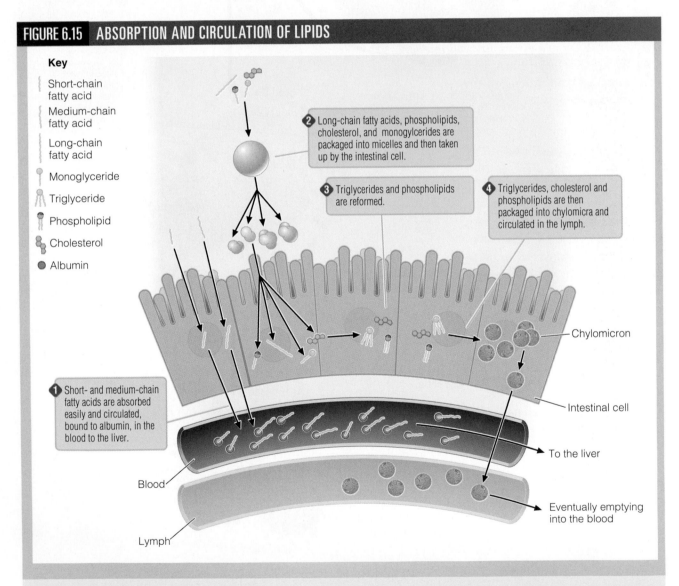

FIGURE 6.15   ABSORPTION AND CIRCULATION OF LIPIDS

Key
- Short-chain fatty acid
- Medium-chain fatty acid
- Long-chain fatty acid
- Monoglyceride
- Triglyceride
- Phospholipid
- Cholesterol
- Albumin

2 Long-chain fatty acids, phospholipids, cholesterol, and monogylcerides are packaged into micelles and then taken up by the intestinal cell.

3 Triglycerides and phospholipids are reformed.

4 Triglycerides, cholesterol and phospholipids are then packaged into chylomicra and circulated in the lymph.

1 Short- and medium-chain fatty acids are absorbed easily and circulated, bound to albumin, in the blood to the liver.

Chylomicron

Intestinal cell

To the liver

Eventually emptying into the blood

Blood

Lymph

The process by which a lipid is absorbed and circulated depends on how hydrophilic it is. In general, it is easier for the body to absorb and circulate hydrophilic lipids (like short-chain fatty acids) than hydrophobic ones (like long-chain fatty acids).

More Hydrophilic Lipids Are Circulated in the Blood

Because short- and medium-chain fatty acids are relatively water soluble, they can be put directly into the blood. Attached to a protein called albumin, these fatty acids are circulated directly from the small intestine to the liver. Once in the liver, short- and medium-chain fatty acids can be metabolized or rerouted for delivery to other cells in the body.

Less Hydrophilic Lipids Are Circulated in Lymph

Compared to what happens with the short- and medium-chain fatty acids, circulation of larger, less hydrophilic lipids away from the gastrointestinal tract is more involved. First, long-chain fatty acids and monoglycerides must be reassembled into triglycerides inside the intestinal cell. These hydrophobic lipids, along with cholesterol and phospholipids, are then incorporated into **chylomicron** particles, which are released into the lymphatic system for initial circulation. Chylomicra (the plural form of the word *chylomicron*) package their hydrophobic lipids, such as triglycerides, within a hydrophilic exterior shell formed mainly from phospholipids and proteins (see Figure 6.16). The lymphatic system eventually delivers the chylomicra into the blood, where they travel to cells that take up their contents. The chylomicron is an example of a **lipoprotein**, a type of particle that your body makes to transport lipids. In this particular case, chylomicra circulate dietary lipids (exogenous lipids), whereas other lipoproteins circulate lipids made by the liver (endogenous lipids). You will learn more about other lipoproteins in the following section.

The enzyme **lipoprotein lipase** enables chylomicra to deliver their dietary fatty acids to cells. Lipoprotein lipase is produced in many tissues (especially adipose and muscle tissues) and resides within the lumens of the blood vessels surrounding the tissues that produce this specialized enzyme. As chylomicra circulate in the

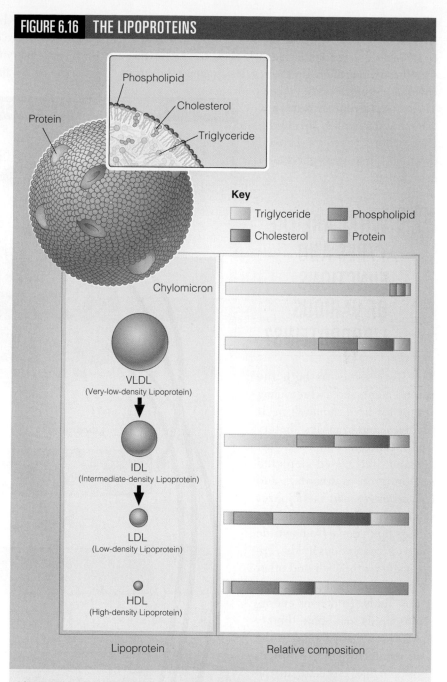

FIGURE 6.16 THE LIPOPROTEINS

Lipoproteins are composed of varying amounts of triglycerides, phospholipids, cholesterol, and proteins. The ratio of lipids to proteins determines the lipoprotein's density and, in most cases, its name.

Adapted from: Christie WC. AOCS Lipid Library. Lipoproteins. Available from: http://lipidlibrary.aocs.org/content.cfm?ItemNumber=39342.

chylomicron A particle, composed of relatively hydrophobic lipids, cholesterol, and phospholipids, that transports dietary lipids in the lymph for circulation. Note that the plural form of *chylomicron* is *chylomicra*.

lipoprotein A type of particle that transports lipids throughout the body.

lipoprotein lipase An enzyme that enables chylomicra and other lipoproteins to deliver fatty acids to cells.

blood, they are attacked by lipoprotein lipase, releasing fatty acids that are then taken up by the surrounding cells. After delivering dietary fatty acids to various tissues throughout the body, the residual fragments of the chylomicra remain in the blood. These **chylomicron remnant** fragments are taken up by the liver where they are broken down and their contents reused or recycled.

are distinguished by whether the particular lipid came directly from the diet, was made in or packaged by the liver for delivery to other cells, or is being transported from the body's tissues to the liver.

6-6 WHAT ARE THE TYPES AND FUNCTIONS OF VARIOUS LIPOPROTEINS?

The liver serves as both the central command and recycling center for lipid metabolism. It receives and recycles dietary lipids (and as you have just learned, fatty acid–depleted chylomicron remnants), and it synthesizes and metabolizes other lipids as needed. Whereas dietary lipids come from outside the body (exogenous lipids), lipids made by the liver and other tissues are referred to as endogenous lipids. To deliver endogenous lipids and some dietary lipids throughout the body, the liver produces a series of lipoproteins that circulate in the blood. A summary of the origins and functions of the various lipoproteins (including the chylomicra) is presented in Figure 6.17. As you will see, there are three general pathways by which lipids are transported in the body, and these pathways

chylomicron remnant A residual fragment of a chylomicron that remains in the blood until it is taken up by the liver.

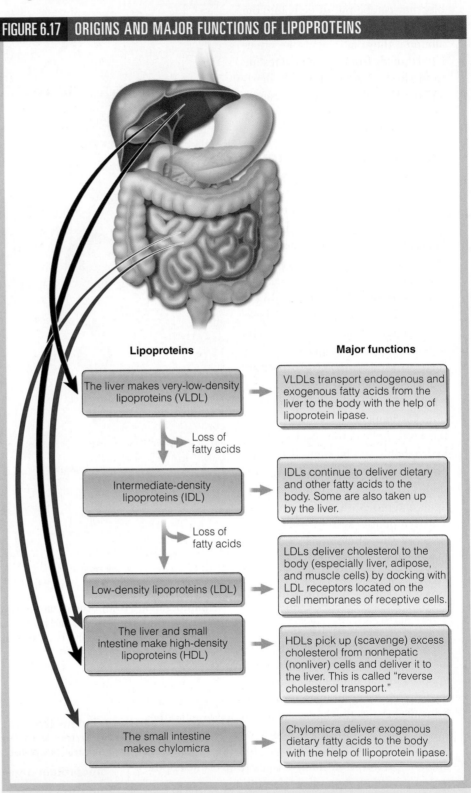

FIGURE 6.17   ORIGINS AND MAJOR FUNCTIONS OF LIPOPROTEINS

Lipoproteins

The liver makes very-low-density lipoproteins (VLDL)

Loss of fatty acids

Intermediate-density lipoproteins (IDL)

Loss of fatty acids

Low-density lipoproteins (LDL)

The liver and small intestine make high-density lipoproteins (HDL)

The small intestine makes chylomicra

Major functions

VLDLs transport endogenous and exogenous fatty acids from the liver to the body with the help of lipoprotein lipase.

IDLs continue to deliver dietary and other fatty acids to the body. Some are also taken up by the liver.

LDLs deliver cholesterol to the body (especially liver, adipose, and muscle cells) by docking with LDL receptors located on the cell membranes of receptive cells.

HDLs pick up (scavenge) excess cholesterol from nonhepatic (nonliver) cells and deliver it to the liver. This is called "reverse cholesterol transport."

Chylomicra deliver exogenous dietary fatty acids to the body with the help of llipoprotein lipase.

Both the liver and small intestine make lipoproteins that circulate lipids in the body.

6-6a Lipoproteins

Because lipids are relatively hydrophobic, their transport in the water-rich blood is somewhat complex. To aid in the process, the liver produces lipoproteins that function to transport lipids in the blood. These lipoproteins are different from the one you have already learned about—the chylomicron—which is the largest and the least dense member of the lipoprotein family. Whereas chylomicra are produced in the small intestine and transport *dietary* lipids in the bloodstream exclusively, other lipoproteins are produced in the liver and carry lipids originating from the liver, itself, as well as various other tissues and organs and (indirectly) from the diet. This also includes lipids released from storage in adipose tissue. Lipoproteins are complex globular structures that contain varying amounts of triglycerides, phospholipids, cholesterol, and special proteins called apoproteins. Embedded within the outer shell of a lipoprotein, **apoprotein** (or **apolipoprotein**) molecules enable a lipoprotein to circulate in the blood and interact with the cells that require its contents. Like the chylomicra made in the small intestine, the lipoproteins made in the liver are constructed so that their hydrophilic components (proteins and phospholipids) face outward and their hydrophobic components (such as triglycerides) face inward (recall Figure 6.16).

Because lipid is less dense than protein, the density of a lipoprotein depends on its relative amounts (or percentages) of lipids and proteins. Lipoproteins with more lipid relative to the amount of protein have lower densities than those with more protein relative to the amount of lipid. With the exception of chylomicra, lipoproteins are named according to their densities, so the relative densities give each lipoprotein its name.

Very-Low-Density Lipoproteins Deliver Fatty Acids
One type of lipoprotein made by the liver is a **very-low-density lipoprotein (VLDL)**. VLDLs are similar to chylomicra in that they contain triglycerides and cholesterol in their cores and are surrounded by phospholipids and proteins on their surfaces. However, VLDLs have a lower lipid-to-protein ratio than do chylomicra, making them smaller and denser. Like chylomicra, the primary function of VLDLs is to deliver fatty acids to cells. But unlike chylomicra, which *only* transport dietary fatty acids, VLDLs also transport fatty acids

HDLs

...are smaller and denser than...

LDLs

...are smaller and denser than...

IDLs

...are smaller and denser than...

VLDLs

...are smaller and denser than...

Chylomicra

derived from liver and other tissues such as adipose tissue. When VLDLs in the blood circulate past cells that produce lipoprotein lipase, the enzyme cleaves off fatty acids that are in turn taken up by the surrounding cells. In this way, tissues in need of fatty acids are able to obtain them directly from the bloodstream.

Low-Density Lipoproteins Deliver Cholesterol
The removal of fatty acids from VLDLs causes them to become smaller and denser. The resulting lipoprotein, which is smaller and less dense than a VLDL, is referred to as an **intermediate-density lipoprotein (IDL)**. Some IDLs are taken up by the liver, whereas others remain in circulation where they continue to give up additional triglycerides. Eventually, an IDL becomes a cholesterol-rich **low-density lipoprotein (LDL)**, which is smaller and denser still. An **LDL receptor** is a specialized type of protein located on cell membranes. LDL receptors bind to the apoproteins (called apolipoprotein B) embedded in the surface of the LDLs, allowing the LDLs to be taken up and broken down by the cell. In this way, much of the cholesterol is removed from the blood and used by cells to maintain their cell membranes and synthesize other compounds such as steroid hormones and vitamin D.

Sometimes, however, cholesterol-containing LDL particles penetrate walls of arteries, such as those supplying the heart with blood. When this happens, the lipids within the LDLs become oxidized, and the apolipoprotein B molecules are modified in such a way that they attract immune cells to the area. In an effort to

apoprotein (or **apolipoprotein**) A protein embedded within the outer shell of a lipoprotein that enables it to circulate in the blood and interact with the cells that require its contents.

very-low-density lipoprotein (VLDL) A lipoprotein, made by the liver, that has a lower lipid-to-protein ratio than do chylomicra. It delivers fatty acids to cells.

intermediate-density lipoprotein (IDL) A lipoprotein that is slightly less dense than a VLDL and that delivers fatty acids to cells.

low-density lipoprotein (LDL) A lipoprotein that delivers cholesterol to the body's cells.

LDL receptor A specialized type of protein, located on cell membranes, that binds to the apoproteins embedded in the surface of an LDL, allowing it to be taken up and broken down by the cell.

repair any damage that has been done to the blood vessel, these immune cells take up and dismantle the LDL particles. This, in turn, initiates a cascade of inflammatory responses that can further damage the blood vessel and can result in the buildup of a fatty substance called **plaque**, which is comprised of cholesterol, fatty substances, cellular waste products, calcium, and fibrin. Eventually, the accumulation of plaque can slow or even block blood flow.[11] Epidemiologic studies suggest that high levels of LDL cholesterol in the blood, or at least cholesterol contained in some types of LDL, are related to increased risk for cardiovascular disease.[12] Thus LDL cholesterol is often referred to as "bad cholesterol." The relationship between cholesterol and cardiovascular disease is addressed in greater detail later in this chapter.

High-Density Lipoproteins Eliminate Excess Cholesterol The liver and to a lesser extent the small intestine, also make a type of lipoprotein called **high-density lipoprotein (HDL)**. Compared to other lipoproteins, HDLs have the lowest lipid-to-protein ratios, and therefore they have the highest densities (hence, *high*-density lipoprotein). HDLs circulate in the blood to collect excess cholesterol from cells and transport it back to the liver, a process referred to as **reverse cholesterol transport**. Because a high level of HDL cholesterol in the blood is generally associated with a lower risk of cardiovascular disease, HDL cholesterol is often referred to as "good cholesterol." There are several types of HDL, however, and not all forms are equally effective in removing excess cholesterol. Different HDLs have different proteins (apoproteins) on their surfaces, resulting in somewhat different functions. The presence of particular apoproteins makes some HDLs more or less efficient at cholesterol removal, and researchers continue to investigate why different people have different apoproteins associated with their HDL particles.

In conclusion, there are three overarching lipid transport pathways, each involving different lipoproteins, as summarized here.

- One pathway exclusively transports dietary (exogenous) lipids and involves chylomicra.

- Another primarily transports endogenous lipids originating in the liver and adipose tissues and exogenous lipids and involves VLDLs, IDLs, and LDLs.

- Another transports excess cholesterol back to the liver and involves HDL (reverse cholesterol transport).

 ## 6-7 HOW ARE DIETARY LIPIDS RELATED TO HEALTH?

After lipids have been digested, absorbed, and circulated, they are ready to be used by your body for myriad purposes. As described previously, fatty acids and their by-products regulate metabolic processes within your cells and orchestrate a variety of physiological responses. Phospholipids are vital components of cell membranes, aid in lipid digestion and absorption, and are needed for many cellular functions. Cholesterol is incorporated into cell membranes, is a precursor for vitamin D and many hormones, and is involved in lipid digestion and absorption via its role in bile.

Although many lipids are vital to good health, too much of some and certain types of other lipids are associated with poor health. For example, an excessive dietary intake of some lipids is related to increased risks for cardiovascular disease and certain cancers. Of course, consuming too much lipid-rich foods can lead to excess calories, and subsequently, unwanted weight gain.

6-7a High-Fat Foods and Obesity

Obesity is defined as the overabundance of body fat. Although obesity is a complex condition, excess energy intake is a major factor in its development. Because fatty acids provide more than twice the calories per gram compared to carbohydrates and proteins, high fat intake is likely an important piece of the obesity puzzle. Whatever its causes, obesity is a major public health concern worldwide because of its association with increased risks for diseases such as cardiovascular disease, type 2 diabetes, and some forms of cancer. Because there is an intense interest in decreasing obesity rates, experts recommend that adults limit their energy intake—especially from high-fat foods and refined carbohydrates, which both contribute substantial calories without providing many of the essential nutrients we need. You will learn much more about lipid consumption and its relationship to obesity in Chapter 9.

plaque A fatty substance that builds up within the walls of blood vessels.

high-density lipoprotein (HDL) A lipoprotein, made primarily by the liver, that has the lowest lipid-to-protein ratio. It circulates in the blood to collect excess cholesterol from cells and transport it back to the liver through a process called reverse cholesterol transport.

reverse cholesterol transport A process whereby high-density lipoproteins remove excess cholesterol from cells and carry it back to the liver.

Although consumption of a high-fat diet can increase risk for obesity, there are many other factors, such as genetics and inadequate exercise, that contribute to this health problem.

6-7b Dietary Lipids and Cardiovascular Disease

Like obesity, **cardiovascular disease**, defined as a disease of the heart or vascular system, is caused by a complex web of factors, including genetics, physical inactivity, tobacco use, and dietary factors. The most common types of cardiovascular disease are **heart disease**, a slowing or complete obstruction of blood flow to the heart, and **stroke**, which is a slowing or complete obstruction of blood flow to the brain. In addition, cardiovascular disease can affect peripheral arteries that circulate blood to the extremities (arms, hands, legs, and feet). When this blood supply is disrupted, amputation is sometimes necessary. When blood flow is restricted, cells do not receive adequate oxygen and nutrients, ultimately causing them to die. Restriction of blood flow is generally caused by a condition called **atherosclerosis**, characterized by a narrowing and stiffening of the blood vessels (often due to plaque buildup). A **blood clot** (a small, insoluble particle made of clotted blood and clotting factors) or **aneurysm** (the outward bulging of the vessel due to weakness) may also restrict blood flow. These conditions are illustrated in Figure 6.18.

Both total lipids and specific types of lipids can influence a person's risk for developing atherosclerosis, blood clots, and aneurysms—and therefore, cardiovascular disease. For example, high intakes of

FIGURE 6.18 CAUSES OF CARDIOVASCULAR DISEASE

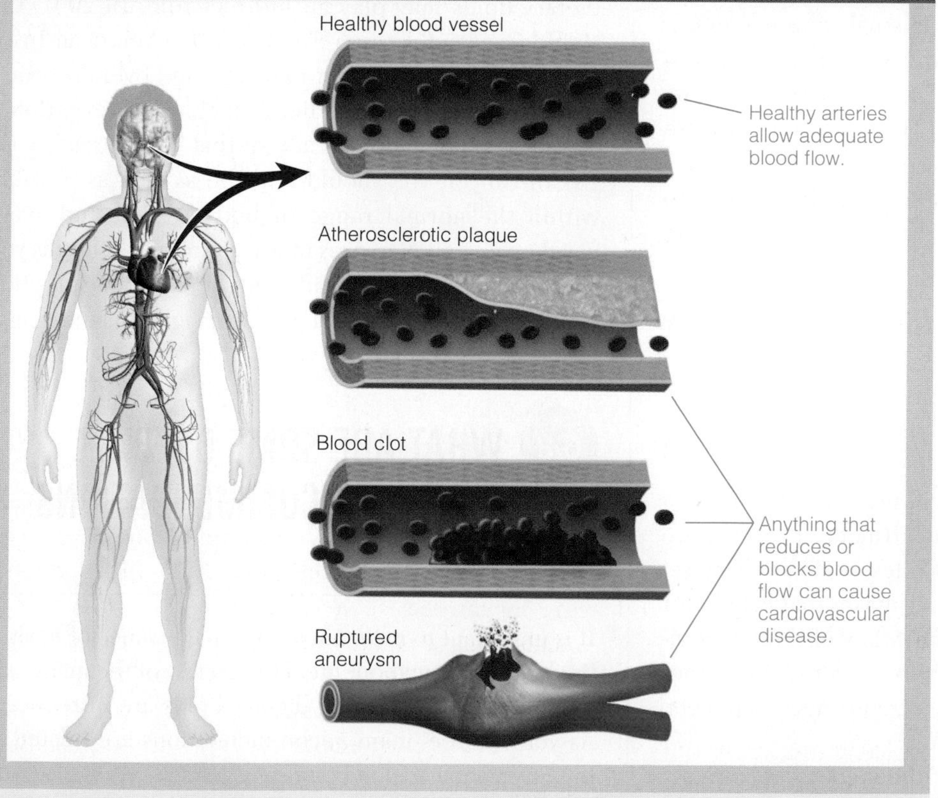

Healthy blood vessel

Healthy arteries allow adequate blood flow.

Atherosclerotic plaque

Blood clot

Anything that reduces or blocks blood flow can cause cardiovascular disease.

Ruptured aneurysm

Atherosclerosis, blood clots, and aneurysms can all reduce or stop blood flow, leading to cardiovascular disease.

cardiovascular disease A disease of the heart or vascular system.

heart disease A slowing or complete obstruction of blood flow to the heart.

stroke A slowing or complete obstruction of blood flow to the brain.

atherosclerosis A narrowing and stiffening of the blood vessels due to plaque buildup, causing the restriction of blood flow.

blood clot A small, insoluble particle made of clotted blood and clotting factors.

aneurysm An outward bulging of a blood vessel due to weakness in the vasculature.

total fats, certain SFAs, or *trans* fatty acids may increase risk for cardiovascular disease in some people, whereas MUFAs may have the opposite effect. Excessive consumption of carbohydrates, particularly refined carbohydrates and fructose, may also increase risk of cardiovascular disease by a mechanism that involves LDLs. Researchers believe that consumption of these types of sugars cause LDL particles to become smaller and less dense, and that these subclasses of LDL are especially atherogenic.[13]

A person's genetic makeup can also interact with diet to influence health, and this holds especially true for lipids and cardiovascular disease. For instance, many studies have shown that the consumption of a diet high in cholesterol can increase blood cholesterol (a risk factor for cardiovascular disease) in some people.[14] However, some people can eat very high amounts of cholesterol without experiencing this effect. This inconsistency may arise because some people absorb dietary cholesterol efficiently, while others may naturally excrete more cholesterol in their feces.[15] Genetic variation can also influence a person's HDL and LDL levels and functioning. For example, the ability of immune cells in blood vessel walls to take up LDL particles (and thus form plaques) may be influenced by variations in the genes that code for LDL receptor proteins.[16] Sometimes, inherited genetic variation in the LDL receptor makes it completely nonfunctional. This condition, referred to as *familial hypercholesterolemia*, results in extremely high levels of blood cholesterol that can only be lowered using medications.[17] As scientists learn more about how genetic makeup interacts with diet to influence health, health professionals will be better able to prescribe the most heart-healthy diet given a person's individual genetics.

Nutritional Guidelines for Cardiovascular Health The multifaceted and complex relationship between diet and cardiovascular risk is yet another example of how dietary variety, moderation, and balance are essential to health and well-being. To promote optimal cardiovascular health, agencies such as the American Heart Association, the U.S. Department of Agriculture (USDA), and the National Institutes of Health have set forth guidelines regarding dietary intakes that help lower risk for heart disease and stroke. Many of these recommendations do not relate to dietary fat per se, but rather to the many different types of nutrients needed to keep the cardiovascular system healthy.

First and foremost, moderate your overall energy intake and balance your macronutrients. The primary goal set forth in many dietary recommendations related to energy consumption and cardiovascular disease is to consume only enough calories to maintain a healthy body weight. Being overweight or obese profoundly increases a person's risk for heart disease and stroke. A person who wants to lose weight is advised to decrease energy intake and increase energy expenditure to achieve a specific bodyweight goal. This is best accomplished by following the Institute of Medicine's recommendations for energy intake (recall the Estimated Energy Requirement formulas from Chapter 2) and Acceptable Macronutrient Distribution Range (AMDR) recommendations for balancing energy intake from the various macronutrient categories. In addition to following these recommendations, making sure that at least half of one's daily grain choices are whole grains helps increase intake of dietary fiber, which is known to lower risk for cardiovascular disease. Whole-grain foods are also rich sources of vitamins, minerals, and heart-healthy lipids.

6-7c Dietary Lipids and Cancer

Although some studies show a link between high-fat diets and a risk for cancer, the data are inconclusive. Obesity, however, is a risk factor for several types of cancer, including colon and breast cancers.[18] As obesity is often associated with consumption of high-fat diets, dietary lipids may play an indirect role. In 2007, the World Cancer Research Fund and the American Institute for Cancer Research jointly issued overall recommendations concerning diet and cancer prevention.[19] These recommendations advise that, to decrease your risk of cancer, you should both be as lean as possible within the normal range of body weight and avoid weight gain and increases in waist circumference as you age. They also recommend that you limit consumption of energy-dense foods and only sparingly consume high-fat fast foods if at all.

 ## 6-8 WHAT ARE SOME OVERALL DIETARY RECOMMENDATIONS FOR LIPIDS?

It is important to maintain an optimal balance of healthy lipid intake throughout life. Here, a few of the many recommendations regarding dietary lipids are introduced. As you will see, many recommendations are related to the associations among dietary lipid intake, obesity, and cardiovascular disease. These recommendations will help you choose the right dietary lipids to help promote optimal health.

6-8a Consume Adequate Amounts of the Essential Fatty Acids

It is important to consume adequate amounts of the essential fatty acids; the Dietary Reference Intakes (DRIs) for these substances are found on the Dietary Reference Intakes card provided with this book. For example, Adequate Intake (AI) levels for linoleic acid are 17 and 12 g/day for adult men and women, respectively. This is the amount of linoleic acid contained in approximately 1¾ tbsp (or 26 mL) of corn oil. For linolenic acid, the AIs are 1.6 and 1.1 g/day for adult men and women, respectively. This is the amount contained in about 1½ tbsp (or 22 mL) of soybean oil.

6-8b Pay Particular Attention to the Long-Chain ω-3 Fatty Acids

In addition to specific recommendations for the essential fatty acids, the USDA and other health organizations suggest that you consume at least two servings of fish high in ω-3 fatty acids, such as salmon, each week, as well as other foods rich in linolenic acid, such as flaxseed and canola oils. People with cardiovascular disease are advised to consume about 1 gram of ω-3 fatty acids daily from fish. This is the amount contained in approximately 2 ounces (57 g) of Atlantic salmon. This recommendation has also been endorsed by the American Heart Association.

6-8c Limit Saturated Fat and *Trans* Fat

Scientists have long known that certain SFAs are related to an increased risk of cardiovascular disease for some people. As such, many dietary guidelines suggest decreasing one's SFA intake. For example, the Institute of Medicine recommends that "intake of SFAs should be minimized while consuming a nutritionally adequate diet."[20] The Dietary Guidelines suggest that SFAs should constitute no more than 10 percent of a person's total caloric intake, and the American Heart Association recommends that SFA intake represent 7 percent or less of one's total calories. Reading food labels such as those illustrated in Figure 6.19 can help you keep track of your SFA intake.

Both the Dietary Guidelines and the Institute of Medicine recommend that people "minimize their intakes of *trans*

FIGURE 6.19 READING NUTRITION FACTS PANELS

Nutrition Facts
Serving Size: 1 cup (244g)

Amount per serving

Calories 122   Calories from fat 45

% Daily Value*

Total fat 5g	7%
Saturated fat 3g	15%
Trans fat 0g	
Cholesterol 20mg	7%
Sodium 100mg	4%
Total carbohydrates 11g	4%
Dietary fiber 0g	0%
Sugars 12g	
Protein 8g	

Vitamin A 9%		Vitamin C	1%
Calcium 29%		Iron	0%

*Percent Daily Values are based on a 2,000 calorie diet.

You can tell how much total fat, saturated fat, and *trans* fat is in a food by checking its Nutrition Facts panel.

fatty acids."[21] Note that these guidelines apply to commercially produced *trans* fatty acids—not to naturally occurring ones.

No DRIs are established for cholesterol. However, the American Heart Association recommends that you consume less than 300 milligrams of cholesterol daily. This is the amount found in one to two eggs, or two servings of beef. The Institute of Medicine simply recommends that "cholesterol intake should be minimized." However, the 2015 Dietary Guidelines for Americans do not recommend limiting cholesterol consumption. This is because their analysis of the available scientific literature suggested that (at least for the vast majority of Americans) increased consumption of cholesterol, per se, does not lead to cardiovascular disease.

Unlike previous versions, the 2015 Dietary Guidelines for Americans does not recommend limiting cholesterol intake.

6-8d Guidelines for Total Lipid Consumption

The Institute of Medicine has not established RDA or AI values for total lipid intake, except during infancy, when AIs are set at approximately 30 g/day. However, the AMDRs recommend that healthy adults consume 20 to 35 percent of their energy from lipid. Based on a caloric requirement of 2,000 kilocalories/day, a person should consume between 400 and 700 kilocalories of lipid. Considering that 1 gram of lipid contains about 9 kilocalories, the daily lipid intake for most adults should be between 44 and 78 grams. This amount is easy to obtain: a typical day's menu that includes three servings of low-fat milk, a bagel with cream cheese, a peanut butter sandwich, spaghetti with meatballs, and a salad with ranch dressing collectively contains about 56 grams of lipids.

Even high-fat foods, if consumed in moderation, can be part of a healthy diet.

© Georgy Markov/ Shutterstock.com

NUTR
ONLINE

PREPARE FOR TESTS ON
THE STUDYBOARD!

● CORRECT

● INCORRECT

● INCORRECT

● INCORRECT

Personalize Quizzes from Your StudyBits

Take Practice Quizzes by Chapter

CHAPTER QUIZZES
▶ Chapter 1
Chapter 2
Chapter 3
Chapter 4

4LTR
PRESS

Access NUTR ONLINE at www.cengagebrain.com

7 | The Vitamins

LEARNING OUTCOMES

7-1 Summarize several historical achievements related to the discovery of vitamins.

7-2 Discuss factors that the water-soluble vitamins have in common.

7-3 Understand the functions and sources of the B vitamins.

7-4 Describe the functions and sources of vitamin C.

7-5 Discuss factors that the fat-soluble vitamins have in common.

7-6 Describe the functions and dietary sources of vitamin A and the carotenoids.

7-7 Understand the functions and sources of vitamin D.

7-8 Understand the functions and sources of vitamins E and K.

7-9 Describe the advantages and disadvantages associated with obtaining vitamins from dietary supplements versus food.

© leonori/Shutterstock.com

After finishing this chapter go to **PAGE 182** for **STUDY TOOLS.**

7-1 HOW DID SCIENTISTS FIRST DISCOVER VITAMINS?

As described in previous chapters, your body uses carbohydrates, proteins, and lipids (the macronutrients) for a multitude of functions. But it also requires micronutrients—vitamins and minerals—albeit in much smaller amounts. Both vitamins and minerals are needed to maintain a healthy body, but the two categories of micronutrients are quite different. Whereas minerals are simple inorganic elements, vitamins are complex organic compounds. You will learn about vitamins in this chapter and then about minerals in Chapter 8.

Over a century ago, scientists only had the capability and knowledge to classify vitamins based on their solubility in water and fat. *Fat-soluble* vitamins were collectively called vitamin A, while *water-soluble* vitamins were referred to as vitamin B.[1] Researchers soon discovered, however, that vitamins A and B actually comprised several unique substances. As it was discovered, each of these new substances was designated with its own classification letter (such as C, D, and K). Researchers subsequently learned that what was initially called vitamin B was not a single substance, but a group of several different interrelated ones. The B vitamins were not all renamed with new letters, but, instead, they were differentiated by numbers such as B_1, B_6, and B_{12}. When grouped together, the B vitamins are often called *B-complex vitamins*.

Today, scientists know of at least 13 essential fat- and water-soluble vitamins. Beyond their letter designations, many vitamins have common names such as *thiamin* (B_1) and *riboflavin* (B_2). Most vitamins also have chemical names; vitamin C, for example, is referred to as *ascorbic acid*. Because it is possible for a single vitamin to have numerous names, distinguishing one from another can sometimes be confusing. In this chapter, you will learn about the essential vitamins' nomenclatures, dietary sources, functions, deficiencies, toxicities, and recommended intakes. You will also learn what dietary supplements are, where to find reliable information about them, and when you might want to consider taking them.

7-2 WATER-SOLUBLE VITAMINS

Each category of nutrients you have studied thus far has distinct characteristics that make it chemically unique. For instance, all proteins are made from nitrogen-containing amino acids, and all carbohydrates are made from monosaccharides. This trend ends with vitamins, however. Unlike carbohydrates, proteins, and lipids, there is no single molecular structure that groups vitamins together and distinguishes them from other nutrients. In other words, aside from the fact that they are all organic molecules, each vitamin is chemically distinct from the next. Nonetheless, there are basic biological properties that define and characterize vitamins.

7-2a Commonalities among the Water-Soluble Vitamins

Even though they are not structurally similar, water-soluble vitamins tend to share three critical biological attributes. The first and perhaps most basic is that all water-soluble vitamins dissolve in water (unlike fat-soluble vitamins). This single attribute concretely distinguishes water-soluble vitamins from fat-soluble

vitamins. Second, although there are exceptions, the body generally absorbs and transports water-soluble vitamins in a similar way. Water-soluble vitamins are absorbed in the small intestine and, to a lesser extent, the stomach. Bioavailability is influenced by many factors, such as nutritional status, other nutrients and substances in foods, medications, age, and health status. All of the water-soluble vitamins are circulated in the blood (as opposed to the lymph), and they are transported from the gastrointestinal tract directly to the liver. Finally, because the body retains only small amounts of water-soluble vitamins, their overconsumption generally does not lead to toxic effects.

Water-Soluble Vitamins Can Be Destroyed by Heat and Improper Storage A balanced and varied diet can provide all of the vitamins needed to maintain health. Unfortunately, no matter how carefully you choose your foods, some preparation, cooking, and storage methods destroy vitamins—especially water-soluble vitamins. Whereas some are lost through exposure to water, air, heat, and/or light, others are unavoidably affected by acidity. Some vitamin loss is to be expected, but you can prevent excessive loss by properly preparing and storing foods.

For example, overcooking some foods can result in the loss of some water-soluble vitamins, and excessive sunlight can degrade vitamin C and certain B vitamins. Information about how each of the water-soluble vitamins can best be protected is addressed throughout the chapter.

7-2b Fortification and Enrichment of Food

Although all foods contain a variety of naturally occurring nutrients, some also have nutrients added during processing. As you learned in Chapter 1, this is called *fortification*. Because processing can destroy some vitamins, fortification is generally used to improve the nutritional value of processed foods. For example, because the portion of rice that contains most of the grain's thiamin is removed during the refinement process, this vitamin is often added to milled rice, making it a *fortified* food. Adding vitamins through fortification helps make many refined products such as white (polished) rice more nutritious. In some cases, a fortified food is even more nutritious than its original version. For instance, orange juice typically does not contain calcium. Therefore, calcium-fortified orange juice is actually more nutritious than its unfortified form.

Fortified food products that meet specified nutrient levels suggested by the U.S. Food and Drug Administration (FDA) can be labeled "enriched." **Enrichment** is the fortification of a select group of foods (rice, flour, bread or rolls, farina, pasta, cornmeal, and corn grits) with specified levels of B vitamins thiamin, niacin, riboflavin, folate, and the mineral iron. The main goal of the U.S. national enrichment program is to provide nutrients that tend to be underconsumed by Americans at levels predicted to improve overall population health.

enrichment The fortification of a select group of foods with FDA-specified levels of thiamin, niacin, riboflavin, folate, and iron.

thiamin (vitamin B$_1$) An essential water-soluble vitamin involved in energy metabolism and the synthesis of DNA and RNA.

coenzyme A vitamin that facilitates the function of an associated enzyme.

beriberi A life-threatening condition caused by thiamin deficiency. There are four forms: wet, dry, infantile, and cerebral beriberi.

Unless it has been fortified, processed white rice is not a good source of micronutrients such as thiamin.

© Somchai Som/Shutterstock.com

Although they differ chemically, the B-complex vitamins tend to serve similar biological functions within the body, especially in terms of energy metabolism. In this section, you will learn about each of these essential nutrients.

7-3a Thiamin

Thiamin (vitamin B$_1$) is an essential water-soluble vitamin involved in energy metabolism and the synthesis of the genetic material DNA and RNA. It is found in a wide variety of foods, such as pork, peas, fish, legumes, soymilk, enriched cereal products, and whole-grain foods (see Figure 7.1). As you will notice throughout this chapter, whole-grain foods tend to be good sources of many water-soluble vitamins. Thiamin is sensitive to heat and is easily destroyed during cooking. Shorter cooking times and lower temperatures can decrease this loss, although it is important to cook meat until it is done.

Several factors influence thiamin bioavailability. In general, its absorption increases when the body needs more of it, and decreases when thiamin status is adequate. Compounds found in raw fish, coffee, tea, berries, Brussels sprouts, cabbage, and alcohol can interfere with thiamin's absorption, and sulfites, often added to processed foods as preservatives, can destroy thiamin.[2] Conversely, vitamin C can increase thiamin bioavailability.

Thiamin Is Critical to ATP, Protein, and Triglyceride Production Although thiamin is not an energy-yielding nutrient per se, it is intimately involved in ATP production. To aid in energy synthesis, thiamin acts as a **coenzyme**, meaning that an enzyme will not function unless thiamin is present. As a component of a coenzyme, thiamin plays a role in the synthesis of DNA, RNA, and triglycerides. Without thiamin acting as a coenzyme, the enzymes that make DNA and RNA cannot work, and protein synthesis is halted. As a result, cell division is impeded because the genetic material (DNA and RNA) cannot be replicated.

Thiamin Deficiency Causes Beriberi As early as 2600 B.C.E., Chinese physicians described **beriberi**, the life-threatening condition of thiamin deficiency. Although now uncommon in the United States, beriberi is still prevalent in regions of the world that rely heavily on unfortified, refined rice as a major source of energy. There are four forms of beriberi. *Dry beriberi*, found mostly in adults, is characterized by severe muscle wasting, leg cramps, tenderness, and decreased feeling in the feet and toes. *Wet beriberi*

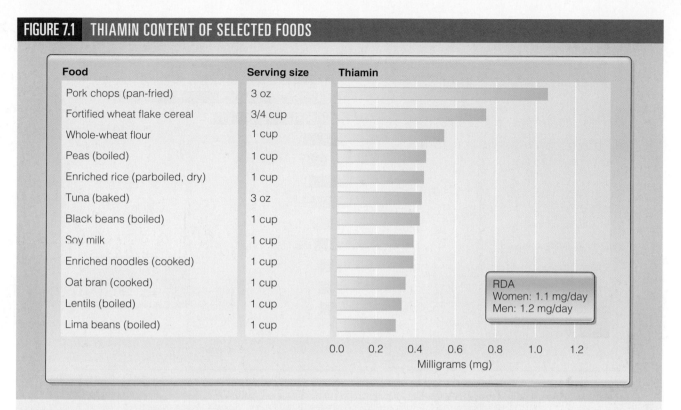

FIGURE 7.1 THIAMIN CONTENT OF SELECTED FOODS

Food	Serving size	Thiamin
Pork chops (pan-fried)	3 oz	
Fortified wheat flake cereal	3/4 cup	
Whole-wheat flour	1 cup	
Peas (boiled)	1 cup	
Enriched rice (parboiled, dry)	1 cup	
Tuna (baked)	3 oz	
Black beans (boiled)	1 cup	
Soy milk	1 cup	
Enriched noodles (cooked)	1 cup	
Oat bran (cooked)	1 cup	
Lentils (boiled)	1 cup	
Lima beans (boiled)	1 cup	

RDA
Women: 1.1 mg/day
Men: 1.2 mg/day

Milligrams (mg)

Source: U.S. Department of Agriculture, Agricultural Research Service. 2004. USDA National Nutrient Database for Standard Reference, Release 16-1. Available at: http://www.ars.usda.gov/nuteintdata

is characterized by severe edema in the arms and legs. An enlargement of the heart and respiratory problems often result in heart failure. *Infantile beriberi*, found in babies breastfed by thiamin-deficient mothers, can also cause heart problems. *Cerebral beriberi* (or *Wernicke-Korsakoff syndrome*) is characterized by abnormal eye movements, poor muscle coordination, confusion, and short-term memory loss. In the United States, cerebral beriberi is typically associated with alcoholism both because alcohol decreases thiamin absorption and because some alcoholics have very poor diets.

Recommended Thiamin Intake The Recommended Dietary Allowances (RDAs) for thiamin are 1.2 and 1.1 mg/day for adult men and women, respectively. Because thiamin toxicity is rare, there is no Tolerable Upper Intake Level (UL). A complete list of vitamin Dietary Reference Intakes (DRIs) is provided on the Dietary Reference Intakes card located at the back of this book.

7-3b Riboflavin

Riboflavin (vitamin B₂) is an essential water-soluble vitamin involved in energy metabolism, the synthesis of a variety of vitamins, nerve function, and protection of biological membranes. Riboflavin is named for its ribose structural component and its yellow color—*flavus* means "yellow" in Latin. Riboflavin is found in a variety of foods, such as liver, meat, dairy products, enriched cereals, and other fortified foods (see Figure 7.2). Fruits and vegetables contain only marginal amounts of riboflavin, while whole-grain foods contain slightly more. Riboflavin is relatively stable during cooking but is quickly destroyed when exposed to excessive light. This is why milk is generally packaged in cardboard, opaque, or translucent (cloudy) containers to protect it from light and preserve its riboflavin content. As with thiamin, riboflavin bioavailability increases when the body needs more of it. Riboflavin found in animal-based foods is somewhat more bioavailable than that found in plant-based foods, and alcohol can inhibit its absorption.

Riboflavin Assists in Reduction-Oxidation (Redox) Reactions
Your body uses riboflavin to produce two coenzymes needed to transfer oxygen

riboflavin (vitamin B₂) An essential water-soluble vitamin involved in energy metabolism, the synthesis of a variety of vitamins, nerve function, and protection of biological membranes.

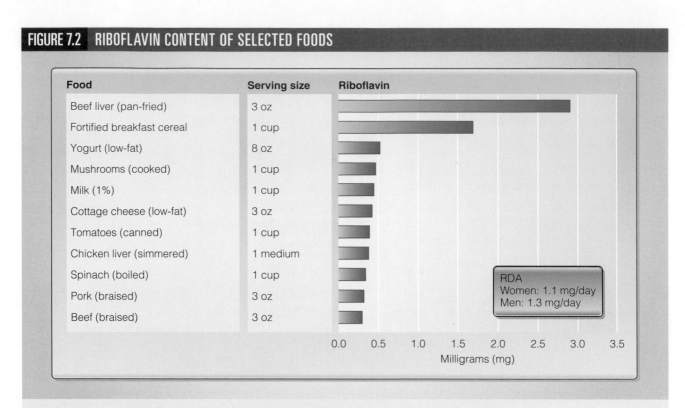

FIGURE 7.2 RIBOFLAVIN CONTENT OF SELECTED FOODS

Food	Serving size	Riboflavin
Beef liver (pan-fried)	3 oz	
Fortified breakfast cereal	1 cup	
Yogurt (low-fat)	8 oz	
Mushrooms (cooked)	1 cup	
Milk (1%)	1 cup	
Cottage cheese (low-fat)	3 oz	
Tomatoes (canned)	1 cup	
Chicken liver (simmered)	1 medium	
Spinach (boiled)	1 cup	
Pork (braised)	3 oz	
Beef (braised)	3 oz	

RDA
Women: 1.1 mg/day
Men: 1.3 mg/day

Milligrams (mg)

Source: U.S. Department of Agriculture, Agricultural Research Service. 2004. USDA National Nutrient Database for Standard Reference, Release 16-1. Available at: http://www.ars.usda.gov/nuteintdata

© Dani Vincek/Shutterstock.com
© David Crockett/Shutterstock.com
© mama_mia/Shutterstock.com
© Valentyn Volkov/Shutterstock.com

Riboflavin is especially abundant in meat, fortified cereals, and dairy products.

or electrons (negatively charged particles) from one molecule to another. This type of reaction, called a **reduction-oxidation (redox) reaction**, is critical for ATP production. Recall from Chapter 3 that an atom or molecule with a net positive or negative charge is called an ion. Ions are formed by the loss or gain of one or more electrons. Charged molecules, and certain ions, play an important role in redox reactions. In general, negatively charged ions (or molecules) donate oxygen or electrons to positively charged ions (or molecules). When this happens, the positively charged particle is *reduced*, and the negatively charged particle is *oxidized*, resulting in a redox reaction. Redox reactions assist in the synthesis of many bodily substances, such as collagen (a component of your skin, muscle tissue, and hair), carnitine (needed for fatty acid metabolism), and several neurotransmitters and hormones.

Beyond its role in redox reactions, riboflavin is also necessary to convert vitamin A and folate (a B vitamin) to their active forms, to synthesize niacin (a B vitamin) from tryptophan (an amino acid), and to form vitamins B_6 and K. Finally, riboflavin is critical to the metabolism of several neurotransmitters, and it is involved in reactions that protect biological membranes from oxidative damage.

reduction-oxidation (redox) reaction A transfer of oxygen or electrons from one molecule to another.

Riboflavin Deficiency Causes Ariboflavinosis
Riboflavin deficiency causes a condition called **ariboflavinosis**, which does not usually occur on its own, but is associated with general malnutrition. The signs and symptoms of ariboflavinosis include **cheilosis**, sores on the outside and corners of the lips, **stomatitis**, inflammation of the mouth, and **glossitis**, inflammation of the tongue, as well as muscle weakness and confusion. Riboflavin deficiency is rare in the United States but can occur in some alcoholics who consume inadequate diets and in people with diseases that interfere with riboflavin utilization, such as thyroid disease.[3]

Recommended Riboflavin Intake
The RDAs for riboflavin are 1.3 and 1.1 mg/day for adult men and women, respectively. Because even at very high doses there are no known toxic effects of riboflavin consumption, no ULs have been established.

7-3c Niacin

Niacin (vitamin B₃) is an essential water-soluble vitamin involved in energy and vitamin C metabolism and synthesis of fatty acids and proteins. Niacin is usually obtained through the diet, but can also be synthesized in the body from tryptophan, an essential amino acid. As such, both niacin and tryptophan are considered dietary sources of niacin. The **niacin equivalent (NE)** is a unit of measure for the combined amounts of niacin and tryptophan in a given food. The NEs of selected foods are illustrated in Figure 7.3. Both niacin and tryptophan are found in a variety of foods, such as liver, poultry, fish, tomatoes, beef, and mushrooms. Whole-grain foods, enriched cereal products, and other

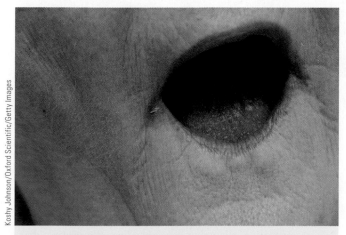

One of the signs of ariboflavinosis includes cheilosis and stomatitis.

Koshy Johnson/Oxford Scientific/Getty Images

fortified foods are also important sources of niacin, which is quite stable and not easily destroyed by preparation or exposure to light.

While the bioavailability of niacin is greater from animal- than plant-based products, treating plant-based foods with alkaline (basic) substances such as baking soda can increase niacin bioavailability. In some traditional cultures, corn and cornmeal are soaked in lime water (an alkaline solution) to increase niacin availability. This practice is historically common in Central American countries, where corn and corn tortillas have traditionally comprised a large part of many peoples' diets. As illustrated in Figure 7.4, lime is still added to some forms of cornmeal. Foods such as tamales prepared with these products are therefore good sources of niacin.

ariboflavinosis A condition caused by riboflavin deficiency whereby cheilosis, stomatitis, glossitis, muscle weakness, and confusion occur.

cheilosis A condition characterized by sores on the outside and corners of the lips.

stomatitis Inflammation of the mouth, often caused by dietary deficiencies.

glossitis Inflammation of the tongue.

niacin (vitamin B₃) An essential water-soluble vitamin involved in energy and vitamin C metabolism and the synthesis of fatty acids and proteins.

niacin equivalent (NE) A unit of measure for the combined amounts of niacin and tryptophan in food.

FIGURE 7.3 NIACIN CONTENT OF SELECTED FOODS

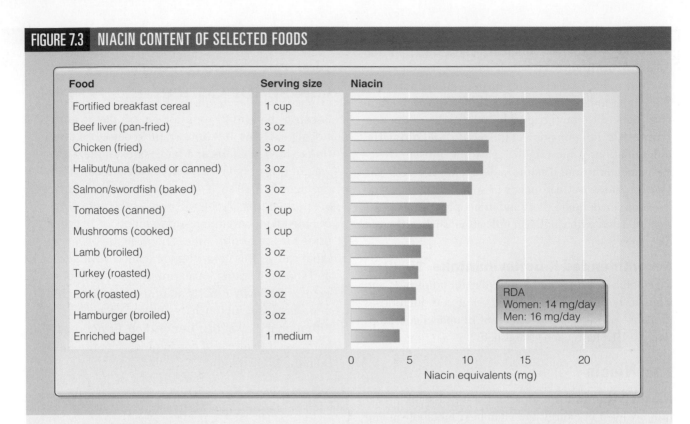

Food	Serving size	Niacin
Fortified breakfast cereal	1 cup	
Beef liver (pan-fried)	3 oz	
Chicken (fried)	3 oz	
Halibut/tuna (baked or canned)	3 oz	
Salmon/swordfish (baked)	3 oz	
Tomatoes (canned)	1 cup	
Mushrooms (cooked)	1 cup	
Lamb (broiled)	3 oz	
Turkey (roasted)	3 oz	
Pork (roasted)	3 oz	
Hamburger (broiled)	3 oz	
Enriched bagel	1 medium	

RDA
Women: 14 mg/day
Men: 16 mg/day

Niacin equivalents (mg)

Source: U.S. Department of Agriculture, Agricultural Research Service. 2004. USDA National Nutrient Database for Standard Reference, Release 16-1. Available at: http://www.ars.usda.gov/nuteintdata

Niacin Is Involved in Reduction-Oxidation Reactions

Like thiamin and riboflavin, niacin is needed to synthesize ATP because niacin's coenzyme forms are critical to redox reactions related to energy metabolism. Niacin's coenzyme forms are also important for the synthesis of several important non-energy-related compounds, such as fatty acids, cholesterol, steroid hormones, and DNA. Finally, niacin's coenzyme forms are needed for the metabolism of vitamin C and folate.

In addition to its role as a coenzyme, niacin is important for the maintenance, replication, and repair of DNA, and it may also play a role in protein synthesis, glucose homeostasis, and cholesterol metabolism. Little is known, however, about how these functions occur.

Niacin Deficiency Results in Pellagra

Niacin deficiency causes **pellagra**, a condition that results in dermatitis, dementia, diarrhea, and death (sometimes referred to as the *four Ds*). Dermatitis, which is characterized by thick, rough, darkened, and

Wondering why there is no vitamin B_4? Substances designated vitamins B_4, B_8, B_{10}, and B_{11} have been declassified as unique vitamins for various reasons, leaving gaps in the numbering system.

© Brian Mueller/Shutterstock.com

sometimes red skin, is often accompanied by neurological problems including depression, anxiety, irritability, dementia, and an inability to concentrate. Associated gastrointestinal disturbances cause loss of appetite, diarrhea, and a characteristically red and swollen tongue. If not treated, severe pellagra can cause death.

pellagra A condition caused by niacin deficiency whereby dermatitis, dementia, diarrhea, and/or death may occur.

FIGURE 7.4 INCREASING NIACIN BIOAVAILABILITY WITH LIME

Ingredients: Specially ground and dehydrated whole kernel corn and **lime**. No preservatives added.

Tamales

Corn tortillas

Atole

Ground corn is often treated with lime (from limestone, not lime juice) to increase the bioavailability of niacin found naturally in corn.

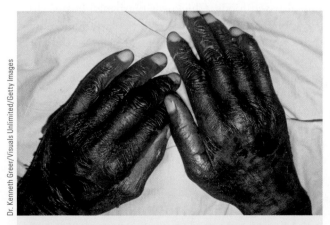

Pellagra can cause serious skin irritation.

Recommended Niacin Intake RDAs for niacin are 16 and 14 mg/day NE for adult men and women, respectively. Because large doses of supplementary niacin can have ill effects, such as increased plasma glucose and liver damage, a UL of 35 mg/day NE has been set for niacin obtained from supplements and fortified foods. Note that this value does not apply to niacin that occurs naturally in foods.

7-3d Pantothenic Acid

Pantothenic acid (vitamin B₅) is a nitrogen-containing water-soluble vitamin involved in energy metabolism, hemoglobin synthesis, and phospholipid synthesis. Because it is found in almost every plant and animal tissue, pantothenic acid is named for the Greek word *pantos*, meaning "everywhere." Pantothenic acid is found in a diverse array of foods, including mushrooms, organ meats (such as liver), and sunflower seeds (see Figure 7.5). Dairy products, turkey, fish, and coffee are also good sources. High temperatures can destroy pantothenic acid. So, cooking food at moderate temperatures is advisable to retain the pantothenic acid it contains. As with other B vitamins, pantothenic acid bioavailability changes depending on the amounts needed by the body.

Pantothenic Acid Is Needed for Carbohydrate, Protein, and Lipid Metabolism

Pantothenic acid functions as a coenzyme in a variety of metabolic reactions—especially those involved in ATP production. The metabolism of glucose, amino acids, and fatty acids depends on pantothenic acid's coenzyme form.

Beyond its coenzyme functions, this vitamin is critical for the synthesis of many important compounds, such as heme, cholesterol, bile, phospholipids, fatty acids, and some of the reproductive hormones.

pantothenic acid (vitamin B₅) A nitrogen-containing water-soluble vitamin involved in energy metabolism, hemoglobin synthesis, and phospholipid synthesis.

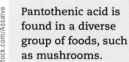

Pantothenic acid is found in a diverse group of foods, such as mushrooms.

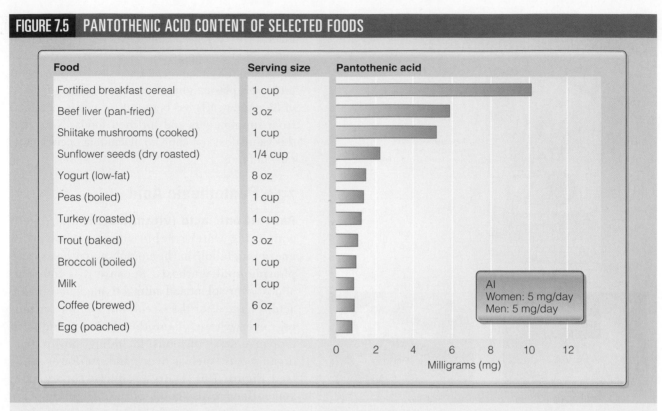

FIGURE 7.5 PANTOTHENIC ACID CONTENT OF SELECTED FOODS

Food	Serving size	Pantothenic acid
Fortified breakfast cereal	1 cup	
Beef liver (pan-fried)	3 oz	
Shiitake mushrooms (cooked)	1 cup	
Sunflower seeds (dry roasted)	1/4 cup	
Yogurt (low-fat)	8 oz	
Peas (boiled)	1 cup	
Turkey (roasted)	1 cup	
Trout (baked)	3 oz	
Broccoli (boiled)	1 cup	
Milk	1 cup	
Coffee (brewed)	6 oz	
Egg (poached)	1	

AI
Women: 5 mg/day
Men: 5 mg/day

Milligrams (mg)

Source: U.S. Department of Agriculture, Agricultural Research Service. 2004. USDA National Nutrient Database for Standard Reference, Release 16-1. Available at: http://www.ars.usda.gov/nuteintdata

Pantothenic Acid Deficiency May Cause Burning Feet Syndrome

Because pantothenic acid is found in almost every food, deficiencies are rare. Nonetheless, some researchers believe that a condition called **burning feet syndrome** may be caused by severe pantothenic acid deficiency. As its name implies, burning feet syndrome is characterized by a tingling sensation in the feet and legs, as well as fatigue, weakness, and nausea. Note that these same sensations can be caused by factors other than pantothenic acid deficiency.

Recommended Pantothenic Acid Intake

There is insufficient information to establish RDAs for pantothenic acid, but an Adequate Intake (AI) level of 5 mg/day has been set for adults. Overconsumption of pantothenic acid is not fatal, but very high intakes have been associated with nausea and diarrhea. Because there is no evidence of pantothenic toxicity, a UL has not been set for this essential vitamin.

7-3e Vitamin B₆

Vitamin B₆ is a water-soluble vitamin involved in the metabolism of glucose, proteins, and amino acids; synthesis of neurotransmitters and hemoglobin; and regulation of steroid hormone function. Like the other B vitamins, vitamin B₆ is found in a variety of plant- and animal-based foods. Chickpeas (garbanzo beans), fish, liver, potatoes, and chicken are particularly good sources of this vitamin, as are fortified breakfast cereals and bakery products (see Figure 7.6). Unlike other B vitamins, vitamin B₆ does not have a widely used common name. The vitamin exists in seven known forms, the most common of which are *pyridoxine, pyridoxal, pyridoxamine,* and *pyridoxal phosphate*. Vitamin B₆ is somewhat unstable and is destroyed by extreme or prolonged heat and cold. For instance, as much as one-half of the vitamin B₆ found in fresh produce can be lost when it is cooked.

Vitamin B₆ Is Critical to Nonessential Amino Acid Synthesis

Like several other B vitamins, vitamin B₆ functions as a coenzyme—its pyridoxal phosphate

burning feet syndrome
A condition believed to be caused by pantothenic acid deficiency whereby tingling in the feet and legs, fatigue, weakness, and nausea may occur.

vitamin B₆ A water-soluble vitamin involved in the metabolism of proteins and amino acids, the synthesis of neurotransmitters and hemoglobin, glycogenolysis, and regulation of steroid hormone function.

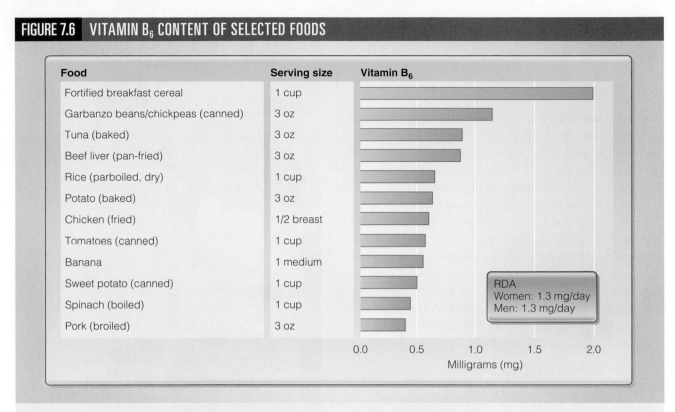

FIGURE 7.6 VITAMIN B₆ CONTENT OF SELECTED FOODS

Food	Serving size	Vitamin B₆
Fortified breakfast cereal	1 cup	
Garbanzo beans/chickpeas (canned)	3 oz	
Tuna (baked)	3 oz	
Beef liver (pan-fried)	3 oz	
Rice (parboiled, dry)	1 cup	
Potato (baked)	3 oz	
Chicken (fried)	1/2 breast	
Tomatoes (canned)	1 cup	
Banana	1 medium	
Sweet potato (canned)	1 cup	
Spinach (boiled)	1 cup	
Pork (broiled)	3 oz	

RDA
Women: 1.3 mg/day
Men: 1.3 mg/day

Milligrams (mg)

Source: U.S. Department of Agriculture, Agricultural Research Service. 2004. USDA National Nutrient Database for Standard Reference, Release 16-1. Available at: http://www.ars.usda.gov/nuteintdata

form is involved in numerous chemical reactions related to the synthesis of nonessential amino acids from essential amino acids. Vitamin B₆ is also essential for the production of some nonprotein substances, such as heme and the neurotransmitters serotonin and dopamine.

Vitamin B₆ Deficiency and Toxicity Vitamin B₆ deficiency impedes heme production, in turn lowering the concentrations of hemoglobin in red blood cells. This condition, called **microcytic, hypochromic anemia**, results in red blood cells that are small (*microcytic*) and light in color (*hypochromic*). Microcytic, hypochromic anemia can diminish the red blood cells' ability to deliver oxygen to tissues, which in turn impairs ATP production. Vitamin B₆ deficiency also causes cheilosis, glossitis, stomatitis, and fatigue. Because these signs and symptoms are very similar to those of riboflavin deficiency, these two deficiencies are often difficult to distinguish without blood analysis.

Vitamin B₆ deficiency is rare, but, as mentioned previously, cooking foods at high temperatures can destroy some of the vitamin B₆ they contain. This was not well understood until the 1950s, when vitamin B₆ added to infant formula was often destroyed during heat processing. As a result, many formula-fed infants developed

vitamin B₆ deficiency, which caused serious complications such as seizures and convulsions. Formulation practices have since changed so that infant formulas now contain sufficient amounts of this essential nutrient.

Relatively large amounts of pyridoxal phosphate are retained within muscle and liver cells. As such, vitamin B₆ toxicity occurs more frequently than does that of other water-soluble vitamins. Vitamin B₆ toxicity can cause severe neurological problems, such as difficulty walking and numbness in the feet and hands. Vitamin B₆ toxicity is not likely to result from the overconsumption of vitamin B₆–rich foods, but may result from excessive intake of supplements.

Recommended Vitamin B₆ Intake Depending on age and sex, RDAs for vitamin B₆ vary from 1.3 to 1.7 mg/day for adults. To prevent the neurological problems associated with excess vitamin B₆, a UL of 100 mg/day has been established for this essential vitamin. Supplements containing 500 milligrams of vitamin B₆ are widely available, making it relatively easy to exceed the UL value. If you take a vitamin B₆

microcytic, hypochromic anemia A condition characterized by low concentrations of hemoglobin, which causes red blood cells to be small and light in color.

supplement (or B-complex vitamins), be sure to carefully read supplement packaging and labels and make sure that your intake does not exceed the UL value.

7-3f Biotin

Biotin (vitamin B₇) is yet another water-soluble vitamin involved in energy metabolism. However, biotin is somewhat unique among the water-soluble vitamins because it is obtained from both the diet and biotin-producing bacteria in the large intestine. As illustrated in Figure 7.7, sources of dietary biotin include peanuts, tree nuts (such as almonds and cashews), mushrooms, eggs, and tomatoes. Biotin is often bound to proteins in food, and it is released during digestion. Sometimes, biotin is bound too tightly to food proteins, making it difficult to absorb. In fact, biotin was first discovered because of its role in *egg white injury*, a condition whereby biotin's bioavailability is reduced severely because it is consumed with *avidin*, a protein present in large quantities in raw egg whites.

biotin (vitamin B₇) A water-soluble vitamin involved in energy metabolism that is obtained both from the diet and biotin-producing bacteria in the large intestine.

Avidin, a protein found in raw egg whites, binds tightly to biotin and decreases its absorption.

© Kjetil Kolbjornsrud/Shutterstock.com

FIGURE 7.7 BIOTIN CONTENT OF SELECTED FOODS

Food	Serving size	Biotin
Peanuts (dry roasted)	1/4 cup	
Almonds (blanched)	3 oz	
Mushrooms (cooked)	1 cup	
Egg yolk (cooked)	1 medium	
Tomatoes (fresh)	1 cup	
Avocado	1 medium	
Sweet potato (baked)	1 medium	
Cashews (dry roasted)	1/4 cup	
Carrots (fresh)	1 medium	
Salmon (baked)	3 oz	
Banana	1 medium	
Cod (poached)	3 oz	

AI
Women: 30 µg/day
Men: 30 µg/day

0   10   20   30   40   50
Micrograms (µg)

Source: Hands E. *Food Finder Vitamin and Mineral Source Guide*, 3rd ed. Salem, OR: ESHA Research.

Do not let the threat of egg white injury dissuade you from eating cooked eggs, however—heat denatures avidin, allowing biotin to be absorbed normally. Alcohol can also decrease biotin absorption, and high cooking temperatures can destroy biotin in foods.

Biotin Adds Bicarbonate (HCO₃⁻) Subunits

Biotin is a critical component of several coenzymes involved in energy metabolism pathways. Specifically, biotin-dependent reactions shift bicarbonate subunits (HCO_3^-) from one molecule to another. Beyond its role as a coenzyme, biotin serves several functions related to gene expression, cell growth, and development.

Biotin Deficiency Can Cause Depression

Though biotin deficiency is uncommon, it is sometimes caused by genetic disorders and other conditions that impair intestinal absorption, such as inflammatory bowel disease). Smoking may also increase risk of biotin deficiency. Signs and symptoms of biotin deficiency are poorly understood, but they can include depression, hallucinations, skin irritations, infections, hair loss, poor muscle control, seizures, and developmental delays.

Recommended Biotin Intake

Because information regarding biotin intake is still insufficient, RDAs have not yet been set for this important micronutrient. However, an AI of 30 µg/day has been set for adults. Because excessive biotin intake does not carry any known detrimental effects, a UL has not been established.

7-3g Folate

Folate (vitamin B₉) refers to a group of related water-soluble vitamins involved in single-carbon transfers, amino acid metabolism, and DNA synthesis. Good sources of folate include organ meats, legumes (such as lentils and pinto beans), okra, and many green leafy vegetables such

Folate (from an Italian word meaning "foliage") is found in many plant-based foods.

© Africa Studio/Shutterstock.com

as spinach (see Figure 7.8). A common form of folate called *folic acid* is rarely found in foods, but is often included in vitamin supplements and is added during the fortification of food. Because cereal grain enrichment has necessitated folic acid fortification since 1998, enriched foods are always very good sources of this vitamin. Folate is destroyed by excessive heat, and cooked foods often contain less folate than do their uncooked counterparts.

The bioavailability of folate depends on its form in food, genetic factors, and the use of certain medications. In general, the body absorbs folate found in supplements and fortified foods (folic acid) better than it absorbs naturally occurring folate. Because folate absorption is so variable, the amount of folate present in a food is expressed in terms of the **dietary folate equivalent (DFE)**, an approximation of the quantity actually absorbed by the body.

Folate Facilitates Single-Carbon Transfers, Amino Acid Metabolism, and Formation of the Neural Tube

Once folate is absorbed by enterocytes, it combines with four hydrogen atoms, converting to its active form, **tetrahydrofolate (THF)**. In its active form, folate facilitates *single-carbon transfers*, which are needed to synthesize many important organic substances such as amino acids. In a single-carbon transfer, a carbon atom in the form of a methyl group ($-CH_3$) is pulled from a molecule and bound to THF, producing 5-methyltetrahydrofolate (5-methyl THF). 5-methyl THF migrates to another molecule and allows it to bind to the methyl group, effectively transferring a single carbon from one molecule to another.

Folate-requiring single-carbon transfers play a large role in amino acid metabolism. For example, to convert the amino acid homocysteine to the amino acid methionine, a methyl group must be transferred to homocysteine. This transfer results in the production of both THF (because 5-methyl THF loses its methyl group) and the essential amino acid methionine (see Figure 7.9). Because the production of methionine from homocysteine requires both folate and vitamin B₁₂, which you will learn about shortly, a deficiency in either can result in a buildup of homocysteine in the body. High levels of homocysteine are associated with an increased risk of heart disease and may have other serious health consequences.[5]

folate (vitamin B₉) A group of related water-soluble vitamins involved in single-carbon transfers, amino acid metabolism, and DNA synthesis.

dietary folate equivalent (DFE) A unit of measure for the approximate amount of folate in a food that is absorbed by the body.

tetrahydrofolate (THF) The active form of folate.

FIGURE 7.8   FOLATE CONTENT OF SELECTED FOODS

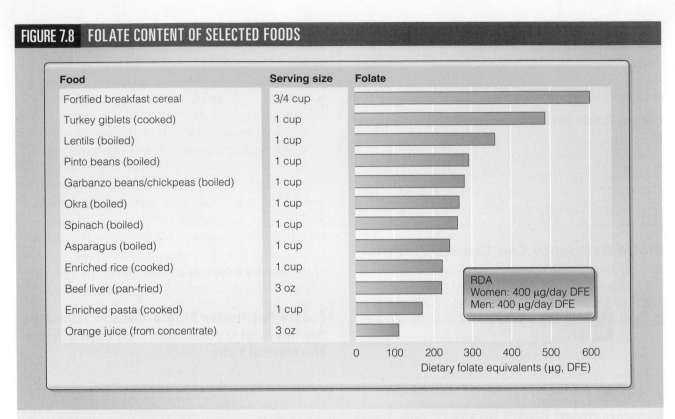

Food	Serving size	Folate
Fortified breakfast cereal	3/4 cup	
Turkey giblets (cooked)	1 cup	
Lentils (boiled)	1 cup	
Pinto beans (boiled)	1 cup	
Garbanzo beans/chickpeas (boiled)	1 cup	
Okra (boiled)	1 cup	
Spinach (boiled)	1 cup	
Asparagus (boiled)	1 cup	
Enriched rice (cooked)	1 cup	
Beef liver (pan-fried)	3 oz	
Enriched pasta (cooked)	1 cup	
Orange juice (from concentrate)	3 oz	

RDA
Women: 400 µg/day DFE
Men: 400 µg/day DFE

Dietary folate equivalents (µg, DFE)

Source: U.S. Department of Agriculture, Agricultural Research Service. 2004. USDA National Nutrient Database for Standard Reference, Release 16-1. Available at: http://www.ars.usda.gov/nuteintdata

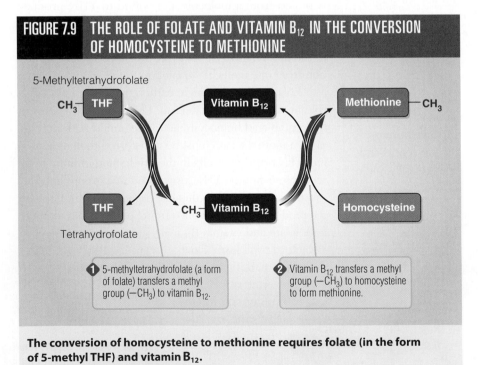

1. 5-methyltetrahydrofolate (a form of folate) transfers a methyl group (—CH₃) to vitamin B₁₂.
2. Vitamin B₁₂ transfers a methyl group (—CH₃) to homocysteine to form methionine.

The conversion of homocysteine to methionine requires folate (in the form of 5-methyl THF) and vitamin B₁₂.

THF is critical to the formation of DNA and RNA, which in turn are essential for the growth, maintenance, and repair of every tissue in the body. The production of DNA and RNA is especially important during periods of rapid growth (such as embryonic and fetal growth) and in cells with very short life spans (such as those that make up the lining of the gastrointestinal tract).

Folate is clearly an important vitamin, but it is particularly important during pregnancy. Folate is essential for the proper formation of the neural tube, which later develops into the spinal cord and brain. Increased maternal intake of folate (both in the form of foods and supplements) has been shown to decrease the risk of a **neural tube defect** in some newborns.[6] A neural tube defect is a malformation whereby neural tissues do not form properly during very early

neural tube defect A malformation whereby neural tissue does not form properly during fetal development.

prenatal development. As a result, the neural tube may not close properly, resulting in exposed tissue in the spinal cord, brain, or both. When this happens, the spinal cord is not protected adequately, and in severe cases, the brain does not form properly. The most common form of neural tube defect is **spina bifida** (Latin for "split spine"), a failure of the neural tube to close properly (see Figure 7.10).[7] Because of the importance of folate to neural tube development, foods labeled as being "enriched" must be fortified with folate. In addition, many other foods are fortified to contain additional folate. The U.S. Centers for Disease Control and Prevention (CDC) report that this public-health approach to increasing folate intake has significantly reduced the number of neural tube defect cases in the United States.[8]

Folate Deficiency Causes Megaloblastic, Macrocytic Anemia

Folate deficiency was once relatively common in the United States, but because enriched cereal products are now fortified with folate, this is no longer the case.[9] Today, folate deficiency occurs most often in alcoholics, people with intestinal diseases, people taking certain medications, and the elderly. Severe folate deficiency causes a condition called **megaloblastic, macrocytic anemia**. Because folate is integral to DNA synthesis and cell maturation, dietary folate insufficiency can cause cells (including red blood cells) to remain large (*macrocytic*) and immature (*megaloblastic*). Folate deficiency is only one cause of this form of anemia, however. Because several nutrient deficiencies can result in megaloblastic, macrocytic anemia, it is sometimes difficult to determine whether suboptimal folate levels are the cause of this potentially dangerous condition.

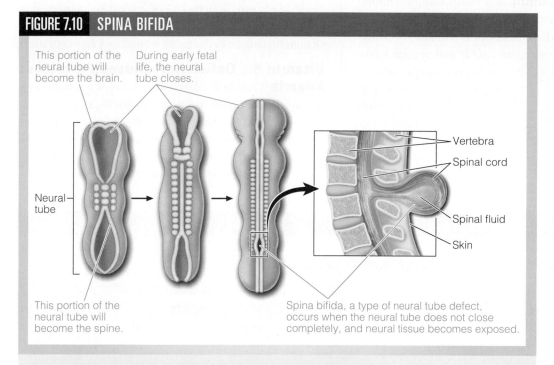

FIGURE 7.10   SPINA BIFIDA

This portion of the neural tube will become the brain.

During early fetal life, the neural tube closes.

Neural tube

This portion of the neural tube will become the spine.

Vertebra

Spinal cord

Spinal fluid

Skin

Spina bifida, a type of neural tube defect, occurs when the neural tube does not close completely, and neural tissue becomes exposed.

Spina bifida is a neural tube defect that occurs early in pregnancy.

spina bifida A failure of the neural tube to close properly during the first months of fetal life.

megaloblastic, macrocytic anemia A condition caused by folate deficiency whereby cells (including red blood cells) remain large and immature.

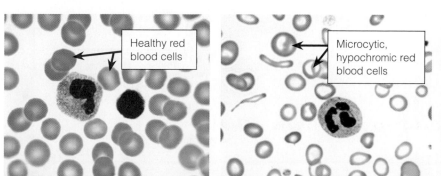

Healthy red blood cells

Microcytic, hypochromic red blood cells

Macrocytic, megaloblastic red blood cells

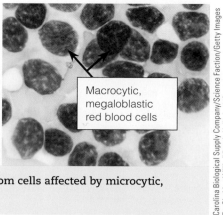

Under a microscope, healthy red blood cells (left) can be distinguished visually from cells affected by microcytic, hypochromic anemia (middle) and macrocytic, megaloblastic anemia (right).

Recommended Folate Intake Adults should consume 400 μg DFE every day. This RDA increases to 600 μg/day for pregnant women. Women capable of or planning to become pregnant are encouraged to consume 400 μg/day DFE in the form of supplements, fortified foods, or both, in addition to a naturally folate-rich diet. Because excess folate intake may make it difficult to detect a vitamin B_{12} deficiency, a UL of 1,000 μg/day DFE has been set for folate derived from fortified foods and/or supplements. There is no evidence that a high intake of naturally occurring folate poses any risk of toxicity.

7-3h Vitamin B_{12}

Vitamin B_{12} (cobalamin) is a water-soluble vitamin involved in energy metabolism and production of methionine (an amino acid). Cobalamin is so-named because it contains the trace element *cobalt* and several nitrogen (or *amine*) groups. Vitamin B_{12} is a unique vitamin because it can *only* be synthesized by microorganisms such as bacteria and fungi. That is, plants and higher animals (like humans) do not synthesize vitamin B_{12}

vitamin B_{12} (cobalamin) A water-soluble vitamin involved in energy metabolism and methionine production.

themselves—they must obtain it from microorganisms living in either their environments or colonized in gastrointestinal tracts. When you eat these plants and animals, you absorb the vitamin B_{12} that they themselves have produced. Good dietary sources of vitamin B_{12} include shellfish, meat, fish, and dairy products; many ready-to-eat breakfast cereals are also fortified with vitamin B_{12}. These and other sources are listed in Figure 7.11.

Vitamin B_{12} Is Involved in ATP and Methionine Production Like several other B vitamins, vitamin B_{12} functions as a coenzyme. It is essential for ATP production and aids folate in amino acid metabolism. Recall that during the conversion of homocysteine to methionine, 5-methyl THF loses a methyl group and is converted to THF (see Figure 7.9). Without adequate vitamin B_{12}, homocysteine blood levels rise, 5-methyl THF cannot be converted to THF, and folate deficiency symptoms appear. In this way, vitamin B_{12} deficiency can cause secondary folate deficiency.

Vitamin B_{12} Deficiency Causes Pernicious Anemia Vitamin B_{12} deficiency is caused by either inadequate dietary intake (primary vitamin B_{12} deficiency) or poor absorption (secondary vitamin B_{12} deficiency). Primary vitamin B_{12} deficiency occurs most often in vegans and infants breastfed by vitamin B_{12}–deficient mothers. Secondary vitamin B_{12} deficiency sometimes occurs when stomach

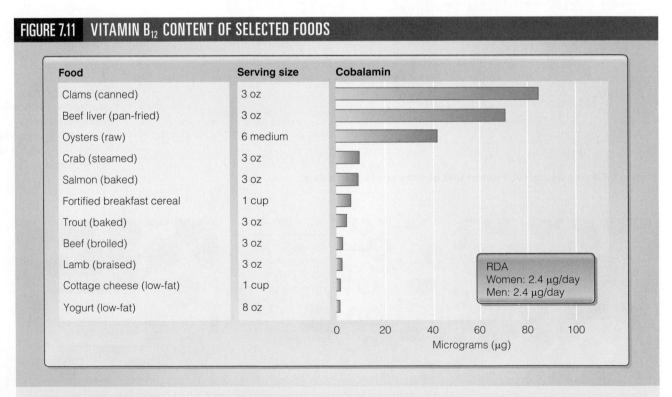

FIGURE 7.11 VITAMIN B_{12} CONTENT OF SELECTED FOODS

Food	Serving size	Cobalamin
Clams (canned)	3 oz	
Beef liver (pan-fried)	3 oz	
Oysters (raw)	6 medium	
Crab (steamed)	3 oz	
Salmon (baked)	3 oz	
Fortified breakfast cereal	1 cup	
Trout (baked)	3 oz	
Beef (broiled)	3 oz	
Lamb (braised)	3 oz	
Cottage cheese (low-fat)	1 cup	
Yogurt (low-fat)	8 oz	

RDA
Women: 2.4 μg/day
Men: 2.4 μg/day

Micrograms (μg)

Source: U.S. Department of Agriculture, Agricultural Research Service. 2004. USDA National Nutrient Database for Standard Reference, Release 16-1. Available at: http://www.ars.usda.gov/nuteintdata

cells fail to produce sufficient amounts of hydrochloric acid or **intrinsic factor**, a protein needed for vitamin B_{12} absorption. The production of either of these substances may slow or cease as a person ages, resulting in poor vitamin B_{12} absorption. Another cause of secondary vitamin B_{12} deficiency is **pernicious anemia**, an autoimmune disease whereby the immune system produces antibodies that destroy the intrinsic factor–producing cells of the stomach lining.[10] People with pernicious anemia or who do not otherwise produce adequate amounts of intrinsic factor cannot simply treat their vitamin B_{12} deficiencies with oral supplements—they must receive vitamin B_{12} by injection.

Although megaloblastic, macrocytic anemia is often associated with both primary and secondary vitamin B_{12} deficiency, this condition is actually caused by an underlying folate deficiency. As you learned, vitamin B_{12} deficiency can cause secondary THF deficiency, even when a seemingly adequate amount of folate is consumed. Because large doses of folate alleviate some of the symptoms of vitamin B_{12} deficiency (such as anemia), such doses can mask a vitamin B_{12} deficiency. However, while anemia may clear up with folate supplementation, other less evident complications associated with vitamin B_{12} deficiency, such as neurologic damage, may not. For this reason, misdiagnosis of a vitamin B_{12} deficiency as a folate deficiency can be dangerous.

Recommended Vitamin B_{12} Intake The RDA for vitamin B_{12} is 2.4 μg/day for adults. Vegans and vegetarians who do not eat any animal-based products should consider taking a supplement or eating foods that have been fortified with vitamin B_{12}. No ULs have been established for this vitamin.

7-4 VITAMIN C

Vitamin C (ascorbic acid) is a water-soluble vitamin that functions as an antioxidant in the body. As illustrated in Figure 7.12 vitamin C is abundant in many fruits and vegetables, such as citrus fruits, peppers, papayas, broccoli, strawberries, and peas. Potatoes are also quite rich in vitamin C. The bioavailability of vitamin C is generally high, although its structure can be destroyed by heat and exposure to oxygen; freshly peeled and/or prepared fruits and vegetables tend to provide more vitamin C than cooked, processed, and/or stored ones.

intrinsic factor A protein produced by the stomach that is needed for vitamin B_{12} absorption.

pernicious anemia An autoimmune disease caused by vitamin B_{12} deficiency, whereby antibodies destroy the stomach cells that produce intrinsic factor.

vitamin C (ascorbic acid) A water-soluble vitamin that serves as an antioxidant within the body.

FIGURE 7.12   VITAMIN C CONTENT OF SELECTED FOODS

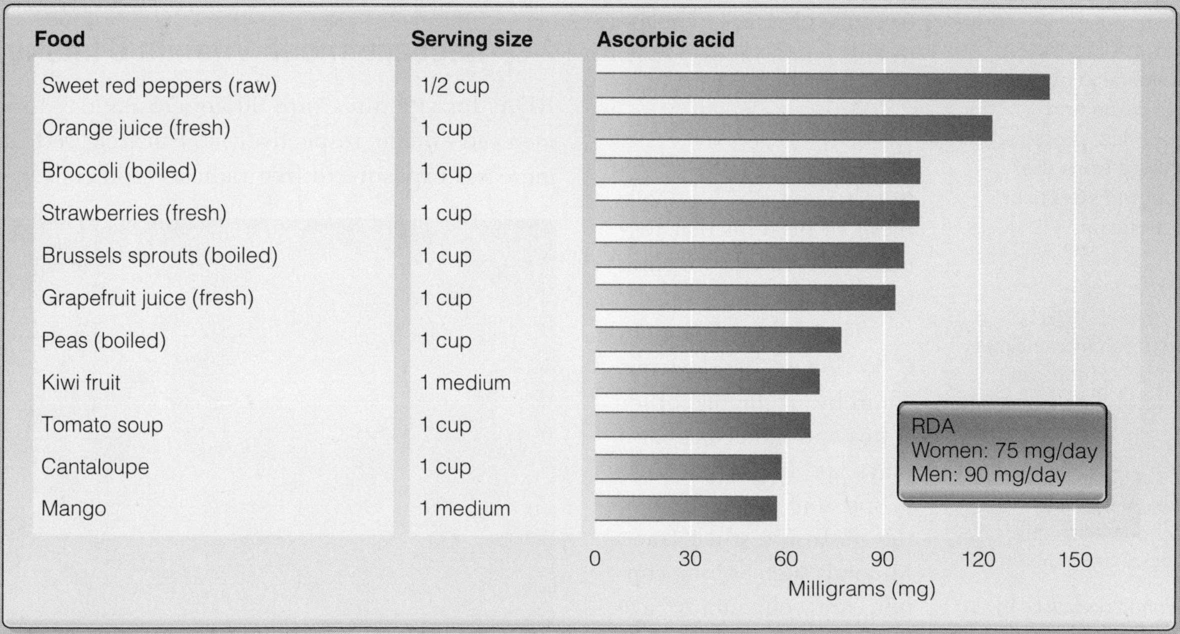

Food	Serving size	Ascorbic acid
Sweet red peppers (raw)	1/2 cup	
Orange juice (fresh)	1 cup	
Broccoli (boiled)	1 cup	
Strawberries (fresh)	1 cup	
Brussels sprouts (boiled)	1 cup	
Grapefruit juice (fresh)	1 cup	
Peas (boiled)	1 cup	
Kiwi fruit	1 medium	
Tomato soup	1 cup	
Cantaloupe	1 cup	
Mango	1 medium	

RDA
Women: 75 mg/day
Men: 90 mg/day

Milligrams (mg): 0 30 60 90 120 150

Source: U.S. Department of Agriculture, Agricultural Research Service. 2004. USDA National Nutrient Database for Standard Reference, Release 16-1. Available at: http://www.ars.usda.gov/nuteintdata

7-4a Vitamin C Is a Potent Antioxidant and May Benefit the Immune System

Unlike the B vitamins, vitamin C is not a coenzyme. Instead, it acts as an **antioxidant**, a compound that donates electrons or hydrogen ions to other substances, inhibiting oxidation. Because vitamin C (like other antioxidants) can easily accept and donate electrons, it is involved in a wide variety of redox reactions. Vitamin C's antioxidant function also protects cells from **free radicals**, molecules that have one or more unpaired electrons and are highly reactive and unstable. Although free radicals are produced in the body during normal cellular metabolism, exposure to intense sunlight, some drugs, and toxic substances, such as smog, cigarette smoke, and ozone, can also generate free radicals. Antioxidants are important because free radicals try to stabilize themselves by taking electrons from other molecules. Because antioxidants have electrons to give, they can effectively neutralize free radicals, in turn protecting other molecules from oxidative damage. If free radicals are not stabilized by antioxidants, they will continue to cause oxidative damage to cell membranes, DNA, and proteins, which can lead to a variety of diseases. For example, damage done to DNA can affect the synthesis of proteins that regulate cell growth, which in turn might lead to cancer.

The ability of vitamin C to donate its electrons (thereby, reducing other atoms and molecules) also plays an important role in nutrient absorption in the gastrointestinal tract. Minerals such as iron, copper, and chromium are better absorbed in their reduced states. Consequently, the consumption of vitamin C can increase the bioavailability of these essential minerals. For example, drinking orange juice with iron-fortified cereal can increase the amount of iron that is absorbed.

Some research suggests that increasing consumption of fruits and vegetables rich in vitamin C may decrease the risk of certain diseases, including the common cold, certain forms of cancer, heart disease, and cataracts.[11] Controlled clinical intervention studies have not consistently demonstrated a protective effect of vitamin C.[12] Growing evidence, however, shows that large doses of vitamin C can benefit the immune system.

7-4b Vitamin C Deficiency Causes Scurvy

Vitamin C deficiency causes **scurvy**, a deadly condition characterized by bleeding gums, skin irritations, bruising, and poor wound healing. In the eighteenth century, British medical doctor James Lind conducted what was perhaps the first controlled nutrition-related experiment in an attempt to determine the cause of scurvy. Lind treated 12 sailors suffering from scurvy with lemons, limes, cider, nutmeg, seawater, or vinegar. He found that consumption of lemons or limes—but not the other treatments—prevented and cured scurvy. Although scurvy was once common, the increased availability of fruits and vegetables has made it relatively rare. However, scurvy cases have been reported in malnourished alcoholics, homeless individuals, and older adults on very restricted diets.[13]

7-4c Recommended Vitamin C Intake

RDAs for vitamin C are 90 and 75 mg/day for adult men and women, respectively. Because cigarette smoke increases exposure to free radicals (and consequently

Vitamin C, found in oranges and many other fruits and vegetables, protects the body from the damaging effects of free radicals.

© Mama_mia/Shutterstock.com

antioxidant A compound that donates electrons or hydrogen ions to other substances, inhibiting oxidation.

free radicals A highly reactive molecule with one or more unpaired electrons; destructive to cell membranes, DNA, and proteins.

scurvy A condition caused by vitamin C deficiency, whereby bleeding gums, bruising, poor wound healing, and skin irritations occur.

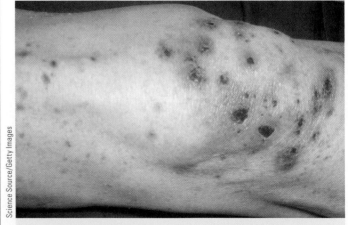

Science Source/Getty Images

One of the signs of advanced scurvy is the presence of red spots on the skin.

the body's need for antioxidants), smokers are advised to increase their vitamin C intake by an additional 35 mg/day. Whether this recommendation also applies to people exposed to secondhand smoke is not clear. To avoid possible gastrointestinal distress, a UL of 2,000 mg/day has been set for supplemental vitamin C.

7-5 FAT-SOLUBLE VITAMINS

Although fat-soluble vitamins share some common characteristics, like water-soluble vitamins, each one is chemically unique. Fat-soluble vitamins are absorbed mostly in the small intestine, a process that requires the presence of both dietary lipids and bile. After they are absorbed, fat-soluble vitamins are circulated away from the small intestine by chylomicra in the lymph. Fat-soluble vitamins eventually circulate to the blood as components of either lipoproteins (such as VLDLs) or transport proteins.

Like many of the water-soluble vitamins, each fat-soluble vitamin has several forms, and some forms are more biologically active than others. Unlike water-soluble vitamins, however, your body can readily store fat-soluble vitamins. Consequently, consuming high doses of some fat-soluble vitamins (especially in supplement form) can result in toxicities—sometimes with serious consequences.

7-6 VITAMIN A AND THE CAROTENOIDS

The term *vitamin A* refers to a series of three compounds referred to as **retinoids (or preformed vitamin A)**: retinol, retinoic acid, and retinal. Although all three forms are important, retinol is the most biologically active and is synthesized in the body from retinal (see Figure 7.13). Retinoic acid can also be synthesized from retinal, but retinoic acid itself cannot be converted to any other retinoid.

Beyond the retinoids, the vitamin A family also includes several **carotenoid**

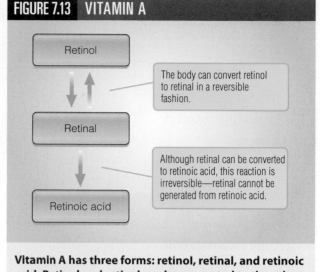

FIGURE 7.13 VITAMIN A

Retinol

The body can convert retinol to retinal in a reversible fashion.

Retinal

Although retinal can be converted to retinoic acid, this reaction is irreversible—retinal cannot be generated from retinoic acid.

Retinoic acid

Vitamin A has three forms: retinol, retinal, and retinoic acid. Retinol and retinal can be converted to the other forms, but retinoic acid cannot.

compounds, which have structures similar to those of the retinoids. Some carotenoids can be converted to vitamin A; one such carotenoid is called a **provitamin A carotenoid**. One of the most common provitamin A carotenoids is **beta-carotene (ß-carotene)**, which the body converts to two retinal molecules. A carotenoid that cannot be converted to vitamin A is called a **nonprovitamin A carotenoid**. Lycopene, astaxanthin, zeaxanthin, and lutein are examples of nonprovitamin A carotenoids and often referred to as phytochemicals.

Good dietary sources of vitamin A or provitamin A carotenoids are listed in Figure 7.14. Because it exists in several forms (each of which has a unique biological identity), the amount of vitamin A in a given food can be

Peppers are excellent sources of provitamin A carotenoids.

© Ruslan Anatolevich Kuzmenkov/Shutterstock.com

retinoids (or preformed vitamin A) A vitamin A compound.

carotenoid A dietary compound with a similar structure to those of the retinoids. Some, but not all, can be converted to vitamin A.

provitamin A carotenoid A carotenoid that can be converted to vitamin A.

beta-carotene (ß-carotene) A provitamin A carotenoid made from two molecules of retinal.

nonprovitamin A carotenoid A carotenoid that cannot be converted to vitamin A.

FIGURE 7.14 VITAMIN A CONTENT OF SELECTED FOODS

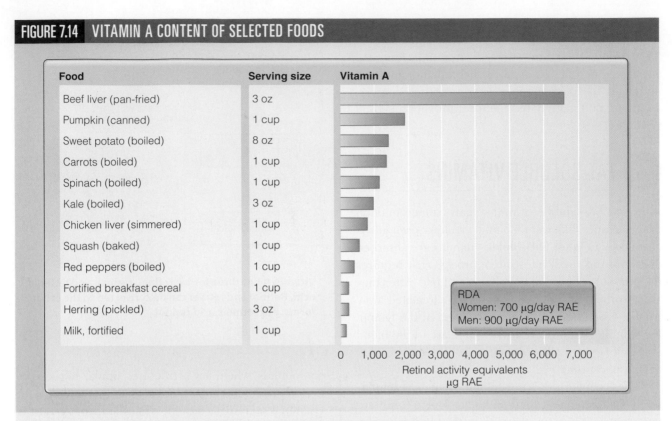

Food	Serving size	Vitamin A
Beef liver (pan-fried)	3 oz	
Pumpkin (canned)	1 cup	
Sweet potato (boiled)	8 oz	
Carrots (boiled)	1 cup	
Spinach (boiled)	1 cup	
Kale (boiled)	3 oz	
Chicken liver (simmered)	1 cup	
Squash (baked)	1 cup	
Red peppers (boiled)	1 cup	
Fortified breakfast cereal	1 cup	
Herring (pickled)	3 oz	
Milk, fortified	1 cup	

RDA
Women: 700 µg/day RAE
Men: 900 µg/day RAE

0   1,000 2,000 3,000 4,000 5,000 6,000 7,000
Retinol activity equivalents
µg RAE

Source: U.S. Department of Agriculture, Agricultural Research Service. 2004. USDA National Nutrient Database for Standard Reference, Release 16-1. Available at: http://www.ars.usda.gov/nuteintdata

difficult to determine. The **retinol activity equivalent (RAE)** is a unit of measure for the combined amounts of preformed vitamin A and provitamin A carotenoids in a food. In general, you consume preformed vitamin A when you eat animal-based foods, such as liver and other organ meats, and fatty fish. Whole-fat dairy products such as whole milk, cheese, and butter are also good sources of vitamin A (especially when they are fortified). Reduced-fat dairy products are not good sources unless they are vitamin A-fortified.

Whereas animal-based foods tend to contain preformed vitamin A, plant-based foods tend to contain provitamin A carotenoids. The yellow and red hues of many carotenoids make carotenoid-rich plant fibers and animal tissues brightly colored. As such, yellow, orange, and red fruits and vegetables, such as cantaloupe, carrots, and peppers, are particularly good sources of the carotenoids, as are brightly colored animal-based foods such as egg yolks, lobsters, crabs, and shrimp. Leafy greens are also good plant-based sources of carotenoids.

retinol activity equivalent (RAE) A unit of measure for the combined amounts of preformed vitamin A and provitamin A carotenoids in a food.

Vitamin A compounds appear to be relatively stable in food. Processing and heating may actually increase the bioavailability of some carotenoids such as β-carotene.

7-6a Vitamin A Is Critical to Vision, Growth, and Reproduction

Anecdotal evidence suggests that ancient Egyptian physicians prescribed vitamin A–rich liver to treat poor vision. It was not until much later, however, that scientists began to understand the mechanisms by which vitamin A–rich foods improved eyesight. As illustrated in Figure 7.15, when light enters your eyes, it passes to an inner back lining called the *retina*. The retina consists of a layer of nerve tissue and millions of cells called *cones* and *rods*. Cones enable you to see color, whereas rods distinguish black from white (a critical aspect of night vision). Rods contain thousands of *rhodopsin* molecules, each of which is composed of *cis*-retinal (a form of vitamin A) and *opsin* (a protein). When light strikes the rhodopsin, the *cis*-retinal is converted to *trans*-retinal and separates from the opsin. This reaction causes a neural signal to be sent to the brain. The light signal is then interpreted by the brain as a recognizable image.

FIGURE 7.15 VITAMIN A AND VISION

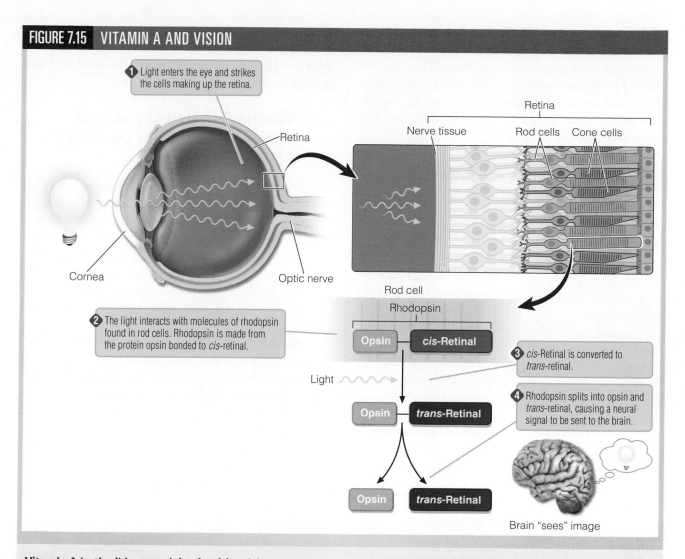

1 Light enters the eye and strikes the cells making up the retina.

Retina

Nerve tissue | Rod cells | Cone cells

Retina

Cornea

Optic nerve

2 The light interacts with molecules of rhodopsin found in rod cells. Rhodopsin is made from the protein opsin bonded to *cis*-retinal.

Rod cell

Rhodopsin

Opsin — *cis*-Retinal

3 *cis*-Retinal is converted to *trans*-retinal.

Light

Opsin — *trans*-Retinal

4 Rhodopsin splits into opsin and *trans*-retinal, causing a neural signal to be sent to the brain.

Opsin — *trans*-Retinal

Brain "sees" image

Vitamin A (retinal) is essential to healthy vision, especially in darkness when the function of rod cells is most important.

Nighttime driving can be dangerous for people with vitamin A deficiency.

Because *cis*-retinal is a form of vitamin A, adequate consumption of vitamin A is needed for this cascade of events to occur. Vitamin A is also important for the health of the eye's outermost tissue layer, the *cornea*, and especially important to vision in low-light environments. This is why adequate vitamin A intake is needed to prevent **night blindness**, a condition whereby vision is impaired in dim light.

Vitamin A has many important functions beyond vision. For example, vitamin A plays an important role in **cell differentiation**, the process by which a nonspecialized, immature cell type

night blindness A condition characterized by an impaired ability to see in low-light environments.

cell differentiation The process by which a nonspecialized, immature cell type becomes a specialized, mature cell type.

becomes a specialized, mature cell type.[14] Vitamin A is also critical for growth and reproduction, and it has a variety of immunological functions. These include the maintenance of protective barriers (such as skin and those that line organs) and the production of immune cells (such as those that produce antibodies). Preformed vitamin A is also essential for healthy bones.

7-6b The Carotenoids Are Important Antioxidants

Adequate intake of the nonprovitamin A carotenoids is associated with reduced risks of heart disease, age-related eye disease, and certain types of cancer (in animal models).[15] Researchers believe that these effects may be attributable to the potent antioxidant functions of some carotenoids.[16] For instance, high circulating levels of lutein and zeaxanthin may decrease one's risk of **macular degeneration**— a chronic and often age-related disease that causes deterioration of the retina.[17]

macular degeneration A chronic disease that causes deterioration of the retina.

vitamin A deficiency disorder (VADD) A spectrum of health-related consequences caused by vitamin A deficiency.

xerophthalmia A condition caused by vitamin A deficiency, whereby the cornea and other portions of the eye are damaged, leading to dry eyes, scarring, and even blindness.

This irreversible, progressive disease is the leading cause of visual impairment in older adults. It is not clear, however, whether increased nonprovitamin A carotenoid intake actually improves vision in older people.

7-6c Vitamin A Deficiency Causes Vitamin A Deficiency Disorder (VADD)

Primary vitamin A deficiency is uncommon in industrialized countries such as the United States. Nonetheless, secondary vitamin A deficiency sometimes occurs in people with diseases that impair lipid digestion and absorption, such as cystic fibrosis. Vitamin A deficiency is also prevalent in malnourished alcoholics with liver damage. Excessive alcohol consumption depletes the body's stores of vitamin A, but the mechanisms of this process are not well understood.

Vitamin A deficiency disorder (VADD) is a spectrum of health-related consequences caused by vitamin A deficiency. VADD is pervasive in developing countries, especially among children. In its mildest form, VADD causes night blindness. Severe forms of VADD can lead to **xerophthalmia**, a condition whereby the cornea and other portions of the eye are damaged, leading to dry eyes, scarring, and even blindness. This complex condition is often accompanied by *Bitot's spots*, white

Vitamin A and International Child Health

Although vitamin A deficiency disorder (VADD) is not common in developed countries, it is still a serious health concern around the world—especially for children. Because vitamin A intake is essential to a healthy immune system, children with VADD are at an increased risk of infection. Protein-energy malnutrition, prevalent in vitamin A–deficient parts of the world, can make vitamin A deficiency even worse. In response to the prevalence of VADD in developing countries, researchers and public agencies have focused their efforts to combat this dangerous condition. For example, the World Health Organization urges health care professionals in the most at-risk countries to provide vitamin A supplements to all children six months of age and older. *Biotechnology* (or *bioengineering*) can also be used to develop new kinds of rice that are naturally enriched with provitamin A carotenoids. With the aid of these and other promising health interventions, fewer children will suffer from vitamin A deficiency in years to come.

International efforts to decrease vitamin A deficiency have improved the health of children worldwide.

Frans Lemmens/Corbis

FIGURE 7.16 VITAMIN A DEFICIENCY

Areas of the world where blindness due to vitamin A deficiency is most common

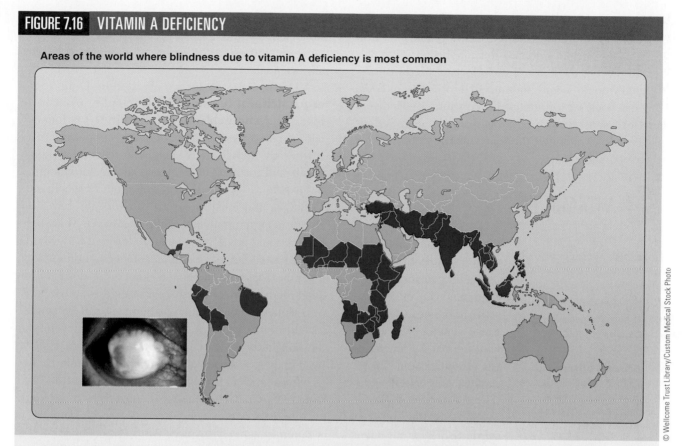

Vitamin A deficiency can cause xerophthalmia, a major cause of blindness around the globe.

spots on the surface of the eye caused by accumulations of dead cells and secretions (see Figure 7.16). Vitamin A deficiency also causes **hyperkeratosis**, a condition whereby skin and nail cells overproduce the protein keratin, causing them to become rough and scaly.

7-6d Vitamin A and Carotenoid Toxicities

Chronic consumption of large doses of preformed vitamin A (three to four times the RDA) can lead to vitamin A toxicity, which can cause **hypervitaminosis A**, a condition characterized by blurred vision, liver abnormalities, and reduced bone strength. Very high doses of naturally occurring and/or synthetic vitamin A can also lead to birth defects such as neurological damage and physical deformities. Isotretinoin (trade name Accutane®), a drug used to treat acne, is a form of vitamin A that can cause severe, life-threatening birth defects. Not only must pregnant women avoid this drug, but sexually active women of child-bearing age must agree to use birth control before it can be prescribed. You can find out more about the dangers of isotretinoin at the National Center for Biotechnology Information's PubMed Health site (www.ncbi.nlm.nih.gov/pubmedhealth/PMHT0010802).

High-dose carotenoid supplementation has been found to increase the risk of lung cancer in some people.[18] Overconsumption of carotenoids can also cause them to accumulate in the skin, turning it a yellow-orange color— a benign condition called **hypercarotenodermia**.

7-6e Recommended Vitamin A and Carotenoid Intake

RDAs for vitamin A intake are 900 and 700 μg RAE/day for adult men and women, respectively. Vitamin A can be consumed either as preformed vitamin A or as provitamin A carotenoids. Because preformed vitamin A is found mostly in animal-based products, vitamin A supplementation may be necessary for vegans and vegetarians who do not consume sufficient amounts of provitamin A

hyperkeratosis A condition caused by vitamin A deficiency, whereby skin and nail cells overproduce the protein keratin, causing them to become rough and scaly.

hypervitaminosis A A condition caused by vitamin A toxicity, whereby blurred vision, liver abnormalities, and reduced bone strength occur.

hypercarotenodermia A condition whereby carotenoids accumulate in the skin, causing it to become yellow-orange.

carotenoids or vitamin A–fortified foods. DRIs have not been set for the nonprovitamin A carotenoids because there is insufficient evidence to support their essentiality. To prevent the known toxic effects of excessive preformed vitamin A consumption, a UL of 3,000 μg RAE/day has been set for adults. The Institute of Medicine advises against dietary carotenoid supplementation for most people, although there are no UL values set for these compounds.

7-7 VITAMIN D

prohormone A compound converted to an active hormone in the body.

ergocalciferol (vitamin D₂) A form of vitamin D found in plant-based foods, fortified foods, and supplements.

cholecalciferol (vitamin D₃) A form of vitamin D that is found in animal-based foods, fortified foods, and supplements and also synthesized in the body.

Vitamin D is considered by many to be both a nutrient and a **prohormone**, a compound converted to an active hormone in the body. But is vitamin D essential, nonessential, or conditionally essential? Some health professionals consider vitamin D a nonessential nutrient because the body can synthesize it in sufficient amounts under normal conditions. However, because some people are not able to produce vitamin D in adequate amounts, some experts consider it to be conditionally essential.

There are two forms of vitamin D in foods: **ergocalciferol (vitamin D₂)** is found in plant-based foods, and **cholecalciferol (vitamin D₃)** is found in animal-based foods. Cholecalciferol is also synthesized in the body. Both forms of vitamin D are found in supplements and fortified foods. Good sources of vitamin D are listed in Figure 7.17. Egg yolks, butter, whole milk, fatty fish, and mushrooms are some of the few foods that contain naturally occurring vitamin D. Many dairy products and breakfast cereals are fortified with vitamin D, and most dietary vitamin D comes from these foods. Vitamin D is quite stable and is not easily destroyed by food preparation, processing, or storage.

Unlike other vitamins, vitamin D can be synthesized through exposure to sunlight; exposing your skin to sunlight for 10–15 minutes three times a week allows your body to make adequate amounts of vitamin D. This is why vitamin D is sometimes referred to as the "sunshine vitamin." This chemical conversion is not quite as simple as it sounds, however. The body's synthesis of vitamin D actually involves two steps, as illustrated in

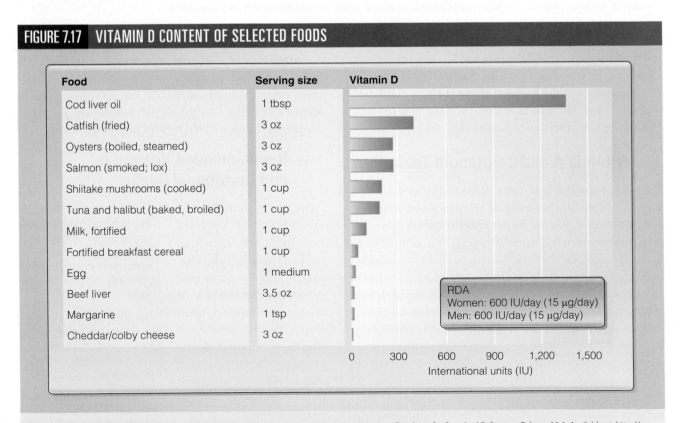

FIGURE 7.17 VITAMIN D CONTENT OF SELECTED FOODS

Food	Serving size	Vitamin D
Cod liver oil	1 tbsp	
Catfish (fried)	3 oz	
Oysters (boiled, steamed)	3 oz	
Salmon (smoked; lox)	3 oz	
Shiitake mushrooms (cooked)	1 cup	
Tuna and halibut (baked, broiled)	1 cup	
Milk, fortified	1 cup	
Fortified breakfast cereal	1 cup	
Egg	1 medium	
Beef liver	3.5 oz	
Margarine	1 tsp	
Cheddar/colby cheese	3 oz	

RDA
Women: 600 IU/day (15 μg/day)
Men: 600 IU/day (15 μg/day)

International units (IU): 0, 300, 600, 900, 1,200, 1,500

Source: U.S. Department of Agriculture, Agricultural Research Service. 2004. USDA National Nutrient Database for Standard Reference, Release 16-1. Available at: http://www.ars.usda.gov/nuteintdata

Because the body can synthesize vitamin D when exposed to sufficient sunlight, this vitamin is sometimes called the *sunshine vitamin*.

Figure 7.18. First, a cholesterol derivative is converted by ultraviolet light to **previtamin D₃** (or **precalciferol**) in the skin. Tanning machines also emit the ultraviolet light that allows the first step to take place, but relying on tanning sessions to obtain enough vitamin D₃ is not recommended. This is because tanning can damage the skin and increase the risk of skin cancer.[19] After the cholesterol derivative is converted to previtamin D₃, this substance is converted in the skin to **vitamin D₃** (or **cholecalciferol**), which then diffuses into the blood and circulates to the liver.

Many environmental, genetic, and lifestyle factors influence the amount of vitamin D₃ produced by the body. In order to synthesize adequate amounts of vitamin D₃, people living in regions with persistent smog, overcast skies, or limited amounts of sunlight likely require more sunlight exposure than do people who live in regions with warm climates and clear skies. People with darker skin may need up to three times more sunlight exposure than do people with lighter skin to produce enough vitamin D₃. For all people, vitamin D₃ production decreases with age. Vitamin D deficiency is also more common in obese individuals, as compared to those who are lean. This may be because, once vitamin D is synthesized, it can be taken up by adipose tissue, making it less available to other tissues of the body.[20] Finally, sunscreen can block the ultraviolet rays needed for vitamin D₃ formation.

Whether consumed through the diet or produced in the skin, vitamin D₃ must be further metabolized before it can be used by the body. This two-step process, illustrated in Figure 7.19, occurs in the liver and kidneys. First, vitamin D₃ is converted to *25-hydroxyvitamin D₃ (25-[OH] D₃)* in the liver. Next, the 25-(OH) D₃ circulates in the blood to the kidneys, where it is converted to *1,25-dihydroxyvitamin D₃ (1,25-[OH]₂ D₃)*, the active form of vitamin D known more commonly as **calcitriol**.

7-7a Vitamin D Is Critical to Blood Calcium Regulation, Bone Health, and Many Other Functions

One of vitamin D's many important functions is the regulation of blood calcium levels. Low blood calcium stimulates the release of **parathyroid hormone (PTH)** from the parathyroid gland. PTH stimulates the conversion of 25-(OH) D₃ to calcitriol in the kidneys. Together, calcitriol and

---

FIGURE 7.18 | **SYNTHESIS OF CHOLECALCIFEROL (VITAMIN D₃) IN THE SKIN**

Cholesterol metabolite

1 Step 1
Exposure of skin to ultraviolet light converts a cholesterol metabolite to previtamin D₃.

Ultraviolet light

Previtamin D₃
(precalciferol)

2 Step 2
Previtamin D₃ is then converted in the skin to vitamin D₃ (cholecalciferol).

Vitamin D₃
(cholecalciferol)

The body can make vitamin D₃ from a cholesterol-like substance when the skin is exposed to ultraviolet light.

---

previtamin D₃(or **precalciferol**) An intermediate product made in the skin during the conversion of a cholesterol derivative to cholecalciferol.

vitamin D₃ (or **cholecalciferol**) The form of vitamin D that is made in the skin and diffuses into the blood and circulates to the liver.

calcitriol The active form of vitamin D produced in the kidneys. Also known as 1,25-dihydroxyvitamin D₃ (1,25-[OH]₂ D₃).

parathyroid hormone (PTH) A hormone released from the parathyroid gland that stimulates the conversion of 25-(OH) D₃ to calcitriol in the kidneys.

FIGURE 7.19 ACTIVATION OF VITAMIN D

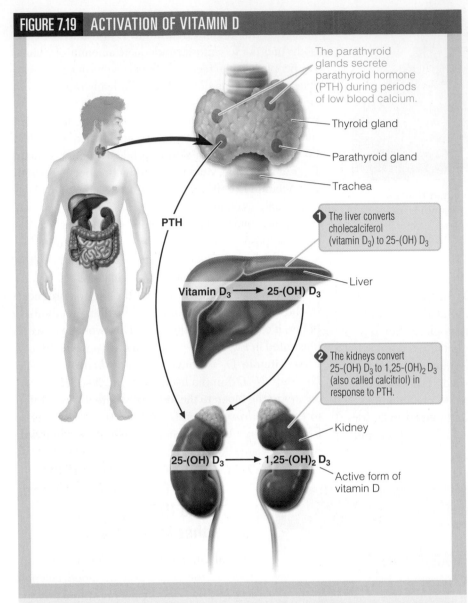

The parathyroid glands secrete parathyroid hormone (PTH) during periods of low blood calcium.

— Thyroid gland

— Parathyroid gland

— Trachea

PTH

1 The liver converts cholecalciferol (vitamin D_3) to 25-(OH) D_3

— Liver

Vitamin D_3 ⟶ 25-(OH) D_3

2 The kidneys convert 25-(OH) D_3 to 1,25-(OH)$_2$ D_3 (also called calcitriol) in response to PTH.

— Kidney

25-(OH) D_3 ⟶ 1,25-(OH)$_2$ D_3

Active form of vitamin D

The conversion of cholecalciferol to calcitriol takes place in the liver and kidneys. This conversion is stimulated by parathyroid hormone (PTH) during periods of low blood calcium.

PTH increase calcium absorption in the small intestine, decrease calcium excretion in the urine, and facilitate the release of calcium from bones (see Figure 7.20).

In addition to the regulation of blood calcium levels and bone health, vitamin D plays important roles in blood pressure regulation, muscle contraction, and nerve function. Vitamin D is also needed for cell differentiation and maturation. Like vitamin A, vitamin D activates selected genes, which in turn prompt cells to synthesize specific proteins that allow the cells to become both specialized

rickets A childhood condition caused by vitamin D deficiency, whereby slow growth and bone deformation occur.

and functional. Although the connection warrants further investigation, some studies have suggested that vitamin D's role in cell regulation may help prevent certain types of cancers, such as those of the colon, breast, skin, and prostate.[21]

7-7b Vitamin D Deficiency Causes Rickets, Osteomalacia, and Osteoporosis

In infants and children, vitamin D deficiency can result in improper bone mineralization—a disease called **rickets**. Rickets is a significant public health concern in some parts of the world.[22] Children with rickets experience slow growth and have characteristically bowed legs or knocked knees that develop because their long leg bones cannot support the stress of weight-bearing activities such as walking. Rickets can also cause the breastbone to protrude outward and the rib cage to become narrow, often resulting in cardiac and respiratory problems.

Rickets was common in the United States during the early 1900s. Vitamin D–fortified milk and infant formulas nearly eradicated rickets in this country—until recently, that is. In 2000, a medical research group published a disturbing report documenting a rise in the incidence of rickets, especially among dark-skinned, breastfed babies.[23] Further research suggested that some infants nourished entirely by human milk may be at increased risk of inadequate vitamin D intake. This phenomenon may be attributable to their mothers' insufficient exposure to sunlight, which results in diminished amounts of vitamin D in their milk. Infants' own insufficient exposure to sunlight may also contribute to the recent rise in rickets, especially among those with darker skin. To counter the potential development of rickets, the American Academy of Pediatrics recommends that all

FIGURE 7.20 REGULATION OF CALCIUM

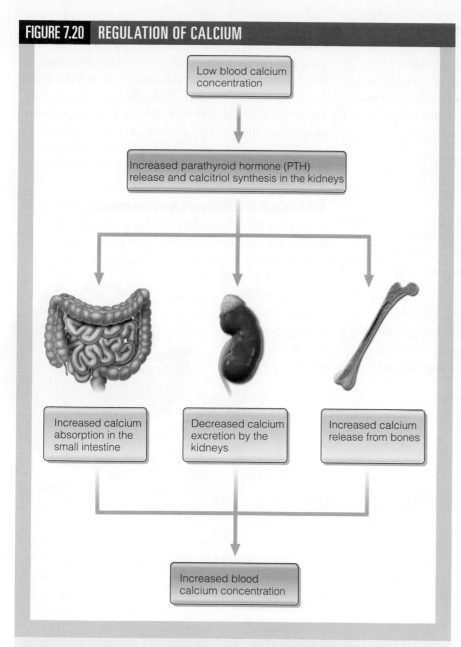

Parathyroid hormone (PTH) and calcitriol work together to increase blood calcium when it is low.

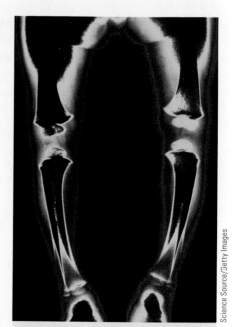

Childhood calcium deficiency can cause rickets.

chronic disease—especially in the elderly—that affects more than 28 million Americans. To help prevent both osteomalacia and osteoporosis, adults over 50 years of age are advised to get at least 15 minutes of sun exposure each day, consume adequate amounts of vitamin D–rich foods, and, in some cases, take vitamin D supplements.

7-7c Recommended Vitamin D Intake

The Institute of Medicine has established an RDA of 600 IU/day for adults.[24] This is approximately the amount of vitamin D in one liter (4.7 cups) of vitamin D–fortified milk. To help diminish bone loss in the elderly, the RDA increases to 800 IU/day at age 71.

Vitamin D toxicity is uncommon unless the diet is supplemented with large amounts of vitamin D. Excess vitamin D can cause calcium levels in the blood and urine to increase, which in turn can cause calcium to be deposited in soft tissues

breastfed infants be supplemented with vitamin D until they can consume adequate amounts of vitamin D–rich foods such as vitamin D–fortified cow milk or infant formula.[24]

When vitamin D intake is inadequate in adults, bones can become soft and weak—a condition called **osteomalacia**. Symptoms of osteomalacia include diffuse bone pain and muscle weakness. People with osteomalacia are also at an increased risk of bone fractures. Inadequate vitamin D intake can also result in the demineralization of previously healthy bone, ultimately leading to **osteoporosis**. Osteoporosis is a serious

osteomalacia An adult condition, caused by vitamin D deficiency, whereby bones become soft and weak.

osteoporosis A serious disease, caused by vitamin D deficiency, whereby bones become weak and porous.

such as the heart and lungs. **Hypercalciuria**, or high levels of calcium in the urine, is also associated with kidney stone formation and renal (kidney) failure. Vitamin D toxicity promotes bone loss and when severe enough, can be fatal. Because of these health concerns, the Institute of Medicine has established a UL of 4,000 IU/day for vitamin D in supplemental form.

7-8 VITAMINS E AND K

Beyond vitamins A and D, adequate intakes of vitamins E and K are needed to meet your fat-soluble vitamin requirements. The functions and importance of these essential micronutrients are discussed next.

7-8a Vitamin E

The term *vitamin E* refers collectively to eight different compounds that have similar chemical structures. Of these, **α-tocopherol** is the most biologically active.

Vitamin E is found in both plant- and animal-based foods, but it is especially abundant in vegetable oils, nuts, and seeds. Some dark green vegetables such as broccoli and spinach contain vitamin E as well. Good dietary sources of vitamin E are listed in Figure 7.21. Vitamin E is easily destroyed during food preparation, processing, and storage. And vitamin E is particularly sensitive to degradation by heating. Proper storage in airtight containers is also important for preserving a food's vitamin E content.

Vitamin E Is a Potent Antioxidant As described in Chapter 6, biological membranes, such as cell membranes, are composed of a bilayer of phospholipid molecules. Naturally, maintenance of these membranes is vital to the stability and function of cells and their contents. Vitamin E, an antioxidant, plays a major role in this maintenance by

© Nata-Lia/Shutterstock.com

hypercalciuria A condition characterized by elevated urine calcium levels.

α-tocopherol The most biologically active form of vitamin E.

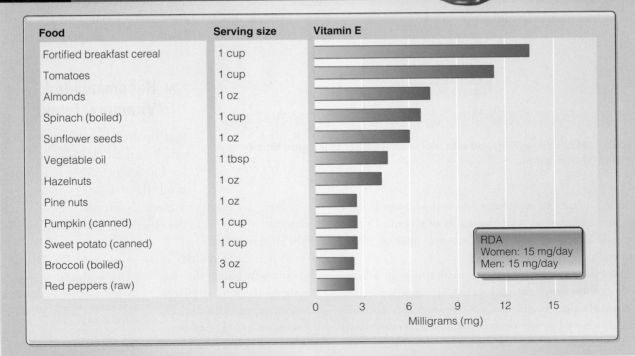

Nuts and seeds are good sources of vitamin E.

FIGURE 7.21 VITAMIN E CONTENT OF SELECTED FOODS

Food	Serving size	Vitamin E
Fortified breakfast cereal	1 cup	
Tomatoes	1 cup	
Almonds	1 oz	
Spinach (boiled)	1 cup	
Sunflower seeds	1 oz	
Vegetable oil	1 tbsp	
Hazelnuts	1 oz	
Pine nuts	1 oz	
Pumpkin (canned)	1 cup	
Sweet potato (canned)	1 cup	
Broccoli (boiled)	3 oz	
Red peppers (raw)	1 cup	

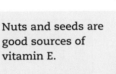

RDA
Women: 15 mg/day
Men: 15 mg/day

Milligrams (mg): 0 3 6 9 12 15

Source: U.S. Department of Agriculture, Agricultural Research Service. 2004. USDA National Nutrient Database for Standard Reference, Release 16-1. Available at: http://www.ars.usda.gov/nuteintdata

protecting the fatty acids embedded in biological membranes from free radical–induced oxidative damage. This protection is especially important for cells that are exposed to high levels of oxygen, such as those in the lungs and red blood cells. The ability of vitamin E to act as an antioxidant is enhanced by the presence of other antioxidant micronutrients, such as vitamin C and the mineral selenium.

Because antioxidants protect DNA from cancer-causing free radical damage, some researchers believe that vitamin E might prevent or cure certain types of cancer. Indeed, diets high in vitamin E are associated with decreased cancer risk, but there is little experimental evidence that vitamin E itself is responsible.[26] In other words, eating foods rich in vitamin E is likely better for you than taking supplementary vitamin E.

Vitamin K was discovered by Danish researcher Dr. Henrik Dam. Dam named the vitamin in honor of its importance in coagulation (*koagulation* in Danish). Dam received a Nobel Prize in Physiology or Medicine in 1943 for the discovery.

Vitamin E Deficiency Causes Neuromuscular Problems Vitamin E deficiency is uncommon; cases have only been observed in infants fed formulas lacking sufficient amounts of vitamin E, people with genetic abnormalities, and people with diseases that cause fat malabsorption. Smoking may also increase the risk of vitamin E deficiency. Nonetheless, vitamin E deficiency is characterized by a variety of symptoms, such as neuromuscular problems, loss of coordination, and muscular pain. Vitamin E deficiency can also damage the retina of the eye.

Recommended Vitamin E Intake The RDA for vitamin E is 15 mg/day—an amount easily obtained from a balanced diet. Unlike the other fat-soluble vitamins, vitamin E rarely causes toxicity, even when large amounts of vitamin E supplements are consumed. This may be because the supplemental form of vitamin E is less biologically active than naturally occurring vitamin E. In some people, however, very high doses of vitamin E supplements can cause dangerous bleeding or hemorrhaging. As such, it is best to be cautious when taking vitamin E supplements. A UL of 1,000 mg/day has been established for vitamin E, regardless of its source.

Broccoli is a good dietary source of vitamin K.

7-8b Vitamin K

The term *vitamin K* refers to three compounds that have similar structures and functions. Vitamin K is found naturally in plant-based foods in a form called **phylloquinone**. This form is also found in some vitamin K supplements. Another form of vitamin K called **menaquinone** is produced by bacteria in the large intestine. Because this bacterial production does not produce sufficient amounts of vitamin K to sustain health, vitamin K is considered an essential nutrient. A third form of vitamin K called **menadione** is neither found naturally in food nor synthesized by intestinal bacteria, but is produced commercially.

In general, dark green vegetables such as kale, spinach, broccoli, and Brussels sprouts are good sources of dietary vitamin K. You can also obtain this vitamin from fish and legumes. Good dietary sources of vitamin K are listed in Figure 7.22. Like many of the vitamins, vitamin K is destroyed by exposure to excessive light and/or heat.

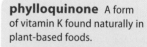

phylloquinone A form of vitamin K found naturally in plant-based foods.

menaquinone A form of vitamin K produced by bacteria in the large intestine.

menadione A form of vitamin K produced commercially.

FIGURE 7.22 **VITAMIN K CONTENT OF SELECTED FOODS**

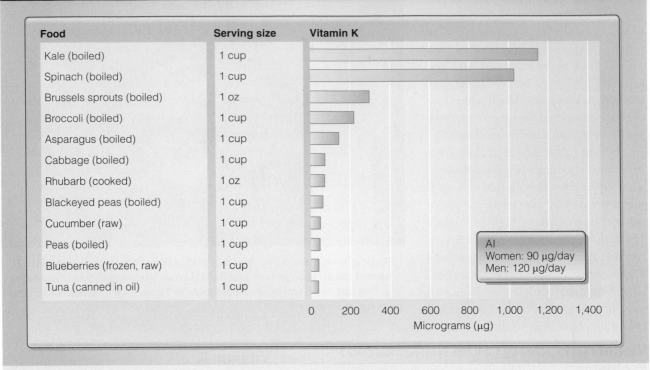

Food	Serving size	Vitamin K
Kale (boiled)	1 cup	
Spinach (boiled)	1 cup	
Brussels sprouts (boiled)	1 oz	
Broccoli (boiled)	1 cup	
Asparagus (boiled)	1 cup	
Cabbage (boiled)	1 cup	
Rhubarb (cooked)	1 oz	
Blackeyed peas (boiled)	1 cup	
Cucumber (raw)	1 cup	
Peas (boiled)	1 cup	
Blueberries (frozen, raw)	1 cup	
Tuna (canned in oil)	1 cup	

AI
Women: 90 µg/day
Men: 120 µg/day

Microgram (µg): 0 200 400 600 800 1,000 1,200 1,400

Source: U.S. Department of Agriculture, Agricultural Research Service. 2004. USDA National Nutrient Database for Standard Reference, Release 16-1. Available at: http://www.ars.usda.gov/nuteintdata

Vitamin K Is Critical to Coagulation Vitamin K functions as a coenzyme in a variety of reactions that ultimately constitute the life-or-death process by which blood clots form. Without this process, called **coagulation**, you might bleed to death after even a minor scrape. For your blood to coagulate and form a clot, a cascade of chemical reactions must first take place. After a cut or scrape occurs, various clotting factors are activated by vitamin K, allowing the next reactions in the cascade to take place. These reactions ultimately result in the production of *fibrin*, a protein that forms a web-like clot that stops the bleeding. Beyond its coagulation functions, vitamin K is also essential for the synthesis of proteins needed for bone and tooth formation.

Vitamin K Deficiency Causes Severe Bleeding Although rare in healthy adults, vitamin K deficiency occurs in some infants. Because a newborn's large intestine completely lacks vitamin K–producing bacteria at birth and human milk often contains very low levels of this essential nutrient, babies receive only minimal amounts of vitamin K during their first few weeks of life. Although this does not present a problem for most infants, some develop severe vitamin K deficiency. This leads to a condition called **vitamin K deficiency bleeding**, originally called *hemorrhagic disease of the newborn*, which is characterized by uncontrollable internal bleeding. To counter the possibility of this condition, the American Academy of Pediatrics recommends that all newborns be given vitamin K injections.[27] Vitamin K deficiency also sometimes occurs in children and adults with diseases that impair lipid absorption, which again causes bleeding. Finally, prolonged

coagulation The process by which blood clots are formed.

vitamin K deficiency bleeding A disease caused by vitamin K deficiency, whereby uncontrollable internal bleeding occurs.

Vitamin K is essential to coagulation, which helps heal your body and protect it from infections.

© iStock.com/Smithore

Vitamin K and Blood-Thinning Drugs

Individuals with cardiovascular disease are often prescribed blood-thinning medications such as warfarin (trade name Coumadin®). Many of these drugs work by interfering with the actions of vitamin K. Because they decrease the formation of blood clots, these drugs can cause serious health problems if taken improperly. People taking blood-thinning drugs are at increased risk of excessive bleeding (even after sustaining a minor injury). Consequently, they should take care to avoid engaging in hazardous activities. Also, because a sudden increase or decrease in vitamin K could interfere with the effectiveness of these medications, those taking such medications should strive to keep their vitamin K intakes constant. It is important to understand drug–nutrient interactions within your body, especially if you are taking potentially life-saving medications.

use of antibiotics can kill bacteria residing in the large intestine, resulting in vitamin K deficiency in all age groups.

Recommended Vitamin K Intake Although RDAs have not been established for vitamin K, AIs of 120 and 90 mg/day have been set for men and women, respectively. Because even very high amounts of vitamin K are rarely toxic, a UL has not been established for this vitamin.

 7-9 TAKING DIETARY SUPPLEMENTS

Now that you know about the essential vitamins, you should be able to choose foods that contain adequate amounts of each of these important dietary components. In fact, the 2015 Dietary Guidelines for Americans recommend that we get all of our essential nutrients from foods, not supplements. Still, you may be wondering about vitamin supplements and the conditions under which they might be taken. Understanding some basic concepts about supplementation can help you make wise choices about what to take—and what not to take.

7-9a Dietary Supplements Can Contain Many Substances

The U.S. Food and Drug Administration (FDA) defines a **dietary supplement** as a product intended to supplement the diet that contains vitamins, minerals, amino acids, herbs or other plant-derived substances, or a multitude of other compounds.[28] Not all dietary supplements contain vitamins. In fact, some do not even contain nutrients.

As with most considerations related to your health, it is important to make informed decisions about which supplements to take and which to avoid. It can be difficult to determine whether a claim regarding vitamin supplementation and health is accurate. Before you take any supplement, consult a reliable scientific source such as the Office of Dietary Supplements (part of the National Institutes of Health). The Office of Dietary Supplements maintains an excellent user-friendly website (http://ods.od.nih.gov) that provides information about most dietary supplements, as well as a wealth of news, studies, and recommendations. For example, before buying and/or using any supplement, contemplate the following tips developed by the Office of Dietary Supplements.

- **Consider safety first.** Some supplement ingredients can be toxic—especially in high doses. Do not hesitate to check with a health professional before taking any dietary supplement.

- **Think twice about following the latest headline.** Sound health advice is generally based on research over time, not a single study touted by the media. Be wary of results claiming a quick fix. If something sounds too good to be true, it probably is.

- **More may not be better.** Some products can be harmful when consumed in high amounts, and/or for a long time. Do not assume that more is better—it might be toxic.

- *Natural* **does not always mean** *safe.* The term *natural* simply means that something is not synthetic or human-made. Do not assume that this term ensures wholesomeness or safety.

dietary supplement Product intended to supplement the diet that contains vitamins, minerals, amino acids, herbs or other plant-derived substances, or a multitude of other compounds.

7-9b When to Consider Taking a Supplement

There are no "hard and fast" rules about when to take a dietary supplement. However, if you have difficulty consuming a variety and balance of healthy foods in adequate amounts, a dietary supplement may help you achieve the essential nutrients' recommended intakes. The following situations often call for dietary supplementation.

- When your food availability or variety is limited by time or cooking constraints (as sometimes happens in college).

- When you decide not to consume certain foods (such as red meat or dairy products).

- During periods of rapid growth and development (such as infancy, childhood, adolescence, and pregnancy).

- When the consequences of normal aging (such as the loss of calcium from bone) make it difficult to consume adequate amounts of a nutrient from food.

- When you consume a low-calorie diet for weight loss.

- If you suffer from a health condition that increases your nutrient requirements or decreases the bio-availability of nutrients you eat.

After deciding to take a dietary supplement, you must then determine which type of supplement to take. This can sometimes be a long and difficult process. Again, one of the best resources in this regard is the Office of Dietary Supplements' website, which contains dozens of *Supplement Fact* sheets. Carefully reading these sheets and consulting with your health care provider should provide valuable information to help you determine which type of supplement best fits your needs.

© Triff/Shutterstock.com

NUTR
ONLINE

REVIEW FLASHCARDS
ANYTIME, ANYWHERE!

**Create Flashcards
from Your StudyBits**

**Review Key Term
Flashcards Already
Loaded on the
StudyBoard**

4LTR
PRESS

Access NUTR ONLINE at www.cengagebrain.com

8 | Water and the Minerals

LEARNING OUTCOMES

8-1 Appreciate the importance of water.

8-2 Differentiate and understand the functions of major minerals in the body.

8-3 Describe the functions of calcium in the body.

8-4 Understand the role of sodium, chloride, and potassium: the major electrolytes in the body.

8-5 Describe the functions of phosphorus and magnesium; the other electrolytes in the body.

8-6 Explain the role of sulfur in human nutrition.

8-7 Differentiate among and understand the functions of trace minerals in the body.

8-8 Differentiate and understand the other trace minerals that can affect health.

©Galyna Andrushko/Shutterstock.com

After finishing this chapter go to **PAGE 219** for **STUDY TOOLS.**

8-1 WATER

"Water, water, everywhere, Nor any drop to drink." This famous passage from Samuel Taylor Coleridge's *The Rime of the Ancient Mariner* relates the lamentation of an experienced seafarer tormented by thirst during a long ocean voyage. Although he is surrounded by water, not a drop is fit for the mariner to consume. Like the ancient mariner, your very existence depends on a plentiful supply of drinkable water. But have you ever wondered why the human body needs water to stay alive? Because water is the most abundant molecule in the human body, a constant supply is essential to replenish the liquids lost through normal activities and physiological processes. Humans can survive for weeks—even months—without food, but not without water. Water is so vital to the human body that a person cannot survive for more than three to five days without it. Truly, water is the essence of life.

For its seemingly simple elegance, water is rarely a pure substance. In fact, it has much greater chemical complexity than you might think. With the exception of distilled water, the water you drink contains a variety of dissolved inorganic minerals that are also vital to your health. Although drinking water provides some of the life-sustaining minerals in your diet, the vast majority of these important nutrients are obtained from food.

Water is an indispensable macronutrient that is involved in myriad physiologic functions. Scientists have long marveled at water's unique physical and chemical properties, such as its ability to dissolve a large variety of chemical compounds. Because it can dissolve so

A molecule of water is composed of two hydrogen atoms and one oxygen atom.

© skyfish/Shutterstock.com

many substances, water is sometimes referred to as the *universal solvent*. Not only does water act as a biological solvent, it facilitates chemical reactions and helps regulate body temperature.

8-1a Distribution of Water in the Body

In adults, approximately 50 to 70 percent of total body weight is composed of water—about 11 gallons (42 liters) total. In infants, water may account for as much as 75 percent of total body weight. Because lean tissue (muscle) retains more water than does fat, men generally have greater water volumes than women. For the same reason, physically fit individuals (with more muscle mass) tend to have greater water volumes than their sedentary counterparts.

As the main component of biological fluid, water is distributed both inside and outside of cells. As illustrated in Figure 8.1, fluid located inside of a cell is referred to as **intracellular fluid**, whereas **extracellular fluid** is located outside of a cell. Extracellular fluid that fills spaces between or surrounding cells is referred to as **intercellular fluid**. Extracellular fluid, a component of your blood and lymph, is referred to

as **intravascular fluid**. You may be surprised to learn that the majority of bodily fluid (roughly 6 gallons, or 23 liters) is located inside cells.

Water Balance Cells rupture if they contain too much water and collapse if they contain too little. To ensure stable fluid balance, the movement of water molecules across cell membranes is carefully regulated. Recall from Chapter 3 that osmosis is the process whereby water molecules move from a region of low solute concentration across a cell membrane to a region of high solute concentration. A **solute** is a substance (e.g., a protein or electrolyte) that is dissolved in a fluid. While water molecules can move freely across most cell membranes, solutes generally cannot. A high concentration of solutes generates a force that attracts molecules of water, diluting the higher concentration of solutes. Thus, the driving

intracellular fluid Fluid located inside of a cell.

extracellular fluid Fluid located outside of a cell.

intercellular fluid Extracellular fluid that fills spaces between or surrounding cells.

intravascular fluid Extracellular fluid located in blood and lymph.

solute A substance dissolved in a fluid.

The pickling of fruits and vegetables depends on osmosis. When a cucumber is placed in a salty solution (brine), water moves out of the cucumber, which then shrivels and turns into a pickle.

force behind osmosis is the difference in the concentration of solutes in bodily fluids. The movement of water ceases when the concentration of solutes is the same on both sides of the cell membrane (see Figure 8.2). As an example of osmosis in action, consider what happens when vegetables are pickled. When a cucumber is placed in a salty brine solution, water moves out of the cucumber in the direction of the greater solute (salt) concentration. As water leaves the cucumber, it shrivels. When the right spices are added to the brine, the cucumber takes on a unique, delicious flavor and is referred to as a pickle.

hydrolysis reaction A chemical reaction whereby a chemical bond is broken by the addition of a water molecule.

condensation reaction A chemical reaction whereby a chemical bond joins two molecules together, releasing water in the process.

8-1b Water's Functions Are Critical to Life

Although water does not provide the body with energy, it does play an active role in hundreds of chemical reactions involved in energy metabolism. Beyond its metabolic functions, this versatile molecule provides protection and serves as an important solvent and lubricant. Finally, water helps maintain a stable internal body temperature, even when the surrounding environment is very cold or very hot.

Hydrolysis and Condensation Reactions

Water is essential to many chemical reactions. A **hydrolysis reaction** occurs when a chemical bond is broken by the addition of a water molecule. You have already learned about many kinds of hydrolytic reactions (e.g., those required for carbohydrate and triglyceride digestion) in preceding chapters. In contrast to hydrolytic reactions, a chemical reaction that results in the formation of a water molecule is called a **condensation reaction**.

FIGURE 8.1   FLUID COMPARTMENTS

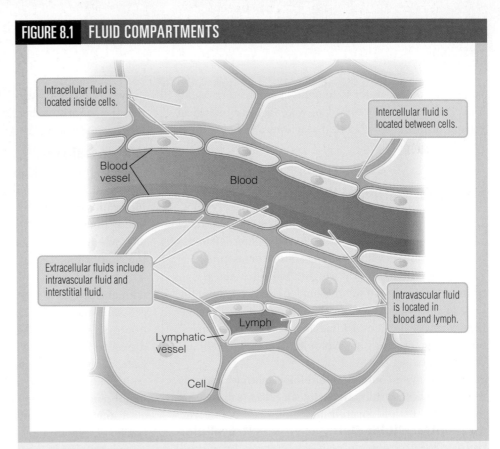

Intracellular fluid is located inside cells.

Intercellular fluid is located between cells.

Blood vessel

Blood

Extracellular fluids include intravascular fluid and interstitial fluid.

Intravascular fluid is located in blood and lymph.

Lymph

Lymphatic vessel

Cell

Water is distributed both inside and outside of cells. Extracellular fluids include fluids surrounding cells and fluid in blood and lymph. Intracellular fluid is located inside cells.

FIGURE 8.2 OSMOSIS

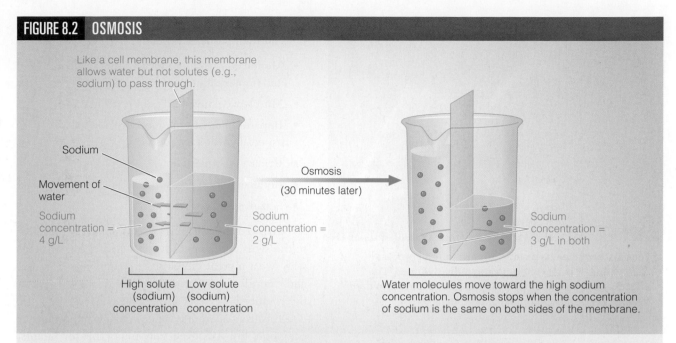

Like a cell membrane, this membrane allows water but not solutes (e.g., sodium) to pass through.

Sodium

Movement of water

Sodium concentration = 4 g/L

Sodium concentration = 2 g/L

High solute (sodium) concentration

Low solute (sodium) concentration

Osmosis (30 minutes later)

Sodium concentration = 3 g/L in both

Water molecules move toward the high sodium concentration. Osmosis stops when the concentration of sodium is the same on both sides of the membrane.

Osmosis is the movement of water molecules across a selective membrane. The direction in which water molecules move is determined by the concentration of dissolved substances (solutes).

Such reactions join molecules together through the formation of a chemical bond. The typical adult produces about approximately 1 cup (or 200–300 mL) of water every day from condensation and other metabolic reactions. Condensation and hydrolysis can be thought of as opposite processes—*make it* and *break it* reactions, respectively.

Water as a Solvent, Transport Medium, and Lubricant
A **solvent** is a substance that can dissolve other substances to form a **solution**. In the body, water is the primary solvent in bodily fluids such as blood, saliva, and gastrointestinal secretions. Blood and lymph, for example, are solutions that consist of water, cells, and a variety of dissolved substances (solutes) such as nutrients and metabolic waste products. These water-based solutions serve as important mediums for transporting myriad substances in the body. For example, nutrients and other dissolved substances in blood move from inside the blood vessel into the watery environment within and around tissues (extravascular fluid). Conversely, waste products produced by cells are dissolved in water, released into the blood, and subsequently eliminated from the body. Without water to dissolve these substances, such critical processes would not be possible.

Water also acts as a lubricant in the GI tract, respiratory tract, skin, and reproductive system, which produce digestive juices, mucus, sweat, and reproductive

fluids, respectively. The ability of the body to incorporate water into its secretions is vital to good health. As an example of its lubricating functions, water is an essential component of mucus produced in the lungs. Mucus both protects the lungs from harmful substances and lubricates the sensitive lung tissue so that it can remain moist and supple. When the amount of water is insufficient or the ability to regulate water balance across membranes is impaired, the production of bodily secretions can become compromised.

Water Regulates Body Temperature
Body heat is generated when energy-yielding nutrients are metabolized, such as during physical activity. Despite metabolic and environmental fluctuations, the average human is able to maintain a relatively stable and comfortable internal temperature of 98.6°F (37°C). The stability of this temperature is critical because even a slight increase or decrease in the body's internal temperature can disrupt normal physiologic functions. To prevent overheating, excess heat must be released from the body. This is accomplished by a process that involves both the skin and sweat glands.

Sweating is your body's most effective method of cooling itself. Water can both absorb and release

solvent A substance that can dissolve other substances to form a solution.

solution A liquid in which a solute is uniformly distributed in a solvent.

Because body heat is dissipated through the process of sweating, it is important to stay fully hydrated. The effectiveness of evaporative cooling can become impaired by exercising in hot, humid conditions.

large amounts of heat without itself changing appreciably in temperature, making it an ideal medium for heat transfer. In order for the body to rid itself of excess heat, blood vessels located near the surface of the skin dilate, releasing beads of sweat. When sweat evaporates from the skin, it takes heat with it. This process, called **evaporative cooling**, is an effective and efficient method of dissipating heat from (and thus cooling) the body. The efficiency of evaporative cooling is one reason why **dehydration**, a condition characterized by an insufficient amount of water in the body, impairs the process of body temperature regulation. Without adequate hydration, the body is unable to produce enough sweat to eliminate excess heat. When a dehydrated person exercises in hot, humid conditions, evaporative cooling is further diminished, making it extremely difficult to dissipate excess heat. This is why, it is particularly important for athletes to stay fully hydrated.

8-1c Dehydration

In its early stages, dehydration often results in thirst and a dry mouth—conditions that prompt you to drink water. Failure to respond to these initial symptoms of dehydration can have serious consequences. Compared to the other macronutrients, your body is least tolerant of water loss. Even a 2 percent loss in water weight can lead

evaporative cooling The process whereby sweat evaporates from the skin, taking heat with it.

dehydration A condition characterized by an insufficient amount of water in the body.

to serious complications. The following are common signs and symptoms associated with dehydration:

- Mental confusion, reduced attention span, and impaired memory
- Muscle weakness and poor coordination
- Impaired body temperature regulation, especially when exercising
- Urinary tract infections
- Reduced blood pressure
- Seizures and coma

To avoid complications, athletes must be sensitive to the signs and symptoms of dehydration. Athletes who sweat a great deal lose large amounts of water, and, therefore, are at particularly increased risk of dehydration.

Ensuring adequate water intake is important for everyone, but it is especially important for infants, children, and the elderly. Because of their small body size, infants and children are more susceptible to dehydration than adults. Thus, it takes very little water loss before the signs and symptoms of dehydration become apparent. Similarly, older adults have less water in their bodies than younger adults. Of particular concern is that signs and symptoms of dehydration in the elderly can easily be mistaken for those of dementia or cognitive decline. This is why it is especially important to assess hydration status in aging adults.

The Body's Response to Dehydration To meet your body's need for water, it is important to balance daily intake of water with daily water loss (see Figure 8.3).

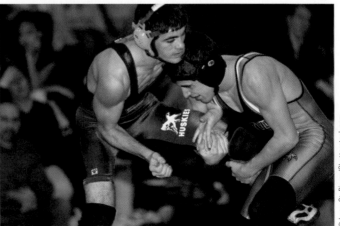

Amateur wrestlers and other athletes sometimes induce dehydration intentionally to lose water weight and meet strict weight class limits. Nutrition experts advise against this approach to weight loss.

FIGURE 8.3 DAILY WATER BALANCE

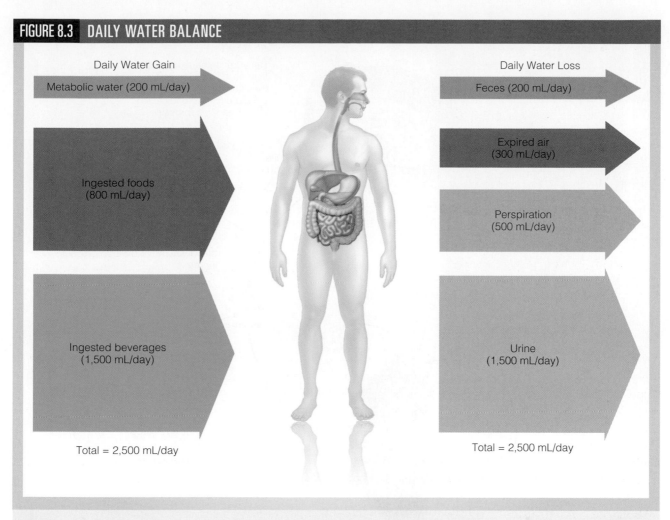

Daily Water Gain

Metabolic water (200 mL/day)

Ingested foods (800 mL/day)

Ingested beverages (1,500 mL/day)

Total = 2,500 mL/day

Daily Water Loss

Feces (200 mL/day)

Expired air (300 mL/day)

Perspiration (500 mL/day)

Urine (1,500 mL/day)

Total = 2,500 mL/day

The body gains and loses water every day. For the body to function properly, average daily water intake must equal average daily water loss.

Adults excrete approximately 1½ quarts (1500 mL) of fluid every day as urine. Water is also expelled as water vapor in expired air (300 mL/day), as perspiration (500 mL/day), and in feces (200 mL/day). In total, the body loses approximately 10.6 cups (about 2,500 mL) of fluid every day. To counter this loss, adults typically need to consume approximately 10 cups (2,300 mL) of fluid daily through a combination of foods and beverages. If you were wondering about the discrepancy between water loss and consumption, remember that cellular metabolism generates approximately 200–300 mL of fluid every day.

When a person becomes dehydrated, several homeostatic control mechanisms work together to restore fluid balance (see Figure 8.4). These

mechanisms involve two hormones: antidiuretic hormone and aldosterone. A progressive loss of body fluids results in low blood volume. In response, the concentration of solutes in extracellular fluids such as the blood

© Patrick Poendl/Shutterstock.com

CHAPTER 8: Water and the Minerals 189

FIGURE 8.4 REGULATION OF BLOOD VOLUME AND SOLUTE CONCENTRATION

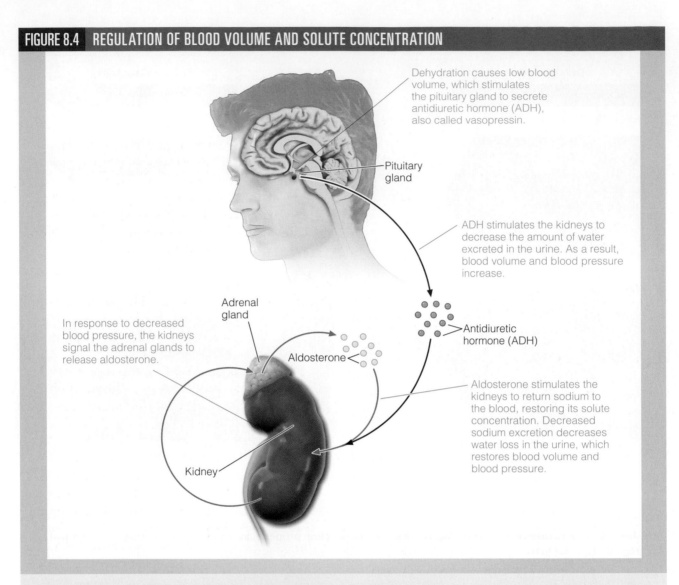

Dehydration causes low blood volume, which stimulates the pituitary gland to secrete antidiuretic hormone (ADH), also called vasopressin.

Pituitary gland

ADH stimulates the kidneys to decrease the amount of water excreted in the urine. As a result, blood volume and blood pressure increase.

Adrenal gland

In response to decreased blood pressure, the kidneys signal the adrenal glands to release aldosterone.

Aldosterone

Antidiuretic hormone (ADH)

Aldosterone stimulates the kidneys to return sodium to the blood, restoring its solute concentration. Decreased sodium excretion decreases water loss in the urine, which restores blood volume and blood pressure.

Kidney

Blood volume and solute concentration are regulated by the brain and kidneys. This process involves two hormones: antidiuretic hormone and aldosterone.

increases. This stimulates the pituitary gland to release **antidiuretic hormone** (**ADH**, or **vasopressin**) into the blood. ADH circulates to the kidneys, where it stimulates water conservation by decreasing the amount of water excreted in the urine. The increased water retention helps restore blood volume to healthy levels.

As the pituitary gland releases ADH in response to low blood volume, low blood pressure (also caused by dehydration) stimulates the adrenal glands to release the hormone **aldosterone**. Aldosterone signals the kidneys to reduce the amount of solutes (sodium) excreted

in the urine and instead return the solutes to the blood. Remember that where solutes go, water follows via osmosis. Together, the complementary actions of ADH and aldosterone restore blood volume to a healthy level.

8-1d Recommendations for Water Intake

In 2004, the Institute of Medicine released its first recommendations for water intake. These guidelines advise women to consume about 11 cups (2.7 liters) of water every day. The recommended water intake for men is 16 cups (3.7 liters) a day. These amounts may sound exorbitant, but they include the water present in all beverages and foods consumed. In general, about 80 percent of total water intake comes from beverages, while the remaining 20 percent comes from foods. Upon careful

antidiuretic hormone (**ADH**, or **vasopressin**) A hormone released by the pituitary gland during periods of low blood volume that decreases the amount of water excreted in the urine.

aldosterone A hormone produced by the adrenal glands in response to low blood pressure.

Does Coffee Cause Dehydration?

Millions of people start the day with a steaming cup of coffee. Indeed, coffee is among the most widely consumed beverages in the world. But is there more to coffee than the pick-me-up that it provides? Aside from being a stimulant, coffee is also a diuretic. Although caffeine does stimulate urination, studies have shown that caffeine-containing beverages do not appear to cause a net loss of total body water.[1] That is, the amount of water consumed while drinking caffeinated beverages adequately compensates for the total amount of water excreted from the body. In addition, there is more to coffee than many people realize. Coffee is a rich source of disease-fighting antioxidants, and may help lower risk of heart disease, certain types of cancer, and type 2 diabetes.

Although caffeine increases urine production, studies have shown that caffeine-containing beverages do not appear to cause a net loss of total body water.

amenic181/Shutterstock.com

consideration, the DRI committee concluded that caffeinated beverages such as coffee and cola contribute to total water intake to the same degree as noncaffeinated beverages.[2] This conclusion stands in contrast to previous recommendations, which suggested that caffeinated beverages (which have diuretic properties) could not be relied upon to supply water to the body.

Physical activity increases a person's need for water, especially when performed in a warm environment. Studies show that sweat rates in athletes competing in hot, humid conditions can be as much as 2 quarts (approximately 2 liters) of water per hour.[3] Specific guidelines have not been established for these individuals, but the Institute of Medicine recommends that very active people take special care to consume enough fluids every day. Consuming sports drinks instead of water during endurance training and competition can help athletes prevent serious dehydration-related complications and replenish electrolytes and energy stores. For less active people, however, water should be the drink of choice for simple hydration and rehydration.

MINERALS

In nutrition, the term **mineral** refers to inorganic substances other than water. Because the body requires only small amounts of dietary minerals (less than 1 g/day), they are considered micronutrients. The body cannot synthesize minerals from other compounds, and therefore are classified as essential nutrients. In other words, you must rely on the foods you eat to provide adequate amounts of these important nutrients. Minerals can be neither created nor destroyed. Even if a food is completely combusted (burned), the minerals will remain as ash. Although minerals comprise only a small fraction of total body weight, they play prominent roles in numerous bodily structures and functions. In the body, minerals are often found in the form of salts (electrolytes), which can separate into charged ions when dissolved. The most prominent electrolytes (mineral salts) in the body are sodium, potassium, calcium, magnesium, and phosphate.

Minerals are classified as either major or trace, depending on the amounts required for normal body functions. A **major mineral** is one required in amounts greater than 100 mg/day. Your body requires seven major minerals: calcium, phosphorus, magnesium, sodium, chloride, potassium, and sulfur. A **trace mineral** is one required by the body only in minute amounts—less than 100 mg/day. Humans require at least eight trace minerals: iron, copper, iodine, selenium, chromium, manganese, molybdenum, and zinc. In addition to the major and trace minerals, several minerals are not currently considered essential nutrients, but may be reclassified as we learn more about them. These minerals include fluoride, arsenic, boron, lithium, nickel, silicon, and vanadium. The functions of these minerals in the body are not fully understood.

8-2a Minerals Serve Diverse Roles

Minerals serve both structural and functional roles. In fact, minerals serve so many roles that they influence virtually every physiological system in the body. While most minerals have distinct functions in the body, others work in concert to carry out interrelated tasks. For example, calcium, magnesium, and phosphorus function together to form and strengthen the structure of

> **mineral** An inorganic substance other than water that is required by the body in small amounts.
>
> **major mineral** A mineral required in an amount greater than 100 mg/day.
>
> **trace mineral** A mineral required in an amount less than 100 mg/day.

Is Bottled Water Better?

More than one-half of all Americans drink bottled water, and one-third drink it regularly.[4] Despite its popularity, there is no evidence that bottled water offers any health advantages over water that runs freely from the tap. In fact, because most

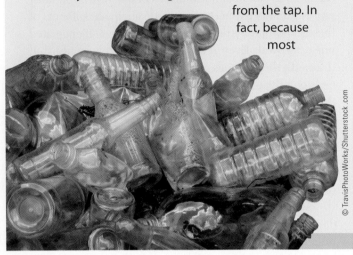

municipalities add fluoride (which helps keep teeth strong) to their water supplies, drinking bottled water instead of tap water may increase a person's risk of tooth decay. Furthermore, by choosing tap water instead of bottled water, Americans could eliminate nearly 2 million tons of plastic waste per year. In fact, most experts agree that the environmental costs associated with bottled water far outweigh any potential nutritional benefits. While some developing countries' water supplies are contaminated (thus necessitating bottled water), American tap water is as wholesome as bottled varieties and is certainly healthier for the environment. However, for those with concerns regarding the purity and/or taste of tap water, a water-filtering system can often remedy the problem.

Americans buy nearly 30 billion bottles of water per year; 90 percent of bottles are not recycled and end up in landfills. It takes thousands of years for plastic to decompose.

the skeleton. Some enzymes need to be activated by a mineral to function. When a mineral such as copper, magnesium, or zinc combines with and activates an enzyme, the mineral is called a **cofactor** and the enzyme is called a **metalloenzyme**. Some minerals are involved in the metabolic breakdown of energy-yielding nutrients, whereas others are essential to nerve function, muscle contraction, and the regulation of bodily fluids. Although nutritional scientists have a good understanding of how the major and trace minerals function in the body, there is still much to learn.

8-2b Minerals in Food

Minerals are abundant in both plant- and animal-based foods. In general, animal-based products have higher mineral contents than do plant-based foods. For example, a one-cup serving of milk contains approximately 300 milligrams calcium, while a one-cup serving of broccoli contains approximately 43 milligrams calcium. The location where a plant grows and where an animal grazes can also influence mineral

Mineral water, promoted for its purity and fresh taste, contains naturally occurring and/or added minerals such as sulfur and calcium. Natural mineral water is both consumed and bathed in, and natural mineral springs are often popular tourist attractions because of their perceived curative properties.

contents of foods. For instance, a plant grown in selenium-deficient soil will have a lower selenium content than one grown in selenium-rich soil. Not surprisingly,

cofactor A mineral that activates an enzyme by combining with it.

metalloenzyme An enzyme that is activated when it combines with a mineral.

selenium deficiency is more common in people living in geographic regions with low-selenium soil. The extent to which a food is processed can also influence its mineral content. For example, cereal grains contain minerals such as copper, selenium, and zinc, but these important nutrients can be lost during processing. Although food manufacturers sometimes fortify their products with minerals lost during processing, most health care professionals recommend choosing foods made with whole grains whenever possible.

As with foods, the types and amounts of minerals found in drinking water vary by region. *Hard water* has appreciably high levels of dissolved minerals such as calcium and magnesium. The types of minerals present in water can give it a particular taste and smell. Also, some minerals can collect on hard surfaces of pipes, sinks, and bathtubs, causing problems to homeowners.

8-2c Mineral Availability in the Body

Like most nutrients in the body, optimal and safe mineral levels are maintained through adjustments in absorption and excretion. Changes in these regulatory processes optimize mineral availability and prevent toxicity. In addition to the regulatory processes of absorption and excretion, the body is capable of storing small amounts of certain minerals in the liver, bones, and other tissues. Because these three processes (absorption, excretion, and storage) are tightly regulated, toxicity is rare for most minerals. Still, despite the body's effective mineral management system, genetic disorders and overconsumption of mineral-containing supplements or medications can have serious outcomes.

In some cases, humans can absorb nearly all of the minerals present in food. This is not the case for all minerals, however. Mineral bioavailability is often influenced by factors such as genetics, age, maturity, nutritional status, and interactions with other compounds in food. During periods of growth, for instance, mineral requirements are high, and absorption of some minerals increases. Conversely, mineral absorption sometimes decreases as a person ages. In addition to these factors, mineral bioavailability can also be influenced by the body's need for it. For example, calcium absorption increases when circulating concentrations of calcium are low.

In some instances, the presence of one mineral can interfere with the absorption of another. This is particularly true for minerals that have similar charges. For example, calcium, iron, copper, magnesium, and zinc all carry a positive (2^+) charge. High intake of any one of these nutrients may reduce absorption of the others.

Healthy Plants Require Healthy Soil

The 1980 eruption of Mount St. Helens wreaked havoc throughout the United States Pacific Northwest. The volcano's eruption blanketed the region in a thick layer of volcanic ash and destroyed many of the area's agricultural crops. Apple farmers were particularly devastated as they watched their entire orchards wither away.[5] Although the eruption initially seemed catastrophic to the apple industry, something unexpected happened the following year. Apple trees were covered not only with spring blossoms, but also with plentiful clusters of large, succulent fruit in late summer. This bountiful harvest occurred because the fine powder ash that spewed from the eruption was extremely rich in minerals, which reinvigorated the soil.

Another dietary factor that can decrease the efficiency of mineral absorption is the presence of binding factors, such as those found in some nuts, grains, and vegetables. These foods often contain substances that bind to minerals in the GI tract, inhibiting their absorption. Despite these nutrient interactions, experts do not recommend reducing the consumption of these otherwise nutritious foods.

 CALCIUM

Calcium (Ca) is the most abundant mineral in your body, making up about 2.2 pounds (1 kg) of your weight. You may already know that calcium plays a critical structural role in bones and teeth. In fact, more than 99 percent of your body's calcium is located in your skeleton. The rest is located in the blood and other tissues, where it participates in vital functions such as muscle contraction, energy metabolism, neural signaling, blood clot formation, and blood pressure regulation.

calcium (Ca) A major mineral that has an important structural role in bone and teeth; as well as regulating vital body functions such as blood clot formation, muscle contraction, neural signaling, energy metabolism, and blood pressure regulation.

8-3a Calcium: Structural Role

The skeletal system makes up the basic architecture of your body. Although you might think of the skeleton as an inert structure, your bones and teeth are in fact composed of living tissue. Skeletal calcium is a component of a large crystal-like molecule called **hydroxyapatite**, which combines with other minerals such as fluoride and magnesium to form the structural matrix of your bones and teeth. Hydroxyapatite also functions as a storage depot for calcium. Bone tissue is complex, and is composed of two different kinds of bone cells: osteoblasts and osteoclasts. An **osteoblast** promotes bone formation, whereas an **osteoclast** promotes the breakdown of older bone by a process called **bone resorption**. These cells work in concert to keep your bones healthy and strong, largely by synthesizing and breaking down calcium-containing hydroxyapatite as needed. To facilitate bone resorption, osteoclasts break down small pockets of old and damaged bone, releasing calcium, phosphorus, and other substances into the blood. In order to maintain stable bone mass, osteoblasts fill the dissolved pockets with new bone.

This continuous process referred to as **bone remodeling** (or **bone turnover**) is illustrated in Figure 8.5. Bone remodeling makes it possible for bones to grow and adapt to mechanical stress (exercise) by stimulating bone remodeling in areas with the greatest load. Like muscles, bones also respond to regular exercise by becoming stronger. Equally important, exercise-induced microscopic changes in bone may also help improve bone architecture, the pattern of trabeculae and associated structures.

8-3b Calcium: Regulatory Functions

Aside from its importance to structural maintenance, calcium is also involved in the regulation of several vital body functions. Calcium works with vitamin K to stimulate blood clot formation and facilitates both muscle contraction and nerve impulse transmission. Calcium supports healthy vision, the regulation of blood glucose, and cell division. Although it is not used directly as an energy source, calcium

hydroxyapatite
A large crystal-like molecule that combines with other minerals to form the structural matrix of bones and teeth.

osteoblast A bone cell that promotes bone formation.

osteoclast A bone cell that promotes the breakdown of older bone.

bone resorption
The process whereby osteoclasts break down bone, releasing minerals into the blood.

bone remodeling
(or **bone turnover**)
The continuous process by which older and damaged bone is replaced by new bone.

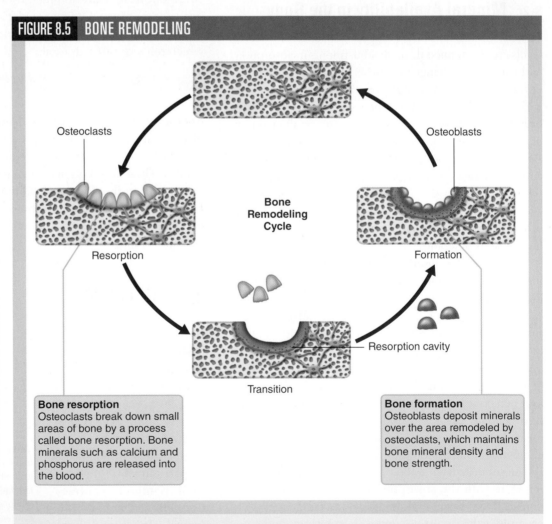

FIGURE 8.5 BONE REMODELING

Osteoclasts

Osteoblasts

Bone Remodeling Cycle

Resorption

Formation

Resorption cavity

Transition

Bone resorption
Osteoclasts break down small areas of bone by a process called bone resorption. Bone minerals such as calcium and phosphorus are released into the blood.

Bone formation
Osteoblasts deposit minerals over the area remodeled by osteoclasts, which maintains bone mineral density and bone strength.

Bone mass is kept stable through a process called bone remodeling. This process involves cells that build bone (osteoblasts) and cells that break bone down (osteoclasts).

is a cofactor for several enzymes needed for ATP synthesis. In addition to the multitude of functions already discovered, scientists are just beginning to understand additional roles that calcium might play in the body. For example, there is growing (albeit controversial) evidence that adequate calcium consumption may help reduce the risk of cardiovascular disease, some forms of cancer, and obesity.[6]

Hormonal Regulation of Blood Calcium

Calcium deficiency affects many tissues in the body. For instance, because calcium plays a role in muscle contraction and nerve function, low blood calcium levels can cause muscle pain, muscle cramping, and a tingling sensation in the hands and feet.

While low calcium intake can cause health problems, excessive calcium intake is equally dangerous. Overconsumption can cause calcium deposits in soft tissues such as muscle and kidney and can interfere with the bioavailability of other nutrients, such as iron and zinc. High calcium intake is also associated with impaired kidney function and the formation of certain types of kidney stones.

The amount of calcium in your blood is tightly regulated by hormonal mechanisms (see Figure 8.6). This well-orchestrated system involves three hormones: calcitriol (vitamin D), parathyroid hormone (PTH), and calcitonin. These hormones work together to maintain healthy blood calcium levels at all times.

FIGURE 8.6 REGULATION OF BLOOD CALCIUM

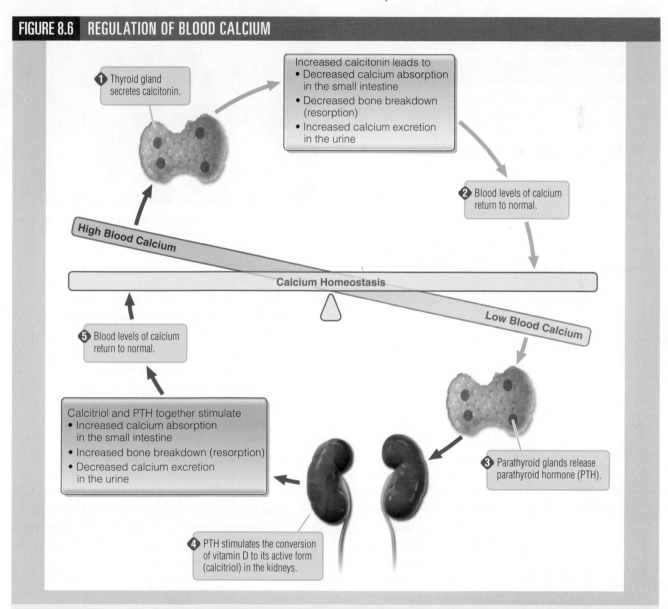

1 Thyroid gland secretes calcitonin.

Increased calcitonin leads to
• Decreased calcium absorption in the small intestine
• Decreased bone breakdown (resorption)
• Increased calcium excretion in the urine

2 Blood levels of calcium return to normal.

High Blood Calcium

Calcium Homeostasis

Low Blood Calcium

5 Blood levels of calcium return to normal.

Calcitriol and PTH together stimulate
• Increased calcium absorption in the small intestine
• Increased bone breakdown (resorption)
• Decreased calcium excretion in the urine

3 Parathyroid glands release parathyroid hormone (PTH).

4 PTH stimulates the conversion of vitamin D to its active form (calcitriol) in the kidneys.

Blood calcium concentration is regulated by three hormones: parathyroid hormone (PTH), calcitriol (vitamin D), and calcitonin.

Recall from Chapter 7 that vitamin D is critical to the regulation of blood calcium levels. When blood calcium levels are low, the parathyroid glands release PTH, which in turn circulates to the kidneys, where it stimulates the conversion of 25-hydroxy vitamin D_3 (25-[OH] D_3) to calcitriol (1,25-[OH]$_2$ D_3, the active form of vitamin D). Working together, PTH and calcitriol increase calcium absorption in the small intestine, reduce calcium excretion in the urine, and signal the release of calcium from bones into the blood. Collectively, these actions increase blood calcium levels back to a normal range.

When blood calcium levels are too high, the parathyroid glands produce less PTH. This decreases the formation of calcitriol, which in turn reduces calcium absorption in the small intestine. Elevated blood calcium also stimulates the thyroid gland to produce a hormone called **calcitonin**, which decreases calcium loss in bone, decreases calcium absorption in the small intestine, and increases calcium levels in the urine. Together, these processes help lower blood calcium levels back to normal.

8-3c Bone Health and Osteoporosis

Most bone growth is completed around age 20, but a small amount of bone (about 10 percent) is deposited between ages 20 and 30. The amount of mineral contained in bone is referred to as **bone mass**. **Peak bone mass**, defined as the point whereby bones have reached their maximum strength and density, is typically attained at about 30 years of age. To determine bone strength, clinicians test **bone density** by measuring the amount of bone tissue (calcium and other minerals) in a segment of bone. Bones with higher densities tend to be stronger and better able to withstand injury, while those with lower densities tend to break more easily.

Because low bone density can lead to negative health consequences later in life, it is important to take the proper steps to ensure that bones remain strong and healthy. You may recall from Chapter 7

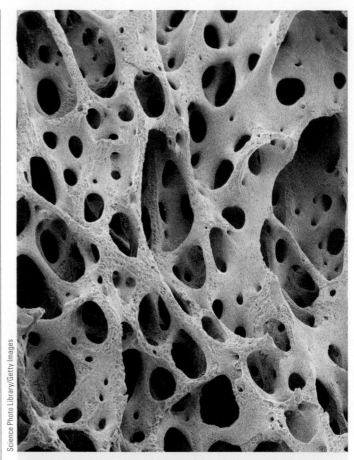

Science Photo Library/Getty Images

Loss of bone matrix can weaken bones and ultimately lead to a condition called osteoporosis. The risk of osteoporosis increases with age.

that rickets, typically caused by a vitamin D deficiency in young children, can also be caused by a lack of calcium in the diet. It is normal for older adults to experience age-related bone loss. However, even moderate bone loss, referred to as **osteopenia**, can cause bones to weaken. Bone loss can progress until bones become fragile and break easily. When this occurs, a person is said to have a condition called **osteoporosis** (meaning "porous bone"). Osteopenia and osteoporosis are especially prevalent in the elderly, for whom even minor falls can result in traumatic fractures that often are difficult to heal. Some people with osteoporosis lose significant height during old age and can develop a curvature in the upper spine, a condition called **kyphosis** (or **dowager's hump**). The National Osteoporosis Foundation estimates that osteoporosis is a major health concern for 54 million U.S. adults age 50 years and older.[7]

A person's risk of developing bone disease depends in part on the peak bone mass achieved in early adulthood. After bone mass peaks, the process of age-related

calcitonin A hormone released by the thyroid gland that decreases calcium loss from bone, decreases calcium absorption in the small intestine, and increases calcium loss in the urine.

bone mass The amount of minerals contained in bone.

peak bone mass The point whereby bones have reached their maximum strength and density.

bone density The amount of bone tissue in a segment of bone.

osteopenia A condition characterized by moderate bone loss in adults.

osteoporosis A condition characterized by brittle and fragile bones caused by a progressive loss of calcium and other bone minerals.

kyphosis (or **dowager's hump**) A curvature of the upper spine.

bone loss begins. There are many reasons for this decline.[8] First, calcium absorption often decreases with age. This is, in part, due to age-related changes in vitamin D synthesis and metabolism, both of which contribute to vitamin D insufficiency. To maintain healthy blood calcium concentrations, bone resorption increases, releasing calcium and other bone-related minerals into the blood. Second, reproductive hormone concentrations naturally decline with age. For example, *estrogen*, a reproductive hormone produced in the ovaries, is important to the maintenance of bone strength in women. As women reach menopause, declining estrogen levels accelerate bone loss. Men also lose bone mass as they age, but the loss is more gradual (see Figure 8.7).

Because bone fractures are a common cause of hospitalization and loss of independence in older adults, it is important to prevent osteoporosis by slowing the progression of bone loss. Many factors are associated with a person's risk of osteoporosis. Biological risk factors include sex, genetics, body size, and advancing age. Although biological risk factors cannot be modified, it is important to be aware of modifiable risk factors that are within a person's control.[9] For example, lifestyle factors such as exercise, diet, and smoking habits can be modified to decrease the risk of osteoporosis. Table 8.1 lists common biological and lifestyle risk factors associated with osteoporosis.

Dietary and Supplemental Sources of Calcium Calcium is found naturally in both plant-and animal-based foods, but the best sources of calcium are dairy products (see Figure 8.8). Other sources of calcium-rich foods

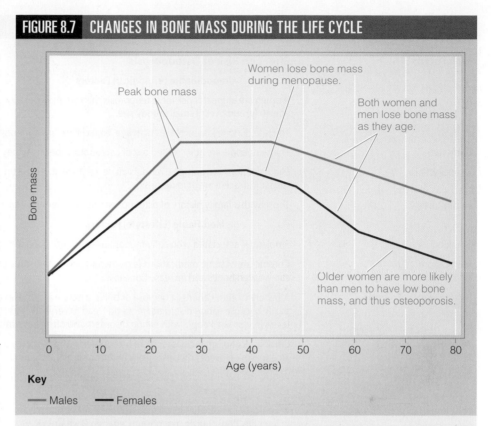

FIGURE 8.7 CHANGES IN BONE MASS DURING THE LIFE CYCLE

Peak bone mass

Women lose bone mass during menopause.

Both women and men lose bone mass as they age.

Older women are more likely than men to have low bone mass, and thus osteoporosis.

Bone mass

Age (years)

Key

— Males — Females

Women are at greater risk of osteoporosis than men because they tend to have lower peak bone mass and are likely to lose more bone mass as they age.

Source: Adapted from Compston JE. Osteoporosis, corticosteroids and inflammatory bowel disease. Alimentary Pharmacology and Therapeutics. 1995, 9:237–250.

Some foods are fortified with calcium.

Bloomberg/Getty Images

TABLE 8.1 FACTORS RELATED TO INCREASED RISK OF OSTEOPOROSIS

Risk Factor	Effect on Risk of Osteoporosis
Unmodifiable (Biological) Factors	
Sex	Women are at greater risk for osteoporosis than are men because estrogen production declines after age 40 (y) and overall smaller body size.
Age	The rate of bone loss increases with age, especially in post-menopausal women.
Body size	Thinner people are at increased risk of osteoporosis because they tend to have lower bone mass.
Ethnicity/Race	People of northern European and Asian descent are at greatest risk, while those of African and Hispanic descent are at lowest risk.
Family history	Those with a family history of osteoporosis are at especially high risk.
Modifiable (Lifestyle) Factors	
Smoking	Smoking is associated with early menopause and subsequent bone loss.
Medication	Chronic use of some medications (e.g., thyroid hormones, some diuretics, and corticosteroid therapy) can weaken bones and increase bone loss.
History of eating disorder	A history of either anorexia nervosa or bulimia nervosa increases risk of low bone density. This is partly because eating disorders frequently lead to very low body fat, which in turn can lower estrogen levels in women. People with eating disorders also often have multiple nutrient deficiencies.
Alcohol consumption	Chronic alcoholism increases risk of osteoporosis. This is mostly due to the negative influence of alcoholism on overall diet, especially in terms of calcium, vitamin D, and protein intake.
Physical inactivity	Regular physical activity, especially weight-bearing exercise, increases bone density and decreases risk of osteoporosis.
Nutritional factors	Chronically low calcium intake can decrease peak bone mass in young adults and increase bone loss in later life. Phosphorus, magnesium, vitamin D, vitamin K, vitamin C, fluoride, and protein are also important to bone health.

FIGURE 8.8 CALCIUM CONTENT OF SELECTED FOODS

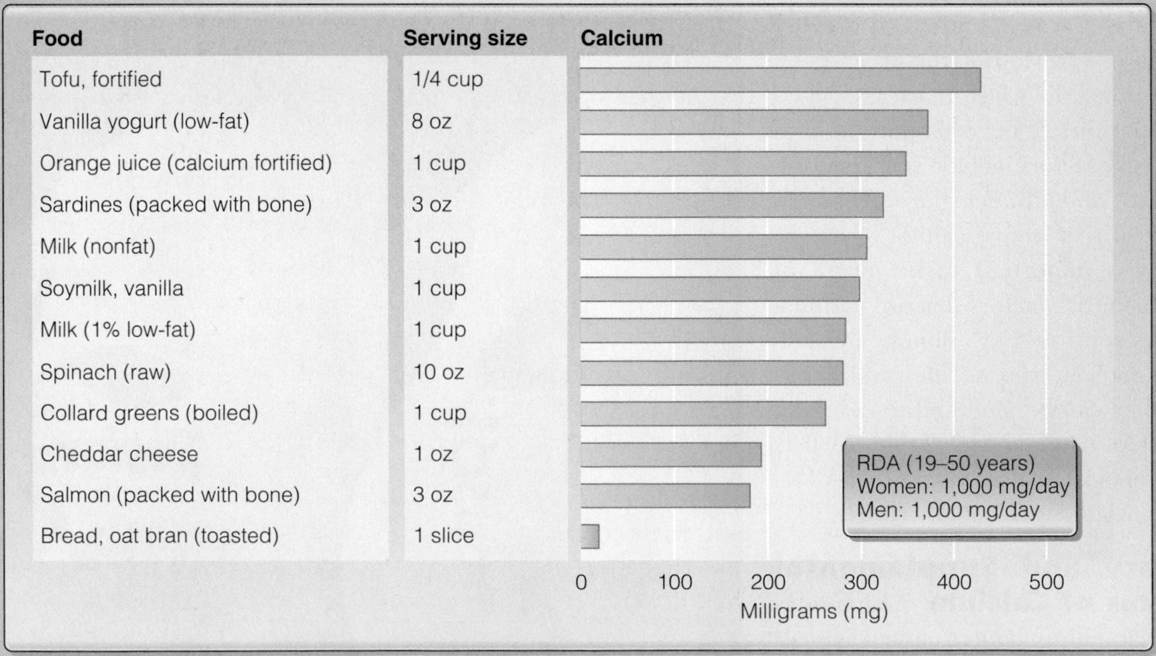

Food	Serving size
Tofu, fortified	1/4 cup
Vanilla yogurt (low-fat)	8 oz
Orange juice (calcium fortified)	1 cup
Sardines (packed with bone)	3 oz
Milk (nonfat)	1 cup
Soymilk, vanilla	1 cup
Milk (1% low-fat)	1 cup
Spinach (raw)	10 oz
Collard greens (boiled)	1 cup
Cheddar cheese	1 oz
Salmon (packed with bone)	3 oz
Bread, oat bran (toasted)	1 slice

RDA (19–50 years)
Women: 1,000 mg/day
Men: 1,000 mg/day

Calcium — Milligrams (mg): 0, 100, 200, 300, 400, 500

Source: USDA Nutrient Database for Standard Reference, Release 28. Available at: https://ndb.nal.usda.gov/ndb/nutrients/index

include dark green leafy vegetables such as collards and spinach, salmon and sardines (with bones), and some legumes. Calcium-fortified foods such as breakfast cereals, orange juice, and soy products also provide considerable amounts of calcium. It is relatively easy to assess the amount of calcium in most packaged foods because Nutrition Facts panels are required to list calcium content. Furthermore, foods fortified with calcium are frequently labeled as such.

If adequate amounts of dietary calcium cannot be obtained, many forms of calcium supplements are available. In general, calcium from supplements is relatively well absorbed and utilized by the body. For maximum absorption, calcium supplements should be taken with food and in amounts no more than 500 milligrams.[10] Although adequate calcium intake is certainly critical to the maintenance of healthy bones, studies suggest that taking calcium and vitamin D supplements produces only small improvements in bone health.[11] Furthermore, limited research also suggests that calcium supplementation might actually increase risk of hip fractures in some people.[12] Clearly, calcium's impact on bone health is a complex subject—one that is impacted by other dietary components.

8-3d Recommended Calcium Intake and Bioavailability

The Recommended Dietary Allowance (RDA) for calcium has been set at 1,000 mg/day for adults. This is approximately the amount of calcium contained in three and a half cups of milk. Because calcium absorption tends to decrease with age, the RDA increases to 1,200 mg/day at 51 years of age for women and at 71 years of age for men. For this reason, it can be difficult for older adults to satisfy their need for this important nutrient by dietary means alone.[13] To decrease the risk of calcium toxicity, a Tolerable Upper Intake Level (UL) of 2,500 mg/day has been established for calcium. At 51 years of age, this UL decreases to 2,000 mg/day.

Many factors influence calcium's bioavailability. For example, *oxalate* and *phytate*, found in some vegetables and whole grains, are compounds that bind calcium in the intestine, hindering its absorption. Because vitamin D is needed for calcium absorption, a person's vitamin D status can also impact calcium bioavailability. This is why many calcium supplements contain vitamin D as well. The acidity of gastric juice improves calcium absorption by making it more soluble in the GI tract. Thus, medications (such as antacids) that inhibit gastric acid production can hinder calcium bioavailability.[14]

8-4 ELECTROLYTES: SODIUM, CHLORIDE, AND POTASSIUM

Recall from Chapter 3 that an electrolyte is a salt that separates into individual charged ions when dissolved in water. When table salt (sodium chloride), one of the most familiar electrolytes, is added to water, it separates into its charged components: **sodium (Na⁺)** and **chloride (Cl⁻)**. These, along with **potassium (K⁺)** are the three most abundant ions in the human body. These 3 charged ions are found in various bodily fluids, and each plays important roles in fluid balance regulation, nerve impulse conduction, and nutrient transport. While sodium is often associated with health problems such as hypertension (high blood pressure), it is important to remember that sodium, like other ions, is essential for many physiological functions within the body.

8-4a Electrolyte Function: Nerve Impulse and Muscle Contraction

Tissues, fluids, and cells all contain varying concentrations of electrolytes. As you previously learned, it is the concentrations of dissolved sodium, chloride, and potassium ions in fluids that direct the flow of water during osmosis. In this way, electrolytes play important roles in regulating fluid balance. A proper distribution of positively and negatively charged ions inside and outside of cells orchestrates many important physiological functions, including nerve cell transmission and muscle contraction.

Nerve impulses and muscle contractions occur when balances of ions shift across cell membranes. When a nerve cell is at rest, positively charged potassium ions are concentrated inside of the cell while positively charged sodium ions and negatively charged chloride ions are concentrated outside of the cell. Upon stimulation, sodium ions rush into the

sodium (Na⁺) A positive ion (cation) found primarily in fluids surrounding cells in the body.

chloride (Cl⁻) A negatively charged ion (anion) found primarily in fluids surrounding cells in the body.

potassium (K⁺) The major positive ion (cation) found inside cells.

cell, and potassium ions rush out. This exchange of ions across the cell membrane produces an electrical signal, or a nerve impulse. Muscle fibers function in much the same manner, but their activation causes calcium (instead of sodium) to flow inside the cell, signaling the muscle to contract. Once calcium has been pumped back out of the muscle cell, it can again relax.

8-4b Electrolyte Function: Blood Pressure Regulation

Blood volume and circulating electrolyte concentrations are intimately linked. A change in blood volume can affect its electrolyte concentrations, and a change in electrolyte concentrations can affect blood volume. Although electrolyte concentrations and blood volume are clearly connected, their regulatory mechanisms are somewhat different. You may recall that a substantial loss of body fluid causes blood volume to decrease. When blood volume decreases, so too does blood pressure. This triggers a series of homeostatic responses that ultimately cause the kidneys to retain sodium and return it to the blood. Again—where electrolytes go, water will follow. Sodium retention increases blood volume, which subsequently restores blood pressure. While this homeostatic response involves antidiuretic hormone (ADH) and aldosterone, a related homeostatic response called the renin-angiotensin-aldosterone system utilizes other substances to assist in blood volume and blood pressure regulation.

As illustrated in Figure 8.9, the kidneys respond to low blood pressure by releasing an enzyme called **renin**. Once in the blood, renin converts the liver-derived protein *angiotensinogen* into *angiotensin I*, which in turn is converted to *angiotensin II* in the lungs. Angiotensin II stimulates the adrenal glands to release aldosterone, which, then, signals the kidneys to retain sodium and return it to the blood. This process assists in the restoration of healthy blood volume by drawing water into the blood via osmosis. Aldosterone also causes blood vessels to constrict, which, along with restored blood volume, re-establishes normal blood pressure. This helps explain why some people with high blood pressure are advised to restrict their sodium intakes. In fact, some drugs control high blood pressure by disrupting this system.

renin An enzyme, secreted by the kidneys, that converts angiotensinogen to angiotensin I.

hyponatremia A condition characterized by low blood sodium concentration.

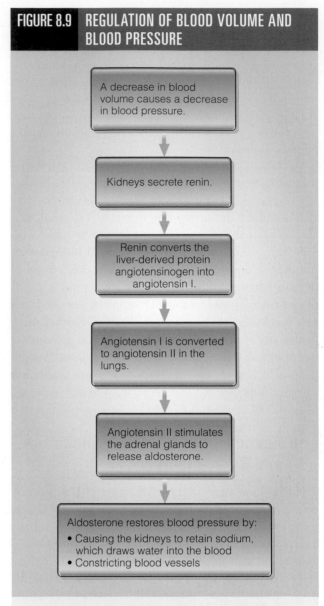

FIGURE 8.9 REGULATION OF BLOOD VOLUME AND BLOOD PRESSURE

A decrease in blood volume causes a decrease in blood pressure.

↓

Kidneys secrete renin.

↓

Renin converts the liver-derived protein angiotensinogen into angiotensin I.

↓

Angiotensin I is converted to angiotensin II in the lungs.

↓

Angiotensin II stimulates the adrenal glands to release aldosterone.

↓

Aldosterone restores blood pressure by:
- Causing the kidneys to retain sodium, which draws water into the blood
- Constricting blood vessels

A loss of body fluids trigger a homeostatic response called the renin-angiotensin-aldosterone system, which helps restore blood volume and blood pressure.

Electrolyte Balance

Sodium, chloride, and potassium are easily absorbed in the small intestine and, to a lesser extent, the colon. Once electrolytes are absorbed, their concentrations in the blood are regulated by the kidneys, which filter excess amounts into the urine. In other words, when blood electrolyte levels are elevated, the excretion of urinary sodium, potassium, and chloride increases. The opposite is true when blood electrolyte levels are low.

Low blood sodium is caused by inadequate sodium intake, loss of sodium through excessive sweating, diarrhea, vomiting, and/or excessive water consumption that is unaccompanied by increased sodium intake. When the concentration of blood sodium falls below a healthy range, a condition called **hyponatremia** develops. Consequences of hyponatremia include nausea, dizziness,

Not All Sports Drinks Are the Same

Some cellular mechanisms associated with intestinal absorption of sodium require the assistance of glucose. In fact, sports drinks designed to replace electrolytes lost during physical activity historically contained both salt and glucose.[15] Today, however, many sports drinks are made with high-fructose corn syrup, a substance that typically contains nearly equal amounts of glucose and fructose. Although it is unclear if the type of sweetener used in a sports drink influences hydration status, sports drinks that contain both glucose and sodium have been found to effectively restore and maintain hydration status after a hard workout.[16] Most experts advise athletes to rely on water for rehydration unless their activity level is especially strenuous or the environment is hot and/or humid.

muscle cramps, and, in severe cases, coma. Hyponatremia can be life threatening—especially in infants and children. Excessive sodium loss associated with sweating is also of particular concern among athletes involved in endurance sports such as marathon running.

Although potassium deficiency is rare, it too can occur because of prolonged diarrhea and/or vomiting. Heavy use of certain diuretics can also cause excessive urinary potassium loss. A **diuretic** is a substance or drug that helps the body eliminate water. Diuretics are often taken to lower blood pressure, but when the body excretes excessive amounts of fluids, it also unavoidably

Most experts recommend that you limit your sodium intake to less than 2,300 mg/day.

excretes electrolytes. This can lead to **hypokalemia**, a condition characterized by a low concentration of potassium in the blood. People with eating disorders that involve vomiting (e.g., bulimia nervosa) are at increased risk of hypokalemia. Potassium deficiency causes muscle weakness, constipation, irritability, and confusion. In severe cases, potassium deficiency can lead to irregular heart function, muscular paralysis, decreased blood pressure, and difficulty breathing.

Dietary Sources of Electrolytes A wide variety of foods contain sodium and chloride. Table salt (sodium chloride), meat, dairy products, and fermented foods such as sauerkraut and pickles are especially good sources of these electrolytes. Highly processed foods such as sandwich meats, chips, fast food, and condiments such as ketchup and soy sauce are often made with high amounts of salt. In general, manufactured foods often contain substantial amounts of sodium, whereas unprocessed foods such as fresh fruits and vegetables contain only small amounts. Dietary sources of potassium include legumes, potatoes, seafood, dairy products, meat, and a variety of fruits and vegetables, such as sweet potatoes and bananas. Common dietary sources of sodium, chloride, and potassium are listed in Figure 8.10.

You can easily determine the salt content of packaged foods by reading Nutrition Fact panels. Because limiting dietary salt intake may decrease the risk of hypertension, the following health claim is sometimes displayed on low-sodium foods: "Diets low in sodium may reduce the risk of high blood pressure, a disease associated with many factors." The following terms are also used on food packaging to describe salt content:

- Salt free: less than 5 milligrams sodium per serving
- Very low salt: less than 35 milligrams sodium per serving
- Low salt: less than 140 milligrams sodium per serving

Sodium Chloride Increases Blood Pressure Aside from making you feel thirsty, an elevated salt intake does not have any demonstrated toxic effects for

diuretic A substance or drug that helps the body eliminate water.

hypokalemia A condition characterized by low blood potassium concentration.

FIGURE 8.10 SODIUM AND POTASSIUM CONTENT OF SELECTED FOODS

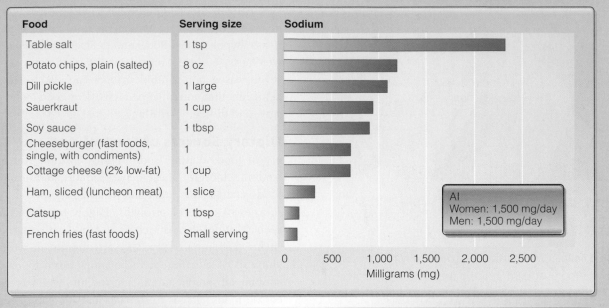

Food	Serving size	Sodium
Table salt	1 tsp	
Potato chips, plain (salted)	8 oz	
Dill pickle	1 large	
Sauerkraut	1 cup	
Soy sauce	1 tbsp	
Cheeseburger (fast foods, single, with condiments)	1	
Cottage cheese (2% low-fat)	1 cup	
Ham, sliced (luncheon meat)	1 slice	
Catsup	1 tbsp	
French fries (fast foods)	Small serving	

AI
Women: 1,500 mg/day
Men: 1,500 mg/day

Milligrams (mg)

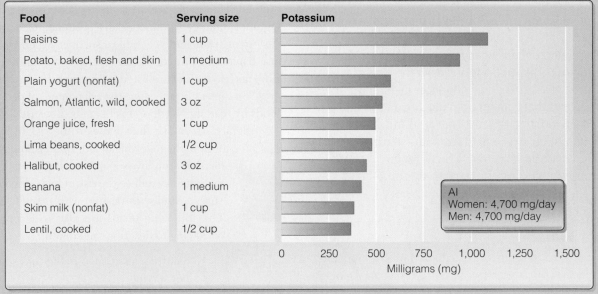

Food	Serving size	Potassium
Raisins	1 cup	
Potato, baked, flesh and skin	1 medium	
Plain yogurt (nonfat)	1 cup	
Salmon, Atlantic, wild, cooked	3 oz	
Orange juice, fresh	1 cup	
Lima beans, cooked	1/2 cup	
Halibut, cooked	3 oz	
Banana	1 medium	
Skim milk (nonfat)	1 cup	
Lentil, cooked	1/2 cup	

AI
Women: 4,700 mg/day
Men: 4,700 mg/day

Milligrams (mg)

Source: USDA Nutrient Database for Standard Reference, Release 28. Available at: https://ndb.nal.usda.gov/ndb/nutrients/index

most people. In some people, however, high intakes of sodium chloride are associated with increased blood pressure, a major risk factor of both heart disease and stroke.[17] Because blood sodium concentration is one of the major regulators of blood volume, high levels of sodium cause blood volume, and thus blood pressure, to rise.

Epidemiologic studies have long suggested that high blood pressure, also called hypertension, is most likely to occur in people who consume large amounts of salt.[18] However, because hypertensive people are also more likely than nonhypertensive people to be overweight and smoke, it is somewhat difficult to discern which specific factors are *causally* related to blood pressure. Even though a definite causal relationship between salt intake and hypertension has not been established for all people, public health organizations have long recommended that individuals lower blood pressure by losing weight, quitting smoking, and restricting salt intake.

Not all people with hypertension benefit from low-salt diets; there is wide variation in how people respond to salt intake. This may be attributable to factors such as genetics, exercise, and the responsiveness of the renin-angiotensin-aldosterone system. Although there is no easy way to determine whether you are salt sensitive, certain populations experience higher incidence of salt sensitivity than do others. Salt sensitivity is most common among the elderly, women, African Americans, and people with hypertension, diabetes, or chronic kidney disease.[19] Regardless of whether you are salt sensitive, one of the best ways to control hypertension is to follow a diet plan that emphasizes reasonable amounts of whole grains, fruits and vegetables, low-fat dairy products, and lean meats.

8-4c Recommended Intakes for Sodium, Chloride, and Potassium

Although the AI for sodium is set at 1,500 mg/day for most adults (19–50 years), the 2015 Dietary Guidelines for Americans recommends a sodium intake of less than 2,300 mg/day. As the majority of dietary chloride is associated with sodium chloride and other salts, the AI for chloride is also set at 2,300 mg/day. Adults age 51 and older, African Americans, and people with hypertension, diabetes, and/or chronic kidney disease should consume less than 1,500 mg/day of sodium. Highly active individuals, such as endurance athletes, and people living in hot, humid climates may require additional sodium in their diets. Recommended intake values have not been established for infants, but health professionals suggest that salt not be added to foods during the first year of life. Most people consume about 1½ tsp (3,400 mg) of sodium every day. Some of the best ways to prevent excessive sodium intake are to avoid highly salted processed foods and to prepare foods with less salt. As with the other electrolytes, an RDA has not been established for potassium. For adults, AIs have been set at 4,700 mg/day. Because potassium toxicity is not associated with dietary sources, a UL has not been set for this mineral.

 8-5 ELECTROLYTES: PHOSPHORUS AND MAGNESIUM

Like calcium, **phosphorus (P)** and magnesium are both major minerals with structural and functional roles. Also, like other electrolytes, if normal levels of phosphorus and magnesium are not maintained, the body does not function properly. Homeostatic mechanisms ensure that the amounts of phosphorus and magnesium in the body are maintained within normal limits so that countless metabolic functions in the body can take place.

8-5a Phosphorus

Phosphorus is essential to the structures of cell membranes, bones, teeth, DNA, RNA, and ATP. Its functional roles include lipid transport, enzyme activation, and energy metabolism. Phosphorus is readily absorbed in the small intestine, and its blood concentration is regulated by the hormones calcitriol, parathyroid hormone (PTH), and calcitonin. When blood phosphorus levels are low, calcitriol and PTH increase both phosphorus absorption in the small intestine and phosphorus release from bone tissue. These actions help return blood phosphorus levels to normal. When blood phosphorus levels are high, calcitonin stimulates osteoblasts to take up phosphorus from the blood to build new bone, thus lowering blood phosphorus levels back to normal.

The primary role of phosphorus in the body is as a component of cell membranes. Recall from Chapter 6 that cell membranes are made from phosophlipids, which contain phosphorus. Because phospholipids also surround lipoproteins, phosphorus is critical to the transport of lipids in blood and lymph. As a component of both DNA and ATP, phosphorus is essential to protein synthesis and energy metabolism. Phosphorus is involved in hundreds of metabolic reactions that involve the activation of molecules (mostly enzymes). Phosphorus is also a key component of the mineral matrix that makes up bones and teeth, and phosphorus-containing compounds help maintain the acid-base (pH) balance in the blood.

Recommended Intake and Dietary Sources of Phosphorus As illustrated in Figure 8.11, good sources of phosphorus include dairy products, meat (including poultry), seafood, nuts, and seeds. Most foods, except for seeds and grains, provide a highly bioavailable source of phosphorus. Generally, the biovaialability of phosphorus from animal-based foods is higher than that of plant-based foods. Because it is added to some foods to promote moisture retention, smoothness, and taste, phosphorus can also be obtained from some processed foods and soft drinks. A typical 12-ounce cola, for example, contains about 50 milligrams of phosphorus. To determine whether a soft drink contains added phosphorus, look for the ingredient *phosphoric acid* on the food label. However, reliance on soft drinks to supply dietary phosphorus is not recommended because these beverages have low nutrient densities and contribute very little to nutrient requirements.

> **phosphorus (P)** A major mineral that is essential to cell membranes, bone and tooth structure, DNA, RNA, ATP, lipid transport, and a variety of processes in the body.

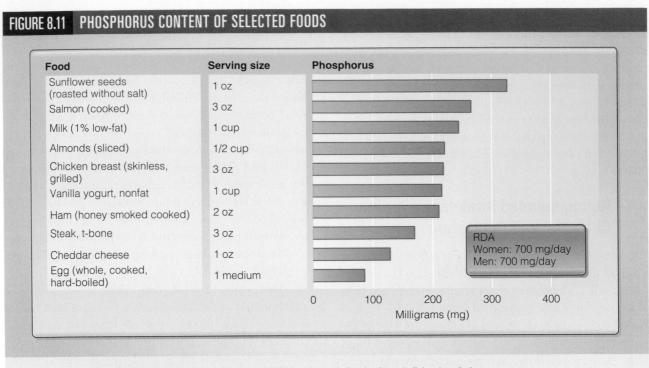

FIGURE 8.11 PHOSPHORUS CONTENT OF SELECTED FOODS

Food	Serving size	Phosphorus
Sunflower seeds (roasted without salt)	1 oz	
Salmon (cooked)	3 oz	
Milk (1% low-fat)	1 cup	
Almonds (sliced)	1/2 cup	
Chicken breast (skinless, grilled)	3 oz	
Vanilla yogurt, nonfat	1 cup	
Ham (honey smoked cooked)	2 oz	
Steak, t-bone	3 oz	
Cheddar cheese	1 oz	
Egg (whole, cooked, hard-boiled)	1 medium	

RDA
Women: 700 mg/day
Men: 700 mg/day

Milligrams (mg)

Source: USDA Nutrient Database for Standard Reference, Release 28. Available at: https://ndb.nal.usda.gov/ndb/nutrients/index

Phosphorus Deficiency, Toxicity, and Recommended Intakes Because of the abundance of phosphorus in most foods—especially those rich in protein—phosphorus deficiency is rare. When it does occur, individuals experience a loss of appetite, anemia, muscle weakness, poor bone development, and, in extreme cases, death. Phosphorus toxicity, on the other hand, is more common, resulting in mineralization of soft tissues such as the kidneys. The RDA for phosphorus is 700 mg/day—approximately the amount you would get from eating two servings of meat and two cups of milk. Because of the dangerous side effects associated with high amounts of phosphorus, a UL of 4,000 mg/day has been established for this mineral.

8-5b Magnesium

Magnesium (Mg) is a major mineral that is important to many physiological processes, such as energy metabolism and enzyme function. Blood magnesium concentrations are regulated primarily by the small intestine and, to a lesser extent, by the kidneys. As with most minerals, magnesium absorption occurs in the small intestine. Absorption increases when blood levels are low and decreases when blood levels are high.

magnesium (Mg) A major mineral that is important to many physiological processes, such as energy metabolism and enzyme function.

The majority of magnesium in the body is associated with bones, where it helps provide structure. Magnesium typically exists as a positively charged ion (Mg^{2+}), which stabilizes enzymes and neutralizes negatively charged ions. For instance, magnesium helps stabilize high-energy compounds such as ATP, and is therefore vital to energy metabolism. All told, magnesium participates in more than 300 chemical reactions. Though perhaps most notable for its role in DNA and RNA synthesis, magnesium also influences nerve and muscle function—especially in heart tissue.

Dietary Sources, Deficiency, Toxicity, and Recommended Intake of Magnesium Good sources of dietary magnesium include whole grains, green leafy vegetables, seafood, legumes, nuts, chocolate, and unprocessed (brown) rice (see Figure 8.12). In general, whole-grain foods contain more magnesium than do those made with refined grains. Magnesium's bioavailability is variable, and it is likely that several dietary factors influence its absorption. Some studies suggest that a diet high in calcium or phosphorus may decrease the bioavailability of magnesium.[20] Protein deficiency and/or high levels of dietary fiber may also decrease the absorption of magnesium.[21]

Magnesium deficiency is rare, but is sometimes observed in malnourished alcoholics experiencing dietary inadequacy and poor overall nutrient absorption.

FIGURE 8.12   MAGNESIUM CONTENT OF SELECTED FOODS

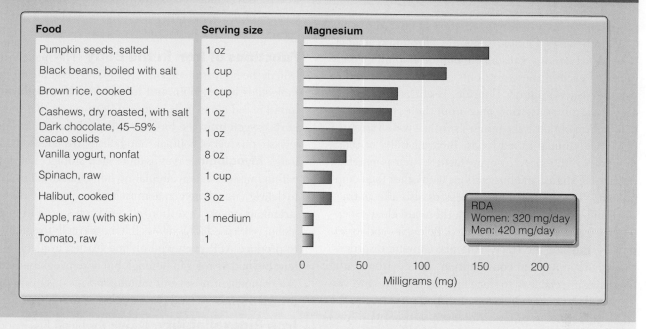

Food	Serving size	Magnesium
Pumpkin seeds, salted	1 oz	
Black beans, boiled with salt	1 cup	
Brown rice, cooked	1 cup	
Cashews, dry roasted, with salt	1 oz	
Dark chocolate, 45–59% cacao solids	1 oz	
Vanilla yogurt, nonfat	8 oz	
Spinach, raw	1 cup	
Halibut, cooked	3 oz	
Apple, raw (with skin)	1 medium	
Tomato, raw	1	

RDA
Women: 320 mg/day
Men: 420 mg/day

Milligrams (mg)

Source: USDA Nutrient Database for Standard Reference, Release 28. Available at: https://ndb.nal.usda.gov/ndb/nutrients/index

Severe magnesium deficiency causes abnormal nerve and muscle function. Magnesium toxicity can occur when large amounts of supplemental magnesium or medications that contain magnesium, such as milk of magnesia (magnesium hydroxide), are consumed. Such medications are often used to treat heartburn, indigestion, and constipation. Symptoms of magnesium toxicity include diarrhea, intestinal cramping, and nausea. Because severe magnesium toxicity can cause heart failure, the consumption of magnesium supplements should be carefully monitored.

RDAs for magnesium are 420 and 320 mg/day for men and women, respectively. These amounts of magnesium are easily obtained by eating three ounces of halibut, a cup of spinach, and a serving of rice. The UL for magnesium is 350 mg/day for adults. This UL is unusual because it is actually similar to or less than the established RDA values. This is because the UL does not apply to magnesium obtained from foods—only that obtained from supplemental and medicinal forms.

Dark chocolate provides many different nutrients, including magnesium.

© Petros Tsonis/Shutterstock.com

8-6 SULFUR

Sulfur (S) is a component of certain amino acids (e.g., cysteine and methionine) in the body. Sulfur-containing amino acids give many proteins structural rigidity. As such, they are necessary for healthy joints, hair, skin, and nails. It is not surprising that protein-rich foods provide the body with generous amounts of sulfur. Sometimes, dissolved sulfur is detectable in drinking water by its egg-like (but harmless) odor and taste. Certain juices, beers, wines, and ciders contain sulfur, and sulfur-containing agents are added to some dried foods to preserve their freshness and color.

Adequate sulfur intake enables the synthesis of compounds that construct healthy connective tissue and facilitate nerve function. Sulfur is also an important component of the B vitamins thiamin and biotin, and is therefore essential to energy metabolism. Sulfur requirements are easily met when adequate amounts of sulfur-containing amino acids are consumed. The DRI committee took this into account when developing intake recommendations for protein and decided not to publish additional DRI values for sulfur.

sulfur (S) A major mineral that is a component of certain amino acids (e.g., cysteine and methionine).

8-7 TRACE MINERALS

Although the body only requires the essential trace minerals (iron, copper, iodine, selenium, chromium, manganese, molybdenum, and zinc) in minute amounts, a daily intake of each of these nutrients is vital to one's health. Another trace mineral, fluoride, is not technically an essential nutrient, but may nonetheless influence health. You obtain trace minerals from a wide variety of plant- and animal-based foods. Bioavailability of these minerals is influenced by many factors, such as genetics, nutritional status, and interactions with other food components. The natural aging processes also affects trace mineral bioavailability. Although there are clear exceptions (e.g., iron and iodine), both deficiencies and toxicities are rare for the trace minerals. Genetic disorders, excessive supplement consumption, and environmental overexposure can lead to trace mineral toxicity, however.

While trace minerals are absorbed primarily in the small intestine, regulation of their respective blood concentrations varies from one mineral to the next. For example, healthy iron status is regulated through iron absorption in the small intestine, whereas other minerals, such as iodine and selenium, are regulated primarily by the kidneys. The liver assists in the regulation of some minerals, such as copper and manganese, by incorporating excess amounts of these trace minerals into bile, which is subsequently eliminated in the feces.

Most trace minerals do not exist in their free forms within cells—they are instead bound to specific proteins. Some trace minerals act as cofactors (enzyme activators), while others function as components of larger nonenzymatic molecules such as hemoglobin. Like several of the major minerals, some trace minerals such as fluoride provide strength and structure to bones and teeth.

8-7a Iron

Iron (Fe) is a trace mineral that is necessary for oxygen and carbon dioxide transport, energy metabolism, the stabilization of free radicals, and the synthesis of DNA. Although iron is likely the most studied trace mineral, iron deficiency remains the most common, yet preventable, micronutrient deficiency in the United States and around the world.[22]

Scientists once believed that iron had just one function in the body, which was to transport oxygen and carbon dioxide in the blood. However, it is now clear that iron serves many other important functions as well.

Functions of Iron in the Body The majority of iron in the body is incorporated into the oxygen-carrying molecules hemoglobin and myoglobin. **Hemoglobin**, found in red blood cells, is a complex protein that transports oxygen from the lungs to cells and carbon dioxide, a waste product of cellular metabolism, from cells to the lungs. **Myoglobin** serves as an oxygen-storage molecule within muscles. When circulating hemoglobin is not able to deliver enough oxygen to muscles (as is the case during strenuous physical exertion), the oxygen stored in myoglobin is released. This allows the muscle cells to continue to produce ATP, even when blood oxygen availability is low. Beyond its roles in hemoglobin and myoglobin, iron is a component of several metalloenzymes, is necessary for DNA synthesis, and plays a role in immunity.

Iron Bioavailability Except for blood loss associated with menstruation, injury, and childbirth, the body loses very little iron once it has been absorbed. The body must therefore be careful not to absorb more iron than it needs. Iron bioavailability is influenced by several factors, such as the type of iron consumed, the presence or absence of other dietary components, and an individual's iron status.[23] Iron status greatly influences the amount of iron absorbed from a food. Specifically, absorption increases when iron stores are low and decreases when they are high. Iron absorption is highest among those with the greatest need for iron such as people with iron deficiency anemia, women with heavy menstrual losses, and pregnant women.

Aside from an individual's iron status, another important factor that influences iron bioavailability is the type of iron consumed. There are two forms of dietary iron: heme iron and nonheme iron. The bioavailability of heme iron is two to three times greater than that of nonheme iron. Because **heme iron** is a part of hemoglobin and myoglobin, it is found only in meat such as beef, pork, poultry, and fish.

iron (Fe) A trace mineral needed for oxygen and carbon dioxide transport, energy metabolism, stabilization of free radicals, and synthesis of DNA.

hemoglobin A complex protein that transports oxygen from the lungs to cells and carbon dioxide from the cells to the lungs.

myoglobin An oxygen-storage molecule located within muscles.

heme iron Iron that is part of hemoglobin and myoglobin and is found only in meat.

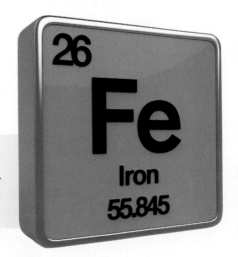

The chemical symbol for iron, Fe, was named for the Latin word for iron—ferrum.

©Nerthuz/Shutterstock.com

Nonheme iron is not part of hemoglobin or myoglobin, and is found mainly in plant-based foods such as green leafy vegetables, mushrooms, and legumes.

Many dietary factors influence the bioavailability of nonheme iron. For instance, nonheme iron exists in two charged forms: *ferric iron* (Fe^{3+}) and *ferrous iron* (Fe^{2+}). The latter is more bioavailable than the former. The following factors can also influence the bioavailability of nonheme iron:

- Vitamin C (ascorbic acid) is one of the best-known enhancers of nonheme iron absorption. Because vitamin C converts ferric iron to ferrous iron in the small intestine, consuming it with nonheme iron in a meal enhances iron absorption.

- **Meat factor** is a compound found in meat that increases the bioavailability of nonheme iron. Consuming even a small amount of meat factor along with nonheme iron-containing grains or vegetables will increase iron bioavailability.

- **Chelator** compounds such as phytates and *polyphenols* bind to nonheme iron in the intestine, making the iron unavailable for absorption. Phytates are found in many vegetables, grains, and seeds, whereas polyphenols are found primarily in plant-based foods such as spinach, tea, coffee, and red wine. If you were to consume one cup of coffee or tea with a nonheme iron-containing meal, iron absorption might decrease by 40 to 70 percent.[24]

Iron Absorption, Transport, and Storage

Although iron deficiency causes many problems, iron toxicity is perhaps even more dangerous. Because of this, complex homeostatic mechanisms regulate the amount of iron absorbed and stored in the body, ensuring that there is neither too little nor too much. Transport proteins begin the process of absorption by escorting iron from the microvilli into the enterocytes. Once iron is absorbed into the intestinal cell, the first level of iron regulation begins. Iron-storing proteins acting as gatekeepers determine the initial fate of iron by either releasing it into the blood or retaining it in the intestinal cell for subsequent elimination in the feces. This dynamic process optimizes iron bioavailability in relation to the body's need for iron (see Figure 8.13).

Dietary iron released from the intestinal cell into the blood binds to a protein called **transferrin**, which delivers the iron to special receptors on cell membranes. The number of iron-binding receptors on a cell membrane changes in response to the cell's need for iron: cells increase the number of

nonheme iron Iron that is not part of hemoglobin and myoglobin and is found primarily in plant-based foods.

meat factor A compound found in meat that increases the bioavailability of nonheme iron.

chelator A compound that binds to nonheme iron in the intestine, making it unavailable for absorption.

transferrin A protein in the blood that binds to dietary iron once it is released from an intestinal cell.

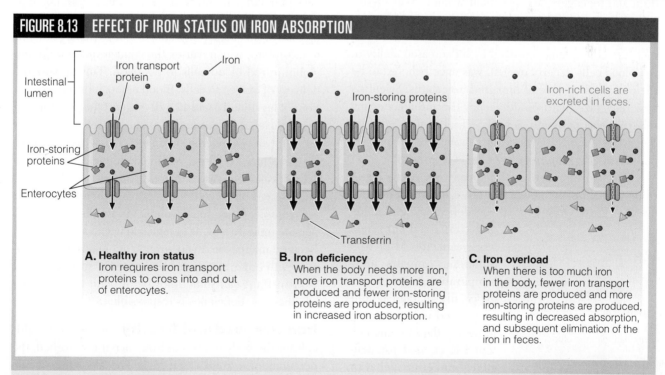

FIGURE 8.13 EFFECT OF IRON STATUS ON IRON ABSORPTION

A. Healthy iron status
Iron requires iron transport proteins to cross into and out of enterocytes.

B. Iron deficiency
When the body needs more iron, more iron transport proteins are produced and fewer iron-storing proteins are produced, resulting in increased iron absorption.

C. Iron overload
When there is too much iron in the body, fewer iron transport proteins are produced and more iron-storing proteins are produced, resulting in decreased absorption, and subsequent elimination of the iron in feces.

The body increases iron absorption when it needs more iron and decreases iron absorption when it has excess amounts.

receptors when they need more iron, and decrease the number of receptors when they need less.

When more iron is asorbed than is needed, small amounts of excess iron are stored in the body. Surplus iron is incorporated into **ferritin**, a protein found primarily in liver, skeletal muscle, and bone marrow cells. Ferritin forms a protein complex that only releases iron when it is needed. In this way, ferritin serves as the body's iron reserve. In fact, measuring blood ferritin levels is one way to assess a person's iron status. Another important regulator of iron is called **hepcidin**, which is made by the liver. Hepcidin regulates the level of iron in blood and its distribution to various tissues of the body by reducing dietary iron absorption.

Women of child-bearing years have the highest risks of developing iron deficiency anemia.

Iron Deficiency When iron-containing red blood cells wear out, the body quickly reclaims the iron contained therein so that it can be used to make new red blood cells. This system of recycling and reusing is not enough to meet the body's requirement for iron, however. If iron intake does not adequately replace iron lost from the body, iron deficiency can develop. Because iron requirements increase during periods of growth and development, iron deficiency is typically observed in infants, growing children and teenagers, and pregnant women. Women are especially at risk for iron deficiency because of regular monthly blood loss associated with menstruation. Anemia, a condition characterized by a decreased ability of red blood cells to carry oxygen, was once believed to be the only consequence of impaired iron status. Scientists now recognize that iron deficiency can influence many other aspects of health as well.[25] If left untreated, iron deficiency anemia increases susceptibility to infections, and can also lead to irregular or rapid heartbeat.

The consequences of impaired iron status begin to occur before the noticeable onset of iron deficiency anemia. Early symptoms of iron deficiency are fatigue and impaired physical work capacity. Mild iron deficiency impairs body temperature regulation—especially in cold conditions—and may negatively influence the immune system.[26] Even mild iron deficiency can affect behavior and intellectual abilities in children.[27] Unfortunately,

some of these effects are irreversible, even with iron supplementation. Studies suggest that mild iron deficiency experienced during pregnancy increases the risk of premature delivery, low birth weight, and maternal death.[28] Because blood ferritin levels (which reflect iron storage) decrease long before anemia develops, measuring blood ferritin concentration allows detection of iron deficiency in its early stages.

Severe iron deficiency is characterized by the presence of microcytic (small), hypochromic (lightly colored) red blood cells. Microcytic, hypochromic anemia develops due to an inability to produce enough heme (and thus hemoglobin), which causes the concentration of hemoglobin in red blood cells to decrease. This impairs the delivery of oxygen to cells. Blood hemoglobin concentrations of less than 13 and 12 g/dL suggest the possibility of iron deficiency in men and women, respectively. Another indicator of severe iron deficiency is low **hematocrit**, the percentage of blood volume comprised of red blood cells. Hematocrit values decrease during iron deficiency because red blood cells are small and therefore take up less volume. Hematocrit values of less than 39 and 35 percent indicate anemia in men and women, respectively. Because microcytic, hypochromic anemia is also associated with other nutritional deficiencies, such as that of vitamin B_6, it can sometimes be difficult to determine whether iron deficiency is responsible.

Iron Overload and Toxicity Because it is difficult for the body to excrete iron once it is absorbed, the potential for iron toxicity is high. The regulation of iron absorption generally protects healthy individuals from iron overload, even when an abundance of iron-rich foods

ferritin A protein complex that stores iron and releases it only when it is needed.

hepcidin A protein released by the liver that regulates the amount of iron circulating in the blood.

hematocrit The percentage of blood volume composed of red blood cells.

are consumed. Iron regulation is not always absolute, however. Ingestion of large amounts of iron (especially in supplemental forms) can overload the body's regulatory mechanisms, resulting in the absorption of too much iron. Signs and symptoms associated with acute iron toxicity include nausea, vomiting, diarrhea, rapid heart rate, confusion, and damage to the liver and other organs. Because iron toxicity is a leading cause of accidental death in young children, parents and care-providers must be sure to store nutrient supplements out of their reach.

Iron toxicity can also be caused by a genetic abnormality called **hereditary hemochromatosis**, which involves a defect in, or overproduction of, one of the iron-regulating proteins. When too much iron is absorbed, it begins to accumulate in the body, eventually causing joint pain, bronze skin color, and organ damage. This condition is easily overlooked in the early states—most people are not properly diagnosed until middle age. One of the most effective treatments for hereditary hemochromatosis is regular removal of small amounts of blood to lower iron levels in the body. Chelation therapy, administering agents that bind excess iron in the blood, is also used to treat this disorder. Hemochromatosis (which affects approximately one out of every 1,000 Americans) is more likely to develop in men than in women, and tends to be more common in Caucasians than in other racial groups.

Dietary Sources and Recommended Intake of Iron

Excellent sources of heme iron include shellfish, beef, poultry, and organ meats such as liver. Enriched foods and other iron-fortified foods are good sources of nonheme iron (see Figure 8.14). Iron RDAs are 8 and 18 mg/day for men and women, respectively. For women, the RDA increases to 27 mg/day during pregnancy. Because a 3-ounce serving of beef contains only about 3 milligrams of iron, it can be difficult for some women to meet their RDAs—especially while pregnant. For this reason, iron supplements are often recommended during pregnancy. To prevent GI distress and other complications from iron excess, a UL of 45 mg/day has been set for iron.

Special Recommendations for Vegetarians and Endurance Athletes

Because only meat contains substantial amounts of highly bioavailable iron, vegetarians may have difficulty consuming adequate amounts of iron. As such, the Institute of Medicine estimates that dietary iron requirements for vegan vegetarians are 80 percent higher than those of omnivores.[29] To take advantage of meat

hereditary hemochromatosis A genetic abnormality whereby too much iron is absorbed, causing it to accumulate in the body.

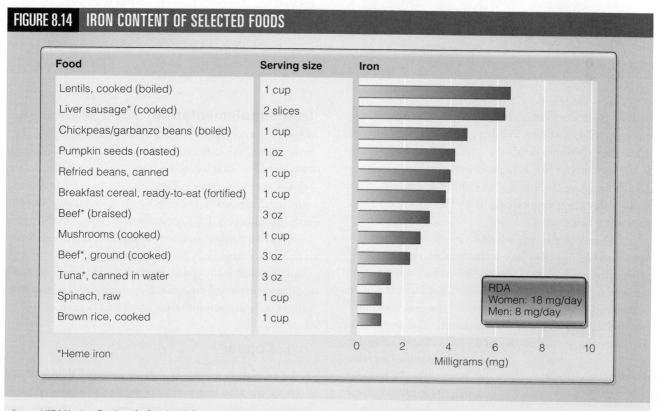

FIGURE 8.14 IRON CONTENT OF SELECTED FOODS

Food	Serving size	Iron
Lentils, cooked (boiled)	1 cup	
Liver sausage* (cooked)	2 slices	
Chickpeas/garbanzo beans (boiled)	1 cup	
Pumpkin seeds (roasted)	1 oz	
Refried beans, canned	1 cup	
Breakfast cereal, ready-to-eat (fortified)	1 cup	
Beef* (braised)	3 oz	
Mushrooms (cooked)	1 cup	
Beef*, ground (cooked)	3 oz	
Tuna*, canned in water	3 oz	
Spinach, raw	1 cup	
Brown rice, cooked	1 cup	

RDA
Women: 18 mg/day
Men: 8 mg/day

Milligrams (mg)

*Heme iron

Source: USDA Nutrient Database for Standard Reference, Release 28. Available at: https://ndb.nal.usda.gov/ndb/nutrients/index

One Lucky Iron Fish

Iron deficiency anemia is one of the most common, yet treatable, nutrition-related health problems in developing countries. Chronic anemia impairs cognitive ability, growth, and development in children. During pregnancy, anemia increases risk of low-birth-weight infants, which compromises the health of future generations. Because iron supplementation is not a realistic, long-term solution in rural, impoverished regions of the world where both access and cost are significant barriers, a new innovation called Lucky Iron Fish™ may be the solution to this growing problem.[31] By placing this fish-shaped cast iron ingot into a cooking pot of boiling water, iron is released. As the ingredients absorb the iron-rich water, the overall iron content of the meal is enhanced. One Lucky Iron Fish provides an entire family with up to 90 percent of their daily iron needs for up to five years. This easy, effective alternative to iron supplementation can improve the health of entire families throughout the world.[32] All it takes is one little fish.

Courtesy of Lucky Iron Fish

Courtesy of Stephanie V. Riley

factor's positive effect on nonheme iron absorption, vegetarians who eat fish should try to consume some fish when they consume nonheme iron sources. Eating even small amounts of fish alongside iron-fortified foods can greatly increase the amount of nonheme iron absorbed. Some fish-eating vegetarians may still require iron supplements, however.

Athletes engaged in endurance sports such as long-distance running may also have increased dietary iron requirements. Increased blood loss in feces and urine and/or chronic rupture of red blood cells caused by the impact of running on hard surfaces may be why endurance athletes have increased iron requirements. The Institute of Medicine suggests that the iron requirements of endurance athletes may increase by as much as 70 percent, although specific DRI values have not been established for this group.

copper (Cu) An essential trace mineral that acts as a cofactor for nine enzymes involved in redox reactions.

Iron Supplementation Iron supplementation is sometimes necessary when diet alone cannot sustain healthy iron status. Supplemental iron is available in both ferrous iron and ferric iron forms. Of these, the ferrous forms (ferrous fumarate, ferrous sulfate, and ferrous gluconate) are most easily absorbed. The U.S. Centers for Disease Control and Prevention (CDC) recommends taking 50–60 mg of iron, the approximate amount in one 300 mg tablet of ferrous sulfate, twice daily for three months to treat iron deficiency anemia. Starting with one-half this dose and gradually increasing to the full dose may help minimize potential side-effects such as gastrointestinal upset and dark stools.

8-7b Copper

Copper (Cu) is an essential trace mineral that acts as a cofactor for nine enzymes involved in reduction-oxidation (redox) reactions. Perhaps best known for its role as a cofactor, copper performs a host of other functions in the body as well. Similar to iron, copper is found in two distinct charged states: the *cupric form* (Cu^{2+}) and the *cuprous form* (Cu^{1+}). This is not the only commonality that copper and iron share, however. Copper and iron also have similar food sources, absorption mechanisms, and functions in the body.

Copper Bioavailability, Absorption, and Functions Copper is absorbed primarily in the small intestine. As with iron, copper absorption increases when stores are low and decreases when stores are high. The mechanisms by which this occurs are poorly understood. Antacids decrease copper bioavailability by forming insoluble copper-containing complexes that cannot be absorbed. Because iron and copper compete for the same transport proteins in the small intestine, excessive iron intake can decrease copper absorption.

Once absorbed, copper circulates in the blood to the liver, where it is bound to its primary transport protein, **ceruloplasmin**. The majority of copper in the body does not circulate freely, but is bound to various transport proteins, storage proteins, and copper-containing enzymes. Excess copper is not stored in the body—it is instead incorporated into bile and eliminated in the feces.

As a cofactor for at least nine metalloenzymes, copper plays an important role in energy metabolism, iron metabolism, neural activity, antioxidant defense, and connective tissue synthesis. **Superoxide dismutase**, a particularly important copper-containing enzyme, acts as an antioxidant, stabilizing the free radical molecules that cause cell and tissue damage in the body. Copper is also required for the syntheses of collagen and norepinephrine, a neurotransmitter that is essential to brain function.

Copper Deficiency and Toxicity Copper deficiency is rare in individuals who eat balanced and varied diets. Still, it occurs occasionally in hospitalized patients and premature infants who receive improper nutritional support. Because antacids impair copper absorption, people who consume large amounts of them can sometimes develop copper deficiencies. Signs and symptoms of copper deficiency include anemia, abdominal pain, nausea, diarrhea, and vomiting. In extreme cases, neurological problems may develop. Copper toxicity may also cause liver damage. Although copper toxicity is rare, people who have consumed large amounts of copper in drinking water and contaminated soft drinks have experienced cramping, nausea, and diarrhea.[33]

Dietary Sources and Recommended Intake of Copper Although organ meats such as liver are likely the best sources of copper, this mineral is also found in shellfish, whole-grain products, mushrooms, nuts, and legumes (see Figure 8.15). An RDA of 900 µg/day has

been established for adults. To prevent possible liver damage caused by copper toxicity, a UL of 10 mg/day has also been established.

8-7c Iodine

Iodine (I) is an essential trace mineral that although affects the body in many ways, has only one essential function—the production of hormones by the thyroid gland. These hormones regulate growth, reproduction, and energy metabolism. Thyroid hormones also influence the immune system and neural development. Technically, most of the iodine in your body exists in the form of *iodide* (I^-). Consistent with much of the literature, however, this mineral is referred to herein and commonly as *iodine*.

Iodine Bioavailability, Absorption, and Functions Iodine is highly bioavailable, and is absorbed mostly in the small intestine. Once absorbed, iodine is circulated to the thyroid gland and used to synthesize the thyroid hormone *thyroxine* (T_4). Thyroxine, which contains four iodine atoms, can be converted to *triiodothyronine* (T_3), a more active form that contains three iodine atoms. T_3 and T_4 help regulate energy metabolism, growth, and development, and are critical to proper brain, spinal cord, and skeletal development during fetal growth. Because thyroid hormones are involved in the regulation of energy metabolism, low levels can cause severe fatigue.

Thyroid-stimulating hormone (TSH), a hormone produced by the pituitary gland, regulates iodine uptake by the thyroid gland (see Figure 8.16). During periods of iodine deficiency, TSH production increases, in turn increasing the percentage of iodine taken up by the thyroid gland. As you might expect, TSH production decreases when excess iodine is consumed. Iodine not needed by the body is excreted in the urine.

A type of dietary compound called a **goitrogen** can decrease the ability of the thyroid gland to utilize iodine. Goitrogens are found in soybeans, cassava (a root eaten worldwide), and cruciferous vegetables such as cabbage, cauliflower, and Brussels sprouts. Goitrogens were named as such because

ceruloplasmin A transport protein that binds to copper and transports it in the blood.

superoxide dismutase A copper-containing enzyme that helps stabilize highly reactive free radical molecules.

iodine (I) An essential trace mineral that serves as a component of the thyroid hormones.

thyroid-stimulating hormone (TSH) A hormone, produced by the pituitary gland, that regulates iodine uptake by the thyroid gland.

goitrogen Substances that interfere with iodine uptake by the thyroid gland, which subsequently impairs the production of thyroid hormones.

FIGURE 8.15 COPPER CONTENT OF SELECTED FOODS

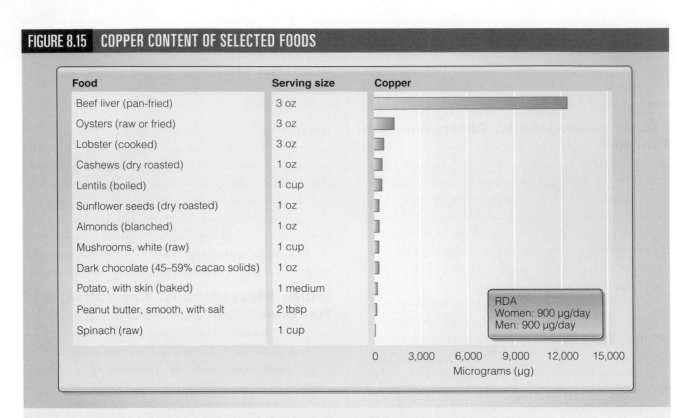

Food	Serving size	Copper
Beef liver (pan-fried)	3 oz	
Oysters (raw or fried)	3 oz	
Lobster (cooked)	3 oz	
Cashews (dry roasted)	1 oz	
Lentils (boiled)	1 cup	
Sunflower seeds (dry roasted)	1 oz	
Almonds (blanched)	1 oz	
Mushrooms, white (raw)	1 cup	
Dark chocolate (45–59% cacao solids)	1 oz	
Potato, with skin (baked)	1 medium	
Peanut butter, smooth, with salt	2 tbsp	
Spinach (raw)	1 cup	

RDA
Women: 900 µg/day
Men: 900 µg/day

0   3,000   6,000   9,000   12,000   15,000
Micrograms (µg)

Source: USDA Nutrient Database for Standard Reference, Release 28. Available at: https://ndb.nal.usda.gov/ndb/nutrients/index

---

they can potentially cause goiter, a disease characterized by enlargement of the thyroid gland. The consumption of goitrogens does not typically pose a problem, except in conditions of very low iodine intake or in people who have thyroid dysfunction.

Iodine Deficiency and Toxicity

Iodine deficiency is a significant public health problem worldwide. It is most prevalent in countries that do not have access to iodized salt and those that are not located near an ocean or sea.[34] Iodine deficiency manifests in a broad spectrum of conditions collectively called **iodine deficiency disorders (IDDs)**. The type of IDD that a person develops depends on factors such as genetics, severity of the deficiency, and age. The two most studied forms of IDD are cretinism and goiter.

iodine deficiency disorders (IDDs) A broad spectrum of conditions caused by iodine deficiency.

FIGURE 8.16 REGULATION OF IODINE UPTAKE BY THE THYROID GLAND

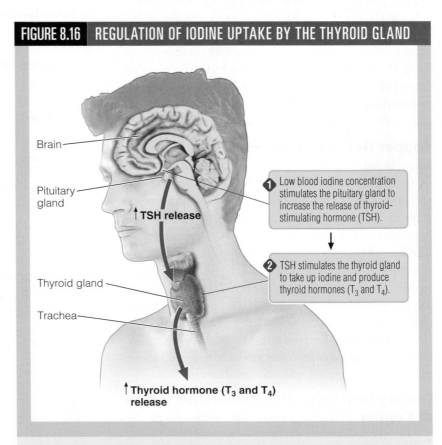

Brain

Pituitary gland

↑ TSH release

Thyroid gland

Trachea

❶ Low blood iodine concentration stimulates the pituitary gland to increase the release of thyroid-stimulating hormone (TSH).

❷ TSH stimulates the thyroid gland to take up iodine and produce thyroid hormones (T_3 and T_4).

↑ Thyroid hormone (T_3 and T_4) release

Produced in the pituitary gland, TSH stimulates iodine uptake in the thyroid gland and the production of thyroid hormones T_3 and T_4. The production of TSH increases during iodine deficiency.

The most severe form of iodine deficiency, **cretinism**, affects babies born to iodine-deficient mothers. During fetal growth, the baby relies on the mother's thyroid hormones for its own growth and development. If a mother is iodine-deficient, she does not produce sufficient thyroid hormones, which affects the baby's growth and development. Cretinism causes severe mental impairment, poor growth, infertility, and increased risk of death. Cretinism is preventable by consuming iodine-rich food or iodine supplements early in pregnancy.

When iodine deficiency occurs, low levels of T_3 and T_4 signal the pituitary gland to secrete increased amounts of TSH in order to boost thyroid hormone production. Eventually, elevated levels of TSH cause the thyroid gland to enlarge, a condition known as **goiter**. When a goiter grows very large, it can obstruct a person's trachea, making breathing and swallowing difficult. In some cases, portions of the thyroid gland are surgically removed to prevent these complications.

A high intake of iodine can cause iodine toxicity, which can take several forms, including both *hypothyroidism* (underactive thyroid activity) and *hyperthyroidism* (overactive thyroid activity). While iodine deficiency is the main cause of goiter, iodine toxicity can cause thyroid gland enlargement as well. The mechanisms by which this occurs are poorly understood. Further complicating the diagnosis of iodine deficiency, both hypothyroidism and hyperthyroidism can be caused by factors other

Iodine Deficiency and Iodization

Iodine deficiency is relatively rare in the United States, but this was not always the case. Prior to salt iodization in the 1920s, iodine deficiency was a major public health problem in some regions of the country. In fact, the region that spanned the Great Lakes and Rocky Mountains was once referred to as the *Goiter Belt*. When iodine deficiency was discovered to be the cause of goiter, state-wide campaigns promoting iodized salt as the cure for goiter were launched. A 75 percent reduction in childhood goiter was noted in just four years, and by the 1950s, less than 1 percent of children were found to be affected by goiter. Clearly, iodized salt has drastically improved the health of many people—a prime example of how scientists can team up with public health officials and food manufacturers to improve the well-being of millions of people through better nutrition.

© Carlos Yudica/Shutterstock.com

than diet. For example, the autoimmune disease Hashimoto's thyroiditis is caused by the production of antibodies that destroy the thyroid hormone-producing cells in the thyroid gland, resulting in hypothyroidism.

Dietary Sources and Recommended Intake of Iodine The amount of iodine contained in a food frequently depends on the iodine content of the soil or water in which the food was grown (see Figure 8.17). Ocean fish and mollusks tend to contain high amounts of iodine because they concentrate iodine from seawater into their tissues. Edible varieties of seaweed such as nori are excellent sources of iodine, as are many dairy products because

©Mediscan/Visuals Unlimited, Inc. / Lonely Planet Images/Getty Images

Children born to iodine-deficient women are at increased risk for cretinism (left), which causes delayed growth, developmental delay, and other abnormal features. Goiter (right), caused by enlargement of the thyroid gland in response to iodine deficiency, affects both children and adults.

cretinism A severe form of iodine deficiency that affects babies born to iodine-deficient mothers.

goiter A sign of iodine deficiency characterized by an enlarged thyroid gland.

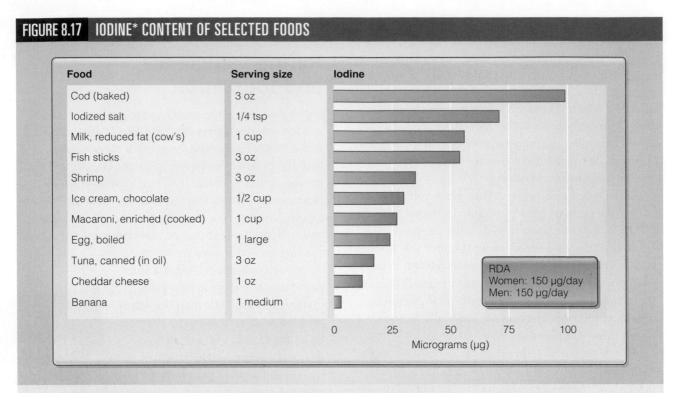

FIGURE 8.17 IODINE* CONTENT OF SELECTED FOODS

Food	Serving size	Iodine
Cod (baked)	3 oz	
Iodized salt	1/4 tsp	
Milk, reduced fat (cow's)	1 cup	
Fish sticks	3 oz	
Shrimp	3 oz	
Ice cream, chocolate	1/2 cup	
Macaroni, enriched (cooked)	1 cup	
Egg, boiled	1 large	
Tuna, canned (in oil)	3 oz	
Cheddar cheese	1 oz	
Banana	1 medium	

RDA
Women: 150 µg/day
Men: 150 µg/day

Micrograms (µg)

*Iodine content of foods is variable; values should be considered approximate.

Source: Pennington JAT, Schoen SA, Salmon GD, Young B, Johnson RD, Marts RW. Composition of core foods of the U.S. food supply, 1982–1991. III. Copper, manganese, selenium, iodine. Journal of Food Composition and Analysis. 1995, 8:171–217. National Institutes of Health. Office of Dietary Supplements. Iodine Fact Sheet for Health Professionals. Available from: https://ods.od.nih.gov/factsheets/Iodine-HealthProfessional/#en10

iodine is used in milk processing. Most of the iodine you consume probably comes from iodized table salt. Approximately one-half of all salt consumed in America is iodized, meaning that it has been fortified with iodine. (Kosher salt and sea salt are not traditionally iodized.) Iodine's RDA is 150 µg/day. To prevent iodine toxicity, a UL of 1,100 µg/day has been established for this mineral.

8-7d Selenium

Selenium (Se) is an essential trace mineral that is critical to reduction-oxidation (redox) reactions, thyroid function, and the activation of vitamin C. The essentiality of selenium was only discovered in the 1950s. Since that time, much has been learned about how the body uses this trace mineral. Optimal selenium intake is believed to decrease the risk of cancer, protect the body from toxins and free radicals, activate vitamin C, and enhance immunity.

selenium (Se) An essential trace mineral that is critical to redox reactions, thyroid function, and the activation of vitamin C.

selenoprotein A protein composed of selenium-containing amino acids.

Keshan disease A disease caused by severe selenium deficiency that mostly affects children and causes serious heart problems.

Selenium Bioavailability, Absorption, and Functions The bioavailability of selenium is high, and its intestinal absorption is not regulated. Once absorbed, selenium circulates in the blood to the body's cells. Some selenium is incorporated into amino acids, which are subsequently used to make **selenoprotein** molecules. There are at least 14 selenoproteins in the body. One group of enzymatic selenoproteins helps protect against oxidative damage and may protect against certain types of cancer, but more research is required to understand the role that selenium plays in this process.[35] The kidneys maintain blood selenium concentrations, and excess amounts are simply excreted in the urine. When consumption is high, selenium can also be expelled in the breath, causing a garlicky odor.

Dietary Sources, Deficiency, Toxicity, and Recommended Intake of Selenium Good sources of selenium include Brazil nuts, meat (especially organ meat), and fish (see Figure 8.18). Severe selenium deficiency results in **Keshan disease**, a potentially fatal disease that causes serious heart problems and mostly affects children. Keshan disease was first documented in the Keshan region of China, which has very low levels of selenium in its soil. Consuming high amounts of selenium

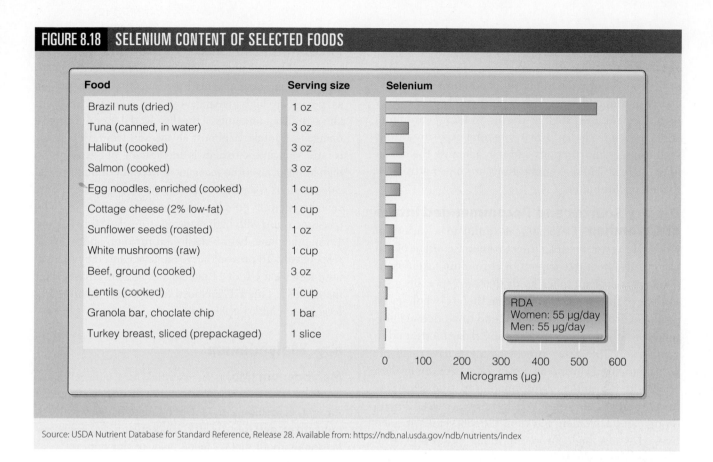

FIGURE 8.18 SELENIUM CONTENT OF SELECTED FOODS

Food	Serving size	Selenium
Brazil nuts (dried)	1 oz	
Tuna (canned, in water)	3 oz	
Halibut (cooked)	3 oz	
Salmon (cooked)	3 oz	
Egg noodles, enriched (cooked)	1 cup	
Cottage cheese (2% low-fat)	1 cup	
Sunflower seeds (roasted)	1 oz	
White mushrooms (raw)	1 cup	
Beef, ground (cooked)	3 oz	
Lentils (cooked)	1 cup	
Granola bar, choclate chip	1 bar	
Turkey breast, sliced (prepackaged)	1 slice	

RDA
Women: 55 µg/day
Men: 55 µg/day

Micrograms (µg)

Source: USDA Nutrient Database for Standard Reference, Release 28. Available from: https://ndb.nal.usda.gov/ndb/nutrients/index

can result in nausea, vomiting, diarrhea, and brittleness of the teeth and fingernails. Selenium's RDA is 55 µg/day, and its UL is 400 µg/day.

8-7e Chromium

Chromium (Cr) is a trace mineral that may be critical to proper insulin function. Chromium was first designated as an essential nutrient because scientists discovered that its deficiency caused a diabetic-like state in certain animals. It was discovered later that chromium might be critical to glucose regulation and insulin function in humans as well. While there is considerable debate as to whether chromium is a required nutrient for humans, it is still classified by some experts as essential.[36]

Chromium Bioavailability, Absorption, and Functions Chromium bioavailability is increased by the presence of vitamin C and acidic medications such as aspirin (acetylsalicylic acid). Conversely, chromium absorption is decreased by the presence of antacids. Regardless of how much is consumed, however, very little (less than 2 percent) chromium is absorbed—most is excreted in the feces. Absorbed chromium is circulated in the blood to the liver, and excess is excreted in the urine. Scientists do not understand why

consuming large amounts of simple sugars can increase urinary chromium excretion.[37] When the body is unable to excrete excess chromium in the urine, it is deposited mainly in the liver.

Some studies suggest that chromium is required for the hormone insulin to function properly—especially in people with type 2 diabetes.[38] Chromium may also be essential to normal growth and development in children. At the very least, it appears to increase muscle mass and decrease fat mass in laboratory animals.[39] Because of this, a form of chromium called **chromium picolinate** has been widely marketed as an ergogenic aid for athletes. Other types of chromium supplements are promoted as products that help regulate blood glucose.

Chromium Deficiency and Toxicity Chromium deficiency has only been observed in hospitalized patients receiving inadequate nutrition support. Chromium deficiency results in elevated blood glucose levels, decreased sensitivity to insulin, and weight loss. Chromium toxicity is also rare, even when

chromium (Cr) A trace mineral that may be critical to proper insulin function.

chromium picolinate A form of chromium taken as an ergogenic aid by some athletes.

supplemental chromium is consumed. This is probably attributable to chromium's low bioavailability. Toxic levels of chromium have been observed in people exposed to high levels of industrially released chromium, however. For example, when stainless steel is heated to very high temperatures, such as during welding, chromium is released into the air. Environmental exposure of this kind causes skin irritations and may increase the risk of lung cancer.[40] These complications are never attributed to dietary chromium.

Dietary Sources and Recommended Intake of Chromium

Chromium is found in a variety of foods. However, because the chromium content of soil can greatly affect the amount of chromium taken up by plants, it is difficult to list foods that are especially good sources of this trace mineral. Nonetheless, whole-grain products, fruits, and vegetables tend to be rich in chromium, whereas refined cereals and dairy foods tend to contain very little. Some processed meats have relatively high amounts of chromium, as do some beers and wines.

There is insufficient information to establish RDAs for chromium. However, AIs have been set at 25 and 35 µg/day for women and men (19–50 years), respectively. Because high chromium intake poses no known adverse effects, the Institute of Medicine has not established ULs for this mineral.

8-7f Manganese

Manganese (Mn) is an essential trace mineral that is a cofactor for metalloenzymes needed for bone formation, glucose synthesis, and energy metabolism. As with chromium, manganese deficiency is rare in humans, and toxicity typically occurs only from excessive environmental exposure.

Manganese Bioavailability, Absorption, and Functions

Regardless of manganese intake, very little (less than 10 percent) is absorbed. Excess manganese is delivered to the liver, where it is incorporated into bile and excreted in the feces. Manganese is a cofactor for numerous enzymes, such as those involved in glucose production and bone formation. Manganese is also important to several enzymes involved in energy metabolism. Like copper and selenium, manganese is a cofactor for an enzyme that protects cells from free radicals.

manganese (Mn) An essential trace mineral that is a cofactor for metalloenzymes needed for bone formation, glucose synthesis, and energy metabolism.

molybdenum (Mo) An essential trace mineral that is a cofactor for several important metalloenzymes needed for amino acid and purine metabolism.

Dietary Sources, Deficiency, Toxicity, and Recommended Intake of Manganese

Good sources of manganese are whole-grain products, pineapples, nuts, and legumes. Dark green leafy vegetables such as spinach are high in manganese, and water can also contain significant amounts of the dissolved mineral. Because manganese intake is almost always sufficient, deficiency is rare. Manganese toxicity is uncommon, but exposure to high levels of airborne manganese (a result of mining) can cause serious neurological problems. Manganese toxicity can also occur in people with liver disease and those who consume water with high manganese levels. AIs of 2.3 and 1.8 mg/day have been established for men and women, respectively. To prevent nerve damage caused by manganese toxicity, a UL of 11 mg/day has been established for manganese. This UL includes manganese consumed as food, water, and supplements.

8-7g Molybdenum

Molybdenum (Mo) is an essential trace mineral that is a cofactor for several important metalloenzymes needed for protein metabolism. Molybdenum deficiency is almost unheard of, and there is little concern among health professionals about dietary inadequacy of this mineral.

Molybdenum Bioavailability, Absorption, and Functions

Molybdenum appears to be absorbed almost completely in the intestine. After circulating in the blood to the liver, molybdenum serves as a cofactor for several enzymes involved in the metabolism of sulfur-containing amino acids such as methionine and cysteine. Molybdenum is also critical to the structural building blocks of DNA. Finally, molybdenum-containing enzymes are essential to the detoxification of drugs in the liver.

Dietary Sources, Deficiency, Toxicity, and Recommended Intake of Molybdenum

Although the molybdenum content of food varies greatly depending on soil molybdenum levels, legumes (e.g., peas and lentils), grains, and nuts tend to be good sources of this mineral. Only one case of primary molybdenum deficiency has been documented. In this case, a hospitalized patient received intravenous nutritional support that lacked molybdenum. The patient's molybdenum deficiency was characterized by abnormal heart rhythms, headache, and visual problems.[41] There are no known adverse effects of high intakes of molybdenum in humans, but toxicity has been shown to harm reproduction capabilities in animals. Molybdenum's RDA is 45 µg/day, and to prevent potential reproductive problems, a UL of 2,000 µg/day has been established for this mineral.

8-7h Zinc

Zinc (Zn) is an essential trace mineral involved in gene expression, immune function, and cell growth. Zinc is a cofactor for more than 300 metalloenzymes and is found in almost every cell of the body. Inadequate intake of dietary zinc can disrupt many important physiological functions, such as growth, reproduction, taste, smell, immunity, DNA replication, and protein synthesis. Even moderate zinc deficiency can cause serious health concerns.

Zinc Bioavailability, Absorption, and Functions Zinc bioavailability is influenced by a variety of dietary factors, many of which are similar to those that influence iron absorption. The bioavailability of zinc is greater in animal-based foods than in plant-based foods. Acidic substances (e.g., vitamin C) increase its absorption. Because zinc absorption relies on the same transport protein as several other similarly charged minerals (e.g., Fe^{2+}, Cu^{2+}, and Ca^{2+}), high intakes of these minerals can decrease zinc bioavailability.

Zinc absorption is highly regulated, requiring at least two proteins. After the first protein transports zinc into enterocytes, the mineral is bound to a second protein, **metallothionine**, which regulates the amount of zinc released into the blood. The synthesis of metallothionine decreases during periods of zinc excess and increases during periods of zinc deficiency. Because intestinal cells are continually shed and replaced, excess zinc bound to metallothionine is excreted in the feces.

Zinc functions primarily as a cofactor. Because enzymes needed for gene expression rely on zinc, this mineral is essential to protein synthesis, cell maturation, growth, and proper immune function. Zinc also acts as a potent antioxidant, and it appears to play a role in the stabilization of cell membranes. Zinc supplementation is often touted as a treatment for the common cold. While most studies do not support this claim, limited data suggest a beneficial effect in children.[42]

Dietary Sources, Deficiency, Toxicity, and Recommended Intake of Zinc High concentrations of zinc are found in shellfish, meat (especially organ meat), dairy products, legumes, and chocolate (see Figure 8.19). Zinc is also sometimes added to fortified cereals and grain products.

Zinc deficiency was first documented in Egypt and Iran, where people tend to eat plant-based diets high in phytates, which chelate zinc in the intestine. Children with zinc deficiencies

zinc (Zn) An essential trace mineral involved in gene expression, immune function, and cell growth.

metallothionine An intestinal protein that regulates the amount of zinc released into the blood.

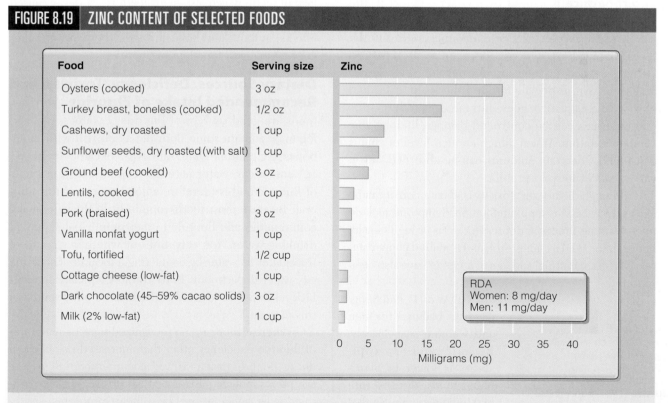

FIGURE 8.19 | ZINC CONTENT OF SELECTED FOODS

Food	Serving size	Zinc
Oysters (cooked)	3 oz	
Turkey breast, boneless (cooked)	1/2 oz	
Cashews, dry roasted	1 cup	
Sunflower seeds, dry roasted (with salt)	1 cup	
Ground beef (cooked)	3 oz	
Lentils, cooked	1 cup	
Pork (braised)	3 oz	
Vanilla nonfat yogurt	1 cup	
Tofu, fortified	1/2 cup	
Cottage cheese (low-fat)	1 cup	
Dark chocolate (45–59% cacao solids)	3 oz	
Milk (2% low-fat)	1 cup	

RDA
Women: 8 mg/day
Men: 11 mg/day

Milligrams (mg)

Source: USDA Nutrient Database for Standard Reference, Release 28. Available at: https://ndb.nal.usda.gov/ndb/nutrients/index

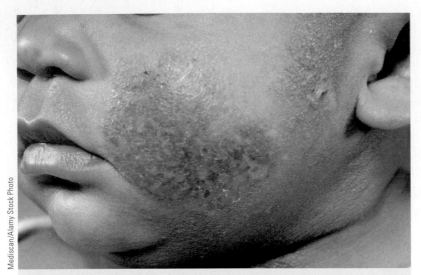

Acrodermatitis enteropathica is a genetic disorder that causes decreased zinc absorption.

exhibit delayed growth, skin irritations, diarrhea, under-developed genitals, and delayed sexual maturation. Zinc supplementation corrected many of these abnormalities, and was therefore deemed an essential nutrient.

Secondary zinc deficiency can result from chronic conditions such as liver disease, impaired kidney function, and diabetes. A condition called **acrodermatitis enteropathica** is caused by a genetic defect in the protein that transports zinc into intestinal cells. Because zinc is not absorbed, excessive amounts are lost in the feces. Babies born with acrodermatitis enteropathica fail to grow properly and usually develop severely red and scaly skin, especially around the scalp, eyes, and feet. While fatal if it goes untreated, acrodermatitis enteropathica can be controlled through lifelong zinc supplementation. When a high enough level of zinc is consumed, adequate amounts can be absorbed—even with a faulty transport protein.

Although dietary zinc toxicity is uncommon, it can be caused by excessive supplementation. Symptoms include poor immune function, depressed levels of high density lipoprotein (HDL) cholesterol, and impaired copper status. Nausea, vomiting, and loss of appetite have also been reported.

RDAs of 11 and 8 mg/day have been set for men and women, respectively. Because zinc bioavailability is lower in plant-based foods, the Institute of Medicine suggests that vegetarians—particularly vegans—consume up to 50 percent more zinc than nonvegetarians. A UL of 40 mg/day has been established for zinc.

8-7i Fluoride

Fluoride (F⁻) is a trace mineral and although it is not actually an essential nutrient, it is important because it strengthens bones and teeth. The dietary sources and functions of fluoride within the body are explained next.

Fluoride Bioavailability, Absorption, and Functions The gastrointestinal tract does not readily regulate fluoride absorption. In fact, almost all fluoride consumed is absorbed by the small intestine. It then circulates in the blood to the liver and is ultimately integrated into bones and teeth. Excess fluoride is excreted in the urine.

Fluoride affects the health of bones and teeth in several ways. First, it strengthens them by being incorporated into their basic mineral structures. Teeth that contain extra fluoride are especially resistant to bacterial breakdown and cavities (dental caries). Second, topical (not dietary) application of fluoride-containing toothpastes and treatments works directly on the enamel surface of teeth, which also helps teeth resist decay caused by bacteria and food debris. Aside from its beneficial effects on tooth and bone strength, there are no other known biological functions of fluoride.

Dietary Sources, Deficiency, Toxicity, and Recommended Intake of Fluoride Very few foods are good sources of fluoride. Potatoes, tea, and legumes contain some fluoride, as do fish with intact bones (e.g., sardines and some types of canned salmon). In some areas, water naturally contains high amounts of fluoride and is therefore a good source of this mineral. Because most foods contain little fluoride, many communities add fluoride to their municipal supply of drinking water. You can find out whether your town fluoridates its water by contacting a local dentist or the city water department. Bottled water typically contains little, if any, fluoride and is therefore a poor source of this mineral.[43]

Whereas there are no recognized signs or symptoms of fluoride deficiency other than increased risk of tooth decay, fluoride toxicity is well documented.[44] Signs and symptoms include gastrointestinal upset, excessive production of saliva, watery eyes, heart problems, and in severe cases, coma. Excessive fluoride intake can cause

acrodermatitis enteropathica A genetic abnormality that causes secondary zinc deficiency.

fluoride (F⁻) A trace mineral that strengthens bones and teeth.

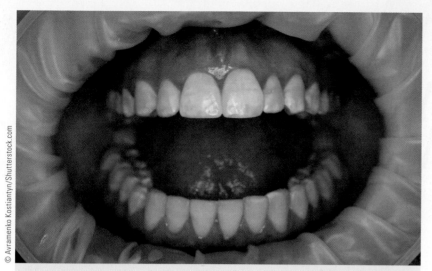
Fluoride toxicity can cause discoloration of the teeth—a condition called dental fluorosis.

- *Silicon* (Si) is important to bone growth in laboratory animals. Dietary sources of silicon include whole-grain products, and it is also used as a food additive.

- *Vanadium* (V) influences cell growth and differentiation, and it may affect thyroid hormone metabolism. Dietary sources of silicon include mushrooms, shellfish, and black pepper.

- *Arsenic* (As) may be a cofactor for a variety of enzymes, and may be important to DNA synthesis. Seafood and whole grains are good sources of arsenic.

- *Boron* (B) is especially important to reproduction in laboratory animals, and deficiency causes severe fetal malformations. Boron is found in fruits, leafy vegetables, legumes, and nuts.

dental fluorosis, pitting and mottling (discoloration) of the teeth, and **skeletal fluorosis**, weakening of the skeleton. Fluorosis is a special concern in small children, who sometimes swallow large amounts of toothpaste on a daily basis. Thus, parents should carefully monitor tooth-brushing routines.

An RDA for fluoride has not been established. However, AIs of 4 and 3 mg/day have been set for men and women, respectively. Fluoride's UL is 10 mg/day.

> **dental fluorosis** Pitting and mottling of teeth caused by excessive fluoride intake.
>
> **skeletal fluorosis** Weakening of the skeleton caused by excessive fluoride intake.

8-8 OTHER IMPORTANT TRACE MINERALS

In addition to those already covered in this chapter, other minerals may also influence your health. Whether the body requires these minerals has yet to be confirmed or denied through scientific experimentation. Still, each of these minerals deserves comment on possible biological functions and food sources. Similar to fluoride, RDAs have not been established for these minerals. Some have ULs, however.

- *Nickel* (Ni) may be critical to protein and lipid metabolism and gene expression. It may also influence bone formation. Nickel is found in legumes, nuts, and chocolate.

- *Aluminum* (Al) may be important to reproduction, bone formation, DNA synthesis, and behavior. It is found in many baked goods, grains, some vegetables, tea, and many antacids.

STUDY TOOLS 8

READY TO STUDY? IN THE BOOK, YOU CAN:

☐ Rip out the Chapter Review Card, which includes key terms and chapter summaries.

ONLINE AT WWW.CENGAGEBRAIN.COM, YOU CAN:

☐ Is caffeine healthy—or no? Watch a short video.

☐ Explore the calcium balance with an animation.

☐ Interact with figures from the text to check your understanding.

☐ Prepare for tests with quizzes.

☐ Review the key terms with Flash Cards.

9 | Energy Balance and Body Weight Regulation

LEARNING OUTCOMES

9-1 Understand the concept of neutral, negative, and positive energy balance.

9-2 Describe the factors that influence energy intake.

9-3 List and describe the components of energy expenditure.

9-4 Understand how body weight and composition measurements can be used to assess health.

9-5 Discuss factors that influence body weight.

9-6 Describe the physiologic mechanisms by which energy balance and body weight are regulated.

9-7 Discuss the components of a successful weight-loss plan.

Westend61/Getty Images

After finishing this chapter go to **PAGE 244** for **STUDY TOOLS.**

9-1 WHAT IS ENERGY BALANCE?

Some count them, some ignore them, while others curse them—but what role do calories actually play in body weight regulation? Moreover, while many people struggle to lose weight, why do some actually find it difficult to gain weight? If you have ever pondered these questions, you are not alone. In this chapter, you will learn about the relationships between energy intake, energy expenditure, and weight gain—in other words, how the foods you eat and the activities you engage in affect your body weight and composition. You will also learn about body weight regulation, the growing prevalence of obesity in the United States, and the various approaches to weight loss and weight maintenance.

What determines whether the energy in the foods you eat is used to fuel your body or is instead stored for later use? It all comes down to balance—specifically, the balance between energy intake and energy expenditure. Unless you are attempting to gain or lose weight, the amount of energy you consume should be equal to the amount of energy your body requires to function. More specifically, energy is needed to support vital body functions (basal metabolism), physical activity, and to process food (thermic effect of food), all of which are discussed in this chapter. When energy intake equals energy expenditure, a person is in a state of **neutral energy balance**, and body weight is relatively stable. When more energy is consumed than the body needs, a person is in a state of **positive energy balance**, which typically results in weight gain. **Negative energy balance** results when energy intake is less than energy expenditure. This results in weight loss. Therefore, under most conditions, body weight is a useful indicator of whether you are in neutral, positive, or negative energy balance. The relationship between energy balance and body weight is illustrated in Figure 9.1.

9-1a Energy Balance and Body Weight

During periods of positive energy balance, muscle and/or adipose tissue increases in size and weight. However, for most adults, increased body weight is primarily associated with increased body fat, and is generally considered

neutral energy balance A condition whereby energy intake equals energy expenditure.

positive energy balance A condition whereby energy intake is greater than energy expenditure.

negative energy balance A condition whereby energy intake is less than energy expenditure.

unhealthy. It is important to remember, however, that weight gain can also be associated with increased muscle mass as well as fluid retention. Positive energy balance is considered desirable and necessary during periods of growth and development, such as infancy, adolescence, and pregnancy.

Negative energy balance occurs during periods of decreased energy intake, increased energy expenditure, or both. Under this condition, stored energy reserves are broken down and utilized as a source of energy. Consequently, negative energy balance usually results in weight loss. One pound of adipose tissue supplies the equivalent of approximately 3,500 kilocalories of energy. However, it is important to recognize that weight loss does not always result exclusively from loss of body fat during periods of negative energy balance. Water and muscle loss contribute to total weight loss too.

9-1b A Closer Look at Adipose Tissue

Recall from Chapter 6 that adipose tissue is a type of connective tissue comprised largely of specialized cells called adipocytes. Each adipocyte contains a lipid-filled core that consists primarily of triglycerides. The number and size of adipocytes determine the amount of adipose tissue in your body. Only recently have researchers fully recognized the complexity of adipose tissue and its relationship to health. Once thought of as a mere passive site for energy storage, scientists now know that adipose tissue is a source of many different hormones and other signaling molecules that link obesity with several chronic, weight-related diseases. Furthermore, where adipose tissue is found in the body (subcutaneous adipose tissue vs. visceral adipose tissue) and types of adipose tissue (white adipose tissue vs. brown adipose tissue) are now known to be associated with overall health and risk for disease.

Adipose tissue consists of cells called adipocytes. The accumulation of fat droplets in adipocytes gives adipose tissue a unique look under the microscope.

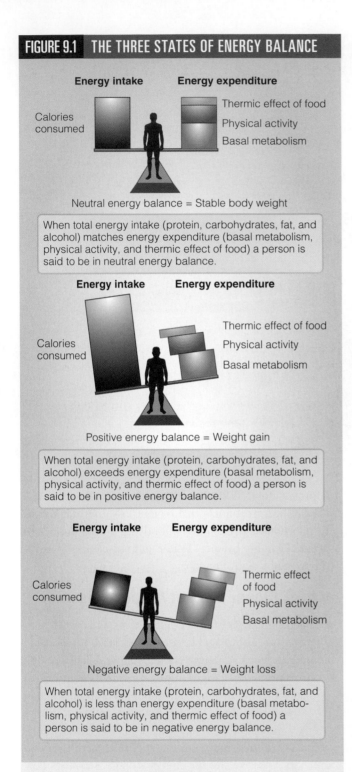

FIGURE 9.1 THE THREE STATES OF ENERGY BALANCE

Energy intake   **Energy expenditure**

Calories consumed

Thermic effect of food
Physical activity
Basal metabolism

Neutral energy balance = Stable body weight

When total energy intake (protein, carbohydrates, fat, and alcohol) matches energy expenditure (basal metabolism, physical activity, and thermic effect of food) a person is said to be in neutral energy balance.

Energy intake   **Energy expenditure**

Calories consumed

Thermic effect of food
Physical activity
Basal metabolism

Positive energy balance = Weight gain

When total energy intake (protein, carbohydrates, fat, and alcohol) exceeds energy expenditure (basal metabolism, physical activity, and thermic effect of food) a person is said to be in positive energy balance.

Energy intake   **Energy expenditure**

Calories consumed

Thermic effect of food
Physical activity
Basal metabolism

Negative energy balance = Weight loss

When total energy intake (protein, carbohydrates, fat, and alcohol) is less than energy expenditure (basal metabolism, physical activity, and thermic effect of food) a person is said to be in negative energy balance.

A change (or lack thereof) in body weight indicates if a person is in neutral, positive, or negative energy balance.

adipokines Signaling proteins secreted by adipocytes that cause a cellular response in another organ.

metabolic syndrome A cluster of conditions that occur together that increase a person's risk of heart disease, stroke, and type 2 diabetes.

Adipose Tissue, Adipokines, and Weight-Related Chronic Diseases Adipose tissue is the body's primary site of energy storage and serves as a buffer during times of energy imbalance. In addition, adipose tissue produces an array of signaling molecules, collectively referred to as **adipokines**. These important hormone-like substances provide a means of communication between adipocytes and other tissues in the body. It is believed that the accumulation of body fat may influence the production and release of certain adipokines. Although some adipokines have important roles in health maintenance, others seem to trigger a plethora of adverse metabolic disturbances. Indeed, it appears that adipokines are key players in the link between obesity and weight-related disturbances such as insulin resistance, inflammation, and other chronic diseases such as hypertension and cardiovascular disease.

Visceral Adipose Tissue and Subcutaneous Adipose Tissue Although adipose tissue is found throughout the body, visceral adipose tissue (VAT) is primarily found in the abdominal cavity (see Figure 9.2). Also referred to as intra-abdominal fat, abnormally high deposits of VAT are associated with an increased risk of a constellation of weight-related health problems, collectively referred to as **metabolic syndrome**. Metabolic syndrome by itself is not a disease. Rather, it is characterized by a cluster of conditions (elevated blood glucose, hypertension, unhealthy blood lipids, and excessive abdominal fat) that increase a person's risk of heart disease, stroke, and type 2 diabetes.

FIGURE 9.2 CROSS SECTION OF THE UPPER BODY

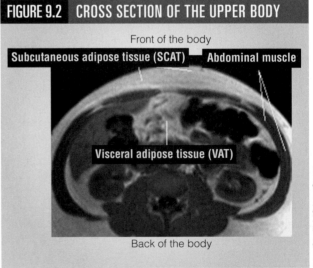

Front of the body

Subcutaneous adipose tissue (SCAT)   Abdominal muscle

Visceral adipose tissue (VAT)

Back of the body

James Cavallini / Science Source

Visceral adipose tissue (VAT) is located between the internal organs in the abdomen. Subcutaneous adipose tissue (SCAT) is found directly beneath the skin.

The majority (approximately 85 percent) of fat in the body is located directly under the skin and is referred to as subcutaneous adipose tissue (SCAT). In addition to insulating the body from heat and cold (thermoregulation), SCAT also plays an important role during times of energy imbalance. Indeed, SCAT represents your body's major energy reserve. Although SCAT is found in most regions of the body, it is most predominant around the thighs, hips, and buttocks.

Although excess body fat in general can be detrimental to health, a surplus of intra-abdominal (VAT) fat is of particular concern. In fact, numerous studies have demonstrated that having a large waist circumference (**central obesity** or **central adiposity**) is of greater concern than having excess fat deposited in the lower regions of the body (hips, thighs, and buttocks). Although excessive VAT and SCAT both impose health risks, it remains unclear as to why people with central obesity are at greater risk for developing weight-related health problems (e.g., type 2 diabetes, hypertension, and cardiovascular disease) than are those who store excess adipose tissue in lower regions of the body (e.g., hips and thighs). However, recent studies suggest that the anatomical location of VAT and the types of adipokines released may account for these differences.[1]

Determining a person's body fat distribution pattern is easier than you might think. For example, **waist circumference** is a good indicator of central obesity. This measurement is easily obtained by placing a tape measure around the narrowest area of the waist (see Figure 9.3). A waist circumference of 40 inches or greater in men and 35 inches or greater in women generally indicates excessive central adiposity.[2] People with central adiposity tend to have waist circumferences that are larger than their hip circumferences and are commonly described as being "apple-shaped." Conversely, people who carry relatively more adipose tissue around the hips and thighs are often referred to as "pear-shaped." Some clinicians prefer to use the ratio of waist circumference to hip circumference (the *waist-to-hip ratio*) to determine body fat distribution, although studies show that waist circumference alone provides a simple yet effective measure of central adiposity.[3] Based on waist circumference measures (>40 inches in men and >35 inches in women), over one-half of U.S. adults have excessive abdominal fat.[4]

White Adipose Tissue and Brown Adipose Tissue
In addition to where adipose tissue is located in the body, it is also important to recognize that there are different types of adipose tissue: **white adipose tissue (WAT)** and **brown adipose tissue (BAT)**.

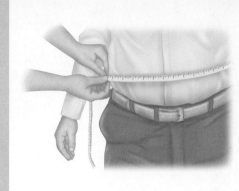

FIGURE 9.3 WAIST CIRCUMFERENCE IS USED TO ASSESS CENTRAL ADIPOSITY

- Person should stand with his feet 6 to 7 inches apart, with his weight evenly distributed to each leg.
- Person should be relaxed, and the measurement should be taken while breathing out.
- The tape measure should be positioned around the narrowest area of the waist.
- The tape should be loose enough to place one finger between the tape and the person's body.
- A waist circumference of 40 inches or greater in males and 35 inches or greater in females indicates central adiposity.

Central adiposity increases a person's risk for weight-related health problems. A person's waist circumference is an important indicator of central adiposity.

People with central obesity are often described as apple-shaped, and those who carry excess adipose tissue around their hips and thighs are described as pear-shaped.

© Vankad/Shutterstock.com

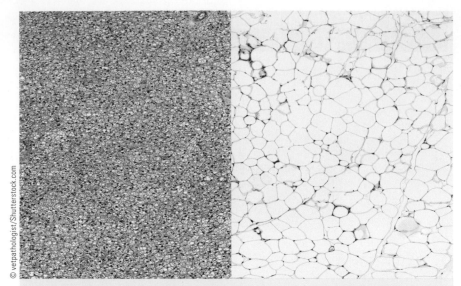

Brown adipose tissue (left) contains multiple small, lipid droplets and numerous iron-containing mitochondria, which gives brown adipose tissue a dark color. White adipose tissue (right) contains single large, lipid droplets, and very few mitochondria. Whereas the primary function of white adipose tissue is to store energy, the primary function of brown adipose tissue is to generate heat.

Whereas WAT serves as the body's primary depot for energy storage, BAT plays an important role in body temperature regulation and basal metabolism. Found primarily in the neck and upper back, BAT contains numerous iron-containing mitochondria that not only give BAT its characteristic brownish appearance, but are also metabolically active, releasing heat (thermogenesis) in the process. Until recently, it was believed that, at least in humans, only newborns had significant amounts of brown adipose tissue. However, more recent studies have demonstrated that this is not the case.[5] Not only do adults have BAT, but both exposure to cold and physical activity, particularly in cold temperatures, appear to increase its metabolic rate. This has important implications in terms of body weight regulation because BAT-related metabolism increases basal energy expenditure.

hunger A basic physiological drive to consume food.

satiety A physiological response whereby a person feels he or she has consumed enough food.

hypothalamus A region of the brain that regulates energy intake by balancing hunger and satiety.

neurotransmitters Hormone-like substances released by nerve cells; transmit electrical impulses from one nerve cell to the next.

9-2 WHAT FACTORS INFLUENCE ENERGY INTAKE?

Although it may seem simple at first, understanding energy balance is more complicated than you might think. The energy your body needs comes from the foods you eat, but understanding what you eat, how much you eat, and why you eat is quite complex. Your food choices are related to many facets of your life, and these choices serve numerous purposes beyond providing your body with nourishment. It is also important to recognize that foods people enjoy as adults may be the very foods they avoided as children. Clearly, eating behavior is complicated, and it is important to consider the psychological, physical, social, and cultural forces that influence it.

9-2a Hunger and Satiety

Hunger is a basic physiological drive to consume food, whereas **satiety** is a physiological response whereby a person feels that enough food has been consumed. Most of the time, people eat to the point of comfort. In other words, they eat until they are satiated. Nonetheless, some people continue to eat even though they feel full, sometimes to the point of discomfort. Regardless of one's level of fullness, however, most people feel hungry again within 3–4 hours after eating.

Some people may think that the stomach is the main determinant of hunger and satiety. Yet, a variety of factors influence these complex physiological sensations. Early experimental work demonstrated that hunger and satiety can be controlled in mice through electrical stimulation of specific regions of the brain. These areas soon became known as the hunger and satiety centers, respectively. Scientists now recognize that a specific, complex region of the brain called the **hypothalamus** regulates energy intake by balancing energy intake (governed by hunger and satiety) with energy expenditure. The hypothalamus receives signals that influence hunger and satiety from different parts of the body, such as the GI tract and from adipose tissue itself (via adipokines). In response to these signals, the hypothalamus releases hormone-like substances called **neurotransmitters**. Whereas some neurotransmitters promote hunger and, in turn, stimulate eating, others promote satiety and meal termination. Not only do these neurotransmitters regulate short-term food intake, they also play an important role in the long-term regulation of body weight.

The Role of the GI Tract in Hunger and Satiety

Understanding the precise mechanisms associated with hunger and satiety is of great interest to researchers. In addition to emotional and social factors, eating behaviors are influenced by a variety of physiological

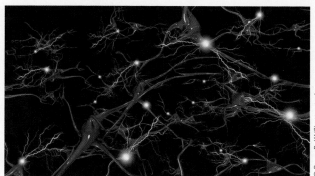

Neurotransmitters released from neurons in the brain play an important role in regulating hunger and satiety.

How Does Weight-Loss Surgery Work?

Obesity can be a chronic condition that, overtime, compromises a person's health. Steadily increasing rates of intractable obesity have prompted some health care professionals to recommend **bariatric surgery** as a safe and effective treatment option when other attempts at weight loss have failed. There are several types of bariatric surgical procedures, all of which cause weight loss by either limiting the amount of food the stomach can hold (restriction), altering digestion and nutrient absorption (malabsorption), or a combination of restriction and malabsorption.[6] The three most commonly performed bariatric procedures include gastric banding, gastric bypass, and sleeve gastrectomy (see Figure 9.4). **Gastric banding** utilizes an adjustable, inflatable band that is carefully placed around the upper portion of the stomach. This restriction creates a small stomach pouch above the band, with the remainder of the stomach below the band. Because the upper pouch is very small, it takes less food to promote satiety. Gastric banding does not impact nutrient digestion and absorption. Rather, this procedure promotes weight loss by reducing hunger and subsequent caloric intake via restriction. Another bariatric procedure, called **gastric bypass**, is the most commonly performed weight-loss surgery. This procedure surgically divides the stomach to create a small stomach pouch. Next, a large segment of the small intestine is bypassed, meaning that the newly formed stomach pouch is surgically connected to the mid-section of the small intestine. However, the bypassed regions of the stomach and duodenum are left in place so that their digestive

secretions can mix with food in the jejunum. This procedure not only restricts food intake, but it also disrupts nutrient digestion and absorption. Although gastric bypass surgery can result in significant long-term weight loss, rerouting the flow of food through the GI tract can lead to vitamin and mineral deficiencies. Thus, people who undergo gastric bypass surgery must adhere to life-long nutrient supplementation regimens. A newly developed weight-loss surgical procedure, called **sleeve gastrectomy**, involves the surgical removal of approximately 80 percent of the stomach to create banana-shaped tubular stomach pouch. Although irreversible, the advantage of this procedure, compared to gastric bypass, is that it does not require rerouting the flow of food, nor, does it impact the release of secretions needed for digestion. However, because food intake is restricted, there is still potential for nutrient deficiencies. For many severely obese individuals, weight-loss surgery can improve quality of life both physically and emotionally. Still, it is important to remember that any type of bariatric surgery is a life-altering procedure, and its effectiveness depends on a person's willingness to eat a healthy, well-balanced diet and exercise regularly.

bariatric surgery Surgical procedure performed to promote weight loss.

gastric banding A type of bariatric surgery whereby an adjustable, fluid-filled band is wrapped around the upper portion of the stomach, dividing it into a small upper pouch and a larger lower pouch.

gastric bypass A surgical procedure that involves reducing the size of the stomach and bypassing a segment of the small intestine so that overall nutrient absorption is decreased.

sleeve gastrectomy A surgical weight-loss procedure that involves the removal of a large part of the stomach, creating a smaller sleeve- or tubular-shaped stomach pouch.

states. One of the most potent physiological satiety triggers is gastric distention (stomach stretching). Gastric distention occurs when food accumulates in the stomach, causing the stomach walls to stretch. This initiates neural signals within the stomach walls, which in turn convey information to the brain. The brain responds by increasing the release of neurotransmitters that elicit the sensation of satiety. Gastric stretching may explain why some foods are better at promoting satiety than others. High-volume foods, such as those with large amounts of water and/or fiber, increase gastric stretching, which in turn helps people feel full and satisfied. For example, two cups of grapes contains approximately 100 kilocalories, the same amount of energy contained in an equivalent amount (1/4 cup) of dried grapes, or raisins. However, because the volume of the grapes is greater than that of the raisins, the grapes cause more gastric stretching, which promotes satiety.

Although gastric stretching helps relieve the sensation of hunger, it is not the only factor associated with satiety. The sensation of fullness is also triggered by the presence of certain nutrients in the blood following a meal. For example, when food intake causes blood glucose levels to increase, the brain responds by releasing neurotransmitters that stimulate satiety, providing a signal to terminate eating. Some studies also show that high-fat meals suppress feelings of hunger longer than low-fat meals with the same number of calories.[7] The signaling of satiation brought on by the presence of certain nutrients in the blood following a meal is an example of a *post-absorptive mechanism* of food regulation.

A number of hormones produced by the GI tract also play a role in regulating hunger and satiety. For example, researchers have recently become interested in **ghrelin**, a hormone dubbed the *hunger hormone*. When the stomach is empty, this potent hunger-stimulating hormone is released by cells in the stomach lining and circulated in the blood to the brain. Elevated levels of ghrelin serve as a pre-meal signal that stimulates the sensation of hunger, but after food is consumed, ghrelin levels decrease. Some obese people may produce too much ghrelin, which could explain why they do not experience a feeling of satiety following a meal.[8] While ghrelin serves as a physiological reminder to eat, there is also evidence that other GI hormones such as cholecystokinin (CCK), which is released from the small intestine, provide equally powerful

ghrelin A hormone secreted by cells in the stomach lining that stimulates hunger.

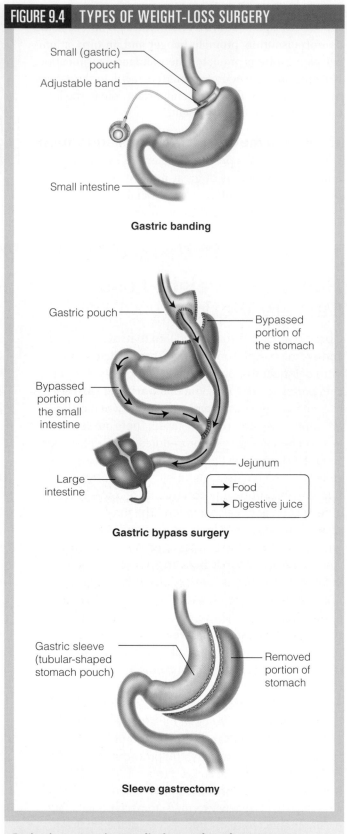

FIGURE 9.4 | TYPES OF WEIGHT-LOSS SURGERY

Small (gastric) pouch
Adjustable band
Small intestine

Gastric banding

Gastric pouch
Bypassed portion of the stomach
Bypassed portion of the small intestine
Large intestine
Jejunum

→ Food
→ Digestive juice

Gastric bypass surgery

Gastric sleeve (tubular-shaped stomach pouch)
Removed portion of stomach

Sleeve gastrectomy

Bariatric surgery is a medical procedure that promotes weight loss. The three most common types of bariatric surgery are gastric banding, gastric bypass surgery, and sleeve gastrectomy.

FIGURE 9.5 REGULATION OF SHORT-TERM FOOD INTAKE

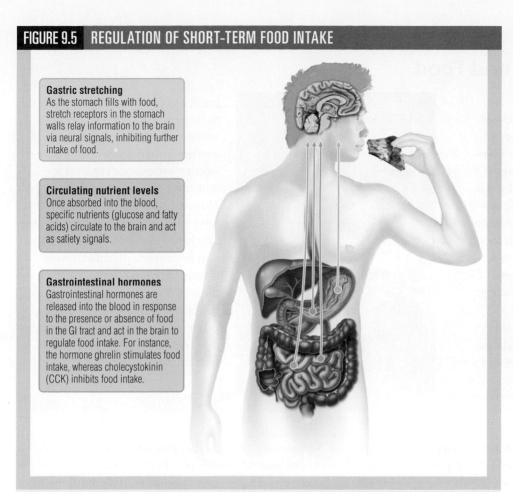

Gastric stretching
As the stomach fills with food, stretch receptors in the stomach walls relay information to the brain via neural signals, inhibiting further intake of food.

Circulating nutrient levels
Once absorbed into the blood, specific nutrients (glucose and fatty acids) circulate to the brain and act as satiety signals.

Gastrointestinal hormones
Gastrointestinal hormones are released into the blood in response to the presence or absence of food in the GI tract and act in the brain to regulate food intake. For instance, the hormone ghrelin stimulates food intake, whereas cholecystokinin (CCK) inhibits food intake.

Signals originating from the GI tract stimulate the brain to regulate short-term food intake by releasing neurotransmitters that influence hunger and satiety.

satiety cues. Some of the factors believed to regulate short-term food intake are summarized in Figure 9.5. Collectively, these signals work together to initiate and terminate food intake. As you will learn, other signals such as the hormone leptin are primarily involved in long-term body weight regulation.

9-2b Psychological Factors

Eating when you are hungry and stopping when you are full are two of the most important keys to maintaining a healthy body weight. However, people often eat for reasons other than hunger or may continue to eat after they feel satiated. Whereas hunger and satiety are physiological cues that regulate food intake, psychological factors also play a role. The term **appetite** refers to the psychological longing or desire for food, whereas **food aversion** refers to a strong psychological dislike of particular foods. The term **food craving** is used when a person experiences a strong desire for a particular food. In addition to psychological origins, there is evidence

that food cravings may also be related to hormonal fluctuations. The most commonly craved foods tend to be calorie rich, such as cookies, cakes, potato chips, and chocolate.[9] Appetite, food aversions, and food cravings are all important psychological determinants of what, when, and how much you eat.

Appetite can easily be aroused by sensory factors such as the appearance, taste, and/or smell of food. For instance, the pleasing smell of baked bread makes many people want to eat, regardless of whether they are hungry. Conversely, unpleasant odors can spoil an appetite, even if a person is hungry. Emotional states can also dramatically affect appetite, but the effect varies from person to person. Some people respond to emotions such as fear, depression, disappointment, excitement, and stress by eating, while others respond by not eating at all. Clearly, emotional states can have a profound influence on appetite, sometimes overriding the physiological cues of hunger and satiety.

9-3 WHAT DETERMINES ENERGY EXPENDITURE?

As you have learned, a complex interplay of factors influences when, what and how much you eat. However, energy intake is only one component of energy balance—the other is energy expenditure. The body expends energy to maintain physiological functions, support physical activity, and process food. Collectively, these components of energy use by

appetite A psychological longing or desire for food.

food aversion A strong psychological dislike of a particular food.

food craving A strong desire for a particular food.

Food Cravings and Food Aversions

Are food aversions the flip side of food cravings? Even the experts are unsure. The reality is that there are no definitive answers when it comes to explaining food cravings and aversions. A food craving is a powerful, irresistible, intense desire for a particular food, whereas food aversions develop when certain foods are viewed as repugnant. Although certain emotional states can provoke food aversions, most are conditioned responses resulting from a paired association between physical discomfort and a

particular food. Food aversions tend to be persistent and long lasting, whereas food cravings can come and go. During pregnancy, many women experience cravings for and/or aversions to certain foods. Although some experts believe that food cravings help pregnant women satisfy their needs for nutrients and that food aversions may provide protection from harmful substances, there is very little scientific data to support either of these claims. In other words, a craving for ice cream does not necessarily mean that a woman is calcium deficient.

Rubberball Productions/Getty Images

---

the body are referred to as **total energy expenditure (TEE)**. As illustrated in Figure 9.6, TEE has three main components:

1. basal metabolism,
2. physical activity, and
3. thermic effect of food.

Total energy expenditure also includes two minor components: **adaptive thermogenesis** and **nonexercise activity thermogenesis**. Adaptive thermogenesis is a temporary expenditure of energy that enables the body to adapt to temperature changes in the environment and physiological conditions such as trauma and stress. Shivering in response to cold weather is an example of adaptive thermogenesis.

Nonexercise activity thermogenesis is expenditure of energy associated with spontaneous movement such as fidgeting and maintaining posture. The contributions of adaptive thermogenesis and nonexercise activity thermogenesis to TEE have yet to be determined, but they are likely significant for some people.

total energy expenditure (TEE) The collective sum of energy used by the body.

adaptive thermogenesis A temporary expenditure of energy that enables the body to adapt to temperature changes in the environment and physiological conditions.

nonexercise activity thermogenesis An expenditure of energy associated with spontaneous movement such as fidgeting and maintenance of posture.

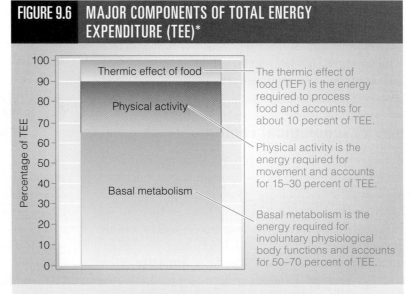

FIGURE 9.6 MAJOR COMPONENTS OF TOTAL ENERGY EXPENDITURE (TEE)*

The thermic effect of food (TEF) is the energy required to process food and accounts for about 10 percent of TEE.

Physical activity is the energy required for movement and accounts for 15–30 percent of TEE.

Basal metabolism is the energy required for involuntary physiological body functions and accounts for 50–70 percent of TEE.

*Energy associated with adaptive thermogenesis and nonexercise activity thermogenesis also contributes slightly to TEE, although its exact contribution is not known.

Source: Adapted from: Levine JA. Non-exercise activity thermogenesis. Proceedings of the Nutrition Society. 2003, 62:667–79.

9-3a Basal Metabolism

Basal metabolism is the energy expended to sustain basic, involuntary physiological functions such as respiration, beating of the heart, nerve function, and muscle tone. **Basal metabolic rate (BMR)**, the amount of energy expended per hour (kcal/hour) to carry out these functions, accounts for most of TEE—approximately 50–70 percent. **Indirect calorimetry** is the most commonly used method for estimating a person's BMR. This involves measuring oxygen consumption and carbon dioxide production following an overnight fast, while a person is at complete rest. Indirect calorimetry can also be used to measure energy expenditure in other situations, such as when exercising.

Factors Influencing Basal Metabolic Rate Basal metabolic rate is not static. This means that your BMR changes in response to a variety of factors. The major factors influencing BMR include body shape, body composition, age, sex, nutritional status, and genetics.[10] All other factors being equal, tall, thin people tend to have higher BMRs than do short, stocky people of equal weight and body composition. This is partly because the former have more surface area, resulting in greater loss of body heat. Similarly, people with high proportions of muscle tend to have higher BMRs than do people with more body fat. This is because it takes more energy to maintain lean tissue (muscle) than body fat. Age can also influence BMR: after the age of 30, BMR tends to decrease by about 2–5 percent each decade. Scientists believe that this decrease is mainly caused by age-related loss of muscle (fat-free mass). This is largely attributed to a decline in levels of growth hormone, which plays an important role in the maintenance of body tissues. It has been estimated that the secretion of growth hormone begins to decline after age 30 at a rate of about 1 percent per year, which over time can contribute to decrease muscle size and strength.[11] Women tend to have a lower BMR than men do because women tend to have proportionately less muscle. Perhaps the most striking factor influencing BMR is severe energy restriction, which can cause the loss of muscle.[12] The impacts of these and other factors on BMR are summarized in Table 9.1.

TABLE 9.1 FACTORS AFFECTING BASAL METABOLIC RATE (BMR)

Factor	Effect on BMR
Age	After physical maturity, BMR tends to decrease with age.
Body composition	Because muscle requires more energy to maintain than does adipose tissue, people with more lean tissue have a higher BMR than do people of equal weight and more adipose tissue.
Body shape	Tall, thin people have higher BMRs than do short, stocky people of equal weight and body composition.
Body temperature	Increased body temperature causes a transient increase in BMR.
Body weight	BMR increases with increased body weight, and decreases with loss of body weight.
Energy restriction	Loss of body tissue associated with fasting and starvation decreases BMR.
Growth	BMR is higher during periods of growth.
Lactation	Milk synthesis increases BMR.
Pregnancy	BMR increases during pregnancy.
Sex	Men tend to have a higher BMR than do women of equal size and weight.
Stress	Chronic stress increases BMR.
Thyroid function	Elevated levels of thyroid hormones increase BMR, while decreased levels of thyroid hormone decreases BMR.

9-3b Physical Activity

After basal metabolism, the energy expended to support voluntary physical activity is the second most significant component of TEE. The amount of energy required for physical activity is quite variable, but for most people, it accounts

basal metabolism An expenditure of energy to sustain vital, involuntary physiologic functions such as respiration, beating of the heart, nerve function, and muscle tone.

basal metabolic rate (BMR) The amount of energy expended per hour (kcal/hour) so that the body can carry out basic, involuntary physiologic functions.

indirect calorimetry A technique that provides an estimate of energy expenditure based on oxygen consumption and carbon dioxide production.

The amount of energy required for physical activity is quite variable. Some elite athletes may require as many as 2,000 to 3,000 extra kilocalories each day to support the demands of physical activity.

for 15–30 percent of TEE.[13] Sedentary people are at the lower end of this estimate, whereas physically active people are at the upper end. Some elite athletes require as much as 2,000 to 3,000 extra kilocalories each day to support the demands of physical activity. Many factors affect the amount of energy expended for physical activity. Rigorous activities such as biking, swimming, and running have higher energy costs than less-demanding activities such as walking. Body size also affects energy expended for physical activity. For example, larger people have more body mass to move than smaller people do and therefore expend more energy to accomplish any given task.

9-3c Thermic Effect of Food

Another component of TEE is the **thermic effect of food** (**TEF**, or **diet-induced thermogenesis**). TEF is the energy required to digest, absorb, and metabolize nutrients following a meal. In other words, it is the cost associated with your body's utilization of the foods you eat. The amount of energy associated with TEF depends on the amount of food consumed and the types of nutrients present in that food. In general, more energy is required to process large amounts of food, and high-protein foods have higher TEFs than do fatty foods. Considering that meals generally contain a mixture of nutrients, TEF is estimated to be about 5–10 percent of total energy intake.[14] In some ways, this component of total energy expenditure is like a caloric sales tax on energy intake. For example, after consuming a 500-kilocalorie meal, a person typically expends 25–50 kilocalories (TEF) just to digest and utilize the nutrients it contains. It is sometimes rumored that a diet consisting solely of "negative-calorie" foods, those that take more energy to digest than the number of calories they provide, will facilitate weight loss. Not only is there no scientific evidence to support this claim, it is unreasonable to think that this approach to weight loss is sustainable over time.

thermic effect of food (**TEF,** or **diet-induced thermogenesis**) The energy needed to digest, absorb, and metabolize nutrients following a meal.

overweight Excess weight for a given height.

obese Excess body fat.

body mass index (BMI) An indirect measure of body fat calculated by dividing a person's body weight by their height (squared).

9-4 HOW ARE BODY WEIGHT AND COMPOSITION ASSESSED?

You now understand how the balance between energy intake and energy expenditure affects changes in body weight (energy storage). Clearly, trying to keep energy intake and energy expenditure in balance is key to maintaining a stable body weight. However, this can be difficult for many people, resulting in unwanted weight gain. At what point does added weight gain become unhealthy? And where does one draw the line between a few extra pounds and a serious health concern? To answer these questions, it is important first to understand how body weight and body composition are defined, measured, and interpreted.

9-4a Overweight Is Excess Weight; Obese Is Excess Fat

Although the terms **overweight** and **obese** are often used interchangeably, they have very different meanings. Overweight refers to excess weight for a given height, regardless of whether the extra weight is muscle or adipose tissue. Conversely, obesity refers to excess body fat regardless of weight. It is therefore possible for muscular people such as athletes to be overweight, but not obese. Similarly, inactive people may not be overweight, but may still be obese. However, most excessively overweight people are obese as well, because adult weight gain is usually associated with an increase in adipose tissue rather than an increase in muscle. For this reason, excessive body weight is often used as an indirect indicator of obesity.

Body Weight Assessment: Height–Weight Tables and Body Mass Index Because there is substantial variation in body weight for any given height, defining an *ideal body weight* may not be possible. Consequently, recommended body weights are simply reference values and are not necessarily ideal for all people. The reference standards most commonly used to assess body weight are height–weight tables and **body mass index (BMI)**.

Recommended weight ranges based on height–weight tables were initially developed to predict weight ranges associated with the longest life expectancies. Although these tables have been revised a number of times, their data may not accurately represent diverse population groups. Even though height–weight tables are still used to assess body weight, BMI offers a more reliable and accurate measurement. For this reason, BMI is used more widely than height–weight tables. A person's BMI is easily calculated using either of the following formulas.

$$BMI\ (kg/m^2) = \frac{weight\ (kg)}{height\ (m^2)}$$

$$BMI\ (lbs/in^2) = \frac{weight\ (lb) \times 703}{height\ (in^2)}$$

For example, the BMI of a person who weighs 150 lb (68.2 kg) and is 65 in (1.65 m) tall is 25 kg/m². You can calculate your BMI by using either of these formulas or by using the chart provided with this book. People with a low BMI (<18.5 kg/m²) typically have low amounts of body fat, whereas those with a high BMI (>25.0 kg/m²) tend to have higher amounts of body fat. Because each BMI unit represents 6–8 pounds of body weight for a given height, an increase in just two BMI units represents a 12- to 16-pound increase in body weight.

Because it is based on the ratio of weight to height, BMI is a better indicator of obesity than weight alone. The cut-off values for BMI classifications (underweight, healthy weight, overweight, and obese) are based on the association between BMI and weight-related morbidity and mortality (see Figure 9.7). Most medical organizations, including the U.S. Centers for Disease Control and Prevention (CDC), use the following BMI criteria to assess body weight in adults.

- Underweight: <18.5 kg/m²
- Healthy weight: $18.5–24.9$ kg/m²
- Overweight: $25.0–29.9$ kg/m²
- Obese: ≥ 30 kg/m²

9-4b Obesity Is Related to Excess Body Fat

Although BMI is a useful method for interpreting a person's weight in relation to height, health professionals sometimes want to know a person's body composition. To this end, imagine that the body has two main components: fat mass and fat-free (lean) mass.

A person's fat (adipose tissue) mass consists mostly of stored triglycerides and supporting structures, whereas the fat-free mass (lean body mass) consists mostly of muscle, water, and bone. The amount of fat and fat-free mass a person has is determined by many factors, such as sex, genetics, physical activity, hormones, and diet. An increase in fat mass can occur when adipocytes either increase in number (**hyperplastic growth**), increase in size (**hypertrophic growth**), or a combination of both. The formation of new adipocytes can occur during any stage of the life cycle,

hyperplastic growth A process whereby new adipocytes are formed.

hypertrophic growth A process whereby adipocytes fills with lipid causing them to enlarge.

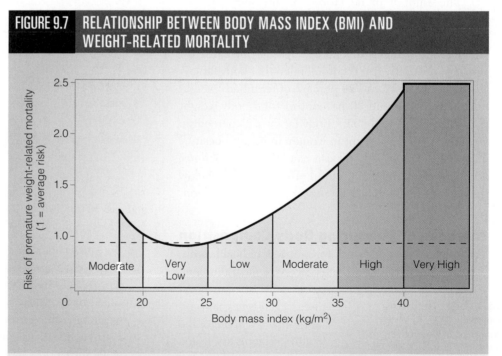

FIGURE 9.7 RELATIONSHIP BETWEEN BODY MASS INDEX (BMI) AND WEIGHT-RELATED MORTALITY

Body mass index between 20 kg/m² and 25 kg/m² is associated with low risk of weight-related mortality. As BMI increases over 30 kg/m², the risk of weight-related mortality increases to moderate. A person is considered very high risk when BMI reaches 40 kg/m².

Source: Adapted from Gray DS. Diagnosis and prevalence of obesity. Medical Clinics of North America. 1989, 73:1–13. Lew EA, Garfinkle L. Variations in mortality by weight among 750,00 men and women. Journal of Chronic Disease. 1979, 32:563–76. Srinivas N, Koka S, Chinnala KM, Boini KM. Obesity: An overview on its current perspectives and treatment options. Nutrition Journal. 2004, 3:3.

FIGURE 9.8 HYPERTROPHIC AND HYPERPLASTIC GROWTH OF ADIPOSE TISSUE

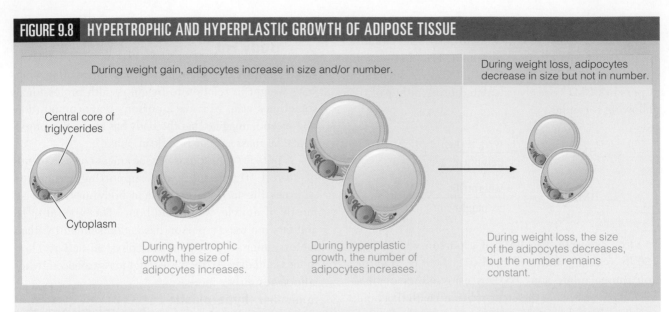

During weight gain, adipocytes increase in size and/or number.

During weight loss, adipocytes decrease in size but not in number.

Central core of triglycerides

Cytoplasm

During hypertrophic growth, the size of adipocytes increases.

During hyperplastic growth, the number of adipocytes increases.

During weight loss, the size of the adipocytes decreases, but the number remains constant.

The amount of a person's adipose tissue depends on the number and size of his or her adipose cells (adipocytes).

and once formed, the total number remains relatively constant. Therefore, when a person loses body fat, only a reduction in cell size is possible (see Figure 9.8).

The amount of fat stored in the body changes throughout the life cycle. For example, nearly 30 percent of total body weight in a healthy 6-month-old infant is fat, whereas this percentage may be cause for concern in adults. It is recommended that body fat levels be between 12 and 20 percent of total body weight for men and 20 and 30 percent of total body weight for women (see Table 9.2). Body fat over 25 percent in men and over 33 percent in women indicates obesity.[15] Just as too much body fat can cause health problems, too little can also have harmful effects; body fat below 12 percent for men and 20 percent for women is considered too low.

Methods of Measuring Body Composition

Obesity increases one's risk for a variety of health problems, including type 2 diabetes, sleep irregularities, joint pain, gout, stroke, heart disease, gallstones, and certain cancers.[16] Some of the health problems associated with obesity are illustrated in Figure 9.9. Clinicians can use a variety of anthropometric methods to estimate body fat. Whereas some of these methods are expensive and complex, others are more readily available

hydrostatic weighing
A method of estimating body composition in which a person's body weight is measured in and out of water.

dual-energy x-ray absorptiometry (DEXA)
A method of estimating body composition in which low-dose x-rays are used to visualize fat and fat-free compartments of the body.

TABLE 9.2 OBESITY CLASSIFICATIONS BASED ON PERCENT BODY FAT

Classification	Body Fat (% total body weight)	
	Males	Females
Healthy	12–20	20–30
Borderline obese	21–25	31–33
Obese	>25	>33

Source: Bray G. What is the ideal body weight? Journal of Nutritional Biochemistry. 1998; 9:489–92.

and easy to use. For illustrations of the common methods used to estimate body composition, see Figure 9.10.

- Body fat is less dense than water, while the fat-free mass has a greater density than water. Because these tissues have differing densities, a procedure called **hydrostatic weighing** (or underwater weighing) can be used to estimate percent body fat. First, a person's weight is measured on land. Next, while sitting on a special scale, a person's weight is measured while submerged under water. The more body fat a person has, the less they will weigh under water. Hydrostatic weighing is usually reserved for clinical evaluations because it requires special equipment and is neither practical nor convenient.

- **Dual-energy x-ray absorptiometry (DEXA)** provides a highly accurate estimate of body

FIGURE 9.9 HEALTH PROBLEMS ASSOCIATED WITH OBESITY

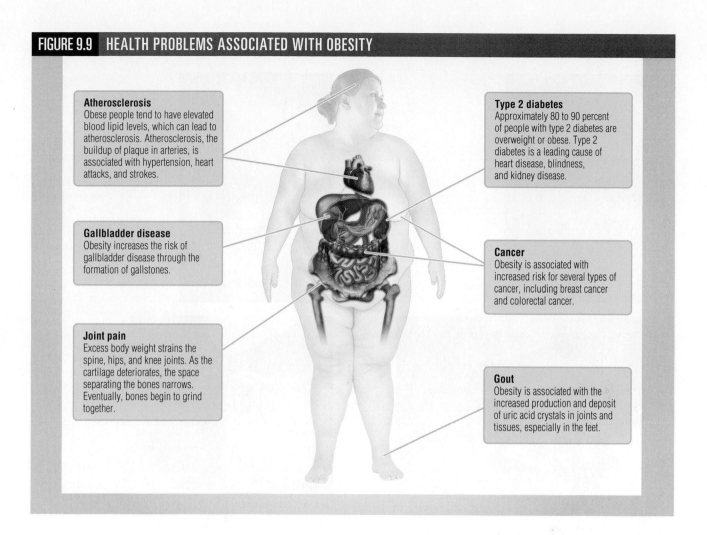

Atherosclerosis
Obese people tend to have elevated blood lipid levels, which can lead to atherosclerosis. Atherosclerosis, the buildup of plaque in arteries, is associated with hypertension, heart attacks, and strokes.

Gallbladder disease
Obesity increases the risk of gallbladder disease through the formation of gallstones.

Joint pain
Excess body weight strains the spine, hips, and knee joints. As the cartilage deteriorates, the space separating the bones narrows. Eventually, bones begin to grind together.

Type 2 diabetes
Approximately 80 to 90 percent of people with type 2 diabetes are overweight or obese. Type 2 diabetes is a leading cause of heart disease, blindness, and kidney disease.

Cancer
Obesity is associated with increased risk for several types of cancer, including breast cancer and colorectal cancer.

Gout
Obesity is associated with the increased production and deposit of uric acid crystals in joints and tissues, especially in the feet.

composition. Originally developed to assess bone-mineral density, DEXA can also be used to measure both total and regional body composition (fat mass and fat-free mass). A DEXA scan involves lying face-up on a table while an x-ray scanning device passes over the entire length of the body. DEXA is considered the gold standard of body composition analysis and is often used by researchers and clinicians.

- Because lean tissue conducts electrical currents better than adipose tissue, **bioelectrical impedance** can be used to estimate body composition. This is because tissues associated with the fat-free mass contain more water than body fat, and therefore more readily conduct the flow of an electrical current. Using a device that emits a weak electrical current, body fat percent can be estimated on the basis of conductivity. This technique is relatively accurate, simple to use, and is considered an acceptable method for estimating body composition in a clinical setting. Bathroom scales with built-in bioelectrical impedance systems are now available for home use, although little is known about their accuracy.

- The **skinfold thickness method** has been used for many years to estimate body fat. An instrument called a **skinfold caliper** is used to measure the thickness of skin and subcutaneous fat at various locations on the body. Body fat is then estimated using mathematical formulas based on the skinfold thickness measurements.

You are now familiar with some of the methods by which weight and obesity are assessed. But to understand what truly causes weight

bioelectrical impedance
A method of estimating body composition in which a weak electric current is passed through the body.

skinfold thickness method
A method of estimating body composition whereby a skinfold caliper is used to measure the thickness of skin and subcutaneous fat at various locations on the body.

skinfold caliper An instrument used to measure the thickness of skin and subcutaneous fat.

FIGURE 9.10 **METHODS USED TO ESTIMATE BODY COMPOSITION**

Hydrostatic weighing
Hydrostatic weighing requires the subject to exhale air from the lungs and be submerged in water. It is important to remain motionless while weight is measured underwater.

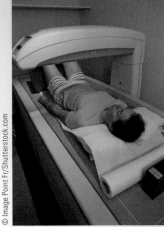

Dual-energy x-ray absorptiometry (DEXA)
While the person is lying on a table, a scanning device passes over the body. The x-ray beams emitted differentiate between the fat mass, lean mass, and skeletal mass. A two-dimensional image of the body is displayed on a computer screen.

Bioelectrical impedance
Electrodes are placed on a person's hand and foot, and a weak electrical current is passed through the body. The conductivity of the current is measured, which provides an estimate of the fat mass and fat-free mass.

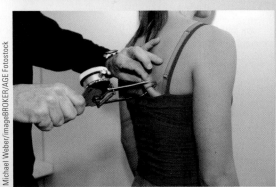

Skinfold thickness
Using a measuring device called a caliper, the health care provider measures the thickness of a fold of skin with its underlying layer of fat at precise locations on the body.

gain and obesity, you must look beyond the individual and examine the overall influences that affect eating habits and activity levels.

9-5 GENETICS VERSUS ENVIRONMENT: WHAT CAUSES OBESITY?

When it comes to our bodies, we tend to be our own worst critics. Although you might wish that your body looked or was shaped differently, it is important to recognize that the issue of obesity is of concern not because of societal norms, but rather because excess weight can harm your health. In the late nineteenth century, only 3 percent of American adults were overweight, and very few were obese. Today, nearly 69 percent of adults (>19 years) are classified as overweight, 35 percent of which are obese.[17] Similar trends are evident in children and adolescents: In the last 20 years, the percentage of overweight American children and adolescents has tripled. An overview of obesity trends in the United States is presented in Table 9.3.

It is important to recognize that obesity is not only on the rise in the United States, but around the world as well.[18] But what is causing the global obesity epidemic? To answer this question, it is important to consider both lifestyle and genetic factors, because a combination of these factors can shift energy balance in favor of positive energy balance and weight gain—especially in today's world.

9-5a Eating Habits

One of the most important factors that influence body weight is the amount and energy density of the foods consumed. It is likely that an increase in energy intake

TABLE 9.3	ESTIMATED PREVALENCE OF OBESITY IN THE 1970s AND 2009–2012	
	Prevalence of Obesity (%)	
Age Group	**1970s**	**2009–2012**
2–5 years	5	10
6–11 years	4	18
12–19 years	6	20
>19 years	15	35
Number of states with adult obesity rate >25%	0	45

Sources: Centers for Disease Control and Prevention. Division of Nutrition, Physical Activity, and Obesity. Available at: http://www.cdc.gov/obesity/data/adult.html. National Center for Health Statistics. Health, United States, 2014: With Special Feature on Adults Aged 55–64. Hyattsville, MD. 2015. Available at: http://www.cdc.gov/nchs/data/hus/hus14.pdf#059

over the last few decades is a major cause of today's obesity epidemic. According to estimates by the U.S. Department of Agriculture, Americans on average consume 500 kcal/day more than they did 30 years ago (2,600 vs. 2,100 kcal/day, respectively).[19] A variety of societal, cultural, and psychological factors influencing what, how much, when, and where people eat may be contributing to this trend.

Societal and Cultural Influences on Eating Habits

Food-related societal and cultural norms affect what and how much food people consume. For instance, eating away from home and consuming large portion sizes of foods—especially energy-dense foods—is becoming more and more common. In the United States, energy-dense, inexpensive, flavorful foods have become readily available and are accepted as a cultural norm. Today, consumption of fast food, especially among children and adolescents, is very common. On any given day, 34 percent of all American children and adolescents consume fast food, which contributes approximately 17 percent of their total daily calories.[20] Although the consumption of fast food is not the sole cause of obesity, it may certainly be a contributing factor because many fast food items are high in fat, refined starchy carbohydrates, and calories. As such, a super-sized value meal might provide more than one-half the calories required in a day. To counter the perception that their food is unhealthy, many fast-food restaurants also offer healthier food choices such as salads, wraps, and sandwiches made with lean meats and whole-grain breads.

Selecting these menu options is one way for individuals and families to eat healthier. However, it is important for consumers to take the time to read the nutrition facts that are provided by most fast-food restaurants. Sometimes, a food may sound like a healthier choice, but in reality, it is not. For example, a grilled chicken wrap at one popular fast-food restaurant has 470 kilocalories, which is more than the caloric content of a typical cheeseburger (300 kilocalories).

Beyond the type of food consumed, studies show that serving size also influences food consumption. That is, when larger food portions are served, people may inadvertently eat more.[21] For example, participants in a study ate 39 percent more M&Ms® when they were given a 2-pound bag than when they received a 1-pound bag.[22] Why does this happen? It appears that some people depend more on visual cues than on physiological cues such as hunger and satiety to judge how much to eat. This phenomenon is particularly important to consider because many restaurants continue to offer super-sized food portions. Portion size distortion can easily translate into excess calories. Therefore, learning how to judge appropriate serving sizes is an important aspect of weight management.

Factors related to food itself can also influence what and how much food a person consumes. When presented with a variety of food choices, people tend to eat more than when they are offered fewer choices.[23] Thus, buffet-style restaurants may be more conducive to overeating than traditional restaurants with more limited menu choices and fixed portion sizes. In addition, when people consume or perceive they have consumed fewer calories at one meal, they tend to reward themselves by eating more at subsequent meals.[24] This may explain in part why increased consumption of reduced-calorie foods and beverages does not necessarily lead to a reduction in total calorie intake over time.

© Somchai Som/Shutterstock.com

How is fat removed from milk? Traditionally, it was removed using gravity. When fresh milk is left to sit and settle, the cream (fat) rises to the top. Skimming off the cream resulted in milk with less fat. Today, this is accomplished by a machine called a centrifugal separator.

When presented with a variety of food choices, people tend to eat more than when they are presented with fewer food choices. Buffet-style eating does not pose a problem when the options are nutritious, but may lead to weight gain when only high-calorie food options are available.

Sociodemographic and Psychological Influences on Eating Habits Factors such as economic status, marital status, and education level can also influence a person's risk for obesity. Although sociodemographic factors do not directly cause obesity, some may indirectly contribute to the problem. Beyond sociodemographic factors, some psychological disorders are also related to a person's likelihood of becoming obese. It is not clear, however, whether obesity predisposes individuals to these disorders or if some psychological profiles lead to obesity. All scientists can say for certain is that some personality types appear to be associated with increased risk for obesity. For example, obese individuals are more likely to experience clinical depression and panic attacks than non-obese people.[25] This may stem from the fact that obesity sometimes lowers self-esteem and confidence. Conversely, individuals who have difficulty coping with stress may turn to food for emotional comfort and gratification, making them more susceptible to obesity.

Research also suggests that people's social network may contribute to their risk for obesity.[26] Those with friends who have experienced weight gain are more likely to gain weight than those with weight-stable friends. It is not clear how social networks are linked to weight gain, but researchers believe that weight gain among close friends and family members might serve as a permissive cue for others to gain weight as well.

9-5b Sedentary Lifestyle

Choosing what and how much to eat is not the only lifestyle choice related to unwanted weight gain and obesity. At the same time they are eating more, Americans are becoming less physically active. A decrease in the availability of jobs that require physical work and an increase in the availability of labor-saving devices make daily life less physically demanding for many people. Despite public education efforts, the majority of Americans do not meet the minimum recommendations for weekly aerobic and muscle-strengthening physical activity. Based on self-reported measures from 2013, only 21 percent of adults engage in any kind of aerobic and muscle-strengthening physical acitivities meeting the 2008 Physical Activity Guidelines for Americans.[27] Together, these changes have likely had a significant negative impact on the nation's health.

It is important to realize that physical activity is not limited to formal exercise. **Physical activity** is any bodily movement that results in energy expenditure, whereas **exercise** is a planned, structured, and repetitive body movement done to improve or maintain physical fitness.[28] Physical activities such as walking up the stairs or mowing the lawn can be just as beneficial as formal exercise. Not surprisingly, the vast majority of studies show that a lack of physical activity increases the risk of being overweight or obese.[29] For most people, walking an extra mile each day— equivalent to taking about 2,000–2,500 extra steps— would increase energy expenditure by 100 kcal/day.[30] If all other factors remained unchanged, a person could lose 1 pound of body weight per month by making this simple change. In addition to helping maintain a healthy body weight, physical activity and exercise also help a person stay healthy and physically fit.

physical activity Any bodily movement that results in energy expenditure.

exercise Planned, structured, and repetitive bodily movement done to improve or maintain physical fitness.

9-5c Genetics

Although there is general agreement that lifestyle factors are a driving force behind the obesity epidemic, it is important to recognize that a person's genetic makeup influences body weight as well. In an environment where energy-dense foods are abundant and physical activity is low, genetic factors make some people more susceptible to weight gain than others.

Discovery of Leptin Provided the First Genetic Clue to Obesity Scientists have long believed that genetic makeup influences body weight. However, direct evidence was lacking until the discovery of a mouse with a gene mutation that made it obese. This mutant animal was referred to as the **ob/ob (obese) mouse**. Ob/ob mice consume large amounts of food (**hyperphagic**), are inactive, and therefore, they gain weight easily. Soon after the ob/ob mouse was discovered, researchers also discovered another mouse with a different gene mutation that also made it obese. They referred to this mouse as the **db/db (diabetic) mouse** because it was used as a model to study obesity-induced type 2 diabetes.[31] In studying these two varieties of mutated mice, scientists learned that the *ob* **gene** codes for a hormone called **leptin**, a potent satiety signal produced in several tissues but especially in adipocytes; in fact, leptin was the first adipokine discovered.[32] Because of a mutation in the *ob* gene, ob/ob mice do not produce leptin, causing them to be hyperphagic. Soon thereafter, researchers discovered that the *db* **gene** codes for the leptin receptor found primarily in the hypothalamus. Db/db mice do not produce the leptin receptor and therefore cannot appropriately respond to leptin's satiety signal.

Many people hoped that leptin—initially touted as the *anti-obesity hormone*—would become the miracle cure for obesity. Yet, further research indicated that the vast majority of obese people produce appropriate or even elevated amounts of leptin. In fact, only a few cases of leptin deficiency in humans have been reported.[33] Still, scientists found that, when severely obese, leptin-deficient individuals are given leptin injections, they experience dramatic weight loss. Although leptin has not proved to be effective for treating human obesity in most cases, its discovery led to important insights into body weight regulation.

A gene mutation made the mouse on the right obese.

9-6 HOW ARE ENERGY BALANCE AND BODY WEIGHT REGULATED?

Although the discovery of leptin has not solved the obesity dilemma, it has provided scientists with a profoundly greater understanding of body weight regulation. Much remains unknown about body weight regulation and possible defects in key energy-regulating activities, but it is now clear that the body plays an active role in influencing energy intake, energy expenditure, and energy storage. This homeostatic body weight regulatory system involves hormonal signaling pathways—some of which are described next.

9-6a Set Point Theory of Body Weight Regulation

The body's ability to adjust energy intake and energy expenditure on a long-term basis serves an important purpose: body weight regulation. If the body had no means of regulating energy balance, the consequences would be catastrophic. Even a slight imbalance in daily energy intake and energy expenditure could result in substantial weight gain or loss over time. To prevent such imbalances, long-term energy balance regulatory signals communicate the body's energy reserves to the brain, which in turn releases neurotransmitters that influence energy intake and/or expenditure.[34] If this system functions effectively, body weight remains relatively stable over time. In other words, neutral energy balance is maintained.

Scientists have suspected for many years that a complex signaling system regulates body weight by adjusting (increasing/decreasing) energy intake and expenditure.[35] To test this theory, researchers observed weight-gain

ob/ob (obese) mouse A mouse with a gene mutation that impairs leptin production, which leads to obesity.

hyperphagic Exhibiting excessive hunger and food consumption.

db/db (diabetic) mouse A mouse with a gene mutation for the leptin receptor, which leads to obesity.

ob gene The gene that codes for the hormone leptin.

leptin A potent hormone (adipokine) produced primarily by adipose tissue that signals satiety.

db gene The gene that codes for the leptin receptor.

and weight-loss cycles in food-restricted mice. When food intake was restricted, the mice lost weight. Not surprisingly, when the mice were fed, consumption increased, and the mice soon returned to their original weights. However, what was surprising to scientists was the fact that, after returning to their original weights, the mice maintained this weight by spontaneously decreasing their intake of food. This phenomenon prompted scientists to propose what was referred to as the **set point theory** of body weight regulation. Proponents of the set point theory speculated that an unidentified hormone circulating in the blood played an important role in body weight regulation. However, it was not until 1994, when researchers first discovered the *ob* gene and the hormone leptin, that there was biological evidence to support this theory.

9-6b Leptin Communicates the Body's Energy Reserve to the Brain

Although the complex mechanisms that regulate long-term energy balance are not fully understood, leptin appears to play an important role by communicating energy reserves (adipose tissue) to the brain (see Figure 9.11). Leptin is produced primarily by adipose tissue and is therefore considered to be an adipokine. When body fat increases, so does the concentration of leptin circulating in the blood. When body fat decreases, leptin production also decreases. Thus, fluctuations in blood leptin levels reflect changes in the body's primary energy reserve, adipose tissue. Leptin does more than convey the message, however—it also triggers a physiological response. Elevated leptin concentrations signal neurons in the hypothalamus, which in turn decrease the release of neurotransmitters that trigger hunger,

set point theory A scientific concept whereby hormones circulating in the blood are theorized to regulate body weight by communicating the amount of adipose tissue in the body to the brain, which adjusts energy intake and expenditure to maintain neutral energy balance.

FIGURE 9.11 LEPTIN AND BODY WEIGHT REGULATION

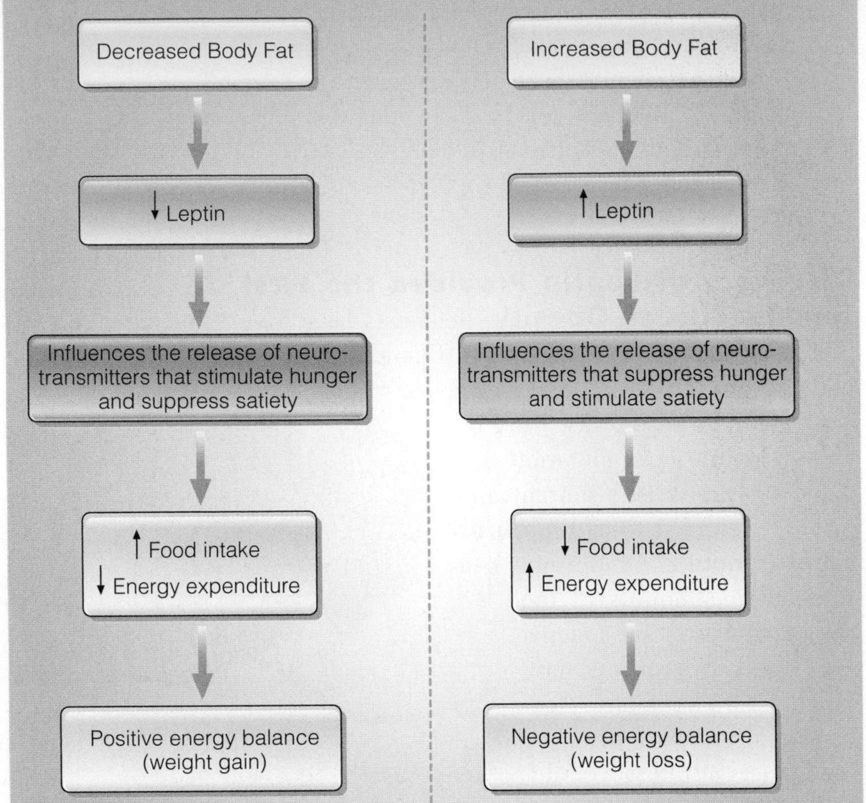

Decreased Body Fat → ↓ Leptin → Influences the release of neurotransmitters that stimulate hunger and suppress satiety → ↑ Food intake ↓ Energy expenditure → Positive energy balance (weight gain)

When body fat decreases, less leptin reaches the brain. This influences the release of neurotransmitters that stimulate hunger and suppress satiety decreasing energy expenditure, which in turn favors weight gain.

Increased Body Fat → ↑ Leptin → Influences the release of neurotransmitters that suppress hunger and stimulate satiety → ↓ Food intake ↑ Energy expenditure → Negative energy balance (weight loss)

When body fat increases, more leptin reaches the brain. This influences the release of neurotransmitters that suppress hunger and trigger satiety increasing energy expenditure, which in turn favors weight loss.

and increase the release of neurotransmitters that trigger satiety. As a result, the body is able to further resist weight gain by prompting a decrease in food intake and/or an increase in energy expenditure. Conversely, a decrease in blood concentrations of leptin has the opposite effect. Neurons in the hypothalamus respond to low leptin levels by releasing neurotransmitters that stimulate hunger, while decreasing the release of neurotransmitters that trigger satiety. In this way, the body protects itself against further weight loss by stimulating food intake and/or decreasing energy expenditure. In this way, leptin is believed to be part of the complex communication loop that helps maintain a relatively stable body weight over time.

Defects in Leptin Responsiveness May Lead to Obesity You may be wondering, "If leptin is a satiety hormone and curbs food intake, why are so

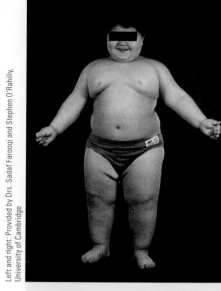
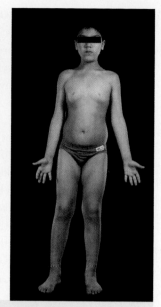

Left and right: Provided by Drs. Sadaf Farooqi and Stephen O'Rahilly, University of Cambridge

This child (left) has a rare genetic condition that causes leptin deficiency. Daily leptin injections had a profound effect on energy intake and energy expenditure, resulting in a marked decrease in body weight (right).

many people obese? If obese people have elevated leptin levels, why do they continue to gain weight?" These are very good questions. A number of researchers believe that some people may have one or more defects in the leptin-signaling pathways that cause a failure to respond to leptin.[36] In other words, the brains of obese people may not be responsive to leptin's appetite suppression signal, regardless of how much leptin is being produced. There are many aspects of this extremely complex signaling pathway that could account for leptin resistance. For example, some researchers believe that a defect in the ability of leptin to be transported into the brain may contribute to leptin resistance, whereas others attribute impaired leptin signaling to defects in leptin receptors found in the hypothalamus.[37] Clearly, there is more to learn about this regulatory system. As scientists continue to study the role of leptin and other adipokines in long-term energy balance, they will gain a better understanding of the underlying mechanisms associated with body weight regulation.

9-7 WHAT IS THE BEST APPROACH TO WEIGHT LOSS?

Although the diet industry would like you to believe otherwise, the truth is clear: there is no quick or easy way to lose weight and keep it off. Approximately one-third of adults in the United States are on a special diet to lose weight. There are plenty of diets to choose from—the obesity epidemic certainly cannot be attributed to a lack of weight-loss advice. Beyond advice and meal planning, a variety of nonprescription weight-loss products are available and marketed aggressively to people trying to lose weight. Far from helping, some of these products may be hazardous to a person's health. Although the U.S. Food and Drug Administration has recently approved several new drugs to aid in weight loss, these too have safety concerns. A pharmaceutical approach such as hunger suppressant drugs may kick start weight loss, but it is not a long-term solution. Clearly, weight loss and weight-loss maintenance require lasting changes that include a sensible diet and exercise program.

Most popular weight-loss plans have specific rules to follow. Some restrict the types of food that can be eaten, others recommend a strict exercise regimen, and still others advocate periods of fasting or cleansing. The real

Chris Hondros/Getty Images

The obesity epidemic certainly cannot be attributed to a lack of weight-loss advice. Amazon.com lists over 22,000 books related to weight-loss diets.

issue that many weight-loss plans fail to address is not what it takes to lose weight but what it takes to keep it off. Rather than succumbing to a fad diet or quick-fix weight-loss approach, consider what actual weight-loss experts have to say.

9-7a Healthy Food Choices Promote Overall Health

Although there are many reasons why a person might want to lose weight, the most important is to improve health. Achieving and maintaining weight loss requires lasting lifestyle changes. A person must evaluate and, if necessary, permanently modify their food choices and their level of physical activity. The extent that Americans have made progress towards satisfying the U.S. Dietary Guidelines and Physical Activity Guidelines remains low. Based on Healthy Eating Index scores, which measures how closely diet aligns with the Dietary Guidelines, the prevalence of overweight/obese Americans is not surprising. Greater adherence to the 2015 Dietary Guidelines for Americans could help millions of Americans achieve health weights, which in turn would lower risk of weight-related chronic disease (Figure 9.12).

Most health experts suggest that people focus less on weight loss and more on healthy eating patterns and overall fitness. Unfortunately, many misguided weight-loss efforts present food as the enemy rather than as a means to good health. People who decide to lose weight at any cost often sacrifice their health in the process. Simply put, to lose weight and keep it off successfully and in a healthy way, a person must choose to eat a balanced diet of nutrient-dense foods and maintain a moderately high level of physical activity.[38] A healthy weight-loss and weight-maintenance program consists of three components:

1. Setting reasonable goals,
2. Choosing nutritious foods in moderation, and
3. Increasing energy expenditure through daily physical activity.

Setting Reasonable Goals Setting reasonable and attainable goals is an important component of any successful weight-loss program. For instance, an obese person's realistic initial weight-loss goal might be to reduce their current weight by 5–10 percent. For someone weighing 180 pounds, this would amount to an initial weight loss of approximately 9–18 pounds. Studies show that even a modest reduction in weight can improve overall health.[39] When it comes to weight loss, slow and steady is the way to go, and weekly weight loss should not exceed 1–2 pounds. This can be achieved by decreasing energy intake by 100–200 kilocalories each day or by walking 1–2 miles each day. Over time, these small changes can result in significant reductions in body weight. Rather than making dramatic dietary changes, small changes such as reducing portion sizes and cutting back on energy-dense snack foods can make a big difference in overall energy intake.

Once a person's body weight stabilizes and the new lower weight has been maintained for a few months, the next decision is whether additional weight loss is needed. Some people benefit from joining weight-maintenance programs such as Weight Watchers® or Take Off Pounds Sensibly®. These types of programs provide long-term support and motivate people to maintain healthy diets and lifestyles. There is also evidence that cell phone applications (apps) and other wearable tracking devices that record and monitor food intake and physical activity

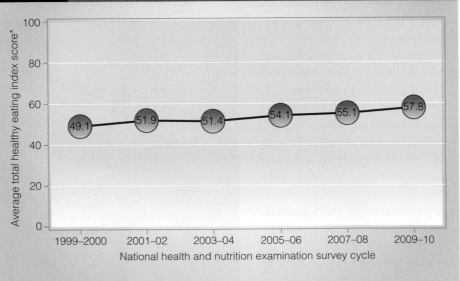

FIGURE 9.12 ADHERENCE OF THE U.S. POPULATIONS TO THE U.S. DIETARY GUIDELINES (AGES ≥2 Y)

*A score of 100 indicates that a diet follows recommendations put forth by the Dietary Guidelines for Americans. Lower scores indicate less adherence.

Source: Adapted from Analyses of What We Eat in American, National Health and Nutrition Examination Survey (NHANES) data from 1999–2000 through 2009–2010.

can have a positive influence on weight-loss efforts. These new technologies provide both guidance and feedback that can keep people motivated.[40]

Choosing Nutritious Foods in Moderation

Weight-loss plans that drastically reduce caloric intake and offer limited food choices often leave people feeling hungry and dissatisfied. Instead, those that allow moderate caloric intake and encourage people to eat foods that are healthy and appealing tend to be more successful. Contrary to popular belief, it is not necessary to avoid foods that contain fat in order to lose weight. In fact, many fat-reduced and fat-free foods are high in refined carbohydrates (both starch and added sugars), which can also be detrimental to health. A key recommendation in the 2105 Dietary Guidelines for Americans is for people to develop eating patterns that emphasize a variety of nutrient-dense foods within all food groups, but at an appropriate level of calories. Limiting intake of foods high in saturated fats, *trans* fats, added sugars, and sodium is still advised.

The idea that dairy products and meats are high-fat foods and therefore should be avoided when trying to lose weight is also a common misperception. Again, the keys to good nutrition and weight maintenance are moderation and making smart choices. Switching from whole to reduced-fat milk is one way to lower caloric intake without losing out on the many vitamins and minerals in milk. Fortified soy beverages offer vegans and those with lactose intolerance a nutritious alternative to dairy foods. Likewise, the types of meat you choose and the methods by which you prepare the meat can greatly affect how many calories you consume. Lean meats prepared by broiling or grilling are both nutritious and satisfying. Eggs, legumes (beans and peas), nuts, seeds, and soy products can also be nutritious food choices that not only add variety, but are good alternatives to meat.

Reducing energy intake is best achieved by cutting back on energy-dense foods that have little nutritional value such as potato chips, cookies, and cakes. Aside from their lower energy densities, nutrient-dense foods such as whole-grain breads, legumes, fruits, and vegetables contain many beneficial substances such as micronutrients and fiber. Furthermore, because these foods tend to have greater volumes than do energy-dense foods, they help people feel more satisfied after they eat.

Healthy eating requires a person to pay attention to hunger and satiety cues. However, because visual cues sometimes have a greater influence than do internal cues on the quantity of food consumed, the amount of food served or packaged—rather than hunger and satiety—often determines how much one eats. Some commercially made muffins, for instance, are extremely large and contain as many calories as several slices of bread. Learning to recognize and choose reasonable portions of food is a critical component of successful weight management. In fact, reducing portion sizes by as little as 10–15 percent can lower daily energy intake by as much as 300 kilocalories. One way to limit serving size (and, incidentally, save money) is to consider sharing a meal the next time you eat at a restaurant. These and other behaviors associated with healthy weight management are listed in Table 9.4.

Increasing Energy Expenditure through Daily Physical Activity

In addition to choosing nutritious foods in moderation, you can also help tip the energy balance equation toward negative energy balance by engaging in physical activity. Walking 1 mile each day, which takes people about 15–20 minutes, uses about 100 kilocalories. This adds up to 700 kilocalories per week. Beyond going out of your way to exercise, just being physically active throughout the day can make a big difference in promoting weight loss and weight maintenance. When possible, you might consider taking the stairs rather than the elevator, walking or biking rather than driving, and incorporating chores into your routine daily activities. Even without weight loss, individuals who are physically active show improved physical fitness. Regardless of one's weight, a lack of exercise may prove the greatest health hazard of all.[41]

It is important to realize that overweight people can still be physically fit, and healthy-weight people can be physically unfit. In fact, studies show that some overweight individuals who are physically fit have fewer health problems than do healthy-weight individuals who

TABLE 9.4	BEHAVIORS ASSOCIATED WITH HEALTHY WEIGHT MANAGEMENT

- Focusing attention on an overall-healthy eating pattern that emphasizes nutrient-dense foods.
- Monitoring food intake, body weight, and physical activity to help make oneself more aware of what and how much is eaten and drunk.
- Selecting small-size and lower-calorie options when eating out.
- Preparing and serving smaller food portions, especially foods and beverages that are high in calories.
- Eating nutrient-dense foods to improve nutrient intake and healthy body weight.
- Reducing screen time to 1–2 hours each day; getting adequate, restful sleep (7–8 hours); and increasing physical activity to help with weight loss and weight maintenance.

are unfit.[42] Healthy blood pressure levels, blood glucose regulation, and blood lipid levels are important indicators of physical fitness. Beyond improving these measures, exercise also promotes a positive self-image and helps people take charge of their lives. Although some people resist the idea of starting an exercise program, they rarely regret it once they begin.

Physical activity is an effective strategy for preventing unhealthy weight gain in healthy-weight, overweight, and obese individuals. For substantial health benefits, the 2015 edition of the Dietary Guidelines recommends that adults engage in at least 150 minutes of moderate-intensity physical activity and perform muscle-strengthening exercises on two or more days each week. To stay physically fit, children and teenagers (6–17 years) should engage in least 60 minutes of physical activity each day, including aerobic, muscle-strengthening, and bone-strengthening activities.[43] People who were formerly obese, however, may require more exercise—60 to 90 minutes each day—to maintain a lower body weight. It is important to remember that, after a person loses weight, their total energy requirement is lower than it was before the weight was lost. For example, when a person loses 10 pounds, their basal metabolic rate decreases by 80 kcal/day. Thus, to maintain a 10-pound weight loss, one must further reduce daily energy intake or increase energy expenditure by 80 kilocalories—the amount in a single slice of bread.

9-7b Characteristics of People Who Successfully Lose Weight

One of the best ways to learn what works in terms of weight loss is to study people who have been successful. The National Weight Control Registry is a large, prospective study of individuals who have lost significant amounts of weight and have been successful at keeping it off.[44] To be eligible to participate in the study, individuals must have maintained a weight loss of 30 pounds or more for one year or longer. Researchers interviewed this unique group of successful individuals to learn about

Not all overweight people are physically unfit and/or unhealthy. It is the combination of being both overweight and sedentary that increases a person's risk of developing weight-related health problems.

their weight-loss and weight-management practices. However, identifying common characteristics among study participants proved difficult. Whereas some reported counting calories or grams of fat, others used prepackaged diet foods. Some participants preferred losing weight on their own, and others sought assistance from weight-loss programs. Nonetheless, researchers identified one strikingly common thread: almost 89 percent of successful weight-loss participants reported that both diet and physical activity were part of their weight-loss plan. Only 10 percent reported using diet alone, and only 1 percent used exercise alone. On average, study participants engaged in 60–90 minutes of physical activity daily, which was equivalent to 2,500 kcal/week for women and 3,300 kcal/week for men. In addition to a high level of physical activity, participants ate breakfast regularly and monitored their weight frequently.

9-7c Does Macronutrient Distribution Matter?

With all the weight-loss advice that is available, it can be difficult to sort fact from fiction. Even experts have differing opinions about the best balance of protein, fat, and carbohydrate intake for achieving weight loss. Today, one of the biggest controversies in weight-loss research is the role of dietary carbohydrate and dietary fat in weight loss and weight gain.[45]

Weight-loss diets that are low in fat and high in carbohydrates have long been considered the most effective in terms of weight loss and weight maintenance. Recall that the Acceptable Macronutrient Distribution Ranges (AMDRs) suggest that one consume 45–65 percent of energy from carbohydrates, 10–35 percent from proteins, and 20 to 35 percent from fats. However, in 1972, Dr. Robert Atkins, one of the first pioneers of the low-carbohydrate diet, turned the nutritional world upside down by proposing that too much carbohydrate—rather than too much fat—actually causes weight gain and metabolic disturbances. We have all heard the expression that *a calorie is a calorie*. However, the body processes energy-yielding nutrients in distinctly different ways. Can these differences impact weight loss and weight maintenance? What scientific evidence exists concerning the effects of macronutrient distribution on weight loss and overall health?

High-Carbohydrate, Low-Fat Weight-Loss Diets Many researchers believe that diets high in carbohydrates and low in fat promote weight loss and have an overall beneficial effect on health. These types of weight-loss plans recommend a relatively low intake of fat (10–15 percent of total calories) by limiting most meat, dairy foods, oils, and olives. However, low-fat meats and dairy products can be eaten in moderation. With an emphasis on fruit, vegetables, and whole grains, these weight-reduction and maintenance plans suggest that about 65–75 percent of total calories come from carbohydrates and that protein and fat make up the remainder.

There are several reasons why advocates of low-fat, high-carbohydrate diets believe that such plans help prevent obesity. First, gram for gram, fat has more than twice as many calories as carbohydrate and protein. It is therefore reasonable to assume that consuming less fat leads to lower energy intake, which in turn results in weight loss. Second, fat can make food more flavorful, contributing to overconsumption. Finally, excess calories from fat are stored by the body more efficiently than excess calories from carbohydrate or protein. Converting excess glucose and amino acids into fatty acids requires energy, which contributes to energy expenditure and thus helps promote weight loss.

Long-standing dietary advice aimed at helping people lose weight has consistently focused on reducing dietary fat. Although total energy intake has increased, the percentage of total calories from fat has declined from 45 percent in the 1960s to approximately 33 percent today. On average, 10 to 20 fewer grams of fat are consumed per day.[46] Obviously, decreased fat intake has not resulted in a decreased prevalence of obesity. In fact, obesity rates have increased under the lower-fat regime.[47] The evidence provided by carefully conducted clinical trials does not support the notion that low-fat diets promote weight loss, nor do they have any health advantages.[48]

Although it is not clear precisely which dietary factors contributed to this trend, some researchers believe that a failure to choose healthy high-carbohydrate foods may be, in part, responsible. While experts hoped that Americans would replace unhealthy fatty foods with more nutritious foods such as whole-grain breads, fruits, and vegetables, this was not the case. The ingredient that typically replaces fat in products advertised as "fat-free" is refined carbohydrate (such as white flour or table sugar)—a nutrient-poor, calorie-containing substitute. Because many of these products have the same amount of calories as the original products, the consumption of high-carbohydrate, fat-free snack foods may actually contribute to weight gain. Some researchers believe that eating foods high in refined carbohydrate (and especially those low in fat) makes people hungrier, and therefore can make them heavier.[49] In short, the theory that a low-fat diet is the best defense against weight gain is not without debate, and many health experts believe that there is now enough scientific evidence to lift the ban on dietary fat.[50]

Low-Carbohydrate Weight-Loss Diets On the opposite end of the weight-loss diet spectrum are diets that are low in carbohydrates and higher in both protein and fat, such as the Atkins, Paleolithic, and Zone diet plans (see Table 9.5). Advocates of these weight-loss plans claim that diets high in protein have some weight-loss advantages. This may be attributed to the fact that protein possibly help people feel fuller on fewer calories and that it takes more energy for the body to assimilate protein (higher TEF), as compared to carbohydrates and fat.[51] Indeed, some experts believe that people are more likely to gain weight from excess carbohydrates than from excess fats or proteins because high-carbohydrate foods cause insulin levels to rise, which in turn favor fat deposition. Thus, limiting one's intake of starch and refined sugars should theoretically help a person lose weight.

TABLE 9.5	CALORIC DISTRIBUTION OF TYPICAL HIGH- AND LOW-CARBOHYDRATE DIETS COMPARED TO THE ACCEPTABLE MACRONUTRIENT DISTRIBUTION RANGES (AMDRs)		
Nutrient	**AMDR[a]**	**High-Carbohydrate Diet[b]**	**Low-Carbohydrate Diet[c]**
	Percent of Total Calories		
Fat	20–35	10–15	55–65
Carbohydrate	45–65	65–75	5–20
Protein	10–35	10–25	20–40

[a]Source: Institute of Medicine. Dietary Reference Intakes for energy, carbohydrate, fiber, fat, fatty acids, cholesterol, protein, and amino acids. Washington, DC: National Academies Press, 2005.
[b]Source: Ornish D. Eat more weigh less: Dr. Dean Ornish's life choice diet for losing weight safely while eating abundantly. New York: Harper Collins, 1993.
[c]Source: Atkins RC. Dr. Atkins' new diet revolution, revised. National Book Network, 2003.

The many low-carbohydrate weight-loss diets differ in terms of the types of foods allowed. Some exclude nearly all carbohydrates, while others take a more moderate approach by allowing healthy, carbohydrate-rich foods such as fruits, vegetables, and whole-grain products. Low-carbohydrate diets have become enormously popular in recent years as people have sought out new weight-loss strategies and alternatives to traditional approaches.[52] Studies comparing weight loss associated with low-carbohydrate diets to weight loss associated with low-fat diets show that, at six months, greater weight loss is achieved on low-carbohydrate diets. Although low-carbohydrate diets appear safe in the short term, some experts have raised doubts about their long-term effectiveness.[53]

Perhaps one of the most legitimate concerns regarding low-carbohydrate diets is the restriction of healthy, high-carbohydrate foods such as fruits, vegetables, and whole-grain products—a valid criticism.[54] Although most low-carbohydrate diet plans recommend that people take dietary supplements, these cannot replace the many other substances (such as fiber and phytochemicals) supplied by these restricted foods. To counter potential nutrient deficiencies and gastrointestinal problems, people following low-carbohydrate diets are strongly encouraged to consume fresh fruits and nonstarchy vegetables. Clearly, moderation is the most important factor when it comes to carbohydrate and fat intake.

© Ditty_about_summer/Shutterstock.com

Total Calories versus Macronutrient Distribution The reality is that nobody knows for certain the ideal distribution of macronutrients for weight loss. What is known, however, is that the AMDRs provide a range of relative proportions of calories from macronutrients associated with a healthy diet. It is up to individuals, therefore, to determine the eating patterns that best suit their needs within the parameters of the AMDRs. However, research strongly suggests that reducing caloric intake is the single most critical component of successful weight loss. What is also most important for weight loss is for people to adhere to a dietary plan that is sustainable. This often means making permanent changes in what we eat, how much we eat, and how often we eat.

Most popular diets approach energy balance in a simplistic fashion, as if there is one magic key that can open the door to easy weight control. On the contrary, maintaining neutral or negative energy balance is quite complex. Eating behaviors are shaped by many factors including genetics, the physiological states of hunger and satiety, and the psychological and social determinants of appetite. In the end, only a combination of healthy eating and exercise can lead to successful long-term weight control. This simple formula can be far from easy to follow, however. Influences ranging from social pressures to genetics to the foods available in a given society all affect one's ability to maintain a healthy weight.

STUDY TOOLS 9

READY TO STUDY? IN THE BOOK, YOU CAN:

☐ Rip out the Chapter Review Card, which includes key terms and chapter summaries.

ONLINE AT WWW.CENGAGEBRAIN.COM, YOU CAN:

☐ Learn about Americans and diet habits in a short video.

☐ Explore the connection between hunger and appetite in an animation.

☐ Interact with figures from the text to check your understanding.

☐ Prepare for tests with quizzes.

☐ Review the key terms with Flash Cards.

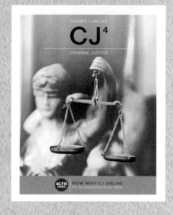

10 | Life Cycle Nutrition

LEARNING OUTCOMES

10-1 Describe the physiological changes that take place throughout the life cycle, and how these changes impact nutrient requirements.

10-2 Describe the major stages of prenatal development.

10-3 Explain how pregnancy and breastfeeding impact the nutrient and energy requirements of women.

10-4 Explain why it is important for mothers to breastfeed their infants.

10-5 Discuss the nutritional needs of infants.

10-6 Discuss the nutritional needs of toddlers and young children.

10-7 Describe how nutritional requirements change during adolescence.

10-8 Understand the effect of aging on nutrient and energy requirements.

After finishing this chapter go to **PAGE 279** for **STUDY TOOLS.**

10-1 WHAT PHYSIOLOGICAL CHANGES TAKE PLACE DURING THE HUMAN LIFE CYCLE?

From beginning to end, the human life cycle is a process of continuous change. The continuum of life encompasses infancy, childhood, adolescence, adulthood, and, for women, the special life stages of pregnancy and lactation. Throughout life, the body changes in size, proportion, and composition. Because the body changes so tremendously throughout life, so too do nutritional requirements. For example, nutritional needs of infants differ vastly from those of adults. Regardless of one's current life cycle stage, an appropriate diet is essential to good health. In addition, nutritional status at an early stage can influence health at later stages. For this reason, the food choices you make today may have far greater consequences on your long-term health than you might think.

10-1a Physiological Changes during the Life Cycle

Cells form, mature, carry out specific functions, die, and are replaced by new cells. In many ways, the life cycle of a cell mirrors human life itself. That is, after a baby is conceived and born, the next 70 to 90 years are characterized by periods of growth and development, maintenance, reproduction, physical decline, and eventually death. The ability to reproduce enables humans to pass their genetic materials (DNA) on to the next generation.

Aging is an inevitable process, and although individuals grow and develop at different rates, physical changes tend to coincide with various stages of the human life cycle. Age-related physical changes affect body size and composition, which in turn influence nutrient and energy requirements. For this reason, the Dietary Reference Intakes (DRIs) recommend specific nutrient and energy intakes for each life-stage group. Newborn and older infants (0 to 6 months and 7 to 12 months); toddlers (1 to 3 years); young children (4 to 8 years);

young and older adolescents (9 to 13 years and 14 to 18 years); young adulthood, middle age adulthood, and adulthood, (19 to 30, 31 to 50, and 51 to 70 years); and older adulthood (over 70 years) each have unique energy and nutrient needs. Beyond these age-related stages, the DRI recommendations also consider the special stages of pregnancy and lactation. The various DRI life-stage groups are illustrated in Figure 10.1. Knowing more about growth and developmental milestones help explain why nutritional needs change throughout life.

Growth and Development Growth refers to a physical change resulting from an increase in either cell size (hypertrophy) or number (hyperplasia), whereas **development** refers to the acquisition of a skill or a function. Growth and development generally occur in a predictable and orderly manner throughout the life cycle. The most rapid rates of growth and development occur during infancy, childhood, adolescence, and pregnancy. In biological terms, human growth and development continue throughout life until physical maturity is reached. Not only does physical maturity coincide with the cessation of growth, but it also signifies the onset of age-related physical changes that gradually lead to a decline in functional skills later in life. Referred to as **senescence**, this

period of the life cycle begins to occur sometime after physical maturity is reached and continues until death.

The most common and useful ways to assess growth are to measure height and weight. The World Health Organization and the U.S. National Center for Health Statistics (NCHS), a division of the Centers for Disease Control and Prevention (CDC), compile height and weight reference standards into growth charts that indicate expected growth for well-nourished infants, children, and adolescents.[1] By comparing a child's weight to the average weight of a group of healthy children of similar age and sex, parents and clinicians can evaluate the adequacy of a child's growth. For example, if the weight of a 5-year-old child is at the sixtieth percentile of an NCHS chart, 40 percent of healthy children of similar age and sex weigh more, and 60 percent weigh less. With the aid of these charts, growth can be monitored over time and used as a general indicator of health throughout these important phases of the life cycle.

growth A physical change that results from an increase in either cell size or number.

development A change in the attainment or complexity of a skill or function.

senescence The gradual physiological deterioration that occurs with age.

FIGURE 10.1 STAGES IN THE HUMAN LIFE CYCLE AND DRI LIFE-STAGE GROUPS

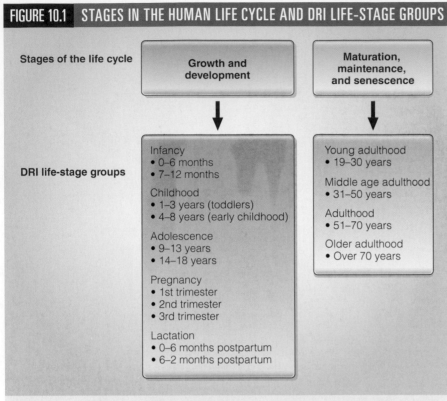

Stages of the life cycle	Growth and development	Maturation, maintenance, and senescence

DRI life-stage groups

Infancy
- 0–6 months
- 7–12 months

Childhood
- 1–3 years (toddlers)
- 4–8 years (early childhood)

Adolescence
- 9–13 years
- 14–18 years

Pregnancy
- 1st trimester
- 2nd trimester
- 3rd trimester

Lactation
- 0–6 months postpartum
- 6–2 months postpartum

Young adulthood
- 19–30 years

Middle age adulthood
- 31–50 years

Adulthood
- 51–70 years

Older adulthood
- Over 70 years

Growth and development affect body size and composition throughout the life cycle. Thus, a person's life stage influences nutrient and energy requirements.

In addition to assessing growth, it is equally important to assess a child's development. Although there is a great deal of variability in terms of human development, developmental patterns are usually predictable. For example, some infants walk as early as 10 months, while others may not take their first steps until several months later. Similarly, infants generally learn to crawl before they learn to walk. Although development varies greatly from child to child, failure to reach major developmental milestones by certain ages is cause for concern and may indicate a problem such as illness or poor nutritional status.

Maturation, Maturity, and Senescence As a person approaches physical maturity, growth and development begin to slow. **Cell turnover**, the cyclical process by which cells form and break down, reaches equilibrium at this time. Once physical maturity is achieved, growth ceases, and the body enters a phase of maintenance during which cell formation equals cell breakdown. As a person ages, the rate at which new cells form decreases, resulting in a loss of some body tissue. Remaining cells become less effective at carrying out their functions, and senescence, the physical changes characteristically associated with aging, gradually become apparent. Senescence brings about a slow decline in physical function and health, which eventually influences a person's nutrient and calorie requirements.

cell turnover The cyclical process by which cells form and break down.

10-2 WHAT ARE THE MAJOR STAGES OF PRENATAL DEVELOPMENT?

Although it is important for all women to meet their nutritional needs, maintaining adequate nutrition is particularly important during pregnancy. Poor nutritional status before and during pregnancy can have serious long-term effects on an unborn child. For example, women with poor nutritional status are at increased risk for having a baby born too early (premature) or too small (intrauterine growth restriction). Because a mother's nutritional choices have such significant effects on the life of her child, early prenatal care and ongoing health assessment are critical.

10-2a Prenatal Development

There may be no other time in a woman's life when her body experiences such extensive changes as during pregnancy. The physiological transformations that a pregnant woman experiences are needed to support the new, emerging life within her. Although every pregnancy is different, much of the growth and development that occurs during pregnancy happens in predictable and organized ways. Prenatal development takes place in two periods—the embryonic period and the fetal period. The **embryonic period** comprises the first 8 weeks of pregnancy and is subdivided into pre-embryonic (0–2 weeks) and embryonic (2–8 weeks) phases.

Embryonic Period Conception takes place when an ovum is fertilized by a sperm, forming a **zygote**. As the zygote moves toward the uterus, it divides repeatedly, eventually forming a dense cellular sphere called a **blastocyst**. Approximately 2 weeks after conception, the blastocyst implants itself into the lining of the uterus. This stage of the embryonic period is referred to as the **pre-embryonic phase**. The **embryonic phase** is the stage of prenatal development that spans from the start of the third week to the end of the eighth week after fertilization. During this phase, the blastocyst is now referred to as an **embryo**. By the end of the embryonic period, the embryo grows to be about the size of a kidney bean. Cell division continues throughout the embryonic period, eventually forming rudimentary structures that eventually develop into specific tissues and organs. By the end of the embryonic period, the basic structures of all major body organs are formed, albeit not fully developed. Each organ follows a precise timetable in terms of development. If specific nutrients are lacking during these critical periods of development, organs may not form properly (see Figure 10.2).

The term **critical period** is often used to describe the time in prenatal development during which adverse effects on growth and development are irreversible. Although critical periods are most likely to occur during the early stages of pregnancy, they can also occur at later stages.

Teratogens **Teratogens** are agents or conditions that can disrupt prenatal growth and development, particularly if exposure coincides with a critical period of organ formation. For example, in the late 1950s, a drug called thalidomide was prescribed to alleviate morning sickness in pregnant women. However, is was soon discovered that thalidomide led to babies being born with severe birth defects such as missing or deformed limbs. This tragedy highlighted that even seemingly safe drugs could have unforeseen harmful effects on the unborn child. Although not all birth defects are caused by teratogens, thousands of substances, including viruses, heavy metals, radiation, environmental pollutants, and nutritional toxicities, are known to be teratogenic. It is, therefore, important for pregnant women to limit their exposure to known teratogens, including alcohol.

It is well known that alcohol consumption during pregnancy can be

Stages of prenatal development from conception to, eventually, a fetus.

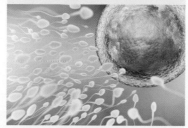
CONCEPTION

BLASTOCYST

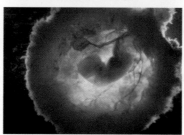
EMBRYO

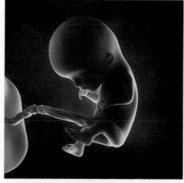
FETUS

© Tatiana Shepeleva/Shutterstock.com

© Sebastian Kaulitzki/Shutterstock.com

Petit Format/Nestle/Science Source/ Photo Researchers, Inc.

© Sebastian Kaulitzki/Shutterstock.com

embryonic period The period of prenatal development that spans from conception through the eighth week of gestation.

zygote An ovum that has been fertilized by a sperm.

blastocyst A dense sphere of cells that implants itself into the lining of the uterus.

pre-embryonic phase The early phase of the embryonic period (0–2 weeks) that begins with fertilization and continues through the formation of the blastocyst.

embryonic phase The stage of the embryonic period (2–8 weeks) during which organs and organ systems form.

embryo A developing human as it exists from the start of third week to the end of the eighth week after fertilization.

critical period A period in prenatal development during which adverse effects on growth and development are irreversible.

teratogens An agent or condition that can disrupt prenatal growth and development.

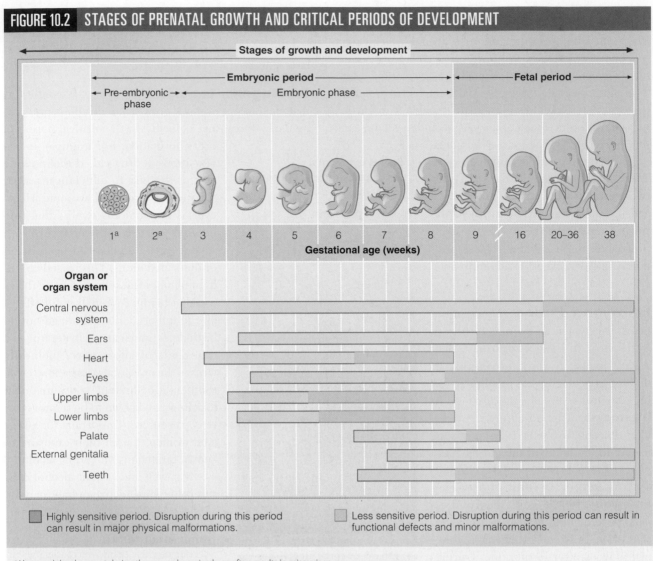

Stages of growth and development

Embryonic period

Pre-embryonic phase

Embryonic phase

Fetal period

| 1ᵃ | 2ᵃ | 3 | 4 | 5 | 6 | 7 | 8 | 9 | 16 | 20–36 | 38 |

Gestational age (weeks)

Organ or organ system

Central nervous system
Ears
Heart
Eyes
Upper limbs
Lower limbs
Palate
External genitalia
Teeth

■ Highly sensitive period. Disruption during this period can result in major physical malformations.

■ Less sensitive period. Disruption during this period can result in functional defects and minor malformations.

ᵃAbnormal development during the pre-embryonic phase often results in miscarriage.

Adapted from: Moore KL, Persaud TVN, Torchia MG. The developing human: Clinically oriented embryology, 10th ed. Philadelphia: W.B. Saunders/Elsevier, 2015.

harmful to the unborn baby, resulting in **fetal alcohol spectrum disorder**. Fetal alcohol spectrum disorder is characterized by a range of physical and behavioral problems that can arise in an infant as a result of alcohol consumption by a woman during pregnancy. For this reason, it is recommended that women completely abstain from drinking alcohol throughout pregnancy.

Fetal Period The embryonic period is followed by the **fetal period**, which starts at the beginning of the ninth week of pregnancy and ends at birth. Now referred to as a **fetus**, a baby in this second stage of pregnancy is characterized by tremendous growth and development. Fetal weight increases by almost 50,000 percent during this period, resulting in a body weight of approximately 7 to 8 lb (or 3.2 to 3.6 kg) and a length of roughly 20 inches (or 51 cm) at birth. Although adequate maternal weight gain and consumption of a healthy diet are important throughout pregnancy, they are particularly important during the fetal period. This is because the fetus undergoes rapid growth and development during the fetal period, and poor nutrition can dramatically hinder these processes. Beyond its direct contribution to fetal growth and development, the food eaten during pregnancy supports dramatic changes in the mother's own body as well.

fetal alcohol spectrum disorder A range of physical and behavioral problems that can arise in a child as a consequence of alcohol consumption by a woman during pregnancy.

fetal period The stage of prenatal development that begins at the ninth week of pregnancy and ends at birth.

fetus A developing human as it exists from the start of the ninth week of pregnancy until birth.

10-2b The Formation of the Placenta

Shortly after the blastocyst implants itself into the lining of the uterus, embryonic and maternal tissues begin to form the **placenta**, an organ that delivers nutrients and oxygen to the developing child. Although the placenta develops early in the pregnancy, it takes several weeks to become fully functional. As illustrated in Figure 10.3, the placenta is a highly vascular structure that serves important functions such as the transfer of nutrients and oxygen from the mother's blood to the fetus via the umbilical cord. Potentially harmful substances can also cross from the mother's blood into the fetal circulation through the placenta. So, pregnant women must be particularly careful about using medications and exposure to substances that might harm the embryo or fetus. Because insects can spread certain viruses that can be harmful to the unborn child, it is important for pregnant women to take certain precautions when traveling to certain countries. For example, pregnant women infected with the Zika virus are at increased risk of having babies born with serious neurological defects that affect brain development. As a result of prenatal exposure to the virus, infants are born with smaller than normal heads, a condition called microcephaly. The placenta not only transports substances to

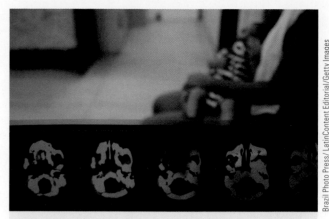

Pregnant women infected with the Zika virus are at increased risk of having babies born with severe birth defects. Officials from the CDC have declared that Zika, which is spread by Aedes species mosquitos, is a teratogen that can cause microcephaly and other serious neurological birth defects in exposed fetuses.

the fetus, but it also carries waste products away. Waste formed by the fetus passes through the placenta into the mother's blood and is ultimately eliminated from her body.

Failure of the placenta to form properly can complicate a pregnancy in many ways. The inability of the placenta to deliver adequate amounts of nutrients and oxygen to the fetus, known as **placental insufficiency**, can impair fetal growth and development. This can also cause the baby to be born too early and/or too small. In severe cases, placental insufficiency can result in fetal death and miscarriage.

FIGURE 10.3 STRUCTURE AND FUNCTIONS OF THE PLACENTA

Placenta
Umbilical cord
Uterus

Fetal blood vessels

Maternal blood vessels

Oxygen, nutrients, and other substances are delivered to the baby from the maternal blood.

Waste products are delivered to the mother from the baby's blood.

Umbilical cord

Placenta

The placenta, made of both fetal and maternal tissues, forms early in the pregnancy.

placenta An organ, made of embryonic and maternal tissues, that supplies nutrients and oxygen to the developing child.

placental insufficiency A complication of pregnancy that can arise when the placenta is unable to deliver adequate nutrients and oxygen to the developing baby.

10-2c Gestational Age

Once a baby is conceived, how long can a woman expect to be pregnant? There are several ways to answer this question. Whereas the *embryonic* and *fetal periods* refer to stages of prenatal development, pregnancy is more commonly described in terms of *trimesters*. The first trimester spans the time from conception through week 13. This includes the entire embryonic period, as well as part of the fetal period.

gestation length The period of time between conception and birth.

gestational age The number of weeks from the first day of a woman's last normal menstrual cycle.

full-term A baby born with gestational age between 37 and 42 weeks.

preterm (or **premature**) A baby born with a gestational age less than 37 weeks.

The second trimester lasts from week 14 through week 26, and the third trimester lasts from week 27 to the end of pregnancy.

The duration of pregnancy—the **gestation length**—is the is the period of time between conception and birth. However, because many pregnant women do not know exactly when they conceived, calculating gestation length can be difficult. A method more commonly used to assess the length of pregnancy is **gestational age**, the number of weeks from the first day of the woman's last menstrual cycle. Based on gestational age, the average length of pregnancy is about 40 weeks. Babies born with gestational ages between 37 and 42 weeks are considered **full-term** infants, whereas those born with gestational ages less than 37 weeks are considered **preterm** (or **premature**), and

Nutrition and Epigeneitc Inheritance

Most people are fully aware that an unhealthy diet during pregnancy can adversely affect the health of a newborn baby. However, mounting evidence also suggests that the consequences of poor nutrition during pregnancy may be more detrimental to the developing fetus than previously believed. In fact, babies born to mothers with poor nutritional status are not only at risk of being born too early and too small, but are also at great risk for health problems that can last a lifetime. Referred to as fetal programming, circumstances such as malnutrition, stress, and alcohol can alter how vital tissues of the body grow and develop *in utero*. As an adult, these changes can make a person more susceptible to diseases such as type 2 diabetes, cardiovascular disease, and hypertension. What is now realized, however, is that some of these traits may be transgenerational, meaning that susceptibility to certain diseases can be passed on to future generations.

Although geneticists have long suspected that something other than changes in our DNA sequence can influence how certain diseases are passed from one generation to the next, the mechanism has largely remained a mystery. But scientists now know that not all inheritable traits are caused by changes in DNA sequence. Rather, some traits can be passed on to future generations by changes in what is called the epigenome. The epigenome consists of complex heritable chemical compounds that influence what proteins

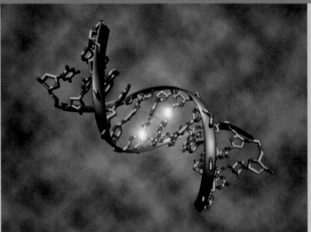

Source: Japan Times

Chemical compounds attached to DNA molecules play an important role regulation of gene expression.

an individual makes (or doesn't make), not by altering the DNA itself, but by modifying gene expression. Changes in the epigenome occur throughout a person's life in response to environmental factors such as smoking, environmental pollutants, stress, and diet, but these changes can be particularly critical when they occur *in utero*. Like DNA, the epigenome is also passed on to offspring. This means that your epigenomic make-up can also affect your children and your grandchildren. This model of inheritance, called epigenetics, is a radical shift from the belief that inherited diseases are caused solely by the genes inherited from your parents. Epigenetic inheritance adds a new dimension to understanding the importance of a mother's diet in shaping the health of her baby, and possibly that of generations to follow.

FIGURE 10.4 CLASSIFICATION OF INFANTS BASED ON GESTATIONAL AGE AND BIRTH WEIGHT

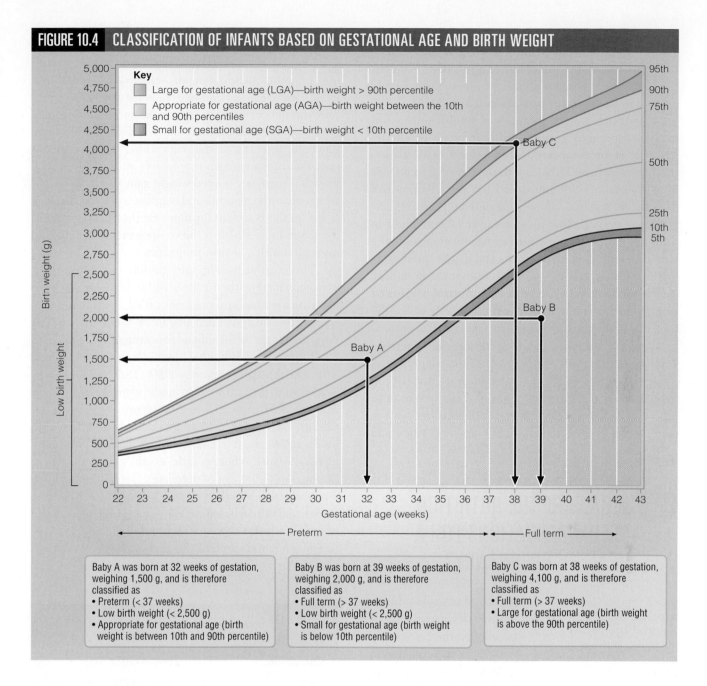

Key
- Large for gestational age (LGA)—birth weight > 90th percentile
- Appropriate for gestational age (AGA)—birth weight between the 10th and 90th percentiles
- Small for gestational age (SGA)—birth weight < 10th percentile

Baby A was born at 32 weeks of gestation, weighing 1,500 g, and is therefore classified as
- Preterm (< 37 weeks)
- Low birth weight (< 2,500 g)
- Appropriate for gestational age (birth weight is between 10th and 90th percentile)

Baby B was born at 39 weeks of gestation, weighing 2,000 g, and is therefore classified as
- Full term (> 37 weeks)
- Low birth weight (< 2,500 g)
- Small for gestational age (birth weight is below 10th percentile)

Baby C was born at 38 weeks of gestation, weighing 4,100 g, and is therefore classified as
- Full term (> 37 weeks)
- Large for gestational age (birth weight is above the 90th percentile)

those born with gestational ages greater than 42 weeks are considered **postterm**. The earlier a baby is born, the greater the risk for complications that can affect the child's survival and long-term health.

Gestational Age and Birth Weight It is not only important that a baby is born full term, but also that the baby is born with a healthy weight. As illustrated in Figure 10.4, growth charts are used to classify infants according to birth weight and gestational age. Babies weighing less than 5 lb, 8 oz (2,500 g) at birth are considered to have **low birth weight (LBW)**. LBW infants are small because they are either preterm or have experienced slow growth *in utero*, also known as

intrauterine growth restriction (IUGR). Babies who have experienced IUGR are often said to be **small for gestational age (SGA)**, meaning that they have birth weights below the tenth percentile for their gestational

postterm A baby born with a gestational age greater than 42 weeks.

low birth weight (LBW) A baby that weighs less than 5 lb, 8 oz (2,500 g) at birth.

intrauterine growth restriction (IUGR) Slow growth while in the uterus.

small for gestational age (SGA) A baby that has a birth weight below the tenth percentile for gestational age.

age. Infants born with birth weights between the tenth and ninetieth percentiles for gestational age are said to be **appropriate for gestational age (AGA)**, whereas those with birth weights above the ninetieth percentile are said to be **large for gestational age (LGA)**.

Low birth weight infants are significantly more likely to experience serious health problems within the first year of life compared to infants born AGA. In fact, premature birth and LBW are the leading risk factors for infant mortality.[2] Not only does LBW put a baby at risk early in life, it may also have profound long-term effects. Evidence suggests that less than optimal conditions in the womb (uterus) may cause permanent changes in the structure and function of organs and tissues, predisposing individuals to certain chronic diseases later in life.[3] This hypothesis, called the **developmental origins of health and disease hypothesis** (formerly called the fetal origins hypothesis), suggests that less than optimal prenatal and early postnatal conditions may increase a person's risk of developing cardiovascular disease, stroke, hypertension, type 2 diabetes, and obesity as adults. You can read more about how early conditions *in utero* can have lasting effects on the health of future generations in the special feature box—Nutrition and Epigenetics.

 10-3 WHAT ARE THE NUTRITION RECOMMENDATIONS FOR A HEALTHY PREGNANCY?

appropriate for gestational age (AGA) A baby that has a birth weight between the tenth and ninetieth percentiles for gestational age.

large for gestational age (LGA) A baby that has a birth weight above the ninetieth percentile for gestational age.

developmental origins of health and disease hypothesis A hypothesis that states that less than optimal conditions in the uterus or during early infancy may cause permanent changes in the structure and function of organs and tissues, predisposing individuals to certain chronic diseases later in life.

Although unavoidable situations can and do affect pregnancy, there are many precautions women can take to help ensure the birth of a healthy child. The recommendations presented throughout the following sections can help decrease the risk of giving birth to a preterm or LBW baby. Of these, it is particularly important for pregnant women to gain an appropriate amount of weight, eat a healthy diet, and refrain from smoking and consuming alcohol.

10-3a Weight-Gain Recommendations

The amount of weight a woman gains during pregnancy is an important determinant of fetal growth and development. Health care practitioners monitor weight gain carefully throughout pregnancy to make sure that a woman gains the appropriate amount of weight—neither too much nor too little. Perhaps surprising to some, weight gained during pregnancy is not solely attributable to fetal growth. Pregnancy-related weight gain also includes tissues such as the placenta and mammary glands (breasts), as well as fluids such as the amniotic fluid that surrounds the fetus. Table 10.1 lists the components of weight gain associated with a healthy pregnancy.

The current weight-gain guidelines for pregnant women, issued by the Institute of Medicine in 2009, are based on a woman's prepregnancy BMI (see Table 10.2).[4] Gaining the recommended amount of weight for her BMI range increases the likelihood of a woman giving birth to a full-term baby with a healthy birth weight. A woman with a healthy prepregnancy BMI (18.5 to 24.9 kg/m²) is advised to gain 25 to 35 pounds, whereas overweight women (BMI from 25.0 to 29.9 kg/m²) are encouraged to gain less—between 15 and 25 pounds. In addition to total weight gain, it is also important to monitor the *rate* of weight gain. Whereas little weight gain is necessary during the early stages of pregnancy, a steady gain of 2 to 4 pounds each month is recommended throughout the second and third trimesters. The ChooseMyPlate website offers a convenient tool to track and evaluate pregnancy weight gain. You can find this tool at http://www.choosemyplate.gov/pregnancy-weight-gain-calculator.

TABLE 10.1	COMPONENTS OF WEIGHT GAIN DURING PREGNANCY	
Component	**Total Weight Gain**
Fetus	7 to 8 lb (3.2 to 3.6 kg)
Placenta	1½ to 2 lb (0.7 to 0.9 kg)
Uterus and supporting structures	2½ to 3 lb (1.1 to 1.4 kg)
Maternal adipose stores	7 to 8 lb (3.2 to 3.6 kg)
Breasts (mammary glands)	1 to 2 lb (0.45 to 0.91 kg)
Body fluids (blood and amniotic fluid)	6 to 7 lb (2.7 to 3.2 kg)
Total weight gain	25 to 30 lb (11.3 to 13.6 kg)

Source: Institute of Medicine. National Research Council. Weight gain during pregnancy: Reexamining the guidelines. National Academies Press. Washington, DC. 2009. Available from: http://nationalacademies.org/HMD/Reports/2009/Weight-Gain-During-Pregnancy-Reexamining-the-Guidelines.aspx

TABLE 10.2 RECOMMENDED RANGES FOR TOTAL WEIGHT GAIN AND RATE OF WEIGHT GAIN DURING PREGNANCY

Pre-pregnancy BMI (kg/m²)	Recommended Total Weight-Gain Range*	Weekly Rate of Weight Gain During Second and Third Trimesters**
Underweight (<18.5)	28 to 40 lb (12.5 to 18 kg)	1 to 1.3 lb (0.44 to 0.58 kg)
Normal weight (18.5–24.9)	25 to 35 lb (11.5 to 16 kg)	0.8 to 1 lb (0.35 to 0.50 kg)
Overweight (25.0–29.9)	15 to 25 lb (7 to 11.5 kg)	0.5 to 0.7 lb (0.23 to 0.33 kg)
Obese (≥30.0)	11 to 20 lb (5 to 9 kg)	0.4 to 0.6 lb (0.17 to 0.27 kg)

*Weight-gain range for singleton pregnancies.

**Calculations assume a 0.5–2 kg (1.1–4.4 lb) gain in the first trimester.

Source: Institute of Medicine. Weight gain during pregnancy: Reexamining the guidelines. National Academies Press. Washington, DC, May 2009. Available from: http://nationalacademies.org/HMD/Reports/2009/Weight-Gain-During-Pregnancy-Reexamining-the-Guidelines.aspx

10-3b Dietary Recommendations during Pregnancy

Dietary recommendations for pregnant women are intended to promote optimal health for both the mother and unborn child. Seeking regular prenatal care from a health care provider is an important first step in establishing a healthy prenatal diet. Pregnant women may also find many useful resources available on the ChooseMyPlate website specifically related to nutrition during pregnancy (see http://www.choosemyplate.gov/moms-pregnancy-breastfeeding). These include an explanation of special nutritional needs during pregnancy, tips for pregnant mothers-to-be, and information about dietary supplements and food safety while pregnant. These resources can be used by pregnant women to create personalized nutrition and physical activity plans to ensure that healthy food choices are being made throughout pregnancy. Beyond those resources, the following guidelines provide key dietary recommendations to promote a healthy and balanced diet that is uniquely suited for pregnancy. A comparison of recommended nutrient intakes for nonpregnant, pregnant, and lactating women is presented in Figure 10.5.

Recommended Energy and Macronutrient Intakes

Adequate weight gain necessitates adequate energy intake. Therefore, it is important for an expectant mother to satisfy her daily energy requirements, as well as those of the growing fetus. Although the energy demands of pregnancy are quite high—about 60,000 kcal over the course of the pregnancy—very little extra energy is needed during the first trimester. During the second and third trimesters, however, rapid fetal growth requires an energy intake beyond that needed to support the mother's own needs. Pregnant women are generally advised to increase their energy intakes by about 340 and 450 kcal/day for the second and third trimesters, respectively. Thus, a woman with an estimated energy requirement of 2,000 kcal/day (when she is not pregnant) requires approximately 2,340 kcal/day during her second trimester and 2,450 kcal/day during her third trimester of pregnancy.

It is important to get enough carbohydrate, protein, and fat to satisfy energy requirements during pregnancy. If the pregnancy is progressing normally, carbohydrates should remain the primary energy source (45 to 65 percent of total calories). An increase of approximately 45 g/day of additional carbohydrates, equivalent to two to three servings of carbohydrate-rich foods such as whole-grain breads or cereals, is optimal. An additional 25 g protein/day is also needed to support the growth of the baby. This is easily obtained by eating sufficient amounts of protein-rich foods such as meat, dairy products, legumes, and eggs.

Dietary fat should make up approximately 20 to 35 percent of total caloric intake during pregnancy. Some essential fatty acids serve other vital roles beyond the provision of energy during pregnancy. For example, linoleic acid and linolenic acid are critical to fetal growth and development. Docosahexaenoic acid (DHA), an ω-3 fatty acid, is essential to fetal brain development and the

© Daniel Rajszczak/Shutterstock.com

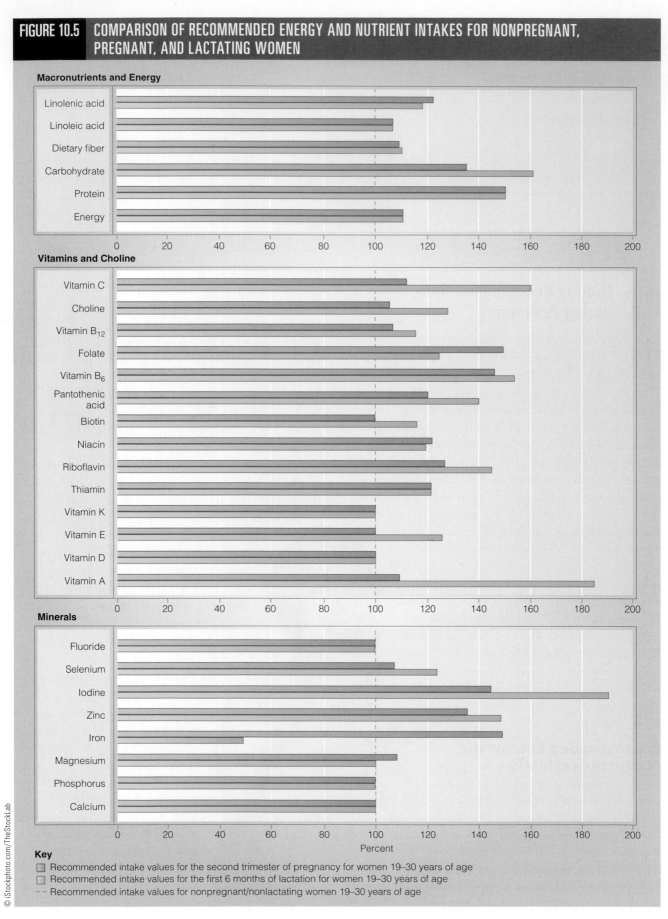

FIGURE 10.5 COMPARISON OF RECOMMENDED ENERGY AND NUTRIENT INTAKES FOR NONPREGNANT, PREGNANT, AND LACTATING WOMEN

Macronutrients and Energy

- Linolenic acid
- Linoleic acid
- Dietary fiber
- Carbohydrate
- Protein
- Energy

Vitamins and Choline

- Vitamin C
- Choline
- Vitamin B$_{12}$
- Folate
- Vitamin B$_6$
- Pantothenic acid
- Biotin
- Niacin
- Riboflavin
- Thiamin
- Vitamin K
- Vitamin E
- Vitamin D
- Vitamin A

Minerals

- Fluoride
- Selenium
- Iodine
- Zinc
- Iron
- Magnesium
- Phosphorus
- Calcium

Percent

Key

- ▢ Recommended intake values for the second trimester of pregnancy for women 19–30 years of age
- ▢ Recommended intake values for the first 6 months of lactation for women 19–30 years of age
- -- Recommended intake values for nonpregnant/nonlactating women 19–30 years of age

formation of the retina. To ensure adequate intake of these important fatty acids during pregnancy, women should eat fish and/or ω-3-rich oils (such as canola or flaxseed oil) several times a week. Because some types of fish contain high levels of mercury, the U.S. Department of Agriculture advises pregnant women to choose fish with low mercury levels (such as salmon, flounder, tilapia, trout, pollock, and catfish) and avoid eating fish believed to have high levels of mercury (such as shark, swordfish, king mackerel, and tilefish).[5] It is also noteworthy that canned "white" tuna (albacore) is higher in mercury than the "light" variety. For this reason, it is recommended that pregnant women limit their consumption of canned white tuna to less than 6 ounces per week. In addition to mercury, fish may contain other harmful chemicals that should be limited or avoided. For this reason, pregnant women should check local advisories to learn about the safety of fish caught in local lakes, rivers, and coastal areas. Advisories may recommend limiting or avoiding eating some types fish caught in certain places. If no advice is available, pregnant women are advised to limit their consumption of local fish to one meal (6 ounces) per week while not consuming any other fish during that week. The 2015 Dietary Guidelines for Americans recommend that pregnant women consume 8 to 12 ounces (227 to 340 g) of seafood per week from a variety of sources while limiting their intake of fish that contain mercury. To obtain needed levels of ω-3 fatty acids, pregnant women should choose fish that are higher in ω-3 fats and lower in mercury, such as salmon, herring, sardines, pollock, and trout. The 2015 Dietary Guidelines for Americans also advise pregnant women to avoid consuming raw and undercooked seafood because it can contain harmful parasites, bacteria or viruses. This includes but is not limited to sushi, sashimi, raw oysters, raw scallops, raw clams, and cured salmon (lox).

Another health concern during pregnancy relates to preventing a disease called *listeriosis*, which can be contracted from eating foods contaminated with the bacteria *Listeria*. For this reason, pregnant women should avoid eating hot dogs, luncheon meats, and other deli meats unless they have been properly reheated. Consuming raw (unpasteurized) milk is also highly discouraged, as is eating store-made salads such as ham and egg salad.

Recommended Micronutrient Intake With few exceptions, the requirements for most vitamins and minerals increase during pregnancy. Perhaps surprisingly, recommended calcium intake does not increase. Although extra calcium is needed for the fetus to grow and develop properly, changes in maternal physiology, such as increased calcium absorption and decreased

urinary calcium loss, accommodate these needs without increasing dietary intake. Therefore, the RDA for calcium for pregnant women, 1,000 mg/day, is the same as that for nonpregnant women.

Unlike calcium, the RDA for iron increases substantially during pregnancy: the recommended intake increases from 18 to 27 mg/day. Iron is essential for both the formation of hemoglobin and the growth and development of the fetus and placenta. Most well-planned diets provide women approximately 15 to 18 mg of iron every day, and, as a result, some pregnant women may have difficulty meeting the recommended intake for iron by diet alone. Therefore, iron supplementation is often encouraged during the second and third trimesters of pregnancy, when iron requirements are the highest.[6]

Adequate folate intake is especially important during pregnancy. Recall from Chapter 7 that folate is critical to cell growth and development, including that of the nervous system. A woman with poor folate status before or in early pregnancy is at an increased risk of having a baby with a neural tube defect, a specific type of birth defect that affects the spinal cord and brain. Because the formation of the neural tube occurs early in the pregnancy—21 to 28 days after conception—the neural tube may already be formed before a woman realizes she is pregnant. Because of the importance of folate to neural tube development, an RDA of 600 μg DFE/day has been set for pregnant women. Women capable of becoming pregnant are advised to consume 400 μg DFE of folic acid as a supplement or in fortified foods in addition to their regular consumption of folate. Examples of folate-rich foods include dark green leafy vegetables, lentils, orange juice, and enriched cereal grain products.

10-3c Staying Healthy during Pregnancy

Every pregnancy is unique. Whereas some women struggle with a wide variety of pregnancy-related discomforts, others experience no problems at all. In most cases, however, women experience physical and/or emotional pregnancy-related changes to some extent. Fortunately, a few simple dietary and lifestyle adjustments can help women feel their best throughout pregnancy.

Pregnancy-Related Physical Complaints Hormonal changes are believed to be the underlying cause of several common physical complaints associated with pregnancy. Morning sickness, fatigue, heartburn, constipation, and food cravings and aversions are often associated with pregnancy-related hormonal changes. Many of these discomforts occur during the early stages of pregnancy, whereas others may persist throughout. Usually these

Not All Neural Tube Defects Are Preventable

Recognizing the importance of folate in the prevention of neural tube defects, the FDA began to require folic acid fortification of all enriched cereal grain products in 1996. In addition, food manufacturers were granted permission to make health claims on appropriate food labels stating that an adequate intake of dietary folate or folic acid (in supplements and fortified foods) may reduce the risk of neural tube defects. Since these nationwide efforts began, folate status in the United States has improved, and the incidence of neural tube defects has decreased.[7] Although folic acid fortification efforts have been successful at decreasing the occurrence of neural tube defects, not all of these disorders can be prevented by increased folate intake. Some defects are *multi-factorial*, meaning that a combination of both environmental and genetic factors contribute to their development. Further investigations into the genetic causes of neural tube defects will hopefully lead to improved methods for early detection and prevention.

discomforts are not serious and can typically be managed with simple diet-related strategies. For example, some women who experience morning sickness, a condition characterized by queasiness, nausea, and vomiting, can find relief by avoiding foods with offensive odors. Other strategies include eating dry toast or crackers, eating small, frequent meals, and eating before getting out of bed in the morning. Heartburn, a common complaint during pregnancy, can often be managed by avoiding spicy or greasy foods, sitting up while eating, and waiting at least 2 to 3 hours after eating before lying down. To alleviate constipation, another common pregnancy-related complaint, women are advised to consume adequate amounts of fruits, vegetables, whole grains, and fluids.

pica The urge to consume nonfood items.

gestational diabetes A form of diabetes that develops when pregnancy-related hormonal changes cause cells to become less responsive to insulin.

Food cravings and food aversions are also common during pregnancy. Powerful urges to consume or avoid certain foods may be caused by hormone-induced heightened senses of taste and smell. Although most food cravings and aversions rarely pose serious problems during pregnancy, some expectant mothers develop powerful desires to consume nonfood items such as laundry starch, clay, soil, and burnt matches. The urge to consume nonfood items, called **pica**, has no known cause and can be potentially harmful to the mother and baby.

Pregnancy-Related Health Concerns Although most minor physical discomforts associated with pregnancy are considered normal, it is important for all expectant women to be aware of changes that could indicate more serious problems. Two common health concerns that can develop during the later stages of pregnancy are gestational diabetes and gestational hypertension.

Approximately 9 percent of pregnant women develop a form of diabetes called **gestational diabetes** during pregnancy—usually around 28 weeks or later.[8] Gestational diabetes occurs when pregnancy-related hormonal changes cause cells to become less responsive to insulin, triggering blood glucose levels to rise. Risk factors associated with gestational diabetes include age (> 25 y), family history of type 2 diabetes, previous miscarriage or delivery of a stillborn, previous birth of a very large baby, history of abnormal glucose tolerance, ethnicity (African, Hispanic, Native American, or Pacific Islander descent), and being overweight or obese. To test for gestational diabetes, most pregnant women are given a routine blood test during the third trimester of pregnancy. Once diagnosed, a healthy diet and exercise regimen can help keep blood glucose levels under control. Some pregnant women who have difficulty controlling blood glucose may require insulin injections, however. Although gestational diabetes disappears within 6 weeks after delivery, nearly 40 percent of all women with gestational diabetes develop type 2 diabetes within the next 10 to 20 years.[9]

To minimize this risk, it is especially important for a woman with a history of gestational diabetes to maintain a healthy weight, make sound food choices, and be physically active. Babies born to mothers with poorly controlled blood glucose (hyperglycemia) can grow very large *in utero*, weighing more than 9 pounds at birth. This can lead to complications during labor and delivery, making it more likely that the baby will be delivered surgically (cesarean section) rather than a vaginal birth. Furthermore, babies born large for gestational age (LGA) are at increased risk to become overweight or obese later in life.

Another problem that can arise during pregnancy is called **pre-eclampsia**, which typically occurs after 20 weeks of gestation. However, this serious hypertensive disorder actually originates early in pregnancy when the placenta is forming. Recall that the placenta is made up of both embryonic and fetal tissues. Defective attachment of the embryonic placental cells to the uterine lining interferes with the ability of maternal placental blood vessels to relax. As the pregnancy progresses, the increased flow of blood through the constricted maternal placental blood vessels causes her blood pressure to increase, sometimes to dangerously high levels.[10] Pre-eclampsia affects about 3 percent of pregnancies, and risk factors include family history of cardiovascular disease, advanced maternal age, obesity, pre-existing hypertension, and a previous pregnancy affected by hypertension. The signs and symptoms associated with pre-eclampsia include a rise in blood pressure; protein in the urine; fluid accumulation (edema) in the face, hands, and feet; headaches; blurred vision; and a sudden, dramatic increase in body weight. Once pre-eclampsia is confirmed, a woman is advised to rest and limit her daily activities for the duration of her pregnancy. Although most women who develop pre-eclampsia deliver healthy babies, some are not as fortunate. In some cases, a woman's blood pressure can increase to dangerously high levels, a condition referred to as **eclampsia**. This serious complication of pregnancy is characterized by dangerous seizures that pose a threat to the life of the mother and unborn child. The only effective treatment for eclampsia is to deliver the baby by a surgical procedure called a caesarian section.

You now understand the important role that nutrition plays before and during pregnancy. However, because breastfeeding requires careful dietary planning to ensure that energy and nutrient needs are satisfied for both mother and child, maintaining good nutrition is equally important after pregnancy as well. The physiological changes associated with milk production and the impact this process has on maternal nutrient requirements are just as vital to consider as prenatal nutrition.

10-4 WHY IS BREASTFEEDING RECOMMENDED DURING INFANCY?

Women often experience noticeable changes in the size and shape of their breasts during pregnancy. These changes are necessary to prepare the *mammary glands* (commonly known as the breasts) for milk production after the baby is born. Human milk is the ideal food for babies: not only does it support optimal growth and development during infancy and early childhood, but the benefits associated with breastfeeding may even extend to later stages of life. Furthermore, because human milk provides immunologic protection against pathogenic viruses and bacteria, it is not surprising that breastfed babies tend to be sick less often than formula-fed babies. Moreover, breastfeeding is beneficial to the mother: it decreases the risks of certain diseases and helps women return to their prepregnancy weight more easily.

10-4a Lactation

During pregnancy, hormones prepare the mammary glands for milk production; this process is called **lactogenesis** and continues until several days after the baby is born. During the early stages of lactogenesis, the number of milk-producing cells in specialized, milk-producing structures called **alveoli** increases and there is an expansion of ducts that will eventually transport milk out of the breast. Women are encouraged to nurse their babies soon after delivery because suckling initiates the final stage of lactogenesis that is required for adequate milk production and release of milk, a process called **lactation**.

Prolactin and Oxytocin Regulate Milk Production The hormones **prolactin** and **oxytocin** regulate milk production in specialized cells that make up alveoli, as well as the release of milk into the surrounding

pre-eclampsia Pregnancy-related condition characterized by high blood pressure, a sudden increase in weight, swelling due to fluid retention, and protein in the urine.

eclampsia A serious complication of pregnancy that is typically preceded by a condition called pre-eclampsia, and is characterized by the onset of seizures.

lactogenesis Structural and functional changes in the mammary glands that begin soon after conception and continue until after the baby is born, enabling mammary glands to produce sufficient amounts of milk.

alveoli Structures made up of milk-producing cells.

lactation The production and release of milk.

prolactin A hormone that stimulates alveolar cells to produce milk.

oxytocin A hormone that causes the muscles around the alveoli to contract.

FIGURE 10.6 NEURAL AND HORMONAL REGULATION OF LACTATION

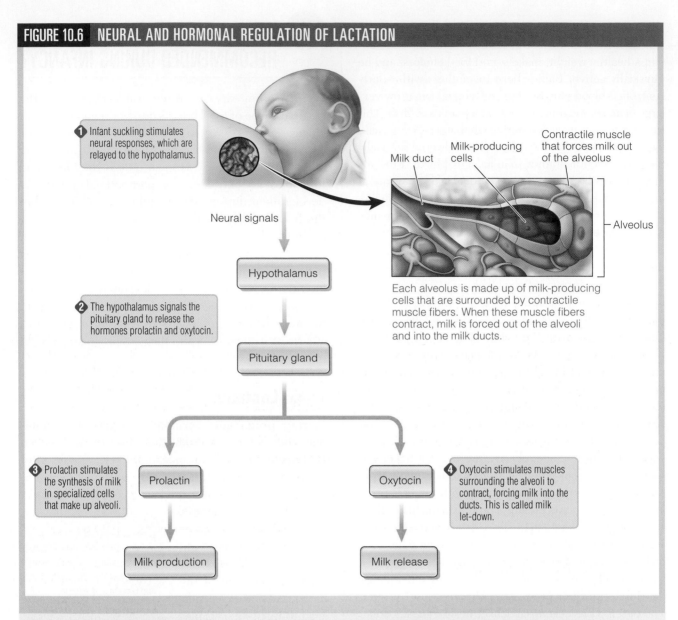

1 Infant suckling stimulates neural responses, which are relayed to the hypothalamus.

Milk duct

Milk-producing cells

Contractile muscle that forces milk out of the alveolus

Alveolus

Neural signals

Hypothalamus

2 The hypothalamus signals the pituitary gland to release the hormones prolactin and oxytocin.

Pituitary gland

Each alveolus is made up of milk-producing cells that are surrounded by contractile muscle fibers. When these muscle fibers contract, milk is forced out of the alveoli and into the milk ducts.

3 Prolactin stimulates the synthesis of milk in specialized cells that make up alveoli.

Prolactin

Oxytocin

4 Oxytocin stimulates muscles surrounding the alveoli to contract, forcing milk into the ducts. This is called milk let-down.

Milk production

Milk release

The human breast is a complex organ composed of many different types of tissues that produce and secrete milk. The hormones prolactin and oxytocin regulate the production and release of milk, respectively.

milk ducts. When a baby suckles, nerves in the nipple are stimulated, signaling the hypothalamus, which in turn signals the pituitary gland to release both prolactin and oxytocin (see Figure 10.6). Prolactin stimulates the synthesis of milk, whereas oxytocin causes the muscles around the alveoli to contract, forcing the milk out of the alveoli into the milk ducts. This active process of milk release is called milk **let-down**. As the baby suckles, the milk moves through the ducts toward the nipple, and into the baby's mouth. Anxiety, stress, and fatigue can sometimes interfere with

let-down The active process whereby milk is forced out of the alveoli and into the milk ducts.

the milk let-down reflex, making breastfeeding challenging. For this and other reasons, it is important for women to seek physical and emotional support during the breastfeeding period.

Milk Production—A Matter of Supply and Demand Whereas milk production is regulated by many physiological factors, the amount of milk produced is largely determined by how much milk the infant consumes. Women who breastfeed exclusively—meaning that human milk is the sole source of infant feeding—produce more milk than do those who supplement breastfeeding with infant formula. On average, women produce around 3 cups (700 mL) of milk per day during the first 6 months postpartum, and 2 ½ cups (600 mL) per day during the second 6 months. The reason women produce less milk during the second 6 months is that most infants are also fed supplemental foods by this age. Because newborns have small stomachs and can only consume small amounts of milk at each feeding, many mothers breastfeed as frequently as every 2 to 3 hours. As a baby grows, more milk can be consumed at each feeding, reducing the need to breastfeed as frequently. The American Academy of Pediatrics (AAP) recommends that women nurse their newborns at least 8 to 12 times each day and feed on demand rather than schedule their babies' feedings.[11]

Breastfeeding requires both proper positioning of the baby at the breast and the ability of the baby to latch onto the nipple. Once these conditions occur, the baby must be able to coordinate sucking and swallowing. Sometimes, parents worry about whether their infant is receiving enough milk. The best indicators of adequate milk production are healthy infant weight gain and appropriate frequency of wet diapers. As a general guideline, parents should expect at least four wet, heavy diapers per day. Most persistent problems associated with breastfeeding can be resolved with the assistance of a medical professional such as a pediatrician or lactation specialist, and all women who experience problems with breastfeeding should seek help. Once breastfeeding is fully established, it is comforting to know that, in most cases, a baby who receives adequate amounts of human milk does not need any other sources of nourishment for the first 4 to 6 months of life.

10-4b Human Milk Is Beneficial to Babies

Over the past 30 years, the number of women who breastfeed has increased steadily in the United States.[12] This trend is in part a response to compelling scientific evidence that human milk is ideally suited to optimal infant growth and development. The AAP recommends exclusive breastfeeding for the first 4 to 6 months of life and breastfeeding coupled with complementary foods (excluding cow milk) from the sixth month until at least 1 year.[13]

Shortly after a mother gives birth, her breasts produce a thick fluid called **colostrum**, which nourishes the newborn and helps prevent disease (colostrum contains an abundance of substances that fight infections). Over the first several days that follow birth, a gradual transition from the production of colostrum to that of mature milk occurs. Like colostrum, mature human milk has many important health benefits. Indeed, human milk is a rich source of nutrients, antimicrobial factors, and bioactive components that provide a vast number of immunological, nutritional, and developmental benefits. Scientists

Most women find it convenient to breastfeed their babies. Healthcare practitioners recommend that mothers breastfeed on demand.

have also recently discovered that human milk even contains bacteria, which are thought to be important in colonizing the infant's gastrointestinal tract. Researchers continue to be intrigued by the complexity of human milk. Not only is its nutrient composition uniquely suited for optimal infant growth and development, but researchers are equally interested in how biologically active components influence the infant's immune system and intestinal microbiota. The importance of establishing healthy gut microbiota early in life has gained universal acceptance among health professionals. In fact, two factors that greatly impact the GI microbiota of infants are mode of delivery (vaginal birth vs. caesarean) and diet during infancy (human milk vs. infant formula). According to the AAP, when compared to infants fed infant formula, breastfed infants have lower occurrences of respiratory illnesses, ear infections, gastrointestinal disease, and allergies, including asthma, eczema, and dermatits. These health benefits may largely be attributed to the various components in human milk that promote early maturation of the newborn's intestinal lining, while establishing a healthy array of organisms in the GI tract.[14]

Nutrient Composition of Human Milk The nutrient composition of human milk perfectly matches the nutritional needs of the healthy, full-term infant. In addition to nutrients, human milk contains enzymes and other compounds that make certain nutrients easier to digest and absorb. The protein contained in human milk, which is both present in the right amount and is easily digested, has an amino acid profile that is uniquely suited to support optimal growth and development during infancy. More than one-half of the calories

colostrum The first secretion from the mammary glands released after giving birth that provides nutrients and substances that help prevent disease.

in human milk come from lipids. Although there are dozens of fatty acids in human milk, scientists are particularly interested in docosahexaenoic acid (DHA) because of its importance in brain and eye development during infancy. Human milk also has an abundance of cholesterol, which serves as an important component of cell membranes. The primary carbohydrate contained in human milk is lactose. Not only is lactose an important source of energy, but it also facilitates the absorption of other nutrients. The vitamins and minerals present in human milk are also found in amounts that promote optimal infant growth and development. Given all of its benefits, health care professionals agree that human milk provides the perfect combination of nutrients and disease-fighting substances for infants.

10-4c Breastfeeding Is Beneficial to Mothers

Most people are aware that breastfeeding is beneficial for infants, but fewer know that it is also beneficial for mothers. Breastfeeding shortly after giving birth stimulates the uterus to contract, minimizing blood loss and shrinking the uterus to its prepregnancy size. Some women also find that breastfeeding helps them return to their prepregnancy weights more easily.[15] Beyond its short-term health benefits after pregnancy, breastfeeding is associated with several long-term maternal health benefits as well. It can delay the return of a woman's menstrual cycle, for example. The span of time between birth and the first postpartum menses allows the mother's iron stores to recover and can reduce the likelihood that she will become pregnant again too soon. Because it is possible to ovulate without menstruating, however, women are encouraged to use contraception until the time is right to have another baby. Finally, breastfeeding reduces a woman's risk of developing some forms of breast cancer, ovarian cancer, and possibly osteoporosis later in life.[16]

10-4d Maternal Energy and Nutrient Requirements during Lactation

Because milk production requires energy, caloric requirements increase during lactation. The amount of additional energy needed during lactation depends on whether a mother is exclusively breastfeeding or feeding by a combination of human milk and infant formula. Because women tend to produce more milk during the first 6 months of lactation than during the second 6 months, additional energy required for milk production is approximately 500 and 400 kcal/day, respectively. However, because some of the energy needed for milk production during the first 6 months should come from stored body fat associated with pregnancy, total dietary energy intake recommended for the first 6 months of lactation is lower than that for the second 6 months. Recommendations for micronutrient intakes during lactation are generally similar to those during pregnancy, though some, such as vitamin A, are somewhat greater and others, such as folate and iron, are somewhat lower. These values are listed on the Dietary Reference Intakes card at the back of this book. Finally, it is important to note that lactating women should consume sufficient amounts of water and other fluids.

Whether or not caffeine should be consumed is a common concern of lactating women. For the most part, drinking 2 to 3 cups of coffee each day (or other caffeine-containing beverages) is generally recognized as safe. However, too much caffeine can make a baby jittery and less likely to sleep.

Similarly, the AAP recommends that alcohol be consumed only in small amounts, if at all. This is because alcohol quickly enters the milk and is therefore consumed by the breastfeeding infant. The MyPlate food guidance system advises that breastfeeding mothers can continue to breastfeed and have an occasional alcoholic beverage if they are cautious and follow these guidelines.[17]

1. Wait until your baby has a routine breastfeeding pattern; at least 3 months of age.
2. Wait at least 4 hours after having a single alcoholic drink before breastfeeding.
3. Or, express breast milk before having a drink and this milk to feed your infant later.

10-4e Infant Formula

Although breastfeeding is usually recommended, there are times when breastfeeding is not possible, such as when the mother is taking chemotherapeutic drugs, infected with human immunodeficiency virus (HIV), using illicit drugs, or has untreated tuberculosis. The only acceptable alternative to human milk is commercial infant formula—it is recommended that infants not be fed cow milk at any time during the first year of life. The nutrient content of cow milk is very different from human milk

© Image Source/Jupiterimages

and infant formula, and substances found in cow milk might cause other health problems in the baby.

When formula is the best option for nourishing an infant, pediatricians recommend that parents use only formulas fortified with iron to help prevent iron deficiency.[18] Parents should also check infant formula for DHA and arachidonic acid fortification for their positive effects on neural and visual development. Although the important health benefits of these two fatty acids are well recognized, it is not mandatory for manufacturers to add them to infant formula.

10-5 WHAT ARE THE NUTRITIONAL NEEDS OF INFANTS?

Rates of growth and development during the first year of life are astonishing. At no other time in the human lifespan—aside from gestaton—do growth and development occur so rapidly. Transitioning from breastfeeding to eating baby food can be challenging, and it is important for parents to be aware of signs that indicate readiness for this next phase. Providing an infant a diet rich in the essential nutrients and an environment that is safe, secure, and engaging helps build a solid foundation for the remainder of life.

10-5a Infant Growth and Development

Not only do major developmental milestones such as walking and self-feeding take place during the first year of life, but this is also a time of rapid growth. Between 4 and 6 months after birth, infant weight doubles and length increases by 20 to 25 percent. After the first 6 months of life, the rate of growth decreases slightly. By the end of the first year, an infant's weight will have tripled.

It is important to monitor growth and development throughout infancy. As such, an infant's weight, length, and head circumference are routinely measured during welll-baby checkups and recorded on growth charts, as illustrated in Figure 10.7. Growth charts enable pediatricians to monitor and assess infant growth over time. Because infants tend to follow a consistent growth pattern, a dramatic change in length or weight could indicate a problem. Poor breastfeeding technique, for example, can prevent an infant from receiving adequate amounts of energy and nutrients, thus slowing or delaying growth.

In addition to gains in weight and length, many developmental changes also occur during infancy. For example, newborns have little control over their bodies. In the first few months of life, however, babies begin to vocalize and are even able to return a friendly smile. Improved muscle control allows developing infants to hold their heads steady, and by 6 months, most infants can sit upright with support. These and other developmental milestones affect how and what babies should be fed. Within this first year of life, infants progress from being fed a diet that consists soley of human milk and/or infant formula, to being able to feed themselves a variety of foods.

10-5b Recommended Dietary Supplementation during Infancy

Are human milk, infant formula, and complementary baby foods adequate sources of nutrition during the first year of life, or should a baby also be given nutritional supplements? The answer to this question depends both on whether the infant is breastfed or formula fed and the infant's age. Supplemental vitamin D, fluoride, iron, and fluids are sometimes recommended for infants, but the decisions to use these supplements should be discussed with a health practitioner. Typical recommendations regarding nutrient supplementation are summarized in Table 10.3.

Vitamin D Supplements Although scientists have long assumed that breastfed babies receive enough vitamin D from human milk and exposure to sunlight, recent evidence indicates that this is not always so. The AAP recommends that breastfed infants and formula-fed infants who consume less than 16 ounces of infant formula a day receive vitamin D supplementation of 400 IU/day beginning in the first few days of life. Supplementation should continue until a baby can obtain sufficient vitamin D from the diet.[19] Because all infant formulas manufactured in the United States are fortified with adequate amounts of vitamin D, there is little risk of vitamin D deficiency in formula-fed infants who consume more than 16 oz (473 mL) of formula daily.

Fluoride Supplements Because fluoride plays an important role in the formation of teeth and the prevention of dental caries later in life, the AAP recommends that breastfed infants receive fluoride supplements starting at 6 months of age if the

© Flashon Studio/Shutterstock.com

FIGURE 10.7 GROWTH DURING THE FIRST THREE YEARS OF LIFE

Birth to 36 months: Girls
Length-for-age and weight-for-age percentiles

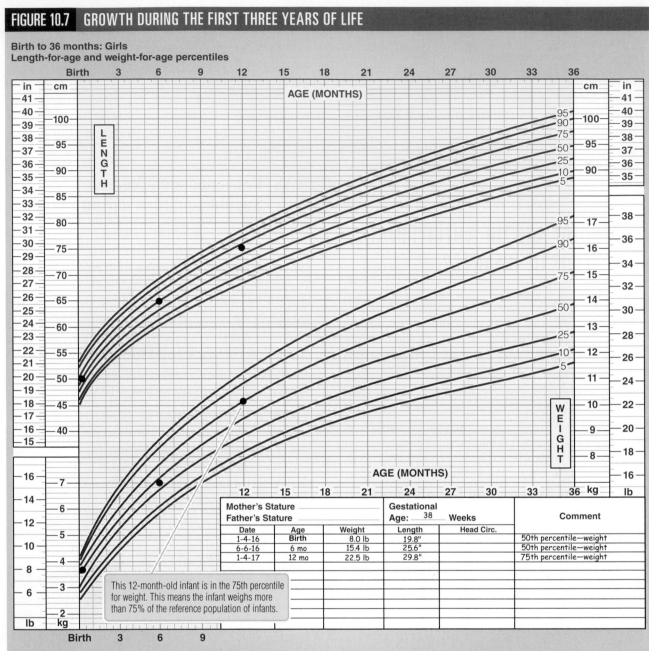

Mother's Stature			Gestational		Comment
Father's Stature			Age: 38 Weeks		
Date	Age	Weight	Length	Head Circ.	
1-4-16	Birth	8.0 lb	19.8"		50th percentile—weight
6-6-16	6 mo	15.4 lb	25.6"		50th percentile—weight
1-4-17	12 mo	22.5 lb	29.8"		75th percentile—weight

This 12-month-old infant is in the 75th percentile for weight. This means the infant weighs more than 75% of the reference population of infants.

By 4 to 6 months, weight doubles, and length increases by approximately 25%.

By 12 months, weight triples, and length increases by approximately 50%.

Weight and length are monitored during infancy and childhood using growth charts.

Source: Developed by the National Center for Health Statistics in collaboration with the National Center for Chronic Disease Prevention and Health Promotion (2000). Available from: http://www.cdc.gov/growthcharts/clinical_charts.htm

local water source has a fluoride concentration less than 0.3 parts per million.[20] Expectant parents should check with local water departments to find out the fluoride content of the drinking water in their communities. However, it is important to know that certain types of water filters (reverse osmosis filters and water distillation devices) can remove almost all fluoride from municipally fluoridated water. If purified bottled water is used to prepare infant formula, it is safe to assume that the infant is not receiving optimal amounts of fluoride. In both cases, fluoride supplements are recommended. The recommended daily dosage of fluoride is 0.25 mg/day for children between 6 months and 3 years of age. Fluoride supplements are not recommended for infants younger than 6 months of age.

TABLE 10.3   RECOMMENDED NUTRIENT SUPPLEMENTATION DURING INFANCY

	Infant-Feeding Method	
Nutrient	Exclusively Breastfed	Iron-Fortified Infant Formula
Vitamin D	• Supplements (400 IU/day) are recommended beginning in the first few days of life and continuing until an infant is consuming at least 16 oz/day of infant formula in addition to or instead of human milk.	• Not needed if an infant is consuming at least 16 oz/day of formula; all formulas in the United States are fortified with vitamin D.
Fluoride	• Supplements (0.25 mg/day) are recommended starting at 6 months if local water has a fluoride concentration of less than 0.3 ppm.	• Not needed if formula is prepared with water that has at least 0.3 ppm fluoride. Supplements (0.25 mg/day) are recommended starting at 6 months if the water used to prepare formula has less than 0.3 ppm fluoride.
Iron	• Supplements (1 mg/kilogram body weight/day) are recommended around 4 to 6 months of age and continued until iron-rich foods such as iron-fortified cereals are introduced.	• Not needed; supplements (1 mg/kilogram body weight/day) are recommended for infants not fed iron-fortified infant formula.
Additional fluids	• Not needed unless an infant has excessive fluid loss due to vomiting and/or diarrhea.	• Not needed unless an infant has excessive fluid loss due to vomiting and/or diarrhea.

Source: American Academy of Pediatrics. Pediatric nutrition handbook, 6th ed., Elk Grove Village, IL; 2008.

Iron Supplements Another nutrient sometimes given to infants via supplementation is iron. A full-term infant is born with substantial iron reserves that help meet iron needs during the first 4 to 6 months of life. After 6 months, an infant consuming iron-fortified formula that contains at least 1 mg iron/100 kcal is likely to maintain adequate iron status. Although human milk contains less iron than infant formula, the iron found in human milk is more easily absorbed. The AAP and CDC recommend iron supplementation (1 mg/kg of body weight/day) for breastfed infants from 4 months of age until iron-rich foods such as iron-fortified cereals are introduced.[21]

Water Because water requirements are likely met throughout the first 6 months of life if adequate amounts of human milk and/or formula are consumed, infants do not need additional fluid supplementation. However, if vomiting and/or diarrhea cause excessive fluid loss, water replacement becomes necessary. The AAP recommends giving an infant plain water rather than juice when it is necessary to replace fluid loss.[22] Although over-the-counter fluid replacement products with added sugars and/or electrolytes are available, such products are not usually needed. In periods of excessive fluid loss, it is important for parents to contact a health care provider.

Iron-fortified cereal is often recommended as a baby's first food.

© Jeanne Provost/Shutterstock.com

10-5c Complementary Foods Can Be Introduced between 4 and 6 Months of Age

Recommendations as to when infants should be introduced to foods other than human milk or infant formula have changed over time. For example, during the 1950s, experts believed that feeding infants solid food early in life was advantageous. As such, parents were encouraged to begin feeding their infants solid foods as early as possible. This is no longer the case. The AAP recommends introducing nonmilk complementary foods (excluding cow milk) when an infant is between 4 and 6 months of age. Until this time, infants are not physiologically or anatomically ready for foods other than human milk and infant formula. When and how to introduce complementary foods depends largely on developmental milestones take place between 4 and 6 months, some of which can help parents determine whether an infant is ready to advance to complementary feeding.[23] Signs that an infant is ready for complementary foods include sitting up with support and good head and neck control.

Because iron status begins to decline at 4 to 6 months, pediatricians typically recommend that an infant's first complementary foods be iron-rich, such as iron-fortified cereal or pureed meat. Rice cereal and other single-grain cereals can be mixed with human milk or infant formula to create a smooth, soft consistency. In addition to sufficient amounts of human milk or iron-fortified infant formula, infants require approximately 25 g (1 ounce) of iron-fortified cereal (or an equivalent source of iron) every day after 6 months of age to meet their iron requirements. Some infants find it difficult to consume food from a spoon, but in time, most become quite skilled at eating pureed food. After spoon-feeding is well established, the consistency of foods can be thickened to make eating more challenging. For example, other foods such as pureed vegetables and fruits can be introduced. During this period, complementary foods are considered *extra* because the infant still needs regular feedings of human milk and/or formula.

Although infants should not be given complimentary foods before they reach 4 to 6 months of age, there is no evidence that delaying introduction beyond this point is beneficial.[24] Nonetheless, it is important for parents to introduce new foods into their infant's diet gradually. After a new food is first introduced, parents should wait 3 to 4 days to make sure the food is tolerated and there are no adverse reactions.[25] Signs and symptoms associated with allergic reactions or food sensitivities include rashes, diarrhea, runny nose, and, in severe cases, difficulty breathing. While parents should be careful when introducing all new foods, there is little scientific evidence that delaying the introduction of highly allergenic foods, such as eggs, peanuts, tree nuts, and fish until early childhood can prevent allergies later in life. In fact, there is now evidence that delaying the introduction of these foods may actually increase the risk of allergy. The AAP recommends that even high-risk infants follow these same guidelines for supplemental foods. However, parents should always consult with their pediatrician at the first sign of an allergic response to a food.[28]

Infants should be fed human milk and/or iron-fortified formula throughout the first year of life. Because cow milk, goat milk, and soymilk are low in iron, their introduction should be delayed until after a baby's first birthday. Consuming large amounts of milk can displace iron-containing foods, increasing an infant's risk

Can Early Introduction of Allergenic Foods Help Prevent Food Allergies?

A swollen face, wheezing, coughing, and hives may not sound life threatening, but when caused by a food allergy, they could mean a trip to the hospital emergency room—or worse. According to the CDC, diagnoses of food allergies have steadily increased in children over the last decade.[26] In 1997, 3.4 percent of U.S. children (0–17 years of age) were diagnosed with a food allergy, but, by 2011, that number increased to 5.1 percent. Today, more than 3 million children are affected with food allergies. While some health care professionals focus on how best to manage food allergies, others focus on how to prevent them. Until recently, most experts believed that the best way to prevent food allergies was to advise parents to delay feeding their children highly allergenic foods (e.g., eggs, soy, dairy, and fish) until 2 to 3 years of age. However, newly released guidelines show a dramatic shift away from this advice. Upon a careful literature review, an expert panel on the diagnosis and management of food allergies concluded that there was no evidence to support the claim that holding off on highly allergenic foods helps protect children against the development of allergies. In fact, researchers concluded just the opposite. That is, early introduction to allergenic foods may actually help to prevent food allergies. In fact, some physicians consider this approach to be a type of "food inoculation." Furthermore, it is no longer recommended that pregnant or lactating women abstain from eating foods regarded as highly allergenic. Perhaps of greatest importance is the recommendation that infants receive only human milk for at least 4, and up to 6 months of age. Although parents with children at high risk for developing allergies should consult with their physicians before introducing new foods, these new guidelines by the American Academy of Allergy, Asthma, and Immunology may help to protect children against food allergies[27]:

- Introduce highly allergenic foods after other complementary foods have been introduced and tolerated.
- Introduce an initial taste of a highly allergenic food at home, rather than at a day care or restaurant.
- Gradually increase the amount of the highly allergenic food if there is no reaction.
- Introduce other new foods at a rate of one new food every 3 to 5 days if no reaction occurs.

Baby bottle tooth decay can result when infants are put to sleep with bottles containing carbohydrate-rich liquids such as juice and milk.

of developing iron deficiency anemia. Similarly, many pediatricians caution parents not to give infants too much fruit juice because it can also displace human milk and iron-fortified formula.[29] Thus, older infants should not be given more than 4 to 6 ounces of fruit juice a day. When they are given juice, the AAP and 2015 Dietary Guidelines for Americans recommend that infants and children be given 100 percent pure fruit juice rather than blends or fruit drinks. Parents may also want to dilute fruit juice with water, and selecting calcium-fortified juices can improve the nutritional value of juice. Parents should be aware that fruit juice that is not pasteurized, such as freshly pressed apple cider, should be avoided because it can harbor harmful bacteria.

Because tooth decay can begin early in life, it is important for parents to establish good dental hygiene practices from the start. Infants allowed to fall asleep with bottles filled with milk, formula, juice, or any other carbohydrate-containing beverage are at risk for developing **baby bottle tooth decay**. The pooling of sugars and other compounds found in carbohydrate-rich beverages in an infant's mouth, while asleep, can damage the newly formed teeth. To avoid baby bottle tooth decay and the dental caries it causes, infants should not be put to bed with bottles that contain anything but water.

As older infants (9 to 10 months of age) become adept at chewing, swallowing, and manipulating food in their hands, they may move toward the next stages of feeding. Certain foods should be avoided during the first few years of life because they pose a risk for choking and are therefore unsafe. These include:

- Popcorn
- Peanuts
- Whole grapes
- Pieces of hot dogs
- Hard candy

In addition, the CDC recommends not feeding honey to children less than 1 year of age because it can contain spores that cause botulism, a serious foodborne illness. Even very low exposure to these spores can make young children sick.

10-6 WHAT ARE THE NUTRITIONAL NEEDS OF TODDLERS AND YOUNG CHILDREN?

In terms of nutrient requirements, childhood is divided into two stages: toddlers (ages 1 to 3 years) and young children (ages 4 to 8 years). Toddlers and young children grow at a steady rate, but one that is considerably slower than that of infants. Growth charts are used to monitor the adequacy of this growth, but unlike infancy, BMI-for-age is used to assess weight in children older than 2 years of age.

Childhood is a time of growing independence, as children gain the ability and confidence to function on their own. Children become more opinionated during this time and often express their likes and dislikes impetuously. Indeed, feeding toddlers and young children can be challenging. Childhood is also a time when attitudes about food are formed; parents play an important role in helping toddlers and young children develop healthy relationships with food. Thus, the ways that parents deal with the challenges of feeding children are very important. Regardless of a child's eating habits and food preferences, it is important for parents to provide enough nutritious food for optional growth and development.

10-6a Feeding Behaviors in Children

Some parents are quite surprised at how quickly mild-mannered infants become willful and opinionated toddlers. The ways that parents respond to feeding challenges can determine whether fussy behaviors persist or fade. Although forcing a child to eat is never recommended, it can be difficult for parents to remain calm during mealtime.

> **baby bottle tooth decay**
> A condition whereby dental caries occur in an infant who habitually sleeps with bottles filled with milk, formula, juice, or any other carbohydrate-containing beverage.

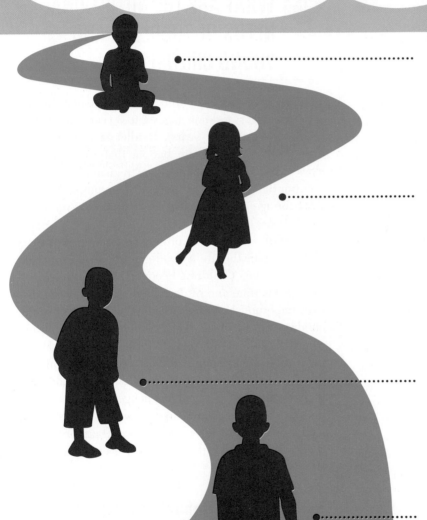

Behavioral Milestones

2-5 year olds!

The preschool years are an important time for developing healthy habits for life. From 2 to 5 years old, children grow and develop in ways that affect behavior in all areas, including eating. The timing of these milestones may vary with each child.

2 YEARS

- Can use a spoon and drink from a cup
- Can be easily distracted
- Growth slows and appetite drops
- Develops likes and dislikes
- Can be very messy
- May suddenly refuse certain foods

3 YEARS

- Makes simple either/or food choices, such as a choice of apple or orange slices
- Pours liquid with some spills
- Comfortable using fork and spoon
- Can follow simple requests such as "Please use your napkin."
- Starts to request favorite foods
- Likes to imitate cooking
- May suddenly refuse certain foods

4 YEARS

- Influenced by TV, media, and peers
- May dislike many mixed dishes
- Rarely spills with spoon or cup
- Knows what table manners are expected
- Can be easily sidetracked
- May suddenly refuse certain foods

5 YEARS

- Has fewer demands
- Will usually accept the food that's available
- Dresses and eats with minor supervision

Go to www.ChooseMyPlate.gov for more information.
USDA is an equal opportunity provider and employer.

Center for Nutrition Policy and Promotion
July 2015

The following six general guidelines regarding common childhood feeding problems can encourage healthy eating:

- *Avoid using food to control behavior.* Most experts agree that using food as a reward or punishment is not a good idea. Although rewarding a child with treats may correct behaviors in the short term, it will likely create serious food issues down the road. For example, giving a child a dessert to reward good behavior may establish a connection between sweet foods and approval. Instead, parents should teach children that food is pleasurable and nourishing—not something to turn to for approval or emotional comfort.

Feeding young children can be challenging for parents. Children often have strong opinions about what and when they want to eat.

- *Model good eating habits.* Studies show that, if parents or siblings enjoy a particular food, a child will be more likely to enjoy it as well.[30] Families that regularly eat meals together are more likely to have children who eat a greater variety of healthy foods than families that do not.[31] Mealtime provides an opportunity for the entire family to be together and share quality time.

- *Be patient.* Childhood food preferences do not always appear rational to adults. Children often judge foods as acceptable or unacceptable based on attributes such as color, texture, and appearance, rather than on taste and nutritional value. Also, children likely experience strong flavors such as onions and certain spices more intensely than do adults. Food likes and dislikes change over time. With patience and encouragement, children are often willing to broaden their food preferences. Although challenging for most parents and caregivers, children usually outgrow their limited food preferences.

- *Introduce new foods.* Making new foods familiar to children is an important first step in food acceptance. Allowing a child to help select and prepare a new food is an effective way to introduce an unfamiliar flavor. Not only does this make the new food more familiar and foster a sense of empowerment, it may also stimulate an interest in healthy food preparation. It is important for parents to remember that accepting new foods takes time. Pressuring children to eat or to try new foods is not recommended.

- *Encourage nutritious snacking.* Because children have small stomachs, they need to eat smaller portions and eat more frequently than do adults. In fact, limiting food consumption to three big meals a day is very difficult for some children. This is why experts recommend that parents and caregivers provide children with nutritious between-meal snacks. Although children who snack frequently are often not hungry at mealtime, this is not usually a problem as long as between-meal snacks are nutrient dense.

- *Promote self-regulation.* It is important that children learn to regulate their food intakes based on internal cues of hunger and satiety. For this reason, serving sizes need to be age appropriate, allowing children to ask for more if desired. Children frequently claim to be too full to eat a certain food. When this happens, experts do not recommend forcing them to eat all the food on their plate. Instead, parents need to set limits: if a child declines to eat healthy meal portions and then asks for dessert, it is reasonable to say no.

Rubberball Productions/Getty Images

©iStock.com/onebluelight

Overweight Children : A Growing Concern

The percentage of overweight children in the United States is on the rise. Health experts estimate that approximately 18 percent of children ages 6 to 11 are obese (meaning that their BMIs are greater than the ninety-fifth percentile).[32] Subsequently, weight-related health conditions such as type 2 diabetes, high blood pressure, and elevated blood lipids are now becoming increasingly prevalent and even commonplace among America's youth. The short- and long-term health and social consequences of childhood obesity are of great concern to parents and health professionals. Because excessive childhood weight gain is likely to continue into adolescence and adulthood, it is important to understand the factors that contribute to this growing trend.

Similar to adults, the behaviors most concretely linked to excessive weight gain in children are unhealthy eating patterns and physical inactivity. Undoubtedly, genetics also plays a role in determining a person's body weight. Although there is much to learn about the specific meal patterns that promote childhood obesity, easy access to high-fat, energy-dense foods is a primary contributing factor. Changing a child's food environment alone is not sufficient to reverse this growing trend. Efforts to encourage children of all ages to participate in healthful physical activities are also important. For many American children, television and other electronic media have largely replaced physical activity.[33] It is not surprising then that an association between obesity and the amount of time spent watching television has been documented.[34] More than one-half of U.S. children watch 2 or more hours of television every day—a figure that does not include additional hours spent playing computer and video games.[35] Also compounding the problem, children often consume calorie-dense foods while watching television.

In our modern world, preventing a child from becoming overweight takes considerable effort at home and at school. Regardless of the many pressures and enticements that children encounter every day, it is essential to teach and model the importance of physical activity and good nutrition. After all, there are as many healthy food choices available today as there are unhealthy ones.

10-6b Recommended Energy and Nutrient Intakes for Children

Although specific nutrient intake recommendations vary by age, dietary recommendations for children of all ages emphasize the importance of healthy food choices that satisfy nutrient and energy requirements. Regular meals and snacks that include fruits, vegetables, low-fat dairy products, lean meats, whole grains, and legumes should provide all of the nutrients needed for proper growth and development, without adding excessive energy to the diet.

Parents and other caregivers are encouraged to use the MyPlate food guidance system (http://www.choosemyplate.gov/children) to determine the number of servings from each food group needed to meet recommended nutrient and energy intakes for toddlers and young children. MyPlate also provides healthy eating materials designed specifically for this age group. The 2015 Dietary Guidelines for Americans and the 2008 Physical Activity Guidelines for Americans stress the importance of regular physical activity to promote physical health and psychological well-being. Parents are encouraged to make sure that children age 6 and older engage in at least 60 minutes of moderate- to vigorous-intensity physical activity every day of the week. Younger children are encouraged to play actively several times each day.

Calcium Calcium-rich foods are particularly important for the development of strong, healthy bones throughout childhood. The RDA for calcium increases from 700 to 1,000 mg/day at 4 years of age. Because milk and other dairy products are good sources of calcium (it would take 4 cups of broccoli to provide the same amount of calcium as 1cup of milk), the 2015 Dietary Guidelines for Americans recommend that children consume dairy or dairy equivalents such as fortified soy beverages. Other products made from plants that are promoted as "milk" (e.g., rice and almond milks) may provide adequate calcium, but are generally not considered to be dairy equivalents because they may lack other important nutrients. For children, the recommended amounts of dairy (or dairy equivalents) are based on age. For children ages 2 to 3 years, 2 cup-equivalents per day are recommended, and for older children (4 to 8 years of age) this increases to 2 ½ cups per day. In general, 1 cup of yogurt, 1.5 ounces of natural cheese, and 2 ounces of processed cheese are all equivalent in calcium contents to 1 cup of milk. Many children do not meet recommended intakes for calcium, which is cause for concern.

Iron Iron is another nutrient for which meeting the recommended intake is critical to a growing body. Iron deficiency is one of the most common nutritional problems observed in childhood. Toddlers and young children require 7 and 10 mg/day of iron, respectively, to meet their needs. Children who drink large amounts of milk and eat limited varieties of other foods are at greatest risk of iron deficiency.[36] Because overconsumption of cow milk and sugar-sweetened juice can displace iron-rich foods and/or beverages from the diet, the AAP recommends that children age 1 to 5 years consume no more than about 3 cups (700 mL) of milk every day, an amount sufficient to meet both calcium and vitamin D requirements. To prevent iron deficiency, parents are encouraged to feed their children a variety of iron-rich foods such as meat, fish, poultry, eggs, legumes, enriched cereal products, whole grains, and other iron-fortified foods. Serving non-meat sources of iron such as cereal and vegetables with vitamin C-rich fruits can help increase iron absorption.

10-7 HOW DO NUTRITIONAL REQUIREMENTS CHANGE DURING ADOLESCENCE?

Toward the end of early childhood, hormones trigger changes in height, weight, and body composition that begin to transform a child into an adolescent. This transition marks the beginning of profound physical growth and psychological development. As with other stages of growth, nutrition plays an important role as an adolescent matures from a child into a young adult. Unfortunately, adolescence is a time when unhealthy eating practices often begin to develop. For example, weight dissatisfaction among teens can lead to inappropriate dieting and other potentially harmful weight-loss behaviors.

10-7a Growth and Development during Adolescence

Adolescence, the bridge between childhood and adulthood, is signified by the onset of **puberty**. Defined as the maturation of the reproductive system, puberty is initiated by hormonal changes that trigger the physical transformation of a child into an adult. The timing of puberty's onset varies, and adolescents of the same age can differ in terms of physical maturation. Consequently, the nutritional needs of an adolescent may depend more on the stage of physical maturation than chronological age. Nonetheless, females tend to enter puberty at an earlier age (around age 9) than do males (around age 11). The first occurrence of menstruation, known as **menarche**, begins in girls around age 13. For reasons that are not clear, more and more American girls are entering puberty at younger ages. Some researchers believe that this trend may be related in part to

the rise in childhood obesity rates, because overweight girls tend to develop physically and experience menarche earlier than do thinner girls.[37]

Considerable physical growth takes place during adolescence. Before the onset of puberty, pre-teens will have attained about 84 percent of their adult height and approximately 42% of their adult weight. During the adolescent growth spurt, females and males grow approximately 9 and 11 inches in height, respectively. Because bone mass increases rapidly during adolescence, it is especially important for teens to consume adequate amounts of the nutrients that promote bone health, such as protein, vitamin D, calcium, and phosphorous.

Changes in linear growth (height) are accompanied by changes in body weight during adolescence. Overall, females and males gain an average of 39 and 52 pounds, respectively. Changes in body composition differ considerably between females and males, however. Whereas females experience a decrease in percentage of lean mass and a relative increase in percentage of fat mass, males experience the opposite. Although good nutrition is important throughout the entire life cycle, the accelerated rates of growth and development experienced during adolescence put this group at particularly high risk for developing diet-related health problems.

10-7b Psychological Issues Associated with Adolescent Eating Behaviors

Adolescence is marked not only by rapid physical changes, but also by numerous psychological and developmental changes. For example, a newfound desire for independence often strains family relationships and can lead to adolescent rebellious behaviors. Furthermore, a strong urge to fit in and be accepted by peers can lead to social anxiety and further familial tension. For these and many other reasons, healthy eating can become a low priority for adolescents. Peers are often more influential than family members in determining adolescent food preferences. As children transform into teens, weight-loss diets, skipping meals, and eating more meals away from home become increasingly

© exopixel/Shutterstock.com

Adolescence is the period between childhood and adulthood.

puberty Maturation of the reproductive system.

menarche The first occurrence menstruation.

common. Although increased body fat is normal and healthy for adolescent females, weight gain can contribute to weight dissatisfaction, which, in turn, can lead to unhealthy dieting and caloric restriction. This can negatively impact growth, development, and reproductive maturation.

10-7c Nutritional Concerns and Recommendations during Adolescence

© Elena Elisseeva/Shutterstock.com

The rapid growth and development associated with adolescence increase the body's need for certain nutrients and energy. Teens can use the MyPlate food guidance system to determine the number of servings from each food group needed to meet recommended nutrient and energy intakes. Using MyPlate's personalized meal planning guide can be particularly advantageous, because an inadequate diet during this stage of life can compromise health and have lasting long-term effects. In addition, there are helpful tips for healthy eating during the teen years on the MyPlate website (see www.choosemyplate.gov/teens).

Food Choices of Adolescents As with adults, the majority of calories in an adolescent's diet should come from healthy, nutrient-dense foods such as fruits, vegetables, lean meats, low-fat dairy products, and whole grains. Average daily intakes of fruits and vegetables, when compared to ranges of recommended intakes, are low among adolescents, for both males and females. It is not surprising then that only a small number of U.S. adolescents satisfy their requirements for dietary fiber. Adolescent males consume on average 18 g of fiber/day compared to 13 g/day for females.[38] Teens average approximately 120–150 total grams of sugars daily, most of which is associated with the consumption of carbonated beverages, sport and energy drinks, snacks and sweets, and caffeine-containing drinks.[39] Between 1977 and 2001, the number of calories consumed from sugar-sweetened beverages increased, while the number of calories obtained from milk decreased by 34 percent.[40]

Important Micronutrients during Adolescence Because too little dietary calcium can compromise bone health later in life, the RDA for calcium

during adolescence is 1,300 mg/day. Getting enough calcium is best achieved by consuming dairy products such as milk, yogurt, and cheese. Average daily intake of dairy and dairy equivalents begins to decline after childhood, and this decline continues throughout adolescence. To establish a healthy eating pattern, it is important to develop strategies to incorporate more calcium-rich foods such as low-fat milk, yogurt, cheese, and fortified soy beverages. These foods not only provide a rich source of calcium, but they also provide other important minerals and vitamins such as vitamins A and D.

In addition to calcium, adolescents require iron to support growth. Because of the iron lost in menstruation, the RDA for iron is higher for girls than for boys during later adolescence (15 and 11 mg/day, respectively). Although the average daily intake of iron among adolescent males is 18 mg, females average approximately 12 mg/day. It appears that most adolescents are well nourished with respect to iron, but approximately 9 percent of U.S. adolescent females have iron deficiency anemia.[41]

The majority of adolescents consume adequate amounts of combined folate and folic acid, which is largely due to folic acid enrichment of food.[42] Given the pregnancy-related health concerns associated with impaired folate status, this is of great importance. Because impaired folate status can increase the risk of neural tube defects, it is imperative for health care professionals to stress the importance of folate-rich foods such as orange juice, green leafy vegetables, and enriched cereals to all females of child-bearing age.

10-8 HOW DO AGE-RELATED CHANGES IN ADULTS INFLUENCE NUTRIENT AND ENERGY REQUIREMENTS?

It is important to make healthy food choices at every stage of life. Implementing the 2015 Dietary Guidelines for Americans can help people transform their eating habits, and establish a healthy eating pattern. Figure 10.8 compares

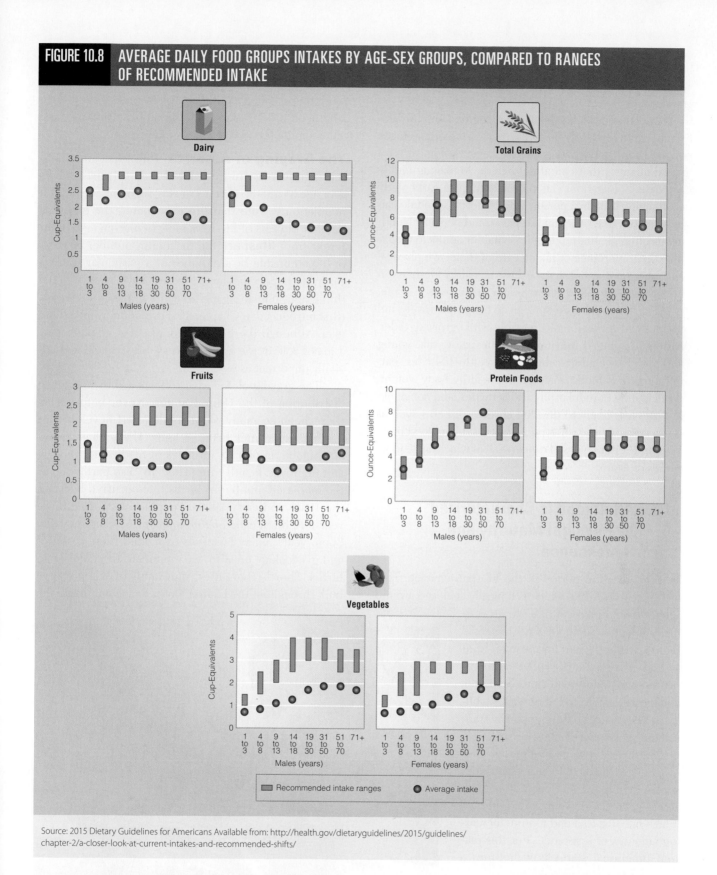

Source: 2015 Dietary Guidelines for Americans Available from: http://health.gov/dietaryguidelines/2015/guidelines/chapter-2/a-closer-look-at-current-intakes-and-recommended-shifts/

average daily food group intakes by age-sex groups to ranges of recommended intake. The ChooseMyPlate website provides many useful guidelines, including those formulated specifically to improve the nutrition and health of students, faculty, and staff on college and university campuses (see http://www.choosemyplate.gov/college).

As a MyPlate Campus Ambassador, you can help faciliate change on your college campus by promoting healthy eating and lifestyles (see http://www.choosemyplate.gov/content/ambassador).

Adulthood spans the largest segment of the life cycle—approximately 60 to 65 years. Like other stages of life, adulthood brings physical, psychological, and social changes that can affect health. Managing family and career obligations can make adulthood a particularly challenging time in a person's life. A hectic schedule can be stressful, free time can be sparse, and personal health needs often go unattended. It is vital for adults to recognize the importance of taking care of their own physical and emotional well-being, because a healthy lifestyle can help ensure a full and active life for many years to come.

There is much a person can do to stay physically and mentally fit for as long as possible while growing older. Although aging is inevitable, studies show that young and middle age adults who adopt healthy lifestyles (eating nutritious foods, maintaining healthy body weight, participating in regular physical activities, and not smoking) are less likely to develop chronic diseases such as cardiovascular disease, type 2 diabetes, high blood pressure, and certain types of cancer—all of which are leading causes of death.

10-8a Adulthood Is Characterized by Physical Maturity and Senescence

Adulthood is the period in the lifespan characterized by physical maturity, which is typically achieved around age 20. Some men continue to grow in their early 20s, and bone mass increases slightly for both men and women until age 30. After achieving physical maturity in young adulthood, adults typically undergo a long period of physical stability (maintenance) before gradually transitioning to senescence. A wide spectrum of health and independence levels is experienced among middle age and older adults. Whereas some people maintain active lifestyles throughout adulthood, others become frail and require increasing amounts of assistance. For this reason, functional status is sometimes a

better indicator of nutritional needs than chronological age during this time. Nonetheless, when establishing DRIs, the Institute of Medicine divided adulthood into four groups: young adulthood (19–30 years), middle age adulthood (31–50 years), adulthood (51–70 years), and older adulthood (>70 years).

The Graying of America Life expectancy continues to increase in the United States. In fact, adults over age 85 are the fastest growing segment of the adult population.[43] Note that life expectancy, the expected number of years of life remaining at a given age, is different from **lifespan**, the maximum number of years of life attainable by a member of a particular species. Both life expectancy and lifespan have risen steadily throughout human history, but the graying of America (first introduced in Chapter 1) is evident now more than ever. Although there is little doubt that genetics play a major role in the aging process, lifestyle choices are vitally important as well.

The graying of America reflects a shift in the age distribution of the United States. Never before have there been so many older adults, and never before has life expectancy been so long. These changes are attributable in part to advances in medical technology and improved health care. They are also partly due to the increased number of births that occurred between the 1940s and early 1960s. People born during these years are often called *baby boomers*. During the baby boom years, many hospitals expanded their obstetric and gynecology units to accommodate the increase in births. Shortly thereafter, thousands of schools were built throughout the United States to accommodate the

© Monkey Business Images/Shutterstock.com

Many adults are physically fit and lead healthy, active lives.

lifespan The maximum number of years of life attainable by a member of a particular species.

growing number of school-age children. Today, there is an increased need for retirement communities and health care practitioners to care for the rising number of older adults.

10-8b Nutritional Concerns and Recommendations during Adulthood

Although adults cannot turn back the hands of time, there is much they can do to keep their bodies strong and healthy. During adulthood, nutrient intake recommendations are intended to both reduce the risk of chronic disease and provide adequate amounts of essential nutrients. The MyPlate food guidance system can assist older adults in planning meals and determining the amount of food from each food group needed to meet recommended nutrient and energy intakes (see http://www.choosemyplate.gov /older-adults). Because older adults generally have lower energy requirements than do younger adults, it is particularly important to consume nutrient-dense foods that help ensure a healthy balance between nutrient and caloric intake. It is also equally important for older adults to stay physically active, to maintain a healthy body weight, and to balance nutrient intake with energy levels. Achieving these three measures of good health can help older adults stay active and maintain good health.

In general, elderly adults comprise an at-risk population for many nutrition-related health problems. Food insecurity, social isolation, depression, illness, hunger, and the use of multiple medications can compromise an older person's nutritional status. Physiological changes associated with aging can also influence the development of nutrition-related health problems. A number of these changes are summarized in Table 10.4.

Optimizing Body Composition and Bone Health during Adulthood
Most adults experience age-related changes in body composition, such as a decrease in lean mass and an increase in fat mass. Although genetics, physical activity, and nutritional status influence these changes, they are likely to occur even when body weight remains stable.[44] Adequate protein intake is important for elderly adults because it helps prevent age-related loss of skeletal muscle. The gradual loss of lean mass, which in turn causes a decline in a person's metabolic rate, can also contribute to age-related weight gain. For this reason, a combination of increased physically activity and/or decreased energy intake is the best approach to maintain a healthy weight. In general, all other factors being equal, a 70-year-old woman

TABLE 10.4	PHYSIOLOGICAL CHANGES TYPICALLY ASSOCIATED WITH AGING

Cardiovascular System
- Elasticity of blood vessels decreases.
- Cardiac muscles weaken.
- Blood pressure increases.

Endocrine System
- Estrogen, testosterone, and growth hormone levels fall.
- Ability to produce vitamin D diminishes.

Gastrointestinal System
- Saliva and mucus production decreases.
- Loss of teeth.
- Difficulty swallowing.
- Production of gastric juice decreases.
- Peristalsis decreases.
- Vitamin B_{12} absorption decreases.

Musculoskeletal System
- Bone mass decreases.
- Lean mass decreases.
- Metabolic rate decreases.
- Strength, flexibility, and agility decrease.

Nervous System
- Appetite regulation is altered.
- Thirst sensation is blunted.
- Ability to smell and taste decreases.
- Sleep patterns change.
- Visual acuity decreases.

Urinary System
- Blood flow to kidneys diminishes.
- Kidney filtration rate decreases.
- Ability to eliminate metabolic wastes decreases.

Respiratory System
- Respiratory rate decreases.

requires approximately 280 fewer kcal/day than does a 30-year-old woman. Likewise, a 70-year-old man requires 380 fewer kcal/day than does a 30-year-old man.

Of course, an older adult's individual energy requirements depend on many interrelated factors such as physical activity, weight, and changes in the relative amounts of muscle and body fat. The 2015 Dietary Guidelines for Americans recommend that all adults prevent gradual weight gain over time by decreasing energy intake and increasing physical activity. Exercise not only decreases fat mass and slows age-related bone loss, but it also helps strengthen muscles and improve coordination. It is recommended that adults, at any age, get at least 2 ½ hours or 150 minutes of moderate-intensity physical activity each week. By finding activities that are enjoyable, older adults can stay active and physically fit. The MyPlate website (www.choosemyplate.gov/older-adults) for older adults provides important recommendations and tips regarding how to eat healthy and how to stay physically fit during this stage of life (Table 10.5).

Over time, age-related bone loss can lead to osteoporosis, which can make bones fragile. Although men can develop osteoporosis, the condition is far more common in women. Many factors influence bone density, but maintaining adequate intakes of the nutrients that support bone health (such as protein, calcium, vitamin D, phosphorus, and magnesium) is essential. To promote lasting bone health, the RDA for calcium increases from 1,000 to 1,200 mg/day at age 51 in women and at age 70 in men. However, older adults often find it difficult to meet this RDA. Lactose intolerance increases with age, inhibiting the ability of many to consume calcium-rich dairy products. To counter this problem, health care providers often recommend lactose-reduced dairy products and/or calcium-fortified foods such as soymilk and supplements. Weight-bearing exercise such as walking is also important because it can help slow the progression of bone loss.

Considerable evidence shows that older adults, especially those who live in northern regions of the United States, are at increased risk of vitamin D deficiency. This condition is likely caused in part by limited exposure to sunlight and by a decreased ability to synthesize vitamin D from cholesterol. For these reasons, the RDA for vitamin D increases from 15 to 20 µg/day at age 70. Good dietary sources of vitamin D include fortified milk, eggs, salmon, and tuna.

Changes in the Gastrointestinal Tract
Age-related changes in the GI tract affect nutritional status, especially in older adults. Aging muscles become less responsive to neural signals, which can impact digestive functions such as swallowing and peristalsis. In fact, difficulty swallowing is a common cause of choking in the elderly. The threat or experience of this frightening and life-threatening event can compel older adults to avoid eating certain foods altogether. Preparing foods so that they are moist and soft can help prevent choking.

Decreased GI motility (peristalsis) can also lead to constipation. Fecal material that remains in the colon for a prolonged period can become hard and compacted, making it difficult to eliminate. Drinking adequate amounts of fluids and eating fiber-rich foods can improve GI function and help prevent this problem. Although the

TABLE 10.5 HEALTHY EATING RECOMMENDATIONS FOR OLDER ADULTS

Recommendations for Nutrients

- Obtain nutrients needed by the body, such as potassium, calcium, vitamin D, vitamin B_{12}, minerals, and fiber.
- Lose weight or maintain a healthy weight.
- Reduce the risk of developing chronic diseases such as high blood pressure, type 2 diabetes, hypertension, and heart disease. If you have a chronic disease, eating well can help to manage the disease.
- Meet individual calorie and nutrition needs.
- Try to maintain energy levels.

Special Nutrition Concerns for Older Adults

- Our daily eating habits change as our bodies get older. Make small adjustments to help you enjoy the foods and beverages you eat and drink.
- Add flavor to foods, using spices and herbs instead of salt, and look for low-sodium packaged foods.
- Add sliced fruits and vegetables to your meals and snacks. Look for presliced fruits and vegetables on sale, if slicing and chopping is a challenge.
- Ask your doctor to suggest other options if the medications you take affect your appetite or change your desire to eat.
- Drink 3 cups of fat-free or low-fat milk throughout the day. If you cannot tolerate milk, try small amounts of yogurt, butter milk, hard cheese, or lactose-free foods. Drink water instead of sugary drinks.
- Consume foods fortified with vitamin B_{12}, such as fortified cereals.

Be Active Your Way

- Adults at any age need at least 2½ hours or 150 minutes of moderate-intensity physical activity each week. Being active at least 3 days a week is a good goal.
- Find an activity that is appropriate for your fitness level. If you are not active, start by walking or riding a stationary bike. Strive for at least 10 minutes of exercise at a time and be as active as possible.
- Include activities that improve balance and reduce your risk of falling, such as lifting small weights. Add strength-building activities at least 2 times per week.
- Being active will make it easier to enjoy other activities such as shopping, playing a sport, or gardening.
- If you are not sure about your level of fitness, check with your doctor before starting an intense exercise program or vigorous physical activity.

Source: U.S. Department of Agriculture ChoolseMyPlate.gov Available from: http://www.choosemyplate.gov/older-adults

recommended fiber intake for older adults is 20 to 35 g/day, older men and women consume on average only 18 and 14 g/day, respectively. There are many reasons why older adults may be reluctant to consume fiber-rich foods, but it is important that they find ways to incorporate fiber into their diets. Good sources of fiber include whole grains, nuts, beans, fruits, and vegetables. Dietary fiber supplements are also sometimes advised.

With increased age comes a decline in the number of stomach cells that produce gastric juice (e.g., hydrochloric acid, intrinsic factor, digestive enzymes). This can reduce the bioavailability of nutrients such as calcium, iron, biotin, folate, vitamin B_{12}, and zinc. For example, without intrinsic factor, vitamin B_{12} cannot be absorbed, leading to vitamin B_{12} deficiency. Symptoms associated with vitamin B_{12} deficiency include dementia, memory loss, irritability, delusions, and personality changes—all of which can easily be overlooked or misdiagnosed in older adults. To avoid the development of vitamin B_{12} deficiency, older adults are often advised to take vitamin B_{12} supplements or to consume adequate amounts of vitamin B_{12}-fortified foods. Foods naturally containing vitamin B_{12} include fish, meat, poultry, eggs, and dairy products.

Effects of Menopause on Women's Nutrition
As a woman ages, she experiences a natural decline in estrogen production. This can lead to a variety of notable physical changes, such as irregular menstrual cycles and sudden feelings of warmth (hot flashes) and insomnia. The decline in estrogen that occurs during this **perimenopausal** stage of life also accelerates the rate of bone loss. By the time a woman reaches **menopause** around her fifth or sixth decade of life, her ovaries are producing very little estrogen, causing her menstrual cycle to stop completely. Declining levels of estrogen cause bone loss to accelerate further, leading some women's bones to become weak and fragile. Women whose bones were dense before the onset of menopause have fewer problems—another addition to the long list of reasons why adequate intake of nutrients related to bone health is important throughout the lifecycle.

Because the monthly blood loss associated with menstruation ceases, menopause can improve iron status. In fact, the RDA for iron decreases from 18 to 8 mg/day for post-menopausal women—the same amount

A Meals on Wheels America volunteer delivers a meal to a home-bound older adult.

as recommended for adult men. Regardless of life stage, adequate iron intake remains a concern among older adults who limit their intakes of meat, poultry, and fish.

Other Nutritional Issues in Older Adults
Other age-related physiological changes can contribute to nutritional deficiencies in older adults. Problems with oral health, missing teeth, or poorly fitting dentures can make food less enjoyable and limit the types of foods a person can eat. Foods that require chewing such as meat, fruits, and vegetables can cause pain, embarrassment, and discomfort for the elderly and therefore may be avoided. It is important for older adults who experience problems with oral health to get proper dental care. Sensory changes in taste and smell can also affect food intake in older adults: the ability to smell diminishes with age, often making food tasteless and unappealing. Furthermore, certain medications can alter taste and diminish appetite. Older adults may find that adding spices to foods makes them more appealing, flavorful, and enjoyable.

The sensation of thirst can also become blunted with age. Because of this, many older adults do not consume enough fluid and are at increased risk of dehydration, which can upset the balance of electrolytes in cells and tissues. A lack of fluid can also disrupt bowel function and exacerbate constipation. Certain medications increase

> **perimenopausal** A life stage characterized by a natural decline in estrogen production.
>
> **menopause** A life stage characterized by very little estrogen production, which causes the menstrual cycle to stop.

water loss from the body, and some elderly people may intentionally limit fluid intake to avoid embarrassment over loss of bladder control. Symptoms of dehydration, which are often overlooked in the elderly, include headache, dizziness, fatigue, clumsiness, visual disturbances, and confusion. Adequate fluid consumption and early detection of dehydration are very important for older adults. To stay fully hydrated, adults are encouraged to consume 8 to 10 cups (1.9 to 2.4 l) of fluids per day. Fluid can come from both beverages and foods.

Finally, the elderly are at particularly high risk of adverse drug–nutrient interactions. Older adults often take multiple medications to treat a variety of chronic diseases. Some drugs can cause loss of appetite, leading to inadequate food intake. Others can alter taste, making food unpleasant to eat. Even nutrient absorption can be affected by certain medications. It is important for older adults to be aware of such problems and seek advice regarding the nutrient-related side effects of their medications.

10-8c Assessing Nutritional Risk in Older Adults

Clearly, many interrelated factors put older adults at increased nutritional risk. For this reason, several national health organizations jointly sponsored the Nutritional Screening Initiative (NSI), a collaborative effort that helped identify risk factors closely associated with poor nutritional status in older adults. The risk factors most closely associated with poor nutritional health were compiled and used to develop a screening tool called the NSI DETERMINE checklist, which is illustrated in Figure 10.9. As you can see, nutritional risk factors are explained and exemplified by the DETERMINE acronym, and a nutritional risk

---

FIGURE 10.9 NUTRITIONAL SCREENING INITIATIVE DETERMINE CHECKLIST

DETERMINE YOUR NUTRITIONAL HEALTH

Circle the number in the "yes" column for those that apply to you or someone you know. Total your nutritional score.

	YES
I have had an illness or condition that made me change the kind and/or amount of food I eat.	2
I eat fewer than 2 meals per day.	3
I eat few fruits or vegetables or milk products.	2
I have 3 or more drinks of beer, liquor or wine almost every day.	2
I have tooth or mouth problems that make it hard for me to eat.	2
I don't always have enough money to buy the food I need.	4
I eat alone most of the time.	1
I take 3 or more different prescribed or over-the-counter drugs a day.	1
Without wanting to, I have lost or gained 10 pounds in the last 6 months.	2
I am not always physically able to shop, cook and/or feed myself.	2
TOTAL	

Total Your Nutritional Score. If it's—

0–2	Good! Recheck your nutritional score in 6 months.
3–5	You are at moderate nutritional risk. See what can be done to improve your eating habits and lifestyle.
6 or more	You are at high nutritional risk. Bring this Checklist the next time you see your doctor, dietitian or other qualified health or social service professional.

DETERMINE: Warning signs of poor nutritional health.

Disease
Any disease, illness or chronic condition which causes you to change the way you eat, or makes it hard for you to eat, puts your nutritional health at risk.

Eating Poorly
Eating too little and eating too much both lead to poor health. Eating the same foods day after day or not eating fruit, vegetables, and milk products daily will also cause poor nutritional health.

Tooth Loss/Mouth Pain
A healthy mouth, teeth and gums are needed to eat. Missing, loose or rotten teeth or dentures which don't fit well, or cause mouth sores, make it hard to eat.

Economic Hardship
As many as 40% of older Americans have incomes of less than $6,000 per year. Having less—or choosing to spend less—than $25–30 per week for food makes it very hard to get the foods you need to stay healthy.

Reducing Social Contact
One-third of all older people live alone. Being with people daily has a positive effect on morale, well-being and eating.

Multiple Medicines
Many older Americans must take medicines for health reasons. Almost half of older Americans take multiple medicines daily.

Involuntary Weight Loss/Gain
Losing or gaining a lot of weight when you are not trying to do so is an important warning sign that must not be ignored. Being overweight or underweight also increases your chance of poor health.

Needs Assistance in Self Care
Although most older people are able to eat, one of every five have trouble walking, shopping, buying food and cooking food, especially as they get older.

Elder Years Above Age 80
Most older people lead full and productive lives. But as age increases, risk of frailty and health problems increase. Checking your nutritional health regularly makes good sense.

The Nutrition Screening Initiative, an effort led by several organizations, developed screening tools to assess risk factors associated with poor nutritional health in older adults.

Adapted from Nutritional Screening Initiative. Report on Nutritional Screening: Toward a Common View. Washington DC: Nutritional Screening Initiative; 1991.

scoring system is outlined. An individual with a score of 6 or more on the NSI DETERMINE checklist is considered to be at high nutritional risk.

Many services are available to help older adults improve nutritional status and overall health. Most communities have congregate meal programs whereby older adults can enjoy nutritious, low-cost meals in the company of others. Most congregate meal programs are subsidized, and total cost is often based on an individual's ability to pay. Another program that provides meals to senior citizens is called Meals on Wheels® America. Relying heavily on volunteers, this program works to deliver low-cost meals to homebound elderly and disabled adults. The mission of Meals on Wheels America is to nourish and enrich the lives of people who are homebound, while promoting dignity and independent living. The federally funded Supplemental Nutrition Assistance Program (SNAP; formerly known as the Food Stamp Program) also assists with the food-related expenses of low-income adults.

It goes without saying that adequate nutrition is an integral and intimate part of life at every stage. Whether a person is receiving Meals on Wheels® Association of America milk from the breast, eating pureed vegetables, snacking at school, feeding a family, or being fed as an older adult, good nutrition provides the nutrients and energy needed to sustain life. By being aware of your own life stage, nutritional needs, and intake amounts, you can take one of the most profound steps toward ensuring a long, healthy, and happy life.

STUDY TOOLS 10

READY TO STUDY? IN THE BOOK, YOU CAN:

☐ Rip out the Chapter Review Card, which includes key terms and chapter summaries.

ONLINE AT WWW.CENGAGEBRAIN.COM, YOU CAN:

☐ Learn about food allergies in children in a short video.

☐ Interact with figures from the text to check your understanding.

☐ Prepare for tests with quizzes

☐ Review the key terms with Flash Cards.

11 | Nutrition and Physical Activity

LEARNING OUTCOMES

 11-1 Understand how physical activity contributes to overall health.

11-2 Explain how the body uses nutrients to fuel physical activity.

11-3 Describe how the body responds to exercise and athletic training.

 11-4 Understand how physical activity influences dietary requirements.

After finishing this chapter go to **PAGE 297** for **STUDY TOOLS.**

 11-1 **WHAT ARE THE HEALTH BENEFITS OF PHYSICAL ACTIVITY?**

Some people exercise to get or stay in shape, others to improve health and physical fitness, and still others because they enjoy the challenge of training and participating in athletic competitions. Regardless of the reason, routine physical activity is essential to a long and healthy life. Even moderate amounts of physical activity impart tremendous benefits in terms of overall health, fitness, and function. Equally important, regular physical activity and exercise contribute to emotional well-being. And yet, although the importance of physical activity may be common knowledge, nearly one-half of all Americans lead sedentary lifestyles.[1] Furthermore, many people are unclear about the types and amounts of physical activity needed to stay healthy and the kinds of foods needed to fuel exercise.

As you learned in previous chapters, being physically active provides many health benefits, and is important throughout a person's life. In this chapter, you will learn how the body responds to the physical demands of exercise, current public health recommendations regarding physical activity, and the overall importance of nutrition during athletic training and competition.

11-1a Physical Activity Improves Health and Fitness

Recall from Chapter 9 that physical activity is any bodily movement that results in increased expenditure of energy. Physical activity includes day-to-day activities such as gardening, household chores, walking, and leisure-time activities. Exercise is a subcategory of physical activity, defined as planned, structured, and repetitive physical movement done to obtain or maintain a certain level of fitness. Whether a person is young, old, sedentary, physically fit, or has some sort of disability, everyone can benefit from physical activity and exercise.

But what does it mean to be physically active? Numerous studies show that as little as 30 minutes of sustained physical activity on most days of the week can substantially improve health and quality of life.[2] Specifically, sustained physical activity can help reduce a person's risk of obesity and certain chronic diseases such as hypertension, stroke, cardiovascular disease, type 2 diabetes, osteoporosis,

and some forms of cancer. Unfortunately, although overwhelming evidence shows that physical activity reduces the risk of many debilitating diseases and health conditions, these benefits quickly disappear if a person becomes inactive.[3] The overall benefits of regular physical activity and exercise on health include the following.

- *Assistance in weight management.* Regular physical activity helps prevent weight gain and, in some cases, facilitates weight loss. This occurs in part because physical activity increases energy expenditure and improves appetite regulation. It also promotes increased muscle mass and decreased body fat.

- *Decreased risk of cardiovascular disease.* Physical activity can improve overall cardiac function, lower resting heart rate, reduce blood pressure, and improve blood lipid regulation. These changes help slow the progression of atherosclerosis and cardiovascular disease.

- *Reduced risk of cancer.* Physical activity can reduce the risk of developing colon, breast, endometrial, lung, and prostate cancer. Furthermore, physical activity after a cancer diagnosis can also assist in recovery and lessen the severity of cancer treatment's side-effects, such as fatigue and depression.

Some people exercise to stay in shape while others exercise to improve overall physical fitness. Regardless of the reason, regular exercise offers many health benefits.

- *Decreased back pain.* Physical activity can increase muscle strength, endurance, flexibility, and overall posture—all of which help prevent injuries and alleviate back pain associated with muscle strain.

- *Decreased risk of type 2 diabetes.* Regular physical activity can decrease blood glucose, insulin resistance, and long-term health complications associated with type 2 diabetes.

- *Optimized bone health.* Weight-bearing physical activity can stimulate bone formation and slow the progression of age-related bone loss.

- *Enhanced self-esteem, stress management, and quality of sleep.* Being physically active provides a sense of achievement and empowerment. This can help reduce depression and anxiety, which in turn fosters improved self-esteem, reduced stress, improved mood, and more restful sleep.

Clearly, the health benefits associated with physical activity are hard to ignore. This is why it is important to exercise regularly in conjunction with eating a healthy diet and making reasonable lifestyle choices.

11-1b Physical Activity Recommendations

Because physical *inactivity* is a major public health concern, several health organizations such as the American College of Sports Medicine (ACSM) and the U.S. Department of Health and Human Services (USDHHS), have released specific recommendations as to how much activity a person needs to stay healthy. Although the many different (and sometimes contradictory) guidelines can seem confusing, both the ACSM and USDHHS (2008 Physical Activity Guidelines for Americans) recommends that most people get at least 2.5 hours (150 minutes) of moderate-intensity aerobic physical activity per week, as well as two sessions per week of muscle-strengthening activity. Table 11.1 provides a summary of physical activity recommendations for different population groups, as provided in the 2008 Physical Activity Guidelines for Americans, which are currently in the process of being updated.

TABLE 11.1 | 2008 PHYSICAL ACTIVITY GUIDELINES FOR AMERICANS

Population Group	Physical Activity Recommendations
Children and adolescents (ages 6–17)	• Children and adolescents should engage, on a daily basis, in at least 1 hour of physical activity primarily consisting of either moderate- or vigorous-intensity aerobic physical activity. • Children and adolescents should participate in both muscle- and bone-strengthening activities at least 3 days per week.
Adults (ages 18–64)	• Adults should strive for 2.5 hours per week of moderate-intensity, or 1.25 hours per week of vigorous-intensity aerobic physical activity. Or, they should participate in an equivalent combination of moderate- and vigorous-intensity aerobic physical activity. • Aerobic activity should be performed in episodes of at least 10 minutes, spread throughout the week. • Additional health benefits can be obtained by increasing exercise to 5 hours per week of moderate-intensity aerobic physical activity, 2.5 hours per week of vigorous-intensity physical activity, or an equivalent combination of both. • In addition to aerobic activities, muscle-strengthening activities that involve all major muscle groups should be performed 2 or more days per week. • If necessary, adults should gradually increase the time spent doing aerobic physical activity and decrease caloric intake in order to achieve neutral energy balance and a healthy weight.
Older adults (ages 65+)	• Older adults should follow the adult guidelines if possible, but individuals with chronic conditions should be as physically active as their abilities allow. • If they are at risk of falling, older adults should also do exercises that maintain or improve balance.
Pregnant and postpartum women	• Pregnant and postpartum women should strive for 2.5 hours of moderate-intensity aerobic activity throughout the week. • Women who regularly engage in vigorous-intensity aerobic activity or high amounts of activity can continue their activity, provided that their condition remains unchanged and they have consulted with their health care providers. • After the first trimester, pregnant women should avoid exercises that involve lying on the back. They should also avoid activities that increase the risk of falling or abdominal trauma, such as contact and/or collision sports.
Children and adolescents with disabilities	• When possible, children and adolescents with disabilities should meet the guidelines for all children—or they should engage in as much activity as conditions allow.
Adults with disabilities	• Adults with disabilities should follow the adult guidelines if possible and be as physically active as their abilities allow.

Source: Adapted from U.S. Department of Health and Human Services. 2008 physical activity guidelines for Americans. Washington, DC: U.S. Department of Health and Human Services, 2008.

11-1c Components of Physical Fitness

Regardless of where a person is on the physical fitness continuum, regular physical activity is one of the most indisputable ways to maintain health, fitness, and overall well-being. The ACSM defines **physical fitness** as a set of measurable health parameters (cardiorespiratory fitness, muscular strength, endurance, body composition, flexibility, balance, agility, reaction time, and power) that enable a person to carry out daily tasks with vigor and alertness, without undue fatigue. To achieve physical fitness, the ACSM emphasizes the importance of a well-rounded physical activity program that helps maintain or improve cardiorespiratory, musculoskeletal, and neuromotor fitness.[4] There are countless types of physical activity, most of which are beneficial to health. However, there are five important outcomes of a well-balanced physical fitness regimen.

- **Cardiorespiratory fitness** refers to the efficiency with which the heart and lungs deliver oxygen and nutrients to working muscles to sustain physical activity. Activities that safely elevate your heart rate, such as walking at a brisk pace, swimming, jogging, and bicycling, can help improve cardiovascular fitness.

- **Muscular strength** refers to the maximal force exerted by muscles during physical activity. Activities such as lifting weights and walking up stairs force muscles to work against resistance, and therefore can help improve muscle and bone strength. Strength-building activities such as weight lifting and working with elastic bands are referred to as **resistance exercise** because they involve working against an opposing force.

- **Muscular endurance** refers to a person's ability to exercise for an extended period of time without becoming fatigued. Sustained activities such as skipping rope, swimming, and bicycling can improve muscular endurance.

- **Muscular flexibility** refers to the range of motion around a joint. Activities such as yoga, stretching, and swimming can help stretch and lengthen muscles, which improves flexibility.

- Body composition (discussed in detail in Chapter 2) refers to the relative amounts of muscle, fat, and bone tissue in the body. Body fat can be decreased through cardiovascular exercise, whereas muscle mass can be increased through resistance-training activities. Weight-bearing activities such as walking and jogging can help increase bone mass.

© ID1974/Shutterstock.com

Sister Madonna Buder, the so-called "Iron Nun" from Spokane, Washington, was recently inducted into the USA Triathlon Hall of Fame. She has competed in hundreds of triathlons, more than 30 Ironman® races, and two Boston Marathons®. In 2005, she became the first woman over the age of 75 to complete an Ironman distance triathlon. At 85 years of age, the Ironman organization created a new age bracket for Sister Madonna.

Lennart Preiss/Getty Images

physical fitness Measurable health parameters (cardiorespiratory fitness, muscular strength, endurance, body composition, flexibility, balance, agility, reaction time, and power) that enable a person to carry out daily tasks with vigor and alertness, without undue fatigue.

cardiorespiratory fitness A measure of the circulatory and respiratory systems' ability to supply oxygen and nutrients to working muscles during sustained physical activity.

muscular strength The maximal force exerted by muscles during physical activity.

resistance exercise Strength-building activities that challenge specific groups of muscles by making them work against an opposing force.

muscular endurance The ability of a muscle or a muscle group to repeatedly exert resistance without becoming fatigued.

muscular flexibility A measure of the range of motion around a joint.

11-1d ACSM Exercise Recommendations for Healthy Adults

Clearly, the benefits of exercise outweigh the risks. Newly released recommendations by the ACSM emphasize the importance of a well-rounded exercise program for healthy adults of all ages. These recommendations can also be adapted for those with certain chronic diseases and for those with physical limitations. To improve and maintain physical fitness, the ACSM promotes a physical activity training program that includes the following 4 exercise components: cardiorespiratory, resistance, flexibility, and neuromotor. According to the ACSM, these exercise guidelines, which are summarized in Table 11.2, form the basis of a sound exercise program by improving one or more aspects of physical fitness.

A varied and regular exercise plan not only contributes to overall physical fitness and mental health, but can also help to slow the progression of age-related physical decline and poor health. In addition to these evidence-based recommendations, the ACSM also emphasizes the importance of paying attention to time spent in idle

TABLE 11.2 THE AMERICAN ACADEMY OF SPORTS MEDICINE EXERCISE RECOMMENDATIONS FOR HEALTHY ADULTS

Exercise Component	Type of Exercise	Recommendation for Healthy Adults
Cardiorespiratory Exercise	Regular, purposeful exercise that involves major muscle groups and is continuous and rhythmic in nature such as brisk walking, jogging, running, dancing, stair climbing, cycling, rowing, cross-country skiing, and swimming.	• At least 150 minutes of moderate-intensity exercise per week. • Can be met through 3–60 minutes of moderate-intensity exercise (five days per week) or 20–60 minutes of vigorous-intensity exercise (three days per week). • One continuous session and multiple shorter sessions (of at least 10 minutes) are both acceptable to accumulate desired amount of daily exercise. • Gradual progression of exercise time, frequency, and intensity is recommended for best adherence and least injury risk. • People unable to meet these minimum recommendations can still benefit from some activity.
Resistance Exercise	Involves major muscle groups such as free-weights, machines with stacked weights, and resistance bands.	• Two to four sets of each exercise will help improve strength and power. • For each exercise, 8–12 repetitions improve strength and power, 10–15 repetitions improve strength in middle-aged and older persons starting exercise, and 15–20 repetitions improve muscular endurance. • Wait at least 48 hours between resistance-training sessions. • Train each major muscle group two or three days each week using a variety of exercise and equipment. • Very light or light intensity is best for older persons or previously sedentary adults starting exercise.
Flexibility Exercise	Involves range of motion movements such as stretching, yoga, ballet, Tai chi, and elastic bands.	• At least two or three days each week to improve range of motion. • Each stretch should be held for 10 to 30 seconds to the point of tightness or slight discomfort. • Repeat each stretch two to four times, accumulation 60 seconds per stretch. • Static, dynamic, and proprioceptive neuromuscular facilitation stretches are all effective. • Flexibility exercise is most effective when the muscle is warm. • Muscles can be warmed by light aerobic activity or a hot bath before stretching.
Neuromotor Exercise (also called functional fitness training)	Multifaceted activities such as Tai chi and yoga that incorporate motor skills can improve balance, coordination, gait, and agility through proprioceptive training.	• At least two or three days per week. • Should involve motor skills (balance, agility, coordination, and gait), proprioceptive neuromuscular facilitation stretches, and multifaceted activities such as Tai chi and yoga to improve balance and agility, which can help prevent falls in older adults.

Source: Garber CE, Blissmer B, Deschenes MR, Franklin BA, Lamonte MJ, Lee IM, Nieman DC, Swain DP. Quantity and Quality for Developing and Maintaining Cardiorespiratory, Musculoskeletal, and Neuromotor Fitness in Apparently Healthy Adults: Guidance for Prescribing Exercise. Medicine & Science in Sports & Exercise. 2011;43:1334-59. Available at: http://journals.lww.com/acsm-msse/pages/articleviewer.aspx?year=2011&issue=07000&article=00026&type=fulltext

activities such as prolonged sitting, watching television, and working on a computer. Sedentary behaviors are considered health risks, regardless of physical activity levels. As the saying goes, "Sitting is the new smoking."

Cardiorespiratory Exercise For most healthy adults, cardiorespiratory exercise (moderate and/or vigorous in intensity, and continuous and/or rhythmic in motion) enhances cardiovascular fitness. Cardiorespiratory exercises, such as walking, jogging, and swimming, can be performed daily in a single, continuous session (> 30 minutes/day), or performed in multiple sessions of at least 10 minutes spread throughout the day. Cardiorespiratory exercise promotes physical fitness by lowering blood pressure, improving blood lipid profiles, enhancing blood glucose regulation, and promoting a favorable ratio of lean-to-adipose tissues. In fact, middle-aged and older adults that exercise regularly have healthier cardiorespiratory fitness and metabolic profiles than do those who are sedentary.[5] For this reason, both the 2008 Physical Activity Guidelines for Americans and the ACSM guidelines concur that healthy adults should strive to achieve targeted levels of cardiorespiratory exercise. However, in order to minimize the risk of musculoskeletal injuries, individuals who have been relatively sedentary should gradually increase the duration, frequency, and/or intensity of their exercise sessions.

Resistance Exercise Resistance exercise engages various major muscle groups and helps improve several parameters associated with muscular fitness. For example, resistance exercise can help slow the progression of age-related muscle loss, a condition known as **sarcopenia**.[6] In addition to increasing muscle strength and mass, resistance exercise also benefits bone mineral density and bone strength.[7] As discussed in previous

If you experience frequent nausea, lightheadedness, and/or blurred vision during exercise, you are likely working outside of a safe intensity level.

chapters, age-related bone loss is a risk factor for both osteoarthritis and osteoporosis. An optimal resistance-training program, which includes a regiment of prescribed frequency, intensity, volume, and rest intervals, helps improve muscle strength, endurance, and power. Many types of resistance exercise such as free weights, muscle isolation machines and resistance bands can improve muscular fitness. Importantly, these types of activities help improve the ability of muscle fibers to shorten and lengthen when needed. The ACSM recommends that resistance exercises be performed a minimum of two non-consecutive days each week, with a prescribed number of sets and repetitions. Targeted major muscle groups include arms, legs, abdomen, shoulders, chest, and the back.

Flexibility Exercise Muscle and joint flexibility tends to decrease with age. Because tight muscles can cause back pain, impair balance, and limit range of motion, the ACSM guidelines emphasize the importance of exercise that improves flexibility and balance. This type of exercise is most effective when it involves stretching muscles and holding a static (not moving) stretch to a point of feeling tightness or slight discomfort. In addition to static stretching, dynamic stretching (slow stretching movements

sarcopenia Age-related loss of muscle tissue.

that favor an increase in reach and range) are also helpful.[8] Flexibility exercises should be done at least two days each week. However, those with existing stiffness and limited range of motion should engage in stretching activities on a daily basis. Participating in activities such as yoga and tai chi are excellent ways to improve flexibility and balance.

Neuromotor Exercise Neuromotor exercise (also called functional fitness training) incorporates a variety of motor skills such as balance, coordination, and agility. This is particularly advantageous for older adults who are at increased risk of falls and related injuries. Short workouts, conducted in 10-minute increments throughout the day, are optimal for improving overall physical fitness. Although few studies have evaluated the efficacy of neuromotor exercise, most health care providers recognize that improved agility and balance in elderly adults help decrease the risk of falling, make older adults less fearful of falling, and can improve overall quality of life.[9] **Proprioceptive neuromuscular facilitation stretching** is a technique that uses natural movement patterns to improve spatial and positional body awareness. As a result of participating in this type of exercise, a person can gain better control of their body movements (balance), which often decline with age as a result of impaired sensory feedback.

 11-2 HOW DOES ENERGY METABOLISM CHANGE DURING PHYSICAL ACTIVITY?

Throughout the day, the body is in a state of perpetual motion. Whether you are standing, sitting, sleeping, walking, or running, your body needs energy. At rest, the body expends approximately 1.0–1.5 kcal/minute. During physical exertion, however, energy expenditure can increase to 36 kcal/minute. As illustrated in Table 11.3, physical activity level can contribute substantially to total expenditure, and therefore energy requirements.

proprioceptive neuromuscular facilitation stretching Exercise that improves equilibrium, agility, and balance by improving spatial and positional body awareness.

exercise intensity The magnitude of the effort required to perform an activity or exercise.

exercise duration The number of minutes or hours an activity is performed in one session.

11-2a ATP Generation during Exercise

Movement requires motor units within skeletal muscle fibers to contract and relax in response to neural signals from the brain. To perform this complex action, muscle fibers convert the chemical energy in ATP to mechanical energy. In this way, ATP provides the energy muscles need for physical exertion. During exercise, ATP is generated from the metabolic breakdown of glucose, fatty acids, and, to a lesser extent, amino acids. The availability of these nutrients is determined by the foods one eats and the nutrients stored in the body.

To fuel physical activity, multiple metabolic pathways work together to provide energy in the form of ATP to muscle cells. Some of these pathways function in aerobic (oxygen-requiring) conditions, whereas others function in anaerobic (oxygen-poor) conditions. Although muscles use anaerobic and aerobic metabolic pathways simultaneously, the intensity and duration of activity determines the relative contribution of each. **Exercise intensity** refers to the magnitude of the effort required to perform an activity or exercise, whereas **exercise duration** refers to the length of time for which an activity is performed (minutes or hours in

TABLE 11.3	THE EFFECTS OF PHYSICAL ACTIVITY LEVEL ON TOTAL ENERGY REQUIREMENTS IN HEALTHY-WEIGHT ADULTS[a]			
	Physical Activity Level (PAL)			
Sex and Age (years)	**Sedentary**	**Low Active**	**Active**	**Very Active**
Females[b]		kcal/day		
20	1,879	2,085	2,342	2,651
30	1,784	1,989	2,247	2,555
40	1,688	1,894	2,151	2,460
50	1,593	1,799	2,056	2,365
60	1,498	1,703	1,961	2,270
Males				
20	2,542	2,770	3,060	3,536
30	2,447	2,675	2,964	3,441
40	2,352	2,579	2,869	3,345
50	2,256	2,484	2,774	3,250
60	2,161	2,389	2,679	3,155

[a]Calculated using Estimated Energy Requirement (EER) equations for a 5-foot 10-inch, 154-pound reference man and a 5-foot 4-inch, 126-pound reference woman.

[b]Estimates for females do not include women who are pregnant or breastfeeding.

Source: The Institute of Medicine. Dietary Reference Intakes for energy, carbohydrate, fiber, fat, fatty acids, cholesterol, protein, and amino acids. Washington, DC: The National Academies Press, 2005.

one session. Short-duration, high-intensity exercise (such as a 100-meter sprint) relies heavily on anaerobic pathways, which deliver immediate but short-lived bursts of ATP to muscles. By contrast, endurance exercise (such as running a marathon) relies heavily on aerobic pathways, which deliver a more sustained supply of ATP. In total, the body uses three energy systems to generate ATP: the phosphagen system, the glycolytic system, and the oxidative system. These three systems work together to ensure adequate energy availability so that the physical demands of exercise can be met under a variety of conditions (see Figure 11.1).

The Phosphagen System

At the onset of exercise, small amounts of ATP already present in muscle tissue provide an immediate source of energy. This ATP reserve is quickly depleted, however. The **phosphagen system** is the simplest and most rapid means by which active muscles generate ATP (see Figure 11.2). Under conditions of low oxygen availability, a high-energy compound called **creatine phosphate** can be rapidly broken down. As a result, a molecule

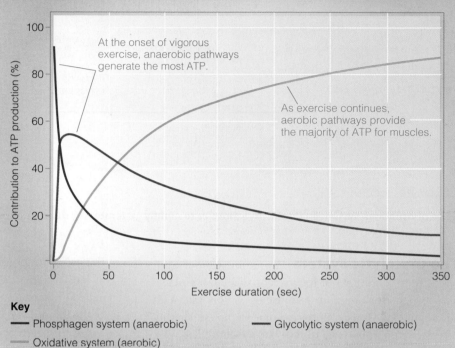

FIGURE 11.1 RELATIVE CONTRIBUTION OF THE BODY'S ENERGY SYSTEMS TO ENERGY EXPENDITURE DURING EXERCISE

At the onset of vigorous exercise, anaerobic pathways generate the most ATP.

As exercise continues, aerobic pathways provide the majority of ATP for muscles.

Key

— Phosphagen system (anaerobic)   — Glycolytic system (anaerobic)
— Oxidative system (aerobic)

Source: Adapted from Gastin PB. Energy system interaction and relative contribution during maximal exercise. Sports Medicine. 2001, 31:725–41.

of phosphate is released, which combines with adenosine diphosphate (ADP) to produce ATP. Only small amounts of creatine phosphate are stored in muscle, and it takes several minutes to replenish that small supply. Therefore, if physical activity continues, muscles must turn to other systems to generate ATP.

Do Creatine Supplements Enhance Athletic Performance?

Recall that the phosphagen system of energy generation depends on an available supply of creatine phosphate. For this reason, some athletes take creatine supplements to enhance performance and endurance. Widely available and sold over-the-counter in most grocery stores and pharmacies, creatine-containing products are promoted as beneficial for muscle mass, strength, and recovery time. Although creatine supplementation does increase creatine levels in muscle tissue, the connection between this increase and improved performance is less clear.

Despite little evidence that they are effective, creatine supplements remain popular among athletes.

© Karefan/Shutterstock.com

phosphagen system The simplest and most rapid energy system; characterized by the use of creatine phosphate to produce ATP.

creatine phosphate A high-energy compound that is broken down to produce ATP.

FIGURE 11.2 THE PHOSPHAGEN SYSTEM

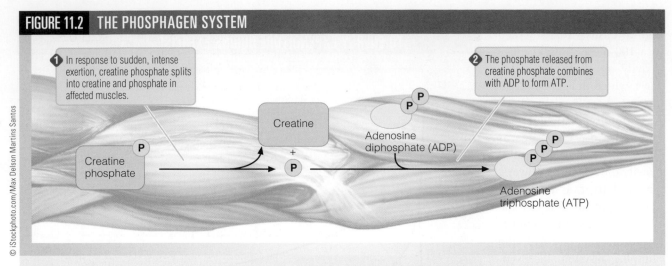

❶ In response to sudden, intense exertion, creatine phosphate splits into creatine and phosphate in affected muscles.

❷ The phosphate released from creatine phosphate combines with ADP to form ATP.

Creatine phosphate

Creatine
+
P

Adenosine diphosphate (ADP)

Adenosine triphosphate (ATP)

The phosphagen system enables active muscles to rapidly generate ATP during periods of low oxygen availability.

The Glycolytic System A continuation of exercise under conditions of low oxygen availability places an even greater demand on muscles to generate ATP. As creatine phosphate stores become depleted, another energy system called the **glycolytic system** takes over. Recall from Chapter 4 that glycolysis is the first step in the metabolic breakdown of glucose, a series of energy-releasing chemical reactions that enzymatically converts glucose to lactate or pyruvate, depending on oxygen availability. Glycolysis can briefly sustain activity after the immediate burst of energy provided by creatine phosphate diminishes. Although glycolysis can generate ATP rapidly, the system is not sustainable. In other words, it can only provide appreciable amounts of ATP for a short duration of time during intense physical activity. For example, in trained athletes, glycolysis can only generate enough ATP to satisfy the energy requirements of intense activity for approximately 3 minutes before their muscles fatigue.[10]

The fate of pyruvate, the end-product of glycolysis, is determined by oxygen availability; if oxygen is available, pyruvate is directed into the oxidative system of energy metabolism. If oxygen availability is limited, as is the case during high-intensity exercise, pyruvate is converted to lactic acid. The majority of lactic acid in the muscle dissociates into lactate and hydrogen ions (H^+), however. Although the accumulation of lactate in muscle is often associated with muscle fatigue, researchers now believe that this may be attributed to the concentration of hydrogen ions, not lactate.[11] In fact, lactate plays an integral role in exercise performance. When lactate produced by muscles is released into the blood, it circulates to the liver where it is recycled into glucose and then released into the blood. This system helps to replenish the supply of glucose for active muscles. In this way, lactate can be viewed as an important reusable fuel source. However, the relationship between lactate production and the generation of hydrogen ions is disadvantageous in terms of muscle fatigue.

The Oxidative System If physical activity continues beyond the capacity of the glycolytic system, muscles must utilize the aerobic **oxidative system** for ATP production. Aerobic pathways are more complex than their anaerobic counterparts, and therefore, they generate ATP more slowly. After several minutes of low- to moderate-intensity activity, breathing gradually becomes faster and harder. The heart begins to beat more frequently and forcefully. These changes in pulmonary and cardiovascular function help deliver much-needed oxygen to muscle tissue throughout the body. As sufficient oxygen becomes available, muscle cells are better able to use aerobic metabolism to produce ATP.

Aerobic metabolism utilizes pyruvate (from glycolysis), fatty acids, and to a lesser extent, amino acids to generate ATP. Because most people have ample amounts of stored body fat (adipose tissue), it is unlikely that the oxidative system could ever completely deplete a person's energy reserve. Indeed, aerobic metabolic pathways can generate a tremendous amount of ATP over an extended period of time—a critical advantage over anaerobic pathways.

glycolytic system An energy system characterized by the anaerobic metabolism of glucose (glycolysis).

oxidative system An aerobic energy system characterized by steady ATP production in an oxygen-rich environment.

Exercise-Related Muscle Soreness

Until recently, many believed that the accumulation of lactate in muscles was the primary cause of delayed muscle soreness following exercise. Although the accumulation of hydrogen ions is now thought to be the source of acute muscle pain and fatigue, ion accumulation does not adequately explain the muscle soreness that often lingers for days after a vigorous workout. Scientists now believe that microscopic tears in muscles may be the cause of this long-lasting muscle soreness. This condition, referred to as *delayed onset muscle soreness*, can occur up to 36 hours after exercise—especially after a new or greater-intensity muscle activity.[12] According to researchers, the tiny tears in muscle tissue can lead to inflammation and localized soreness. The good news is that this process helps muscle fibers to become stronger and more resistant to tearing. Although this type of muscle damage can be quite painful, it indicates that your muscles will be better equipped to handle the same type of exercise in the near future.

11-3 WHAT PHYSIOLOGIC ADAPTATIONS OCCUR IN RESPONSE TO ATHLETIC TRAINING?

As the body acclimates to the rigors of frequent exercise, a series of physiological changes called **adaptation responses** help improve athletic performance and enable the body to recover more efficiently and effectively. The anatomical and physiological changes that result from training are important for improving physical fitness, athletic condition, and overall health. The nature and magnitude of adaptation responses depend on a person's fitness level and the intensity and duration of training. Nonetheless, but everyone who trains regularly experiences these physiological changes to some extent. Training-induced adaptations can cause muscles to undergo **muscle hypertrophy**, meaning that they become larger and stronger. When a person stops training, however, gains in muscle hypertrophy are quickly lost. In fact, muscles that are not frequently challenged by physical activity undergo **muscle atrophy**, meaning that they decrease in size. Another example of an adaptation response is the expansion of blood vessels that occurs in response to cardiovascular and strength-building exercise. This expansion of blood vessels increases a person's ability to take in and deliver oxygen to muscles. Given that adaptation responses differ depending on the frequency, intensity, type, and duration of training, most athletic trainers recommend a varied training regimen that alternates between strength- and endurance-focused workouts. Next, we consider the specific advantages of these different types of exercise regimens.

11-3a Strength Training and Endurance Training

Different types of exercise induce a variety of physical adaptations. **Strength training** (such as weight lifting) challenges muscles with movements that are difficult but brief in duration. These physical adaptations

© Catalin Petolea/Shutterstock.com / Kathy Beerman

The relative contributions of aerobic and anaerobic pathways depend on the intensity of the exercise. Muscles utilize anaerobic pathways when there is a need to rapidly generate ATP, such as during short, high-intensity bursts of activity (left). Activities that entail long periods of endurance rely primarily on aerobic pathways (right).

adaptation responses A series of physiological changes that enables the body to perform and recover more efficiently and effectively in response to training.

muscle hypertrophy An adaptation response to exercise characterized by the enlargement of muscles.

muscle atrophy A process whereby muscles decrease in size.

strength training Exercise that increases skeletal muscle strength, power, endurance, and mass.

Hitting the Wall

Keeping active muscles adequately fueled can become increasingly difficult during prolonged activity (such as running a marathon). In fact, when active muscles become fatigued due to diminishing ATP availability, some athletes lose their stamina, or "hit the wall." Consuming sports drinks, energy bars, or energy gels during exercise can help provide a ready source of glucose, which is easily digested and rapidly absorbed. Sports drinks are particularly helpful for some athletes because they provide not only glucose, but also water and electrolytes (salt). Too much fluid can cause stomach cramps, however. So, it is important for athletes to find out in advance which products work best for themselves.

© paintings/Shutterstock.com

increase maximal oxygen consumption (or VO$_2$ max), which increases muscle size and strength. Conversely, **endurance training** (such as running and swimming) entails steady, low- to moderate-intensity exercise that

endurance training Exercise that entails steady, low- to moderate-intensity exercise that persists for an extended duration.

heart rate The number of heartbeats (contractions) per unit of time.

ventilation rate Breaths expelled per minute.

stroke volume The amount of blood pumped out of the heart (left ventricle) per heartbeat.

maximal oxygen consumption (or **VO$_2$ max**) A measure of the cardiovascular system's capacity to deliver oxygen to muscles.

interval training Exercise that entails alternating short, fast bursts of intense exercise with slower, less demanding activity.

anaerobic capacity The maximum amount of work that can be performed under anaerobic conditions.

persists for a longer duration. Endurance training is beneficial for pulmonary and cardiovascular function, and it can help improve aerobic capacity (the ability of cells to produce ATP in the presence of an oxygen-rich environment). As the intensity and duration of endurance training increases, muscles require more oxygen for aerobic ATP production. This challenges the cardiovascular system to work even harder.

Cardiovascular adaptations to physical training also include decreased resting **heart rate** (the number of heartbeats per minute), decreased **ventilation rate** (breaths expelled per minute), and increased **stroke volume** (the amount of blood pumped out of the heart per heartbeat). As a result, the cardiovascular system pumps blood more efficiently, which facilitates blood flow and the subsequent delivery of nutrients and oxygen to muscles. These adaptive responses increase **maximal oxygen consumption** (or **VO$_2$ max**), a measure of the cardiovascular system's capacity to deliver oxygen to muscles. A high VO$_2$ max increases physical stamina and helps delay the onset of fatigue. Endurance training also increases the ability of muscles to use fatty acids for ATP production. This minimizes glucose use, which in turn delays the onset of muscle fatigue. An ability to conserve glucose is particularly important for endurance athletes, such as marathon runners, cyclists, and distance swimmers, who depend on the utilization of glycogen stores as an energy source over an extended period of time.

Some athletic trainers recommend **interval training**, which entails alternating short, fast bursts of intense exercise with slower, less demanding exertion. For example, a runner might sprint at full speed for 1 minute, then jog at a measured pace for 5 minutes. An interval set is repeated several times during a training session. This type of work-out helps improve both endurance and **anaerobic capacity**, defined as the maximum amount of work that can be performed under anaerobic conditions. Greater anaerobic capacity means that muscles have a higher tolerance for lactate and are better able to perform repetitive, high-intensity activity with little or no rest.

Although it is important to understand how your body responds to the physical demands of exercise, it is equally important to understand how these demands affect dietary needs and energy requirements. These topics will be addressed in the next section.

11-3b Sport Supplements and Ergogenic Aids

Some athletes believe that taking certain dietary supplements and/or **ergogenic aids** can enhance physical performance beyond the natural gains obtained through physical training alone. Ergogenic aids are substances that influence athletic performance with the expectation that they will give athletes a competitive advantage. Substances such as vitamins, minerals, herbal products, caffeine, and extracts from organs and glands are all part of an ever-growing industry of reputed sports-enhancing products. Although these products are largely untested, they are widely available and are popular among athletes. It is important for athletes to be aware that little effort is made to study the safety and efficacy of ergogenic aids.

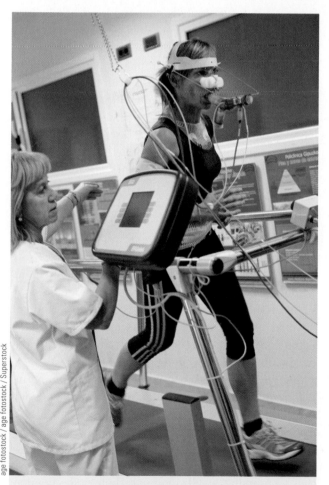

Testing athlete's VO$_2$ max (maximal oxygen uptake) involves measuring oxygen uptake during incremental bouts of exercise on a motorized treadmill.

In fact, the U.S. Food and Drug Administration (FDA) does not require such products to be tested at all before they are placed in the marketplace. This is true for all dietary supplements because they are considered to be food, which also does not need any sort of "testing" to be sold.

 ## HOW DOES PHYSICAL ACTIVITY INFLUENCE DIETARY REQUIREMENTS?

In general, nutritional recommendations are similar for all people—regardless of physical activity level. However, people who participate in athletics should take special care to:

1. consume enough water to replenish fluid loss during training and competition, and

2. consume enough calories to meet the energy demands of physical activity and prevent weight loss.

Although exercise may increase micronutrient requirements slightly, the 2015 Dietary Guidelines for Americans recommend that nutrient requirements be met through an adequate and balanced diet, rather than supplementation.[12] Although a well-balanced diet can easily satisfy nutrient and energy requirements of most physically active adults, competitive athletes may require assistance to ensure that the type, amount, and timing of food and fluid intake meets performance goals.[13]

11-4a Energy Requirements to Support Physical Activity

Athletic individuals must consume enough energy to support normal daily activities, as well as those performed as exercise. To satisfy total energy requirements, it is important to consider the frequency, intensity, type, and duration of a workout. Estimated energy expenditures associated with various activities are listed in Table 11.4. Using this information, you can estimate that a sedentary man who weighs 154 lb (70 kg) needs approximately 2,512 kcal/day simply to maintain body weight. However, running for 1 hour at a pace of 5 miles per hour would expend about 590 kilocalories in addition to those required to support normal daily activities. As a result, approximately 3,102 kilocalories would be needed to maintain body weight. At times, it is important for athletes to consider additional factors that can impact energy requirements. For example, temperature (both hot and cold), stress, altitude, and

ergogenic aids A substance taken to enhance athletic performance.

TABLE 11.4	AVERAGE ENERGY EXPENDITURE FOR SELECTED PHYSICAL ACTIVITIES	
Physical Activity	**Energy Expenditure (kcal/hour)[a]**	
Aerobics, high impact	533	
Aerobics, low impact	365	
Aerobics, water	402	
Backpacking	511	
Basketball	584	
Bicycling, < 10 mph	292	
Golfing, carrying clubs	314	
Hiking	438	
Racquetball	511	
Rope jumping	861	
Running, 8 mph	861	
Skiing, cross-country	496	
Skiing, downhill	314	
Swimming laps, vigorous	715	
Walking, 3.5 mph	314	

[a]Approximate calories used by a 160-lb person (1-hour duration)

Source: Adapted from: Ainsworth BE. Haskell WL, Herrmann SD, Meckes N, Bassett DR, Tudor-Locke C, Greer JL, Vezina J, Whitt-Glover MC, Leon AS. 2011 compendium of physical activities: A second update of codes and MET values. Medicine & Science in Sports & Exercise. 2011; 43:1575.

medications can impact an athlete's energy requirements. Furthermore, decreased energy requirements coincide with reductions in training and decreases in muscle mass.[14]

An athlete who trains on a regular basis most likely has a very high total energy requirement. In fact, some athletes require 4,000 to 5,000 kcal/day. As discussed in Chapter 2, mathematical formulas developed by the Institute of Medicine as part of the Dietary Reference Intakes (DRIs) can be used to estimate total energy requirements for various physical activity levels.[15]

11-4b Recommended Intakes for Macronutrients

As with any diet plan, it is important to consider the ideal distribution of calories, in addition to total energy intake. Recall that the Institute of Medicine's Acceptable Macronutrient Distribution Range (AMDR) suggests that 45–65 percent of total calories come from carbohydrates, 20–35 percent from fat, and 10–35 percent from protein. These recommendations likely hold for all adults, including athletes. Although alcohol is not a nutrient, it is important for athletes to consider alcohol's overall contribution in terms of total energy intake, as well as its effect on athletic performance. Certainly, consuming alcohol in moderation, in conjunction with a well-balanced diet, is not of concern. However, misuse of alcohol can negatively impact athletic training and performance. Consuming alcohol in large amounts can impair regulation of body temperature, hydration, and energy metabolism.[16]

Carbohydrates As previously discussed, exercise is fueled by an integrated system of energy-yielding metabolic pathways. Because glucose is the primary energy-yielding macronutrient that fuels activities with high-energy demand, athletes should pay close attention to carbohydrate intake. Important considerations include consuming a diet that ensures adequate glucose reserves (glycogen) prior to competition, maintaining blood glucose within acceptable levels during competition, and restoring glucose reserves after exercise or competition (recovery). Therefore, in order to optimize athletic performance, nutrition strategies must address carbohydrate requirements before, during, and after physical activity.

Although the Institute of Medicine does not provide special recommendations regarding carbohydrate intake for athletes, the Academy of Nutrition and Dietetics (AND) and the ACSM recommend that daily intake of carbohydrate be personalized to coincide with each athlete's training program and body size. As such, daily carbohydrate recommendations for athletes vary from 3–5 to 8–12 g/kg body weight. For a 200 lb (91 kg) athlete adhering to a moderate-intensity exercise intensity program (1 hour/day), this translates to approximately 455–637 grams of carbohydrate daily.[17] For perspective, one slice of whole-wheat bread contains approximately 17 grams of carbohydrates. While this calculation provides a good estimate for some athletes, total carbohydrate requirements clearly depend on total energy expenditure, exercise intensity, exercise duration, and the type of activity performed.

Nutritional strategies can be used to help maintain blood glucose levels during long, sustained athletic competitions. Because glucose is a key fuel for the central nervous system (including the brain), plummeting blood glucose levels during athletic competition not only compromise performance, but can also impair mental concentration, coordination, and skill. Sports foods such as drinks, bars, and gels offer athletes a rich source of carbohydrates that are easy to consume during competition. Consuming 30–60 grams of carbohydrates per hour is an important tactical approach to enhancing athletic performance because it helps supply muscles and the brain with adequate glucose.[18]

Glycogen is an important energy reserve for athletes participating in distance or endurance sports. To help sustain energy for longer periods of time, some athletes try to increase the amount of glycogen stored in the body (skeletal muscle and liver) using a technique called **carbohydrate loading**. Advocates of this potentially energy-boosting strategy believe that athletes can stimulate their muscles to store more glycogen by combining the right workout intensity with the right level of carbohydrate intake. Although there is not universal support for the effectiveness of carbohydrate loading, proponents recommend that it be carried out in two stages.[19] During the first stage, an athlete consumes approximately 6–8 grams carbohydrate per kg of body weight per day while simultaneously increasing workout intensity. Three to four days prior to competition, the athlete tapers workout intensity and increases carbohydrate intake to 10–12 g per kg of body weight per day. Although athletes participating in high-intensity or endurance competitions often believe that carbohydrate loading gives them a competitive edge, adequate training and a healthy, balanced diet are also important for the improvement of athletic performance. Because glycogen stores are depleted overnight, it is particularly important for endurance athletes to consume carbohydrate-rich foods and/or snacks prior to early-morning workouts and competitions. Low-fiber, low-glycemic-response foods are often recommended because they are thought to cause fewer GI problems than high-fiber, high-glycemic-response foods.[20]

Protein Although considerable amounts of protein are needed to maintain, build, and repair muscle, the debate continues as to whether expressly active people (and even competitive athletes) need additional protein.[21] In 2005, the Institute of Medicine concluded that the estimated protein requirement for a healthy adult (0.8 grams of protein per kg per day) is equally adequate for a physically active individual.[22] However, the AND and ACSM believe that athletes may require as much as 1.2–2.0 grams of protein per kg of body weight per day.[23] Although requirements can fluctuate depending on the amount and rigor of training, athletes should be able to satisfy targeted protein intake goals with a varied diet that includes moderate amounts of high-quality protein sources such as dairy, meat, and eggs. Although protein supplementation is occasionally used to optimize recovery, experts recommend that athletes focus on dietary strategies to satisfy protein requirements. Athletes who adhere to strict vegetarian diets must ensure that protein

requirements are met by consuming high intakes of complementary protein sources such as soy products, nuts, and whole grains.[24] Depending on the extent of dietary restrictions, this can pose practical challenges for athletes in terms of both protein quantity and quality.

Lipids Fatty acids are an important source of energy for physically active individuals. Intake of foods rich in monounsaturated fatty acids (such as olive and canola oils) and polyunsaturated fatty acids (such as vegetable and fish oils) should be emphasized, while intake of saturated and *trans* fatty acids should be limited. Furthermore, because exercise can lead to inflammation, some experts believe that eicosapentaenoic acid- and docosahexaenoic acid-rich ω-3 fatty acids may be especially important for competitive athletes.[25]

11-4c Recommended Intakes for Micronutrients

Vitamins and minerals are essential for many functions, including energy metabolism, the repair and maintenance of body structures, protection from oxidative damage, and prevention of immune suppression related to physical exertion. As long as an individual consumes a varied diet that supplies adequate energy, vitamin and mineral needs will likely be met as well. Nutrient supplementation to enhance athletic performance is controversial and potentially dangerous. Furthermore, there is little evidence regarding the efficacy of this practice.[26] However, athletes who consume foods with low nutrient densities, restrict energy intake, and avoid animal-based products are at increased risk of inadequate intakes of several micronutrients—especially iron, calcium, and antioxidants. As these nutrients are especially important for physical activity, deficiencies may foster severe complications for some athletes. For this reason, coaches and trainers need to be aware of the nutritional issues that affect athletes who follow strict vegetarian diets.[27]

Iron The iron-containing molecules hemoglobin and myoglobin are integral components of red blood cells and muscles, respectively. Recall from Chapter 8 that hemoglobin transports oxygen to cells and removes carbon

carbohydrate loading A technique used by some athletes to increase the amount of glycogen stored in the body (liver and skeletal muscle) so that they can sustain activity for long periods of time.

dioxide, a waste product of energy metabolism. Similarly, myoglobin aids in the delivery of oxygen to muscles. Because iron is vital to oxygen and carbon dioxide transport, iron deficiency with or without anemia can compromise aerobic energy metabolism and thus seriously affect athletic training and performance. Although non-heme iron is available in cereals, grains, and some vegetables, it is not as readily absorbed as the heme iron found in meat. All individuals who do not eat meat—including athletes—are at increased risk of impaired iron status and should have their blood tested periodically. Impaired iron status imposes additional stress on an athlete's body and can limit physical and mental performance. Because it can take up to six months to reverse iron deficiency anemia, it is important for athletes to consume foods rich in bioavailable iron such as red meat. As a second line of defense, iron supplements can help improve iron status, although supplementation may cause GI distress.[28]

Endurance athletes, especially runners, may have higher iron requirements than nonathletes. Increased loss of blood in the feces, iron loss associated with excessive sweating, and the rupturing of red blood cells as a result of the feet repeatedly striking hard surfaces contribute to this phenomenon.[29] Because of the blood loss associated with menstruation, female athletes are also more likely than males to experience impaired iron status. In addition to iron deficiency, some athletes experience **sports anemia**, a temporary type of anemia that occurs at the onset of a training program. Sports anemia is caused by a disproportionate increase in plasma volume as compared to the synthesis of new red blood cells. This physiologic response, referred to as **hemodilution**, can make athletes appear iron deficient even though they are not. Because sports anemia is caused by a healthy physiological response to training rather than poor nutritional status, there is no need for athletes with this condition to add additional iron to their diets.

Calcium Although recommended calcium intake is the same for athletes and nonathletes alike, adequate calcium is essential for the growth, maintenance, and repair of bone tissue. Thus, a diet that is low in calcium increases an athlete's risk of low bone density, which in turn increases the likelihood of stress fractures. Stress fractures are small cracks in bones—especially those in the feet—that occur in response to repeated jarring. Women who have menstrual irregularities are at particularly high risk of bone loss, and those who exercise excessively while

sports anemia A temporary type of anemia that occurs at the onset of a training program due to hemodilution.

hemodilution A disproportionate increase in plasma volume as compared to the synthesis of new red blood cells.

Inadequate calcium intake can lead to bone loss, which can lead to stress fractures. It is important for all athletes, especially those who are post menopausal or have menstrual irregularities, to make sure they meet their recommended calcium intakes.

© i4lcoc l2/Shutterstock.com

restricting their caloric intakes are at risk of developing the female athlete triad, a syndrome characterized in part by bone loss that will be addressed in Chapter 12.

Athletes who restrict their intakes of dairy products may benefit from foods that are fortified with both calcium and vitamin D, as the latter facilitates absorption of the former. Fortified orange juice, soy products, and breakfast cereals are good sources of these two micronutrients. If an athlete's diet is lacking in calcium-rich foods, calcium supplements may be beneficial.[30] Additionally, athletes who train indoors, live in regions with limited sunlight availability, and use lotions with Sun Protection Factor are at increased risk for vitamin D insufficiency. Dietary intervention alone is not considered a reliable means to resolve vitamin D deficiency, and supplementation may be required.[31]

Antioxidants Exercise increases the production of highly reactive molecules called free radicals (also called reactive oxygen species) that are harmful to cells. The body's natural oxidative stress defense system (antioxidants) plays an important role in counteracting the oxidative damage of free radicals. **Oxidative stress** can occur when free radical production exceeds the body's natural ability to stabilize these destructive molecules. For this reason, it is particularly important for athletes to consume foods that provide a good source of antioxidant nutrients (selenium, beta-carotene, vitamin C, and vitamin E). Athletes should also try to incorporate brightly colored (yellow-orange, red, and deep green) fruits and vegetables into their daily diets.[32] Not only are these foods excellent sources of antioxidant nutrients, they also contain numerous phytochemicals that also act as antioxidants in the body. There is little evidence to suggest that antioxidant supplementation is advantageous for athletic performance in terms of endurance, oxygen availability, cardiovascular function, muscle strength, or resistance to fatigue.[33]

11-4d Fluids, Electrolytes, and Dehydration

Although exercise increases body temperature, the body usually dissipates this heat by sweating. However, if the water lost as sweat is not replaced, dehydration can result. To meet the body's need for water, adult men and women should consume approximately 11 and 16 cups (2.6 and 3.8 liters) of water per day, respectively. Athletes often require substantially more than this, however.[34] Although sweat consists mainly of water, it also contains electrolytes such as sodium, chloride, and potassium, and, to a lesser extent, minerals such as iron and calcium. Because extensive sweating can disrupt the body's balance of both fluids and electrolytes, it is important for athletes to stay hydrated.

© VectorLifestylepic/Shutterstock.com

Dehydration Can Cause Serious Illness

Excessive sweating causes blood volume to decrease. As blood flow to the skin diminishes, so too does the body's ability to dissipate heat. As a result, core body temperature can rise, increasing the risk of heat exhaustion and heat stroke. **Heat exhaustion** can occur when as little as 5 percent of an athlete's body weight is lost due to sweating.[35] Heat exhaustion is characterized by feelings of illness, muscle spasms, rapid and/or weak pulse, low blood pressure, disorientation, and profuse sweating (see Table 11.5). If body heat continues to rise, heat exhaustion can quickly progress to **heat stroke**, which occurs when 7–10 percent of body weight is lost due to sweating. This very serious condition is

oxidative stress An imbalance between the body's natural oxidative stress defense system (antioxidants) and free radicals, which results in an impaired ability to prevent cellular oxidative damage.

heat exhaustion A rise in body temperature that occurs when the body has difficulty dissipating heat.

heat stroke A serious condition that can develop if body temperature continues to rise beyond heat exhaustion.

TABLE 11.5	SIGNS AND SYMPTOMS OF MILD TO SEVERE DEHYDRATION
Mild Dehydration	**Moderate to Severe Dehydration**
• Flushed face	• Low blood pressure
• Extreme thirst	• Fainting
• Dry, warm skin	• Disorientation, confusion, and slurred speech
• Impaired ability to urinate; production of reduced amounts of dark, yellow urine	• Severe muscle cramping in the arms, legs, stomach, and back
• Dizziness, made worse when standing	• Convulsions
• Weakness	• Bloated stomach
• Cramping in arms and legs	• Heart failure
• Crying with few or no tears	• Sunken, dry eyes
• Difficulty concentrating	• Skin that has lost its firmness and elasticity
• Sleepiness	• Rapid, shallow breathing
• Irritability	• Fast, weak pulse
• Headache	
• Dry mouth and tongue, with thick saliva	

Source: Adapted from Joseph A Salomone III, MD, associate professor, Department of Emergency Medicine, Truman Medical Center, University of Missouri at Kansas City School of Medicine. Available from: http://www.webmd.com/fitness-exercise/dehydration-topic-overview

characterized by dry skin, confusion, and loss of consciousness. A person experiencing heat stroke requires immediate medical assistance.

Dehydration can cause a number of problems beyond impaired body temperature regulation. For example, excessive sweating or consumption of large amounts of fluid without adequate amounts of sodium can cause blood sodium concentration to decrease. Diminished blood sodium concentration, called **hyponatremia**, is more common in endurance athletes who participate in activities that last several hours, such as marathon and triathlon contestants. A 2005 study of Boston Marathon runners reported that 13 percent of participants developed hyponatremia.[36] Strength- and endurance-focused athletes alike must be particularly careful when exercising in hot or humid weather, as these conditions can increase sweating and thus the loss of water and electrolytes from the body. Because drinking large amounts of water may actually increase one's risk of hyponatremia, many experts recommend that endurance athletes replenish body fluids by consuming drinks that contain both carbohydrates and electrolytes, such as sports drinks.[37]

Preventing Dehydration and Electrolyte Imbalance

To prevent dehydration, athletes must consume adequate amounts of fluids not only during but also before and after exercise. The ACSM recommends that athletes consume 2–4 mL/lb of body weight (5–10 mL/kg body weight) of fluid 2–4 hours before exercising.[38] It is important for athletes to allow sufficient time for the body to void excess fluid prior to training or competition, but additional fluids may be required during prolonged training sessions, particularly in hot environments.[39]

Because dehydration can compromise performance, athletes should also be sure to consume fluids *during* exercise. The purpose of fluid intake during an athletic event is not to rehydrate, but rather to stay hydrated by replacing the fluid lost through sweating. When a workout lasts less than 1 hour, plain water is adequate to replenish body fluids. When a workout lasts more than 1 hour, a fluid that contains small amounts of carbohydrates and electrolytes, such as a sports drink, is likely beneficial.[40] Readily available and well-tolerated by most people during exercise, sports drinks provide roughly 10–20 grams of carbohydrates (40–80 kilocalories) per eight-ounce (237 mL) serving. It is important to be aware, however, that some sports drinks contain caffeine. Whereas some people enjoy the mild stimulatory effect of caffeine, others find that it causes stomach upset, diarrhea, nervousness, and jitters.

hyponatremia Diminished blood sodium concentration.

© Steve Yager/Shutterstock.com

It is important for athletes to prevent dehydration by consuming adequate amounts of fluids before, during, and after exercise. This is why race organizers set up water stations along courses.

11-4e Nutrition Plays an Important Role in Recovery

Hard training is physically demanding, and athletes need to consume the right types and amounts of food to make the most of exercise. Adequate nutrition is also an important aspect of successful post-exercise recovery. A well-balanced and complete dietary recovery plan can help the body repair muscle tissue, replace lost nutrients, and fully replete energy reserves such as glycogen.

Carbohydrates Replenish Glycogen Stores

It takes approximately 20 hours to replenish liver and muscle glycogen stores after a long, intense athletic workout. Although the rate of glycogen synthesis is variable during the post-exercise period, there are several recommendations to promote optimal glycogen resynthesis. First, because the rate of glycogen synthesis is highest within the first 2 hours after exercise, athletes should consume carbohydrate-rich foods as soon as possible after a workout.[41] Consuming 50–70 grams of carbohydrates within the first hour of post-recovery can jump-start this metabolic process.[42] Ideally, a person should consume several small carbohydrate-rich meals instead of a single large one.[43] This promotes a gradual rise in blood glucose rather than a rapid rise followed by a quick drop.

Consuming high-glycemic-index foods later in the recovery period may also help facilitate glycogen resynthesis.[44] Furthermore, the rate of muscle glycogen synthesis may be higher when carbohydrates and proteins are consumed together than when carbohydrates are consumed alone during the post-exercise period.[45] Although an ideal ratio of carbohydrate to protein has not been firmly established, some athletic trainers recommend a ratio of 4 grams of carbohydrates to every 1 gram of protein.[46] This mix of carbohydrates and proteins could promote greater muscle glycogen, but there is no evidence that it improves overall athletic performance.

Protein Promotes Muscle Recovery Because prolonged and intense exercise increases protein breakdown in muscles, high-quality protein sources should be consumed during the post-exercise recovery phase. The first few hours following exercise are often referred to as an *anabolic window*. This period is an important time for an athlete to both reverse the catabolic state associated with physical activity and to increase the rate of muscle protein synthesis. When a person consumes protein-rich foods, the concentration of amino acids in the blood increases. This provides the necessary building blocks for muscle maintenance, recovery, and muscle protein synthesis during the post-workout period.

As previously discussed, experts disagree as to whether athletes require more protein (relative to body weight) than do nonathletes. Some suggests that certain types of protein may be more advantageous in post-workout muscle recovery than others, however. For example, high-quality protein recovery foods such as milk have grown in popularity among athletes in recent years because it can speed tissue repair, which in turn accelerates recovery from exercise-induced muscle damage. This is largely because the proteins in milk are readily digested and it is also abundant in amino acids with high bioavailability.

Water and Electrolyte Replacement Even if athletes consume fluids while exercising, they can still lose up to 6 percent of their body weight during strenuous activity. Therefore, fluid replacement is critical during the post-workout recovery phase. Consuming two to three cups (450–675 milliliters) of fluid for each pound of weight lost during exercise adequately replenishes fluid balance. A simple way for athletes to monitor hydration status is to observe the color of their urine the morning after a hard workout.

© loreanto/Shutterstock.com

Light-colored urine is a general indicator that a person is well hydrated, whereas darkcolored urine may indicate that additional fluids are needed. Although sweat contains significant amounts of sodium, the foods consumed after exercise usually contain enough sodium to restore electrolyte balance. Therefore, it is neither necessary nor recommended that athletes take sodium supplements.[47]

STUDY TOOLS 11

READY TO STUDY? IN THE BOOK, YOU CAN:

☐ Rip out the Chapter Review Card, which includes key terms and chapter summaries.

ONLINE AT WWW.CENGAGEBRAIN.COM, YOU CAN:

☐ Learn about Michelle Obama's "Let's Move" initiative in a short video.

☐ Explore the effect of diet on physical endurance in an animation.

☐ Interact with figures from the text to check your understanding.

☐ Prepare for tests with quizzes.

☐ Review the key terms with Flash Cards.

12 | Eating Disorders

LEARNING OUTCOMES

12-1 Differentiate between disordered eating patterns and the major eating disorders.

12-2 Discuss less-common disordered eating behaviors.

12-3 Understand contributing factors related to eating disorders.

12-4 Discuss the connection between eating disorders and athletics.

12-5 Describe strategies to prevent and treat eating disorders.

Klaus Mellenthin/Getty Images

After finishing this chapter go to **PAGE 315** for **STUDY TOOLS.**

12-1 WHAT IS THE DIFFERENCE BETWEEN DISORDERED EATING AND EATING DISORDERS?

Your body has natural signals that tell you when to eat, what to eat, and how much to eat. Paying attention to and trusting these internal cues helps cultivate a healthy relationship with food—in other words, eating without fear, guilt, or shame. For some people, however, a preoccupation with food and/or weight loss can reach obsessive proportions and disrupt healthy eating behaviors. When this happens, disordered eating behaviors can sometimes develops. **Disordered eating** encompasses a wide range of irregular eating behaviors that are not considered to be eating disorders. Examples of disordered eating patterns include irregular eating, chronic yo-yo dieting,

disordered eating An eating pattern characterized by unhealthy eating behaviors.

eating disorder An extreme disturbance in eating behaviors that can be both physically and psychologically harmful.

excessive calorie counting, frequent weight fluctuations, rigid food and exercise regimes, food preoccupation, emotional eating, and food restriction.[1] Disordered eating behaviors are common and often arise in response to stress, illness, and dissatisfaction with one's appearance. Although disordered eating patterns can be disturbing to others, they typically do not persist long enough to cause serious physical harm. Experts estimate that up to 50 percent of the population, at one time or another, have or will experience disordered eating patterns. For some people, however, disordered eating progresses into a full-blown **eating disorder**—an extreme disturbance in eating behaviors that can be both physically and psychologically harmful. Eating disorders are complex behaviors that arise from a combination of physical, psychological, and social issues. People with eating disorders often feel isolated, and their relationships with family and friends become strained. Eating disorders are not nearly as pervasive as disordered eating patterns, and may affect 1–3 percent of the general population.[2]

The number of people diagnosed with eating disorders has increased over the last 25 years.[3] Although eating disorders are not pervasive in all sectors of the

population, they are more common among women than men. The National Association of Anorexia Nervosa and Associated Disorders estimates that 8 million people in the United States battle some form of eating disorder.[4] Although eating disorders are most prevalent among adolescent girls and young adult women, a pattern of eating disorders has begun to emerge among middle-aged and older women as well.[5] This trend may be developing in response to America's youth-oriented culture, which intensifies insecurities associated with aging.[6]

The prevalence of eating disorders in females is well documented, but far less is known about eating disorders in males. Numbers of eating disorders among males may be difficult to estimate because they may be more reluctant to seek help for what is commonly perceived as a female condition. Among those with eating disorders, approximately 10 percent are male.[7] Although it is not clear which personal and cultural aspects contribute to the development of eating disorders in adolescent boys and young men, possible factors include a history of obesity, participation in a sport that emphasizes thinness, and a heightened cultural emphasis on physical appearance.

As described in Table 12.1, the American Psychiatric Association recognizes three distinct categories of eating disorders. **Anorexia nervosa (AN)** is characterized by an irrational fear of gaining weight or becoming obese; **bulimia nervosa (BN)** is characterized by repeated cycles of bingeing and purging; and **binge-eating disorder (BED)**, which is characterized by frequent episodes of out-of-control eating without purging.[8] The American Psychiatric Association also recognizes that some people experience feeding or eating-related disorders that do not meet the criteria for AN, BN, or BED. These individuals fall within another category referred to as **Other Specified Feeding or Eating Disorders (OSFED)**. Examples of behaviors that fall into this category include recurrent episodes of night eating, bingeing, and purging with lower frequency than that observed with BN, recurrent purging to influence weight in the absence of bingeing, and repeated attempts at weight loss even

anorexia nervosa (AN) An eating disorder characterized by an irrational fear of gaining weight or becoming obese.

bulimia nervosa (BN) An eating disorder characterized by repeated cycles of bingeing and purging.

binge eating disorder Repeated episodes of binge-eating; characterized by out of control consumption of unusually large amounts of food in a short period of time.

Other Specified Feeding or Eating Disorders (OSFED) A category of feeding or eating behaviors that include some, but not all, of the diagnostic criteria for anorexia nervosa, binge eating, and/or bulimia nervosa.

TABLE 12.1 AMERICAN PSYCHIATRIC ASSOCIATION CLASSIFICATION AND DIAGNOSTIC CRITERIA FOR FEEDING AND EATING DISORDERS

Anorexia Nervosa (AN)

A. Persistent restriction of energy intake leading to significantly low body weight (in context of what is minimally expected for age, sex, developmental trajectory, and physical health).

B. Either an intense fear of gaining weight or becoming fat, or persistent behavior that interferes with weight gain even when body weight is extremely low.

C. Disturbance in the way one's body weight or shape is experienced, undue influence of body shape and weight on self-evaluation, or persistent lack of recognition of the seriousness of the current low body weight.

    Subcategories:

- *Restricting type*: Does not regularly engage in binge eating or purging behavior.
- *Binge-eating/purging type*: Regularly engages in binge eating or purging behavior.

Bulimia Nervosa (BN)

A. Recurrent episodes of binge eating. An episode of binge eating is characterized by both of the following:

- Eating, in a discrete period of time (e.g., within any 2-hour period), an amount of food that is larger than most people would eat during a similar period of time and under similar circumstances.
- A sense of lack of control over eating during the episode (e.g., a feeling that one cannot stop eating or control what or how much one is eating).

B. Recurrent inappropriate compensatory behavior in order to prevent weight gain, such as self-induced vomiting, misuse of laxatives, diuretics or other medications, fasting, or excessive exercise.

C. Binge eating and inappropriate compensatory behaviors both occur, on average, at least once a week for three months.

D. Self-evaluation is unduly influenced by body shape and weight.

E. Disturbance does not occur exclusively during episodes of anorexia nervosa.

Binge-Eating Disorder

A. Recurrent and persistent episodes of binge eating. An episode of binge eating is characterized by both of the following:

- Eating, in a discrete period of time (e.g. within any 2-hour period), an amount of food that is definitely larger than most people would eat during a similar period of time and under similar circumstances.
- A sense of lack of control over eating during the episode (e.g. a feeling that one cannot stop eating or control what or how much one is eating).

B. Binge eating episodes are associated with three or more of the following:

- Eating much more rapidly than normal.
- Eating until feeling uncomfortably full.
- Eating large amounts of food despite not being hungry.
- Eating alone because of being embarrassed by eating so much.
- Feeling disgusted, depressed, or very guilty about overeating.

C. Marked distress regarding binge eating.

D. Binge eating occurs, on average, at least once a week for three months.

E. Binge eating is not associated with the recurrent use of inappropriate compensatory behaviors as in Bulimia Nervosa and does not occur exclusively during the course of Bulimia Nervosa, or Anorexia Nervosa methods to compensate for overeating, such as self-induced vomiting.

Source: Adapted from Diagnostic and statistical manual of mental disorders, 5th ed. (DSM-5). Washington, DC: American Psychiatric Association Press, 2013.

though body weight is within normal ranges. To better represent the spectrum of behaviors related to feeding and eating disorders, the latest edition of the Diagnostic and Statistical Manual of Mental Disorders (DSM-5) also now includes diagnostic criteria for pica (persistent eating of nonnutritive substances), rumination disorder (repeated regurgitation of food), and avoidant/restrictive food intake disorder. Each of these categories is divided into subcategories that depend on the presence or absence of specific behaviors. Because behaviors associated with different eating disorders often overlap, it can be difficult to identify the specific disorder that a person has.[9]

12-1a Anorexia Nervosa (AN)

The American Psychiatric Association recognizes two subcategories of AN. In the first, called **anorexia nervosa, restricting type**, low body weight is maintained through food restriction and/or excessive exercise. In **anorexia nervosa, binge-eating/purging type**, low body weight is maintained through both food restriction and periods of **bingeing** and/or purging. Bingeing is defined as uncontrolled consumption of large quantities of food in a relatively short period of time, whereas **purging** is self-induced vomiting, excessive exercise, and/or misuse of laxatives, diuretics, and/or enemas (see Figure 12.1).

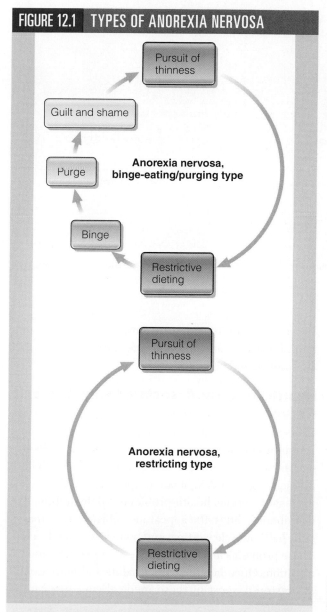

FIGURE 12.1 TYPES OF ANOREXIA NERVOSA

Anorexia nervosa, binge-eating/purging type

Pursuit of thinness → Guilt and shame → Purge → Binge → Restrictive dieting → Pursuit of thinness

Anorexia nervosa, restricting type

Pursuit of thinness → Restrictive dieting → Pursuit of thinness

There are two types of anorexia nervosa, an eating disorder characterized by self-starvation and a relentless pursuit of weight loss.

Plush Studios/Digital Vision/Getty Images

People with distorted body images perceive themselves as fat even if they are dangerously thin.

Those who struggle with AN tend to be very critical of their body shape and size. In fact, there is often a disconnect between actual and perceived body weight and shape such that there is a self-perception of being overweight or obese even when dangerously thin. Because of this distorted point-of-veiw, weight loss becomes an obsession—individuals with AN often take drastic measures to lose weight, spend hours scrutinizing their bodies, and tend to view self-worth in terms of weight and body shape.

Since food restriction (and often, self-starvation) is perceived as an

anorexia nervosa, restricting type
A subcategory of anorexia nervosa characterized by food restriction and/or excessive exercise.

anorexia nervosa, binge-eating/purging type
A subcategory of anorexia nervosa characterized by both food restriction and periods of bingeing and purging.

bingeing Uncontrolled consumption of large quantities of food in a relatively short period of time.

purging Self-induced vomiting and/or excessive exercise, misuse of laxatives, diuretics, and/or enemas.

accomplishment rather than a problem, people with AN often have little motivation to change their behaviors. Although the symptoms of AN center around food, the causes are much more complex. AN and other eating disorders usually stem from psychological issues, and both denial of food and the relentless pursuit of thinness become ways of coping with emotions, conflict, and stress.

Behaviors Associated with Anorexia Nervosa

Typically, people with AN limit their food intake as well as the varieties of foods consumed. Foods are often categorized as either safe or unsafe depending on whether their consumption is believed to cause weight gain. Foods often perceived as unsafe, and typically the first to be eliminated, include meat, high-sugar foods, and fatty foods. As AN progresses, the diet becomes so restricted that nutritional needs are no longer met. Beyond limiting food intake, individuals with AN may develop unusual food rituals to demonstrate self-restraint and control over the urge to eat. These rituals include chewing food but not swallowing, overchewing, obsessively cooking for others, and cutting food into unusually small pieces.[10]

Obsessive behaviors not related to food are also commonly associated with AN.[11] Many individuals with AN maintain rigid exercise schedules—some work out as much as 3–4 hours a day. Furthermore, individuals with AN often keep meticulous daily records to track food intake, amount of time spent exercising, and weight loss.[12] Daily monitoring can provide an anorexic person the motivation and mental stamina to sustain a perpetual state of hunger (see Figure 12.2). A page taken from the journal of a person with AN reflects an inability to recognize problematic behaviors (see Figure 12.3). Finally, it is not uncommon for people with AN to weigh themselves repeatedly throughout the day. These and other obsessive behaviors help quell concerns about being insufficiently

lean and/or excessively fat. Common signs and symptoms, behaviors, and consequences associated with AN are summarized in Table 12.2.

Health Concerns Associated with Anorexia Nervosa

Although AN eventually causes a person to become underweight and appear gaunt, AN-related health problems such as nutrient deficiencies, electrolyte imbalances, and hair loss can develop even before physical signs of AN become apparent. If AN is left untreated, serious health problems—dehydration, cardiac abnormalities, the appearance of lanugo (increased body hair), muscle wasting, and bone loss—can develop. These problems are largely the result of chronic undernutrition. Once healthy eating habits and body weight are restored, most of these medical concerns can be reversed.

In females, a loss of body fat can lead to a decline in estrogen production, which in turn can lead to disrupted menstruation and poor reproductive function.[13] This

FIGURE 12.2 FOOD, WEIGHT, AND EXERCISE RECORD OF A PERSON WITH ANOREXIA NERVOSA

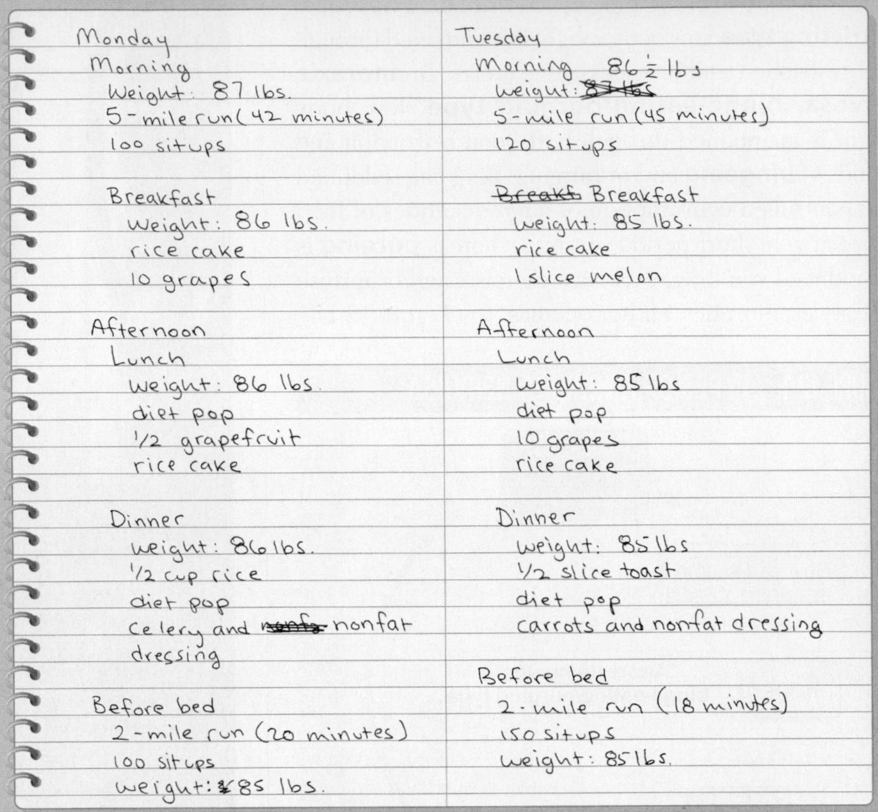

Monday
Morning
  Weight: 87 lbs.
  5-mile run (42 minutes)
  100 situps

Breakfast
  Weight: 86 lbs.
  rice cake
  10 grapes

Afternoon
Lunch
  Weight: 86 lbs.
  diet pop
  1/2 grapefruit
  rice cake

Dinner
  Weight: 86 lbs.
  1/2 cup rice
  diet pop
  celery and ~~nonfat~~ nonfat dressing

Before bed
  2-mile run (20 minutes)
  100 situps
  weight: ~~8~~85 lbs.

Tuesday
Morning 86½ lbs
  Weight: ~~87 lbs~~
  5-mile run (45 minutes)
  120 situps

~~Breakf~~ Breakfast
  Weight: 85 lbs.
  rice cake
  1 slice melon

Afternoon
Lunch
  Weight: 85 lbs
  diet pop
  10 grapes
  rice cake

Dinner
  Weight: 85 lbs
  1/2 slice toast
  diet pop
  carrots and nonfat dressing

Before bed
  2-mile run (18 minutes)
  150 situps
  weight: 85 lbs.

Some people with eating disorders keep meticulous records regarding weight loss, exercise, and diet.

FIGURE 12.3 THOUGHTS AND BEHAVIORS ASSOCIATED WITH ANOREXIA NERVOSA

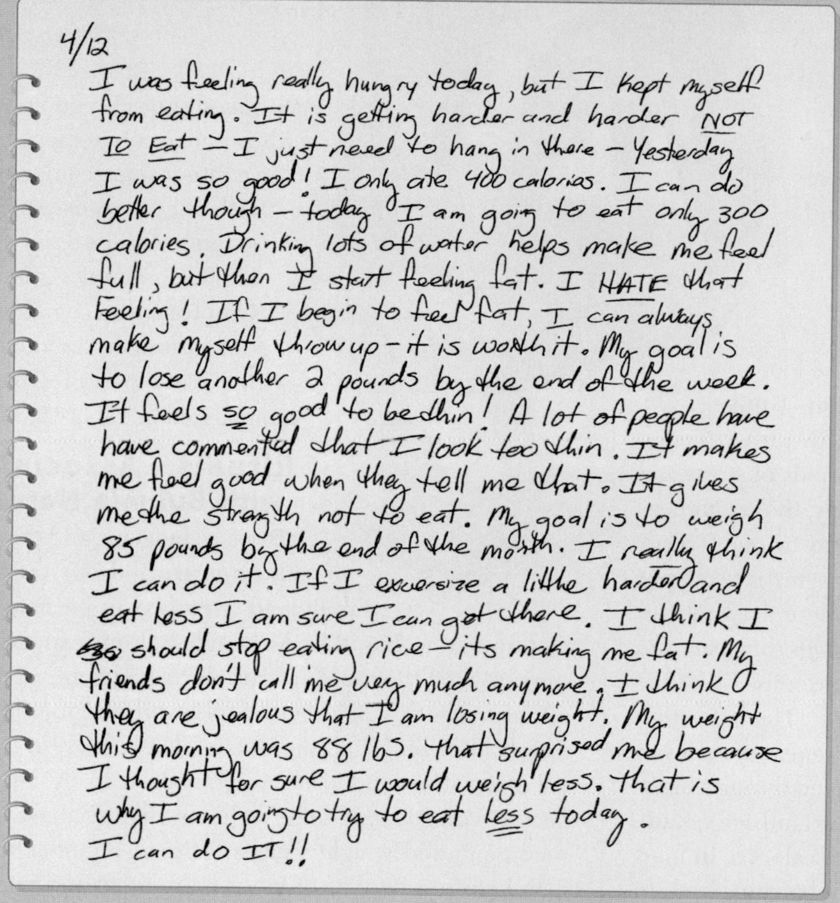

4/12

I was feeling really hungry today, but I kept myself from eating. It is getting harder and harder NOT To Eat — I just need to hang in there — Yesterday I was so good! I only ate 400 calories. I can do better though — today I am going to eat only 300 calories. Drinking lots of water helps make me feel full, but then I start feeling fat. I HATE that feeling! If I begin to feel fat, I can always make myself throw up - it is worth it. My goal is to lose another 2 pounds by the end of the week. It feels so good to be thin! A lot of people have have commented that I look too thin. It makes me feel good when they tell me that. It gives me the strength not to eat. My goal is to weigh 85 pounds by the end of the month. I really think I can do it. If I exercise a little harder and eat less I am sure I can get there. I think I should stop eating rice — its making me fat. My friends don't call me very much anymore. I think they are jealous that I am losing weight. My weight this morning was 88 lbs. that surprised me because I thought for sure I would weigh less. That is why I am going to try to eat less today. I can do IT!!

physiological response most likely evolved as a means to protect women from pregnancy during periods of low food availability. In addition to being critical for reproductive function, estrogen also plays an important role in bone health. The longer a woman goes without menstruating—a condition called *amenorrhea*—the greater the bone loss. Although estrogen levels return to normal when body weight is restored, it is not likely that bones ever fully recover. For this reason, women who have experienced AN-triggered amenorrhea remain at increased risk for osteoporosis even after returning to healthy body weights and eating patterns.[14]

AN is a serious condition that affects a person's physical and mental well-being. It goes without saying then that AN requires immediate medical, psychological, and nutritional intervention.

TABLE 12.2 PHYSICAL SIGNS AND SYMPTOMS, BEHAVIORAL AND EMOTIONAL CHARACTERISTICS, AND HEALTH CONSEQUENCES ASSOCIATED WITH ANOREXIA NERVOSA

Physical Signs and Symptoms	Behavioral and Emotional Signs and Symptoms	Health Consequences
• At or below the 15th percentile of ideal body weight	• Food rituals such as excessive chewing or cutting food into small pieces	• Reproductive problems
• Thin appearance	• Restriction of the amount and type of food consumed	• Bone loss
• Fainting, dizziness, fatigue, overall weakness	• Adherence to rigid daily schedules and routines	• Electrolyte imbalance
• Intolerance to cold	• Refusal to eat	• Irregular heartbeat
• Significant loss of body fat and lean body mass	• Denial of hunger	• Bruises
• Brittle nails	• Excessive exercise	• Injuries such as stress fractures
• Dry skin, dry hair, thinning hair, hair loss	• Depression or lack of emotion	• Impaired iron status
• Low blood pressure	• Food preoccupation	• Impaired immune status
• Growth of lanugo on the body	• Frequent monitoring of body weight	• Slow heart rate
• Irregular or absent menstruation	• Tendency toward perfectionism	• Increased risk of heart failure
	• Rigid attitude	• Increased risk of suicide

Although researchers have reported that AN can lead to death, this is true only for the most severe cases that require hospitalization.[15] Still, the death rate associated with AN is high: between 5 and 20 percent of those diagnosed with AN die from complications within 10 years of the initial diagnosis.[16] The two primary causes of AN-related death are heart attacks and suicide.[17]

12-1b Bulimia Nervosa (BN)

Like people with AN, those with BN use food as a psychological coping mechanism. Unlike those with AN, however, people with BN turn *to* food—rather than *away from* it—during periods of stress and emotional conflict. Perhaps surprisingly, the occurence of BN outnumbers AN by about two to one.[18] The American Psychiatric Association recognizes two types of people with BN: (1) those who vomit or use laxatives, diuretics, or enemas in response to binges (bingeing and purging type), and (2) those who use exercise or fasting in response to binges (nonpurging type). The most common type of BN is characterized by repeated cycles of bingeing and purging.[19] Foods consumed during a binge often include sweets such as cookies and ice cream, and can easily add up to thousands of calories. In most cases of BN, the panic associated with bingeing leads to the drastic measure of induced vomiting. As previously

The term bulimia comes from the Greek word meaning "the hunger of an ox."

mentioned, not all people with BN purge to counteract bingeing; rather some compensate with excessive exercise or extended periods of fasting.[20] Common signs and symptoms, behaviors, and consequences associated with both types of BN are summarized in Table 12.3.

Rituals Associated with Bulimia Nervosa

Because bingeing and purging are often carried out secretly, family and friends may be unaware that someone has BN. Although bulimics may consume several thousand calories during a binge, purging often prevents weight gain. Still, even purging immediately after bingeing cannot completely prevent nutrient absorption, and some weight gain is likely. In fact, people with BN tend to be within or slightly above their recommended weight ranges.[21] Some bulimics alternate between periods of bingeing/purging and periods of food restriction.

TABLE 12.3 PHYSICAL SIGNS AND SYMPTOMS, BEHAVIORAL AND EMOTIONAL CHARACTERISTICS, AND HEALTH CONSEQUENCES ASSOCIATED WITH BULIMIA NERVOSA		
Physical Signs and Symptoms	**Behavioral and Emotional Signs and Symptoms**	**Health Consequences**
• Fluctuations in body weight	• Feelings of guilt or shame after eating	• Erosion of tooth enamel and tooth decay from exposure to stomach acid
• Swollen or puffy face	• Obsessive concerns about weight	• Electrolyte imbalances that can lead to irregular heart function and possible sudden cardiac arrest
• Odor of vomit on breath	• Repeated attempts at food restriction and weight loss	• Inflammation of the salivary glands
• Sores around mouth	• Frequent use of bathroom during and after meals	• Irritation and inflammation of the esophagus
• Irregular bowel function	• Feelings of being out of control	• Hemorrhaging and bleeding after vomiting
	• Moodiness and depression	• Dehydration
	• Laxative abuse	• Weight gain
	• Fear of not being able to stop eating voluntarily	• Abdominal pain, bloating
	• Disappearance of food in household	• Sore throat, hoarseness
	• Eating to the point of physical discomfort	• Broken blood vessels in the eyes
	• Self-induced vomiting after eating	

FIGURE 12.4 THOUGHTS AND BEHAVIORS ASSOCIATED WITH BULIMIA NERVOSA

- 5/9 -

I feel so fat and ugly. Yesterday I saw this
guy that I like walking with another girl—
it made me feel sad. When I got to class,
the teacher passed back our exam. I didn't
get a very good grade. I WISH I WAS PRETTY AND
SMART !!!

Yesterday I felt so bad that I binged
and purged most of the afternoon.
I hate when I do that but I can't help
myself. I didn't intend for it to happen.
I bought all this food thinking that it
would last. It lasted for about an hour. Before
I knew it the cookies and ice cream were gone.
Then I went to the dining hall and ate dinner.
!! As soon as I got back to my room I made
myself vomit again. I was glad that my
roommate wasn't there. She would think I
am so GROSS. That is because I AM GROSS !!!
I wish I could stop this. Tomorrow I am
going to diet. I am going to get thin.
Yeah, right ....

Individuals with BN often experience feelings of shame and regret after episodes of bingeing and purging. This can lead to depression, which increases the likelihood of future binge/purge cycles (see Figure 12.4). These destructive cycles can be difficult to resist because they often bring short-term relief from complex emotional issues. Many people struggle for years before admitting that they have a problem.

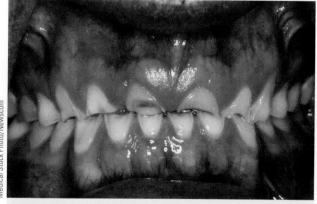

Frequent vomiting can damage tooth enamel. As a result, the teeth of individuals with BN or other disordered eating patterns sometimes appear mottled.

Whereas individuals with AN often feel satisfaction from dieting and weight loss, many individuals with BN feel discontent and are more likely to have low self-esteem. Those with BN also tend to be impulsive, and are prone to other unhealthy behaviors such as substance abuse, self-mutilation, and suicidal tendencies.[22] For these reasons, people with BN are more likely to seek treatment than are those with AN. And sometimes, treatment seeks them: people with BN often only seek help after getting caught in the act of bingeing and purging.

Health Concerns Associated with Bulimia Nervosa Over time, repeated cycles of bingeing and purging damage the body. If a person with BN purges by vomiting too frequently, the delicate lining of the esophagus and mouth can become irritated by repeated exposure to stomach acid. Because the acidity of gastric juice can also harm dental enamel and cause tooth decay, dentists and dental hygienists are often the first to notice the damage caused by repeated vomiting. Furthermore, some people with BN induce vomiting by inserting their fingers deep into their throat, which can cause their hands to become scraped from striking the teeth. Frequent vomiting and/or overuse of laxatives and diuretics can cause dehydration and electrolyte imbalance. This can lead to irregular heart function, and can even result in sudden cardiac arrest.[23] Fortunately, many people with BN seek help and medical assistance before reaching this critical point, and most improve with treatment.

12-1c Binge-Eating Disorder

Although binge-eating disorder (BED) was first described in 1959, it was not until 2013 that it was officially considered a distinct eating disorder characterized by recurrent episodes of binge eating. Like people with other eating disorders, those with BED typically feel shame and guilt associated with binge eating.

Individuals often attempt to conceal these behaviors, but eventually self-loathing triggered by feelings of negativity begin to surface. Repeated episodes of uncontrolled eating become a source of both embarrassment and comfort. Binge eating may even temporarily help relieve negativity, although it often returns with greater intensity.[24] Long-term studies suggest that in terms of severity, persistence, and duration, BED may have more favorable outcomes compared to other types of eating disorders.[25]

The American Psychiatric Association's diagnostic criteria for BED requires that a person experiences recurrent episodes of binge eating, defined as eating larger amounts of food than most people would eat within any 2-hour period. The binge-eating episodes must also be associated with three or more of the following:

1. eating more rapidly than normal;
2. eating until feeling uncomfortably full;
3. eating large amounts of food despite not feeling hungry;
4. eating alone because of being embarrassed by how much one is eating; and
5. feeling disgusted, depressed or very guilty after bingeing.

Additional diagnostic criteria include bingeing at least once a week for three months; bingeing without compensatory behaviors; feeling distressed in response to binge eating; and eating excessively large amounts of food in a 2-hour period at least twice per week for at least six months; and feeling a lack of control over the bingeing episodes.[26] As with BN, binges often typically take place in private and are often accompanied by feelings of shame.

Although most people overeat from time to time, the bingeing associated with BED is distinct in that it serves as a habitual coping mechanism. For many people with BED, binges provide an escape from stress and emotional pain by inducing a psychological numbness and a temporary state of emotional well-being. Anger, sadness, anxiety, and other types of emotional distress often trigger binges. Many people with BED struggle with clinical depression, although it is not clear if depression triggers BED or if BED triggers depression.[27] Some studies indicate that people with BED are likely to have been raised in families affected by alcohol abuse.[28] In the case of BED, however, food rather than alcohol becomes the drug of choice.

The number of people with BED is much greater than the combined numbers of people with AN and BN.[29]

In total, approximately 1–5 percent of the American population has BED.[30] Although people with BED are sometimes of normal weight, most are overweight. In fact, some weight-loss treatment programs estimate that between 20 and 40 percent of obese patients experience BED.[31] Because they are likely to be overweight, people with BED tend to be at increased risk of weight-related health problems such as type 2 diabetes, gallstones, and cardiovascular disease.[32]

12-1d Other Specified Feeding and Eating Disorders (OSFED)

Just because a person does not meet the full diagnostic criteria of an eating disorder does not mean that unhealthy eating patterns should be ignored. For this reason, the American Psychiatric Association in the newly revised DSM-5 created the category of Other Specified Feeding and Eating Disorders (OSFED). Individuals who fall into this category typically have disturbed eating habits that have not yet been uniformly defined or characterized. Nonetheless, the physical, psychological, and behavioral disturbances can create havoc in a person's life. People with OSFED often present with a spectrum of behaviors and traits such as disturbed eating habits, distorted body image, and overall intense fear of gaining weight. Weight fluctuations, food preoccupation, sensitivity to comments related to body shape, low self-esteem, and rigid thoughts about food (good vs. bad) are not unusual among those with OSFED.

 ARE THERE OTHER DISORDERED EATING BEHAVIORS?

Many troublesome disordered eating behaviors exist in addition to the eating disorders formally recognized by the American Psychiatric Association. Not eating in public, situational purging, chewing food but spitting it out before swallowing, and obsessive dieting are some examples of such behaviors. Although insufficient information exists for these disordered eating patterns to be classified as distinct eating disorders, such behaviors can be equally debilitating and disruptive as AN, BN, and BED. Also, if disordered eating behaviors persist, they can easily progress into a full-blown eating disorder. As for any newly recognized pattern of dysfunctional behavior, further research is needed to cultivate a better understanding of these conditions.

Today, some of the most recognized and best understood food-related disturbances are restrained eating, excessive nighttime eating, avoidance of new foods, and obsession with muscularity. Each of these patterns is outlined in the following sections.

12-2a Restrained Eater

Individuals that suppress their desire for food and avoid eating for long periods of time between binges are called **restrained eaters**.[33] Although restrained eaters do not meet the specified diagnostic criteria, for BN (nonpurging type), there are many similarities. Restrained eaters limit their food intakes and fast to lose weight. After extended periods of food restriction, however, restrained eaters find themselves feeling out of control and respond by bingeing. This cycle of fasting and bingeing can be difficult to stop. Many restrained eaters perceive themselves as overweight, and the consumption of large amounts of food generates further feelings of inadequacy and self-contempt. These feelings can cause restrained eaters to turn back to food for emotional comfort. Like others with disordered eating patterns, restrained eaters find themselves in a vicious cycle that results in poor physical and psychological health.

12-2b Nocturnal Sleep-Related Eating Disorder and Night Eating Syndrome

Nocturnal sleep-related eating disorder (SRED) is a disordered eating pattern characterized by eating while asleep without any recollection of having done so.[34] Reportedly, people with SRED leave their beds and walk to their kitchens to prepare and eat food. Individuals may begin to suspect SRED when they notice unexplained missing food and see evidence of late-night kitchen activities. Meals consumed during episodes of SRED are typically high in fat and calories and may involve unusual food combinations. Perhaps unsurprisingly, it is common for those with SRED to experience weight gain. Although both men and women can develop SRED, the disorder is far more common in women.[35] Some researchers believe that stress, dieting, and depression can trigger episodes of SRED and that food restriction during the day may make some individuals more vulnerable to unconscious binge eating at night.[36] Although rates of occurrence are difficult to estimate, 1–3 percent of the American population may experience SRED, and rates as high as 10–15 percent have been reported among individuals with eating disorders.[37]

Night eating syndrome is characterized by an ongoing, persistent pattern of late night binge eating.

© OLJ Studio/Shutterstock.com

Night eating syndrome (NES) is a disordered eating pattern closely related to SRED, except individuals are fully aware of their eating. NES is characterized by a cycle of daytime food restriction, excessive food intake in the evening, and nighttime insomnia. Even though most people report that night eating causes them to feel depressed, anxious, and guilty, those with this type of disordered eating typically consume more than one-half of their daily calories after dinner. Night eating syndrome affects between 1 and 2 percent of American adults and it is more common among women than men.[38] Nine to 15 percent of individuals

restrained eater Individuals that suppress their desire for food and avoid eating for long periods of time between binges.

nocturnal sleep-related eating disorder (SRED) A disordered eating pattern characterized by eating while asleep without any recollection of having done so.

night eating syndrome (NES) A disordered eating pattern characterized by a cycle of daytime food restriction, excessive food intake in the evening, and nighttime insomnia.

participating in weight-loss treatment programs report behaviors associated with NES. Signs and symptoms associated with NES include:

- Not feeling hungry for the first several hours after waking
- Overeating in the evening and consuming more than one-half of daily calories after dinner
- Difficulty falling asleep, accompanied by the urge to eat
- Waking during the night and finding it necessary to eat before falling back to sleep
- Feelings of guilt, shame, moodiness, tension, and agitation—especially at night

12-2c Food Neophobia

It is very normal for parents of young children to complain about picky eating. Indeed, many children are reluctant to try new foods, and for some, even slight changes in food routines can be upsetting. In time, most children outgrow these behaviors and begin to accept new foods and eating patterns. Some do not, however. Children who refuse to try new foods as they grow older sometimes develop irrational food-related fears, which can lead to the development of food neophobia later in life.

Food neophobia is a disordered eating pattern characterized by an irrational fear or avoidance of trying new foods. Individuals with food neophobia typically eat very limited ranges of foods and often adhere to well-defined and unusual food rituals and practices.[39] For example, an individual with food neophobia may refuse to eat foods made from two or more food items. A person may enjoy certain foods when eaten separately, but may find them disgusting when joined together. In adults, these types of behaviors can be socially restricting and embarrassing. In extreme cases, they can lead to nutritional inadequacies, although some individuals with food neophobia take supplements to compensate for their poor eating habits. People with food neophobia tend to be slightly overweight because they often limit themselves to comfort foods that are high in calories, such as hamburgers, French fries, and macaroni and cheese.[40] Perhaps surprisingly, most adults with food neophobia are not interested in therapy or other treatments. Some researchers believe that food neophobia is a type of obsessive-compulsive disorder that requires extensive psychological therapy.[41] Regardless of its classification, clinicians are interested in learning more about food neophobia and how the people who are affected by it react to new foods.[42]

Avoidant/Restrictive Food Intake Disorder (AFRID) This new disordered eating diagnosis referred to as **avoidant/restrictive food intake disorder (AFRID)** describes children, adolescents, and adults who struggle to satisfy their basic nutrient and energy requirements. Although it may originate early in life such as during infancy or childhood, it is best characterized as psychological hurdles that prevent a person from consuming food for reasons other than a drive for thinness.[43] Unlike individuals with food neophobia, people with ARFID often experience substantial weight loss, nutritional deficiency, and physical reactions to foods perceived as perverse. For example, the sight or thought of certain foods may trigger retching, vomiting, or gagging. This food disturbance can be so severe that a person may eventually require tube feeding and other types of nutritional support. In children, ARFID can result in poor growth. The distinguishing characteristic associated with ARFID is the persistent disturbance in eating that leads to weight loss, nutritional problems, and dependency on sources of nourishment other than food (nutrient supplements and/or tube feeding).

12-2d Muscle Dysmorphia

Although societal pressures to be thin are directed more intensely toward females than toward males, it is not uncommon for the latter to experience similar body image issues. Instead of being pressured to be thin, boys and men are more often pushed to be strong and muscular. This societal pressure may be reinforced by the media, which often portrays the ideal male body as having well-defined muscles. Once referred to as *reverse anorexia*, **muscle dysmorphia** is a disorder characterized by a preoccupation with increasing muscularity. Muscle dysmorphia is observed primarily in men who have intense fears of being too small, too weak, and/or too skinny.[44] The term *dysmorphia* comes from the Greek words *dys*, which means "bad" or "abnormal," and *morphos*, which means "shape" or "form." Thus, individuals with muscle dysmorphia are preoccupied with perceived defects in their bodies.

Behaviors associated with muscle dysmorphia include working out for hours each day, allowing exercise to interfere with family and social life, paying excessive

food neophobia A disordered eating pattern characterized by an irrational fear or avoidance of new foods.

avoidant/restrictive food intake disorder (AFRID) An eating disturbance characterized by persistent failure to satisfy nutrition and/or energy requirements due to an aversion to food.

muscle dysmorphia A disorder characterized by a preoccupation with increasing muscularity.

© Lebedev Roman Olegovich/Shutterstock.com

Muscle dysmorphia, often described as the opposite of AN, is characterized by a compulsive preoccupation with increasing muscularity.

attention to diet, and maintaining unusual food rituals and eating practices. People with muscle dysmorphia may also engage in health-threatening practices such as the use of anabolic steroids to gain muscle mass.[45]

Individuals affected by muscle dysmorphia often have personality traits similar to those of people with recognized eating disorders.[46] In fact, one study reported that one-third of men diagnosed with muscle dysmorphia also had a history of eating disorders.[47] Men with muscle dysmorphia tend to have low self-esteem and typically harbor concerns about masculinity.[48] To compensate for these insecurities, they are driven to achieve overtly strong-looking bodies. Affected by obsessive thoughts of being weak, insecure, and vulnerable, men with muscle dysmorphia can have difficulties maintaining personal and social relationships.

Newly released, full-bodied dolls provide a more positive, realistic portrayal of women's bodies than those of the past.

Source: Barbiemedia

 ## 12-3 WHAT CAUSES EATING DISORDERS?

Scientists have many theories as to the factors that contribute to eating disorders, but there are no definitive answers. Undoubtedly, people develop eating disorders for a variety of reasons—why some are more vulnerable than others is not clear. Several factors are likely contributors, however. These include sociocultural characteristics, family dynamics, personality traits, and biological (genetic) factors.[49]

12-3a Sociocultural Factors

Eating disorders are more prevalent in some cultures than in others; when food is abundant and slimness is valued, eating disorders are more likely to develop. This is one reason why eating disorders are more prevalent in industrialized Western countries like the United States than in regions like Africa, China, and the Middle East.[50] Even within American society, differences in sociocultural status influence the development of eating disorders. However, as multicultural youth acculturate to mainstream American values, distinct cultural norms that once protected them from eating disorders may be eroding. For some, a desire to fit prevailing American beauty and social norms may lead to the development of an eating disorder. The impact of these norms has been felt across socioeconomic, cultural, and racial spectrums. For example, data show that the prevalence of eating disorders among Latina and African-American women is catching up to that of Caucasian women in the United States.[51]

The Role of the Media Based on popular media images, the perfect body is tall, lean, and has well-defined muscles. Whereas the average woman is 5 feet 4 inches (1.6 meters) tall and weighs 140 lb (63 kilograms), the

average model is 5 feet 11 inches (1.8 meters) tall and weighs 117 lb (53 kilograms).[52] This discrepancy has become more pronounced as female fashion models, beauty pageant contestants, and actresses have become increasingly thinner over the years. When the Miss America beauty pageant first began in the 1920s, its contestants' body mass indexes (BMIs) averaged 20–25 kg/m². Today, nearly all the participants have BMIs below those considered healthy, and nearly one-half have BMIs consistent with the diagnostic criteria for AN (a BMI less than 18.5 kg/m²).[53] In fact, a recent study reported that

women who participated in beauty pageants as children are more likely to experience body dissatisfaction, interpersonal distrust, and impulsiveness than those who did not.[54]

Major media outlets that glamorize unrealistically thin and overly muscular bodies have long been criticized for evoking a sense of inadequacy in impressionable children and young adults. For example, exceptionally slender celebrities who appear on television and in the movies may skew teenagers' standard of thinness, encouraging them to engage in unhealthy eating practices to achieve a thin and supposedly glamorous appearance.[55] Even the body shapes of popular dolls and action figures given to children during their formative years are unrealistic. In short, media likely play an important role in influencing how people view themselves, and may even predispose susceptible individuals to eating disorders. In response to unrealistic industry standards, some toy manufacturers have taken a bold step—designing dolls with body types that more accurately represent women's physiques. The hope is that full-bodied dolls will instill upon children a more positive, realistic portrayal of female body shapes and sizes.

Because body dissatisfaction is thought to be an essential precursor to the development of an eating disorder, it is important for children to understand that healthy bodies come in many shapes and sizes—despite what the media portrays. It is equally important for older children to be prepared for the physical and emotional changes associated with puberty. The changes in body dimensions and weight gain that occur during adolescence can make girls feel embarrassed and uncomfortable with their maturing bodies. Because such feelings are not always discussed openly among families and friends, girls

The BMIs of beauty contestants have decreased over the last 50 years.

may feel as if they are somehow abnormal. In fact, discomfort with the rapid maturation associated with puberty is extremely common.

Social Networks In addition to the media, a person's circle of peers may also contribute to the development of an eating disorder. Attitudes and behaviors about slimness and appearance are often learned from those with whom we associate. A desire to gain acceptance among one's friends can spark body dissatisfaction and dieting in susceptible adolescents. Unfortunately, many teenagers believe that to be liked and to belong they must be thin.

Although eating disorders frequently begin with a desire to lose weight, relatively few people on weight-loss diets actually develop eating disorders. Likewise, not everyone who lives in an affluent society that stigmatizes obesity and advocates extreme slenderness develops an eating disorder. These factors alone do not cause eating disorders, but, nonetheless, the sociocultural environments of industrialized countries appear to foster the development of eating disorders in individuals who are already vulnerable for other reasons. One such other reason is an unhealthy family dynamic.

Children who participate in beauty pageants may be more likely to dislike their bodies later in life.

AFP/Getty Images

12-3b Family Dynamics

Aside from sociocultural factors, another important issue related to one's risk of developing an eating disorder is having an unhealthy family dynamic. Because parents influence nearly every aspect of a child's life, it should come as no surprise that family dynamics play a decisive role in the development and perpetuation of an eating disorder. Although no one single family type inescapably causes a child to develop an eating disorder, researchers have found that certain distinguishing behaviors and characteristics increase the likelihood of this happening.[56] Such behaviors and characteristics include overprotectiveness, rigidity, conflict avoidance, abusiveness, chaotic family dynamics, and the presence of a mother with an eating disorder.[57]

Enmeshed Families Enmeshment is a style of family interaction whereby family members are overly involved with one another and have little autonomy.[58] An enmeshed family has no clear boundaries among its members. This environment can make it difficult for children to develop independence and individualism. Furthermore, children raised in such families often feel tremendous pressure to please their parents and meet expectations. Rather than doing things for themselves, children with overly involved, protective parents strive to please others. Enmeshed family dynamics promote dependency, which may lay the foundation for the emergence of an eating disorder. Under such circumstances, food often becomes the only component in a child's life over which control can be exerted.

Chaotic Families In contrast to enmeshed families that have exceedingly tight familial structures, chaotic families have remarkably loose structures. A **chaotic family** (or **disengaged family**) employs a style of family interaction characterized by a lack of cohesiveness and little parental involvement.[59] The roles of family members are loosely defined; children often feel a sense of abandonment; and parents may be depressed, alcoholic, or emotionally absent. A child growing up in this type of home may later develop an eating disorder as a way to fill emotional emptiness, gain attention, or suppress emotional conflict.

Mothers with Eating Disorders The presence of a mother with an eating disorder or body dissatisfaction can negatively influence eating behaviors in children.[60] An inability to demonstrate a healthy relationship with food and model healthy eating for one's children is a serious concern. A mother with an eating disorder is

enmeshment A style of family interaction whereby family members are overly involved with one another and have little autonomy.

chaotic family (or **disengaged family**) A style of family interaction characterized by a lack of cohesiveness and little parental involvement.

more likely to criticize a daughter's appearance and to encourage her to lose weight, even if doing so would be unhealthy. As a result, children of women with eating disorders are at increased risk of developing eating disorders themselves.[61]

12-3c Personality Traits and Emotional Factors

Scientists have long believed that certain personality traits make some people more susceptible to eating disorders than others. Some concerning characteristics include low self-esteem; a lack of self-confidence; and feelings of helplessness, anxiety, and depression.[62] An individual with an eating disorder typically has **food preoccupation**, meaning that an inordinate amount of time is spent thinking about food. Another personality trait commonly associated with eating disorders is *perfectionism*.[63] People who exhibit perfectionism often have difficulty dealing with shortcomings in themselves; therefore an imperfect body is not easily tolerated.

12-3d Biological and Genetic Factors

Because certain personality traits and eating behaviors are determined in part by the nervous and endocrine systems, it is logical that brain chemistry and other biological and genetic factors might play a role in the development of an eating disorder. It is also possible, however, that disordered eating may actually disrupt normal neuroendocrine function. For example, studies show that individuals with eating disorders are often clinically depressed.[64] It is difficult to determine whether clinical depression leads to eating disorders or vice versa (or if another factor is at play). In any case, because medication used to treat clinical depression is often effective in the treatment of certain eating disorders, depression is likely a contributing factor.

Scientists are working to identify genes that might influence susceptibility to eating disorders.[65] Studies of identical and fraternal twins have provided evidence suggesting that susceptibility to eating disorders may, in part, be inherited.[66] Although such

> **food preoccupation**
> Spending an inordinate amount of time thinking about food.

studies cannot completely differentiate the contribution of genetics from that of environment, some research suggests that the contribution of genetics may actually be greater than that of the environment. How this occurs remains unclear, however.[67]

12-4 ARE ATHLETES AT INCREASED RISK FOR EATING DISORDERS?

There are more competitive female athletes today than ever before. Perhaps not coincidentally, the number of female athletes with eating disorders has increased as well. Some studies indicate that the prevalence of eating disorders among female student-athletes and nonathletes does not differ, while other studies indicate otherwise.[68] At any rate, losing weight in an unhealthy way can have serious health consequences above and beyond affecting performance for an athlete. It is of utmost importance for coaches and trainers to recognize and act on the early warning signs and symptoms associated with eating disorders.

12-4a Athletics May Foster Eating Disorders in Some People

The prevalence of disordered eating and eating disorders among collegiate athletes is estimated to be somewhere between 15 and 60 percent.[69] Disagreement and inconsistent estimates may be attributable to the reluctance of some athletes to admit that they experience such problems. In addition, some athletes who exhibit disordered eating behaviors may not satisfy all the criteria needed for diagnosis. Regardless of the exact numbers, athletes—especially female athletes—are considered by many experts to be an at-risk group for developing eating disorders.

For some sports, physical performance is determined not only by motor abilities, strength, coordination, but also by body

© JacksColdSweat/Shutterstock.com

Sports that demand slender physiques tend to have more athletes with eating disorders than those for which larger body size is advantageous.

swimming, gymnastics, and diving have long perceived that body size affects how judges rate their performance.

Because athletes tend to be competitive and may equate their self-worth with athletic success, they may be especially willing to engage in risky weight-loss practices to achieve success. Coaches and trainers who believe that excess weight can hinder performance may ignore unhealthy behaviors, and may even encourage them. According to the National Collegiate Athletic Association (NCAA), the sports with the highest numbers of female athletes with eating disorders are cross-country running, gymnastics, swimming, and track and field.[71] Sports with the highest numbers of male athletes with eating disorders are wrestling and cross-country running.

weight. Athletes such as ski jumpers, cyclists, rock climbers, and long-distance runners may deliberately try to achieve low body weight to gain a competitive advantage. Athletes in sports that entail strict weight classes (such as boxing and wrestling) often try to fit the lowest possible weight class to avoid being outsized. Sports that demand a thin physical appearance (such as gymnastics) are also likely to have more athletes with eating disorders than are sports for which greater size may be beneficial.[70] This may be because judges in such sports often consider size and appearance when rating performance. Athletes who participate in dancing, figure skating, synchronized

12-4b The Female Athlete Triad

Because the rigor of athletic training alone is very stressful on the body, athletes with eating disorders are at extremely high risk for developing medical complications. Adequate nourishment is required to meet the physical demands of athletics. This is why serious health problems can arise when an athlete is restricting food intake, bingeing, and/or purging. For example, female athletes are at increased risk for developing a syndrome known as the **female athlete triad**. The female athlete triad is a combination of three interrelated conditions: disordered eating (or eating disorder), menstrual dysfunction, and osteopenia (see Figure 12.5).[72]

The interrelation of the three components of the female athlete triad is complex. Disordered eating can lead to very low levels of body fat, which can cause estrogen levels to decrease. Without adequate estrogen levels, menstrual cycles can become irregular, and in some cases, stop completely. A lack of estrogen

female athlete triad A combination of three interrelated conditions: disordered eating (or eating disorder), menstrual dysfunction, and osteopenia.

FIGURE 12.5 | FEMALE ATHLETE TRIAD

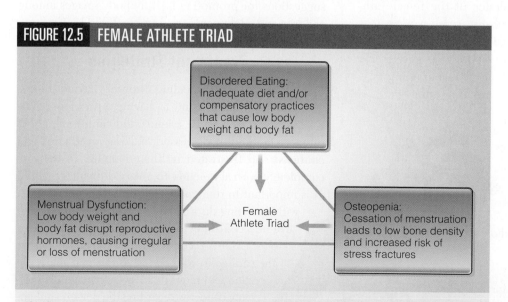

The female athlete triad is a combination of three interrelated conditions: disordered eating (or an eating disorder), menstrual dysfunction, and osteopenia.

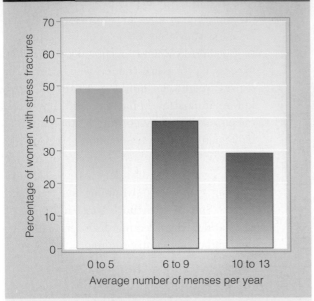

FIGURE 12.6 STRESS FRACTURES AND MENSTRUAL HISTORY IN COLLEGIATE FEMALE DISTANCE RUNNERS

Source: Barrow GW, Saha S. Menstrual irregularity and stress fractures in collegiate female distance runners. American Journal of Sports Medicine. 1988, 16:209–16.

can cause bone loss, which over time weakens bones, resulting in osteopenia. In severe cases, the entire matrix of the bone can begin to deteriorate, leading to osteoporosis. As illustrated in Figure 12.6 a collegiate female runner's risk of stress fractures increases as her menstrual cycle becomes more irregular.[73]

Determining the prevalence of the female athlete triad can be difficult because some women may feel unburdened by the cessation of menstruation and therefore may not be willing to report it. However, these athletes often begin to experience frequent injuries, such as stress fractures, which can draw attention to the fact that they have a problem. It is important for parents, coaches, and health care providers to be aware of the spectrum of disordered eating patterns and eating disorders so that assistance can be provided to athletes who need it.

The risk of stress fractures increases in women when menstrual cycles become irregular or cease.

© Marco Govel/ Shutterstock.com

12-5 HOW CAN EATING DISORDERS BE PREVENTED AND TREATED?

People with eating disorders are at increased risk for serious medical and/or emotional problems. Thus, they need treatment from qualified health professionals. Typically, a treatment team includes mental health specialists who can help address and treat underlying psychological issues, medical doctors who can treat physiological complications, and a dietitian who can recommend healthy food choices. Important treatment goals for individuals with eating disorders are to learn how to enjoy food without fear and guilt and to respond to the physiological cues of hunger and satiety to regulate food intake. For these goals to become possible, specialists help people with eating disorders to recognize and appreciate their own self-worth. While many treatment options are available for people who have eating disorders, the first step is to recognize that there is a problem and seek help.

12-5a Prevention Programs

Educational programs designed to increase awareness and prevent eating disorders in young girls produce varied results. Although it is important to reach children before eating disorders develop, too many prevention programs focus on deterring dangerous eating-disorder behaviors instead of encouraging healthy attitudes toward food, dieting, and body image. To prevent eating disorders from developing, educational strategies must focus on issues related to overall health and self-esteem. Some suggestions for promoting healthy body images among children and adolescents are listed in Table 12.4.

12-5b Treatment Strategies

A person with an eating disorder may not recognize or cannot admit to having a problem. Concerns expressed by friends and family members often go ignored or dismissed, making loved ones feel confused and frustrated by their inability to help—a completely normal reaction to a very difficult situation. It is important to remember that even though a person with an eating disorder may resist help, family and friends continue to play important supportive roles. Experts recommend that they not focus on the eating disorder *per se*, as this may make the person feel more defensive. Instead, expressing concerns regarding the person's unhappiness and encouraging him or her to seek help may be of greatest value.

TABLE 12.4 PROMOTING A HEALTHY BODY IMAGE AMONG CHILDREN AND ADOLESCENTS

- Encourage children to focus on positive body features instead of negative ones.
- Help children understand that everyone has a unique body size and shape.
- Be a good role model for children by demonstrating healthy eating behaviors.
- Resist making negative comments about your own weight or body shape.
- Focus on positive, nonphysical traits such as generosity, kindness, and a friendly laugh.
- Do not criticize a child's appearance.
- Never associate self-worth with physical attributes.
- Prepare a child for puberty in advance by discussing physical and emotional changes.
- Enjoy meals together as a family.
- Discuss how the media can negatively affect body image.
- Avoid using food as a way to reward or punish.

Source: Adapted from Story M, Holt K, Sofka D. Bright futures in practice: Nutrition, 2nd ed. Arlington, VA: National Center for Education in Maternal and Child Health, 2002.

© gajdamak/Shutterstock.com

It is important that the medical team chosen to treat a person with an eating disorder has specific training and expertise in the field of eating disorders. Because eating disorders vary greatly, it is critical that appropriate methodologies and treatment goals are implemented. Treatment goals for people with AN include achieving a healthy body weight, resolving psychological issues (such as low self-esteem and distorted body image), and establishing healthy eating patterns. Because some people with AN are severely malnourished, intensive care provided by inpatient facilities may be necessary.[74] Inpatient facilities are staffed by physicians, nurses, social workers, mental health therapists, dietitians, and other professionals to provide care in all aspects of recovery, including medical support, nutritional programming, and psychological counseling. Like those with AN, people with BN often have nutritional, medical, and psychological issues that must be addressed during recovery as well. In addition to the treatment goals associated with AN, treatment goals for people with BN include the reduction and eventual elimination of bingeing and purging.

Establishing a healthy relationship with food, developing strategies to help resist urges to binge and purge, and maintaining healthy body weight without bingeing and purging are all important.

The sooner a person with an eating disorder gets help, the better the chance of full recovery. Regardless of when that happens, recovery can be a long, trying process. Well-meaning advice such as "just eat" only makes matters worse. It is important for families and friends to know that most people with eating disorders eventually do recover. Treatment often involves counseling for the entire family, which helps everyone heal and move forward in life.

STUDY TOOLS 12

READY TO STUDY? IN THE BOOK, YOU CAN:

☐ Rip out the Chapter Review Card, which includes key terms and chapter summaries.

ONLINE AT WWW.CENGAGEBRAIN.COM, YOU CAN:

☐ Learn about treating eating disorders in a short video.

☐ Interact with figures from the text to check your understanding.

☐ Prepare for tests with quizzes.

☐ Review the key terms with Flash Cards.

13 | Alcohol, Health, and Disease

LEARNING OUTCOMES

13-1 Understand the chemical properties of alcohol, how it is produced, and how it is absorbed and circulated in the body.

13-2 Describe the process of alcohol metabolism.

13-3 Discuss the potential health benefits of alcohol consumption.

13-4 Explain the health risks associated with alcohol consumption.

13-5 Understand the warning signs and consequences of alcohol abuse, and know the current recommendations for responsible alcohol use.

After finishing this chapter go to **PAGE 329** for **STUDY TOOLS.**

13-1 WHAT IS ALCOHOL AND HOW IS IT PRODUCED?

The raising of a celebratory toast is an ancient tradition. Despite a rich history of pageantry and ritual, alcoholic beverages have also brought misery and suffering. Although many people who drink do so without harming themselves or others, long-time abusers know all too well that alcohol can lead to psychological and physical dependency. Today, millions of people around the world are seeking help in their efforts to abstain from alcohol use. There is no easy explanation as to why some people can control their drinking and others cannot. Genetics may predispose some people to alcoholism, but cultural factors play major roles as well.

Scientists have long debated whether alcohol is beneficial or detrimental to health. When it is consumed responsibly, alcohol poses little physical, social, or psychological threat. Studies show that moderate alcohol consumption can even reduce the risk of heart disease in middle-aged and older adults, and it may provide some protection against type 2 diabetes and gallstones.[1] However, when it comes to the benefits of alcohol, more is clearly not better.

If you have ever been around someone who has had too much to drink, you know that alcohol is a drug with mind-altering effects. Indeed, most people act and behave differently when under the influence of alcohol. In excess, alcohol can alter judgment; lead to dependency; and damage the liver, pancreas, heart, and brain. Heavy drinking increases the risk of both accidents and some types of cancer, and maternal alcohol consumption can seriously harm an unborn child. Estimated annual costs associated with alcohol abuse in the United States total more than $180 billion. For these and other reasons, the 2015 Dietary Guidelines for Americans clearly state that people who choose to drink alcoholic beverages should do so sensibly and in moderation.[2] In this chapter, you will learn how alcohol is produced and how it is absorbed and circulated in the body. You will also learn about the many effects that alcohol has on health and disease.

13-1a Alcohol Is Produced through Fermentation

For all of its profound effects on the body, alcohol is a rather simple molecule. From a chemical perspective, the term **alcohol** is used to describe a broad class of organic compounds that have one or more hydroxyl (–OH) groups attached to carbon atoms. There are many types of alcohol, all of which are somewhat similar in terms of chemical structure. Alcohols are quite volatile and tend to be soluble in water. Although the various types of alcohol share certain similarities, there are important differences as well. Whereas **ethanol** (the type of alcohol found in liquor, wine, and beer) can be consumed, most other forms of alcohol are not safe to drink. For instance, methanol which is used to make antifreeze, can be lethal if ingested. While alcohol is not considered a nutrient, it does provide 7 kilocalories per gram. Table 13.1 lists the caloric contents of selected alcoholic beverages.

Alcohol is produced by a process called **fermentation**, which was discovered thousands of years ago. Fermentation occurs when single-cell microorganisms called *yeast* metabolize the sugars found in fruits and grains, ultimately producing ethanol and carbon dioxide. Once the alcohol content of a beverage reaches a certain point, fermentation stops naturally. Alcohol content is often expressed as **alcohol by volume (ABV)**, the percentage of ethanol in a given volume of liquid. For example, if a beverage has an ABV of 14 percent, 100 mL of that beverage contains 14 mL of ethanol. Depending on the strain of yeast used, fermentation usually stops naturally in a barley-based beer when ABV reaches 4–8 percent, and in a grape-based wine when ABV reaches 11–14 percent.

alcohol An organic compound that has one or more hydroxyl (–OH) groups attached to carbon atoms.

ethanol The type of alcohol found in alcoholic beverages.

fermentation The process whereby alcoholic beverages are produced via the addition of yeast to grains or fruit; results in the conversion of sugars into ethanol and carbon dioxide.

alcohol by volume (ABV) The percentage of ethanol in a given volume of liquid.

TABLE 13.1	SERVING SIZES, ENERGY CONTENTS, AND ALCOHOL CONTENTS OF SELECTED ALCOHOLIC BEVERAGES		
Beverage	**Serving Size (oz [mL])**	**Energy (kcal/ serving)[a]**	**Alcohol (g/serving)**
Light beer	12 (355)	103	11
Regular beer	12 (355)	153	13
White wine	5 (148)	121	14
Red wine	5 (148)	125	14
Distilled beverages[b]			
80 proof	1.5 (44)	96	14
90 proof	1.5 (44)	110	16
100 proof	1.5 (44)	123	18
Crème de menthe	1.5 (44)	186	15
Daiquiri	4 (118)	225	14
Whiskey sour	4.5 (133)	217	19
Piña colada	4.5 (133)	245	14

[a] Note that some alcoholic beverages contain energy-yielding nutrients other than alcohol. Therefore, caloric content cannot always be calculated simply by multiplying grams of alcohol by 7 kcal/g.
[b] Distilled beverages include gin, rum, vodka, whiskey, and other liquors.
Source: USDA National Nutrient Database for Standard Reference, Release 23. Nutrient Data Laboratory. Available at: https://ndb.nal.usda.gov/

13-1b Alcohol Absorption

Because alcohol does not require digestion, it is readily absorbed by simple diffusion into the blood. Although some alcohol is absorbed in the stomach, most (80 percent) is absorbed in the small intestine. The actual rate of alcohol absorption is influenced by several factors. For example, alcohol is absorbed more quickly when a person drinks on an empty stomach. The presence of food in the stomach not only dilutes alcohol, it also slows its release into the small intestine. This in turn slows absorption. Although the type of food a person consumes does not have a measurable effect on the rate of alcohol absorption, the alcohol content of a beverage does. In general, the rate of alcohol absorption increases as the alcohol concentration in the beverage increases. However, beverages with high alcohol concentrations (greater than 30 percent ABV) can irritate the stomach lining and increase mucus production. This can delay gastric emptying (the passage of alcohol from the stomach into the small intestine), in turn slowing absorption.[3]

13-1c Alcohol Circulation

Once absorbed, alcohol enters the bloodstream and circulates throughout the body. **Blood alcohol concentration (BAC)** is a unit of measure (grams per deciliter) for the amount of alcohol in the blood. For example, a person with a BAC of 0.10 has one-tenth of a gram of alcohol per deciliter of blood. Within 20 minutes of consuming one standard drink (12 ounces of beer, 5 ounces of wine, or 1.5 ounces of 80-proof distilled liquor), the BAC begins to rise, peaking within 30–45 minutes. Because the rate of alcohol metabolism is generally slower than the rate of alcohol absorption, alcohol accumulates in the blood as more alcoholic beverages are consumed. For this and many other reasons, people who drink should do so slowly and in moderation. In all 50 states, a BAC of 0.08 g/dL is the legal limit for driving (see Figure 13.1).

A process called **distillation** can be used to increase the alcohol contents of some alcohol-containing beverages. During distillation, fermented beverages are heated, which causes the alcohol to become a vaporous gas. The alcohol vapors are collected and cooled until they become liquid again. This process of producing pure liquid alcohol concentrate is used to create distilled alcoholic beverages such as gin, vodka, and whiskey. The alcohol content of a distilled alcoholic beverage, also called *hard liquor*, can be determined from its **proof**. A beverage's proof is twice its alcohol content (ABV). For example, a distilled liquor labeled as 80 proof is 40 percent alcohol by volume.

distillation The process whereby alcohol vapors are condensed and collected to increase alcohol content.

proof A measure of the alcohol content of distilled liquor; proof is twice the percent of alcohol by volume.

blood alcohol concentration (BAC) A unit of measure for the amount of alcohol in the blood, measured in grams per deciliter (g/dL).

Alcohol proof, which is defined as two times the percent of alcohol by volume, is a measure of the amount of alcohol contained in alcoholic beverages.

© mubus7/Shutterstock.com

However, specific laws and penalties related to driving while impaired can vary state-by-state. Although charts such as this help to approximate your BAC levels, they do not take into account other factors such as medication, strength of the alcoholic beverage, and food consumption that can all impact blood alcohol concentration. In addition, alcohol-containing e-cigarettes may also expose people to small amounts of alcohol without their awareness. Because alcohol impairs cognition, decision making, and slows reaction time, the best way to stay under the legal alcohol limit is to abstain from drinking and driving.

Factors Affecting Blood Alcohol Concentration

Alcohol readily disperses in water-filled environments both inside and outside of cells. Because tissues in the body vary greatly in their water contents, some take up alcohol more quickly than do others. For example, muscle tissue contains more water than does adipose tissue, and therefore, it more readily takes up alcohol. In other words, body composition can influence BAC. If two people of similar body weight ingest the same amount of alcohol, the leaner person will likely have a lower BAC. Body size can also influence BAC. A person with a larger body will likely have more blood and body fluids than will someone with a smaller body. The alcohol consumed by a larger person becomes more diluted in the blood, resulting in a lower BAC.

In recent years, caffeinated alcoholic beverages have become very popular. According to many experts, the combination of alcohol and caffeine can be dangerous because the stimulating effects of caffeine can mask intoxication. As a result, an individual may drink more alcohol than intended, which can increase the BAC and lead to hazardous and life-threatening behaviors. In 2010, the U.S. Food and Drug Administration (FDA) warned companies that the addition of caffeine to alcoholic beverages was unsafe. In response to warning letters sent out

FIGURE 13.1 APPROXIMATE BLOOD ALCOHOL CONCENTRATION[a]

Men

Weight (lbs)	1	2	3	4	5	6	7	8	9
100	.04	.08	.11	.15	.19	.23	.26	.30	.34
120	.03	.06	.09	.12	.16	.19	.22	.25	.28
140	.03	.05	.08	.11	.13	.16	.19	.21	.24
160	.02	.05	.07	.09	.12	.14	.16	.19	.21
180	.02	.04	.06	.08	.11	.13	.15	.17	.19
200	.02	.04	.06	.08	.09	.11	.13	.15	.17
220	.02	.03	.05	.07	.09	.10	.12	.14	.15
240	.02	.03	.05	.06	.08	.09	.11	.13	.14

Number of drinks consumed per hour

Women

Weight (lbs)	1	2	3	4	5	6	7	8	9
100	.05	.09	.14	.18	.23	.27	.32	.36	.41
120	.03	.08	.11	.15	.19	.23	.27	.30	.34
140	.03	.07	.10	.13	.16	.19	.23	.26	.29
160	.03	.06	.09	.11	.14	.17	.20	.23	.26
180	.03	.05	.08	.10	.13	.15	.18	.20	.23
200	.02	.05	.07	.09	.11	.14	.16	.18	.20
220	.02	.04	.06	.08	.10	.12	.14	.17	.19
240	.02	.04	.06	.08	.09	.11	.13	.15	.17

Number of drinks consumed per hour

Key

☐ Driving skills impaired   ■ Legally intoxicated

1 drink = 1.5 ounces of 80-proof hard liquor; 5 ounces of wine (12% ABV); 12 ounces of beer (5% ABV).

Estimated blood alcohol concentrations are based on alcohol consumption, body weight, and sex.

[a]Continued consumption of alcohol at these rates (drinks/hour) may result in even higher blood alcohol concentrations.

Source: Adapted from The Pennsylvania Liquor Control Board. Available at: http://www.lcb.state.pa.us/cons/groups/alcoholeducation/documents/form/000340.pdf

The combination of alcohol and caffeine is troubling because the stimulating effects of caffeine can mask intoxication. The U.S. Food and Drug Administration (FDA) has concluded that these products are unsafe.

TABLE 13.2 STAGES OF ALCOHOL INTOXICATION

BAC (g/dL)	Stage of Intoxication	Effects
0.01–0.05	Subclinical	• Behavior that is nearly normal by ordinary observation
0.03–0.12	Euphoria	• Increased sociability, talkativeness, and self-confidence • Decreased inhibition, diminished attention, and altered judgment • Beginning of sensory-motor impairment, loss of efficiency in fine motor skills
0.09–0.25	Excitement	• Emotional instability, loss of critical judgment • Impaired perception, memory, and comprehension • Decreased sensory response, increased reaction time • Reduced visual acuity, peripheral vision, and recovery from flashes of bright light • Impaired sensory-motor coordination and balance • Drowsiness
0.18–0.30	Confusion	• Disorientation, mental confusion, and dizziness • Exaggerated emotional states • Disturbances of vision and perception of color, form, motion, and dimension • Increased pain threshold • Decreased muscle coordination, staggering gait, slurred speech • Apathy, lethargy
0.25–0.40	Stupor	• Extreme lethargy, approaching loss of motor functions • Markedly decreased response to stimuli • Lack of muscular coordination, inability to stand or walk • Vomiting, incontinence • Impaired consciousness, sleep, or stupor
0.35–0.50	Coma	• Complete unconsciousness • Depressed or abolished reflexes • Low body temperature • Impaired circulation and respiration • Possible death
0.45+	Death	Death from respiratory arrest

Source: Adapted from Dubowski KM. Stages of acute alcoholic influence/intoxication. 2006. Available at: http://www.drugdetection.net/PDF%20documents/Dubowski%20-%20stages%20of%20alcohol%20effects.pdf

by the FDA, some manufacturers changed their product formulation to no longer contain caffeine, whereas others removed their products from the marketplace altogether.

13-1d Alcohol Depresses the Central Nervous System

Because it sedates brain activities, alcohol is classified as a central nervous system depressant. This is surprising to many, as consuming small amounts of alcohol often makes people feel euphoric. Pleasant feelings

disinhibition A loss of inhibition.

arise when alcohol selectively depresses the parts of the brain that normally censor social behaviors, aggression, and impulsivity. The inhibitory effect of alcohol on these portions of the brain is called **disinhibition**. In other words, alcohol can cause a temporary loss of inhibitory control, making a person feel relaxed and more outgoing. However, disinhibition also impairs judgment and reasoning. As BAC increases, areas of the brain that control speech, vision, and voluntary muscular movement also become depressed. If drinking continues, a person may lose consciousness, and BAC may reach potentially lethal levels. Table 13.2 lists common responses to different levels of alcohol in the blood.

13-2 HOW IS ALCOHOL METABOLIZED?

Once consumed, alcohol is metabolized into substances that can be safely eliminated from the body. Contrary to popular belief, there is nothing a person can eat or drink to accelerate this process. Small amounts of unmetabolized alcohol are eliminated from the body by the lungs (in expired air), skin (in sweat), and kidneys (in urine). However, the majority of alcohol is chemically broken down (metabolized) by the liver. There is a limit to how much alcohol the liver can metabolize in any given time, however. The average person can metabolize 0.5 ounces (14.8 milliliters) of pure alcohol per hour. In other words, it takes about 1 hour for the body to eliminate the alcohol in a 12-ounce serving of beer. Consequently, if alcohol consumption continues at a rate greater than the liver can metabolize, a person's BAC will increase with time.

13-2a Alcohol Metabolism

During periods of light to moderate drinking, most alcohol is metabolized through a two-step metabolic pathway. The first step requires the enzyme **alcohol dehydrogenase (ADH)**, which is found primarily in the liver. ADH converts ethanol to acetaldehyde, a toxic molecule that the body must quickly break down. If acetaldehyde accumulates, some passes from the liver into the blood, causing the unpleasant side effects associated with heavy drinking (headache, nausea, and vomiting)—otherwise known as a hangover. The second step in the metabolic process, catalyzed by the enzyme **acetaldehyde dehydrogenase (ALDH)**, converts acetaldehyde to acetate.

Genetics Can Influence Alcohol Metabolism
Studies have found that genetic differences (i.e., variation in the DNA) in the enzymes ADH and ALDH can affect a person's ability to metabolize alcohol.[4] For example, a high percentage of Asians have a less functional form of ALDH. As a result, acetaldehyde levels increase quickly, causing dilation of blood vessels, headaches, and facial flushing. Differences in alcohol-metabolizing enzymes have also been found between men and women.[5] Women tend to have lower ADH activity in their stomach cells than do men, which may help explain why some women have lower tolerances for alcohol. These factors may also explain why women are more likely than men to develop alcohol-related health problems.[6]

Tolerance and Cross-Tolerance Over time, chronic alcohol consumption can activate a group of liver enzymes that assist in alcohol metabolism, reducing the amount of time a person remains intoxicated. This is

Cross-tolerance is doubly dangerous.

why some chronic drinkers develop a **tolerance** to alcohol, reducing its effects and allowing them to drink excessive amounts before becoming intoxicated. Because these enzymes also aid in the metabolism of other drugs, a heavy drinker may develop a *cross-tolerance* to these substances as well. In other words, when tolerance to alcohol develops, it is possible to become more tolerant (less responsive) to certain other drugs. In fact, in heavy drinkers, the need to metabolize alcohol can out-compete the need to metabolize other drugs, resulting in drug concentrations that reach dangerously high levels. This is why taking certain drugs in combination with alcohol is not only dangerous, but sometimes deadly.

13-3 DOES ALCOHOL HAVE ANY HEALTH BENEFITS?

Although the detrimental health effects of heavy alcohol consumption have long been known, moderate alcohol consumption may actually benefit health.[7] However, because the amount considered moderate by some might be considered excessive by others, it is important to define exactly what is meant by "moderate alcohol consumption." The 2015 Dietary Guidelines for Americans state that, if alcohol is to be consumed, it should only be consumed in moderation, which is defined as not exceeding one drink per day for women and two drinks per day for men, and alcohol should only be consumed by nonpregnant adults and

alcohol dehydrogenase (ADH) A liver enzyme that converts ethanol to acetaldehyde.

acetaldehyde dehydrogenase (ALDH) A liver enzyme that converts acetaldehyde to acetate.

tolerance A response to chronic drug or alcohol exposure that results in the body being able to metabolize increasingly larger quantities.

those of legal drinking age. In the United States, a standard drink is equivalent to 12 ounces (5 percent alcohol) of beer, 5 ounces (12 percent alcohol) of wine, or 1.5 ounces (80 proof) of distilled liquor, each of which contains approximately 12–14 grams of alcohol. The Dietary Guidelines also stress the importance of recognizing that even moderate consumption of alcoholic beverages contributes calories, which should be incorporated into the calorie limits of a healthy eating pattern. Indeed, alcohol consumption is thought to be a major contributor to unhealthy weight gain.

Many people are surprised to learn that consuming alcohol in moderation can have positive health effects. For example, adults who drink in moderation tend to live longer than do nondrinkers and those who drink heavily.[8] In moderation, alcohol consumption appears to be associated with decreased risks of cardiovascular disease, stroke, gallstones, age-related memory loss, and even type 2 diabetes.[9] However, some evidence suggests that phytochemicals (antioxidants), rather than the ethanol found in alcoholic beverages, may be responsible for these health benefits.[10]

Although moderate alcohol intake may benefit the health of an adult, it provides little (if any) health benefits for young adults. On the contrary, alcohol consumption among young adults is associated with increased risks of injury and death. Studies show that many adults with alcohol-related problems started drinking at a young age. To be more specific, greater frequency of risky drinking behaviors as a young adult is associated with hazardous drinking patterns later in life.[11]

13-3a Moderate Alcohol Consumption and Cardiovascular Disease

The relationship between moderate alcohol consumption and a reduced risk of cardiovascular disease has been confirmed by hundreds of epidemiological studies.[12] More specifically, adults who consume an average of one to two alcoholic drinks a day have a 30–35 percent lower risk of cardiovascular disease than adults who do not consume alcohol (see Figure 13.2).[13] These benefits are lost, however, when alcohol intake becomes excessive. Excessive drinking includes binge drinking, which is defined as four or more drinks for women

resveratrol An antioxidant found in the skin of red grapes.

and five or more drinks for men within 2 hours. The 2015 Dietary Guidelines for Americans define excessive drinking as consumption of eight or more alcohol beverages a week for women and 15 or more alcohol beverages a week for men. Although studies consistently show cardioprotective benefits of light-to-moderate alcohol consumption, the American Heart Association and the Dietary Guidelines discourage individuals from beginning or continuing to drink alcohol for this purpose.[14]

How Moderate Alcohol Consumption May Benefit the Heart You may be wondering how alcohol could lower a person's risk of cardiovascular disease. First, recall that high-density lipoproteins (HDLs) help protect against heart disease and stroke. Studies show that a light-to-moderate daily intake of alcohol is associated with increased HDL-cholesterol levels and therefore may offer some protection from cardiovascular disease.[15] To put this in perspective, one drink per day is associated with a 5 percent increase in HDL, and two to three drinks per day with a 10 percent increase. Light-to-moderate alcohol consumption is also associated with lower circulating triglyceride levels. There is also evidence that alcohol decreases blood clot formation and helps dissolve existing blood clots.[16] Blood clots can block the flow of blood in arteries, and thus, it can cause heart attacks and strokes. Finally, light alcohol consumption may have anti-inflammatory effects, which in turn may lower cardiovascular disease risk.[17]

Additional Health Benefits Associated with Red Wine In addition to health benefits associated with alcohol, biologically active compounds found in grapes and red wine may provide additional health benefits. There is considerable evidence that the antioxidant **resveratrol**, found in the skin of red grapes, helps reduce inflammation and atherosclerosis, thus providing additional cardiovascular disease protection.[18] In fact, there is some speculation that resveratrol might explain the *French paradox*, a term used to describe the observation that French people have a relatively low occurrence of cardiovascular disease despite indulging in diets high in saturated fats and smoking.

A phytochemical found in grapes and red wine called resveratrol may provide some protection from cardiovascular disease.

© Federico Rostagno/Shutterstock.com

FIGURE 13.2 RELATIONSHIP BETWEEN AVERAGE ALCOHOL CONSUMPTION AND CORONARY HEART DISEASE

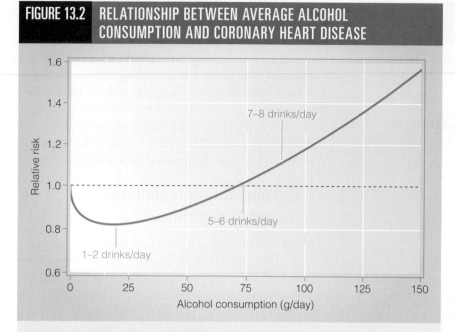

7–8 drinks/day

5–6 drinks/day

1–2 drinks/day

Light-to-moderate alcohol consumption is associated with reduced risk of coronary heart disease. At high levels (7–8 drinks/day), the risk of coronary heart disease increases.

Source: Corrao G, Rubbiati L, Bagnardi V, Zambon A, Poikolainen K. Alcohol and coronary heart disease: A meta-analysis. Addiction. 2000, 95:1505–23. Reprinted with permission of Blackwell Publishing Ltd.

For many adults, alcohol contributes substantially to total daily energy intake (overall 4.7%) which is why even light-to-moderate drinking can result in weight gain.[20]

People who abuse alcohol tend to eat very poor diets, which can quickly lead to nutrient deficiencies or primary malnutrition. When alcohol accounts for more than 30 percent of total energy intake, micronutrient intake is likely to be inadequate.[21] Alcohol can also interfere with digestion, absorption, utilization, and excretion of various nutrients, leading to secondary malnutrition.[22] For example, heavy use of alcohol can interfere with the absorption and metabolism of thiamin, resulting in alcohol-induced brain damage. The impact of heavy alcohol consumption on selected nutrients is summarized in Table 13.3.

13-4 WHAT SERIOUS HEALTH RISKS DOES HEAVY ALCOHOL CONSUMPTION POSE?

In the case of alcohol, more is not better. Although numerous studies suggest that moderate alcohol intake may provide health benefits for middle-aged adults, alcohol is clearly hazardous to one's health when consumed in excess. Over time, heavy drinking can have disastrous consequences on health, leading to impaired nutritional status, liver damage, gout, certain cancers, heart problems, and pancreatic disease (pancreatitis). High-risk drinking also takes a heavy toll on one's family, friends, career, and social life.

13-4a Excessive Alcohol Intake and Nutritional Status

Many studies show that heavy drinking leads to decreased nutrient availability and impaired nutritional status via both primary and secondary malnutrition.[19] Alcoholic beverages have many calories, but very few essential nutrients. In other words, they have low nutrient densities and are sources of empty calories.

13-4b Alcohol Metabolism and the Liver

Chronic alcohol consumption can interfere with normal liver function. For example, it can cause fat to accumulate in and around the liver, a condition referred to as **fatty liver**. This process is reversible if a person stops drinking. In most cases, having a fatty liver does not produce any clinical signs or symptoms. Over time, however, a fatty liver can lead

fatty liver A condition caused by chronic alcohol consumption characterized by the accumulation of triglycerides in the liver.

From left to right, normal, fatty, and cirrhotic livers.

Arthur Glauberman/Arthur Glauberman/Science Source

TABLE 13.3 IMPACT OF HEAVY ALCOHOL CONSUMPTION ON NUTRITIONAL STATUS

Nutrient	Impact of Alcohol	Health Consequences
Water-Soluble Vitamins		
Thiamin	• Impaired absorption • Increased urinary loss • Altered metabolism • Reduced storage	• Paralysis of eye muscles • Degeneration of nerves with loss of sensation in lower extremities • Loss of balance, abnormal gait • Memory loss, psychosis
Vitamin B$_6$	• Decreased activation • Increased urinary loss • Displacement from binding protein	• Anemia • Impaired metabolic reactions involving amino acids
Vitamin B$_{12}$	• Impaired absorption	• Possible neurological damage, numbness and tingling sensations in the arms and legs
Folate	• Decreased absorption • Increased urinary loss	• Anemia • Diarrhea
Riboflavin	• Impaired absorption • Decreased activation	• Deficiency not typical in isolation but occurs in conjunction with other B vitamin deficiencies
Fat-Soluble Vitamins		
Vitamin A	• Decreased conversion to retinal • Reduced absorption	• Impaired vision, night blindness • Liver disease
Vitamin D	• Decreased activation • Impaired absorption	• Bone fractures and osteoporosis
Vitamin E	• Impaired absorption	• Nerve damage • Tunnel vision • Fragility of cell membranes
Vitamin K	• Impaired absorption	• Bruising and prolonged bleeding
Minerals		
Magnesium	• Increased urinary loss	• Muscle rigidity, cramps, and twitching • Irregular cardiac function • Possible hallucinations
Iron	• Increased storage • Impaired absorption • Increased loss in feces	• Both iron deficiency and overload are possible
Zinc	• Decreased absorption • Increased urinary loss • Impaired utilization	• Altered taste, loss of appetite • Impaired wound healing • Night blindness

alcoholic hepatitis Inflammation of the liver caused by obstructed blood flow.

cirrhosis A condition characterized by the presence of scar tissue in the liver.

to the development of a more serious condition called **alcoholic hepatitis**. Alcoholic hepatitis is characterized by inflammation and swelling of the liver, which can lead to medical complications. In addition to alcoholic hepatitis, about 10 percent of heavy drinkers develop a condition called **cirrhosis** of the liver. In cirrhosis, healthy liver tissue becomes damaged and is replaced with scar tissue. If the damage worsens, the ability of the liver to carry out normal physiologic functions becomes impaired. Cirrhosis can eventually lead to liver failure and even death.

13-4c Long-Term Alcohol Abuse and Cancer Risk

The association between alcohol consumption and cancer has been studied extensively. Conclusive evidence shows that heavy drinking (at a rate of approximately four drinks—50 grams pure alcohol—a day) increases a person's risk of developing certain types of cancer, especially those of the mouth, esophagus, colon, liver, and breast.[23] Even a low-to-moderate intake of alcohol may increase a woman's risk of cancer, particularly breast cancer.[24]

Almost 50 percent of cancers of the mouth and esophagus are associated with heavy drinking.[25] It is not clear how alcohol increases the risk of cancer, although it appears that alcohol may act as both a *carcinogen* and a *cocarcinogen*. Carcinogens are compounds that initiate the formation of cancer, whereas cocarcinogens enhance the carcinogenicity of other cancer-causing chemicals. For example, because cigarettes contain multiple carcinogens, heavy drinkers who smoke are at particularly high risk for developing cancers of the mouth, esophagus, and trachea—it is possible that alcohol interacts with cigarette smoke to make it more dangerous.

© Photographee.eu/Shutterstock.com

The combination of smoking and drinking greatly increases risk of cancers of the mouth and esophagus.

Scientists hypothesize that poor nutritional status associated with heavy drinking may also lead to the development of certain cancers. For example, alcohol interferes with folate availability, which may compromise cell division and DNA repair.[26] Similarly, reduced levels of iron, zinc, vitamin E, certain B vitamins, and vitamin A—all nutrients that are commonly deficient in heavy drinkers—have been experimentally linked to certain types of cancer. Because adequate nutrient intake may offer protection against cancer, an overall dietary inadequacy caused by chronic alcohol intake may weaken a person's natural defense mechanisms.

13-4d Alcohol Abuse and the Cardiovascular System

The relationship between alcohol consumption and cardiovascular health is complex. Whereas numerous studies suggest that moderate intakes of alcohol provide some protection against cardiovascular disease, heavy alcohol consumption is clearly detrimental. Thus, it is important to balance the benefits and risks of alcohol intake (Figure 13.3). Recall that the 2015 Dietary Guidelines for Americans define excessive drinking as consumption of more than three drinks on any day or more than seven drinks per week for women, more than four drinks on any day or more than 14 drinks per week for men, or consumption within 2 hours of four or more drinks for women and five or more drinks for men.[27] Cardiovascular consequences associated with long-term heavy drinking include hypertension, stroke, and irregular cardiac function. These detrimental effects can begin to develop in as few as five years.[28]

One of the most serious cardiovascular consequences associated with heavy alcohol consumption is **alcoholic cardiomyopathy**. Over time, chronic exposure to high levels of alcohol causes the heart muscle to weaken. Because the weakened heart cannot contract forcibly, blood flow to vital organs, such as the lungs, liver, kidneys, and brain, is reduced. If drinking continues, alcoholic

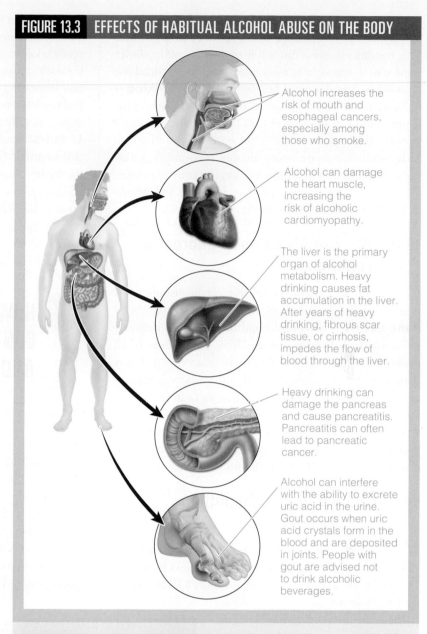

FIGURE 13.3 EFFECTS OF HABITUAL ALCOHOL ABUSE ON THE BODY

Alcohol increases the risk of mouth and esophageal cancers, especially among those who smoke.

Alcohol can damage the heart muscle, increasing the risk of alcoholic cardiomyopathy.

The liver is the primary organ of alcohol metabolism. Heavy drinking causes fat accumulation in the liver. After years of heavy drinking, fibrous scar tissue, or cirrhosis, impedes the flow of blood through the liver.

Heavy drinking can damage the pancreas and cause pancreatitis. Pancreatitis can often lead to pancreatic cancer.

Alcohol can interfere with the ability to excrete uric acid in the urine. Gout occurs when uric acid crystals form in the blood and are deposited in joints. People with gout are advised not to drink alcoholic beverages.

Chronic alcohol intake can lead to health problems such as cancer, heart disease, liver disease, pancreatic disease, and gout.

cardiomyopathy can ultimately result in heart failure. For reasons that are not clear, women are more susceptible to alcoholic cardiomyopathy than are men.

A relationship between heavy drinking and hypertension is also well established: as alcohol intake increases, so does blood pressure.[29] This may be caused in part by alcohol-related liver damage. When a heavy drinker refrains from drinking alcohol, blood pressure often returns to normal. High blood pressure is not the

alcoholic cardiomyopathy
A serious condition whereby the heart muscle weakens in response to chronic alcohol consumption.

only concern, however. Alcohol can also trigger an irregular heartbeat, a condition called **cardiac arrhythmia**. This condition is sometimes referred to as *holiday heart syndrome* because it is associated with binge drinking on weekends and around holidays.[30] **Binge drinking** is defined as the consumption of four or more drinks for women and five or more drinks for men over a 2-hour period. Although its cause is not entirely clear, cardiac arrhythmia due to excessive alcohol consumption may occur as a response to disturbances in electrolyte balance. Whatever its cause, alcohol-induced cardiac arrhythmia can lead to sudden cardiac arrest (heart attack).

13-4e Alcohol Abuse and Pancreatic Function

In addition to its effects on the cardiovascular system, excessive drinking can also damage the pancreas. **Pancreatitis** is a painful condition characterized by inflammation of the pancreas. Up to 90 percent of all cases of chronic pancreatitis are associated with heavy drinking. A chronically high alcohol intake (consuming at least seven drinks a day for more than five years) increases a person's risk of developing this condition. Pancreatitis makes it difficult for the pancreas to release pancreatic juice, which in turn impairs digestion and nutrient absorption in the small intestine.[31]

14-4f Alcohol Abuse, Cognition, and Dementia

While those who drink alcohol responsibly have a lower risk of dementia, heavy consumption appears to be detrimental to brain function. Studies report a faster overall decline in many facets of cognition, particularly memory, in middle-aged adults who consumed 2–3 drinks daily over a 10-year period.[32] While it may not be surprising to learn that alcohol abuse significantly impacts

brain function in older adults, it is troubling to learn that alcohol consumption is particularly harmful to the developing adolescent brain. In fact, one risk factor identified for early-onset dementia is frequent alcohol intoxication beginning in adolescence.[33] A recent report by the National Survey on Drug Use and Health stated that approximately 23 percent of those between the ages of 12 and 20 years reported consuming alcohol within the past month.[34] Underage drinking is a serious problem in the United States, particularly on college campuses. In addition to the direct impact of heavy alcohol consumption on the brain, alcohol can both mask and exacerbate mental illness. Although alcohol *per se* does not cause mental illness, nearly one-third of alcohol abusers have a mental illness.[35]

 13-5 HOW DOES ALCOHOL ABUSE CONTRIBUTE TO INDIVIDUAL AND SOCIETAL PROBLEMS?

Although the majority of people who drink alcohol do so responsibly and without negative consequences, the line between alcohol use and abuse is often blurry. When alcohol is the cause of significant problems in a person's life, drinking is problematic. **Alcohol use disorders (AUDs)**, a spectrum of behaviors that encompasses both recurrent excessive drinking and alcoholism, arises when drinking leads to negative physical, legal, and/or social consequences.[36] AUDs are not discriminatory; they affect men and women, young and old, professional and nonprofessional, and rich and poor.

cardiac arrhythmia An irregular heartbeat that can be caused by a high intake of alcohol.

binge drinking The consumption of four or more drinks for women and five or more drinks for men over a 2-hour period.

pancreatitis A painful condition characterized by inflammation of the pancreas.

alcohol use disorders (AUDs) A spectrum of disorders that encompasses both recurrent excessive drinking and alcoholism.

© william casey/Shutterstock.com

Though they are frequently misdiagnosed, the occurence of AUDs throughout the United States is quite high. Annual costs associated with AUD-related hospitalization, premature death, and alcohol-related illness and injury exceed $249 billion.[37] Habitual alcohol abuse affects virtually every aspect of a person's life, including family relationships and performance at work and at school. Arrests related to public intoxication, driving while under the influence of alcohol, and other alcohol-related crimes often result in substantial legal problems. Despite the myriad problems associated with excessive drinking, many individuals with AUDs are not able to stop or limit their alcohol intake because alcohol use can lead to both physical and emotional dependence.

13-5a Treating Alcohol Abuse

Despite the devastating and pervasive consequences of alcohol abuse, many alcohol-dependent individuals are adept at masking their addiction. Sometimes, it takes tragic circumstances such as an accident or job dismissal to alert family and friends to the problem. Furthermore, many clinicians receive little training in handling substance abuse and therefore may be reluctant to screen patients for drinking-related behaviors and health problems.[38]

Most mental health professionals recognize alcoholism as a complex and widespread disease that can be treated—but not cured. One of the oldest and most reputable organizations for the treatment of alcoholism is **Alcoholics Anonymous® (AA)**. AA offers a 12-step program that provides fellowship and support for individuals who want to achieve sobriety and stay sober. More than 2 million people consider themselves lifelong members of AA. Other organizations such as **Al-Anon®** and **Alateen®** provide support for family members and friends of alcoholics. While Al-Anon is open to all affected family members and friends, Alateen caters specifically to teenage children of alcoholics.

13-5b Alcohol Use on College Campuses

Numerous studies have documented the extent to which alcohol use affects college students.[39] Alcohol abuse, rampant on many college campuses, is associated with a wide range of negative consequences such as vandalism, violence, acquaintance rape, unprotected sex, and death.[40] To give perspective to the magnitude of alcohol-related problems, data from the National Institute on Alcohol Abuse and Alcoholism indicate that four out of five college students drink alcohol, and that approximately 1,825 college students between the ages of 18 and 24 years of age die each year from alcohol-related unintentional injuries.[41] Individuals between 21 and 29 years of age have the highest percentages of fatal alcohol-related traffic accidents.[42] If you find yourself wondering if you or someone you know has a problem with alcohol, you can use the self-assessment tool shown in Table 13.4.

Although 20 percent of college students report that they do not drink alcohol, about one-half describe themselves as binge drinkers.[43] Indeed, college students are more likely to binge drink than their peers who do not attend college.

HANK MORGAN/Getty Images

Alcoholics Anonymous is a fellowship of men and women who share their experiences to help themselves and others recover from alcoholism and to maintain sobriety.

Alcoholics Anonymous® (AA)
An organization that offers a 12-step program that provides fellowship and support for individuals who want to achieve sobriety and stay sober.

Al-Anon® An organization that provides support for family members and friends of alcoholics.

Alateen® An organization that provides support for teenage children of alcoholics.

TABLE 13.4 THE JOHNS HOPKINS UNIVERSITY HOSPITAL ALCOHOL SCREENING QUIZ

1. Do you lose time from work due to drinking?
2. Is drinking making your home life unhappy?
3. Do you drink because you are shy with other people?
4. Is drinking affecting your reputation?
5. Have you ever felt remorse after drinking?
6. Have you had financial difficulties as a result of drinking?
7. Does your drinking make you careless of your family's welfare?
8. Do you turn to inferior companions and environments when drinking?
9. Has your ambition decreased since you started drinking?
10. Do you crave a drink at a definite time daily?
11. Do you want a drink the next morning?
12. Does drinking cause you to have difficulty sleeping?
13. Has your efficiency decreased since drinking?
14. Is drinking jeopardizing your job or business?
15. Do you drink to escape from worries or trouble?
16. Do you drink alone?
17. Have you ever had a loss of memory as a result of drinking?
18. Has your physician ever treated you for drinking?
19. Do you drink to build up your self-confidence?
20. Have you ever been to a hospital or institution on account of drinking?

If you answered three or more of the questions with a "Yes," there is a strong possibility that your drinking patterns are detrimental to your health and that you may be alcohol dependent. Under these circumstances, you should consider getting an evaluation of your drinking behavior by a health care professional.

Source: Office of Health Care Programs, Johns Hopkins University Hospital. Available at: http://www.alcohol-addiction-info.com/Alcohol_Addiction_Self_Assessment_Tools.html

In fact, most college students who binge drink report that these behaviors started after they entered college.[44] Students who binge drink are more likely than those who do not to miss class, have lower academic rankings, get in trouble with campus law enforcement, and drive while intoxicated.[45] Clearly, more research is needed to understand how the collegiate environment encourages risky drinking behaviors and what can be done to help prevent alcohol-related problems on university campuses.

Colleges and Universities Must Create a Culture that Discourages High-Risk Drinking
The issue of underage drinking on university campuses continues to perplex academic administrators, who have come to little agreement about how to address this problem effectively.

Alcohol abuse among college students is a problem that affects the entire campus community. Clearly, traditional educational approaches that focus on distribution of pamphlets and the provision of one-day alcohol awareness programs are not enough—colleges and universities must work together with communities to create a culture that discourages high-risk drinking.[46] To this end, many colleges are developing and implementing comprehensive programs designed to curb the problem of alcohol abuse in unique ways.[47] For example, by increasing campus recreational activities that do not involve drinking, many schools hope to dissuade students from habitually going to bars and clubs. Community-oriented initiatives such as discouraging price discounts on alcohol, two-for-one drink specials, inexpensive beer pitcher sales, and other types of happy hour promotions among campus-adjacent bars are also being implemented.

Binge drinking is a major problem among college students.

Recommendations for Responsible Alcohol Use Both health benefits and risks are associated with alcohol consumption among adults. Although most Americans who drink alcohol do so safely and responsibly, this is certainly not the case for everyone—especially adolescents and young adults. For this reason, one of the goals identified in Healthy People 2020 is to protect health, safety, and quality of life for all—and especially for children—by reducing alcohol abuse.[48] In 1998, the number of alcohol-related deaths was 5.9 per 100,000. Healthy People 2020 aims to reduce the number of alcohol-related deaths to 4 per 100,000.

The 2015 Dietary Guidelines for Americans also addresses the harmful effects of alcohol when consumed in excess; they state that adults who consume alcohol should do so in moderation. Again, moderation is defined as up to one drink per day for women and up to two drinks per day for men. Although these guidelines are generally safe, some people should avoid alcohol completely, such as those younger than the legal drinking age, those who have difficulty restricting their alcohol intakes, those taking medications that can interact with alcohol, and those with specific medical conditions. And, as previously stated, mixing alcohol and caffeine is not considered safe and should be avoided. Furthermore, all people should abstain from drinking when driving, operating machinery, or taking part in activities that require attention, skill, or coordination. Women who are or may become pregnant should not drink. This is because even moderate drinking during pregnancy can cause behavioral and developmental problems for the baby. Excessive drinking during pregnancy can result in serious problems for the baby, including physical malformations and brain damage. The MyPlate website recommends that breastfeeding women can have an occasional alcoholic beverage if they are cautious and follow these guidelines.

- Wait until the baby has a routine breastfeeding pattern, around 3 months of age at a minimum, before consuming alcohol.

- Wait at least 4 hours after having a single alcoholic drink before breastfeeding.

- Or, express breast milk before having a drink and use the expressed milk to feed the infant later.

Because breastfeeding provides many benefits to both mothers and infants, women should not stop breastfeeding altogether just because they would like to have an occasional drink.[49]

Consuming alcohol while pregnant is never advised.

STUDY TOOLS 13

READY TO STUDY? IN THE BOOK, YOU CAN:

- ☐ Rip out the Chapter Review Card, which includes key terms and chapter summaries.

ONLINE AT WWW.CENGAGEBRAIN.COM, YOU CAN:

- ☐ Learn about consequences to lowering the drinking age in a short video.
- ☐ Explore the effects of alcohol in an animation.
- ☐ Interact with figures from the text to check your understanding.
- ☐ Prepare for tests with quizzes.
- ☐ Review the key terms with Flash Cards.

14 | Keeping Food Safe

LEARNING OUTCOMES

14-1 Discuss the infectious and noninfectious agents that cause foodborne illness.

14-2 Understand how noninfectious agents can cause foodborne illness.

14-3 Describe how food manufacturers prevent contamination.

14-4 Take steps to reduce foodborne illness at home and when eating out.

14-5 Understand how to reduce foodborne illness while traveling.

After finishing this chapter go to **PAGE 348** for **STUDY TOOLS.**

14-1 WHAT CAUSES FOODBORNE ILLNESS?

Most Americans are fortunate to have an abundant supply of healthful food. Still, there are times when even healthy food can make a person sick. Although most people understand the importance of avoiding food that is spoiled or stored improperly, food that appears, smells, and tastes safe to eat can still cause illness. Every effort is made to ensure that food is healthful and not harmful, but food safety remains an important public health concern. The U.S. Centers for Disease Control and Prevention (CDC) estimates that 48 million Americans, or one out of every six people, suffer from foodborne diseases annually. Of them, some 128,000 are hospitalized and 3,000 die.[1] In this chapter, you will learn about common disease-causing agents and how to avoid them, so that you can minimize your risk of illness.

You are exposed to thousands of microscopic organisms (microbes) every day. Microbes populate the world you live in, and many even serve useful purposes. For example, microbes in your GI tract assist in food digestion, produce vitamins and other health-promoting substances, and may even help prevent some diseases. Other microbes are pathogenic (disease causing), however. These microbes can make you sick, and consumption of them in foods and beverages is the main cause of foodborne illness. Generally speaking, a **foodborne illness** is a disease caused by the ingestion of unsafe food. Foodborne illness is sometimes referred to as *food poisoning*. You can contract a foodborne illness by ingesting either infectious or noninfectious agents. An **infectious agent** (or **pathogen**) is a living microorganism such as a bacterium, virus, mold, fungus, or parasite. A **noninfectious agent** is an inert (nonliving) substance such as a nonbacterial toxin, a chemical residue from processing, pesticides, or antibiotics; or a physical hazard such as glass or plastic.

foodborne illness A disease caused by the ingestion of unsafe food.

infectious agent (or **pathogen**) A living microorganism that can cause illness.

noninfectious agent An inert (nonliving) substance that can cause illness.

14-1a Basic Microbiology Related to Food Safety

Although most foodborne illnesses are caused by microbes, not all microbes are harmful. In fact, even related pathogenic microbes can cause vastly different types of illnesses (or have no effect at all). Therefore, to understand how to prevent foodborne illness, you must first have a basic knowledge of microbiology, and you must understand how various organisms can make you sick.

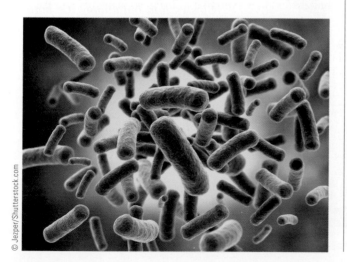

To begin with, a group of closely related microorganisms can vary in their genetics, or DNA code. Each microbial variety is called a **serotype** (or **strain**). Whereas some serotypes are harmless, others cause disease. For example, some serotypes of the bacterium *Escherichia coli* (*E. coli*) live safely in your GI tract, whereas *E. coli* O157:H7 can cause severe illness.[2] Moreover, different serotypes of a single type of pathogenic bacterium can cause illness in different ways. Some pathogenic *E. coli* serotypes cause mild intestinal discomfort within one to three days, while others cause more severe symptoms that take up to eight days to develop. The time elapsed between the consumption of a contaminated food and the emergence of sickness is called the **incubation period**. Because different pathogens (and serotypes thereof) have different incubation periods, health care providers often consider this variable when trying to determine what caused a person to become sick. Quickly identifying which serotype of a microorganism is involved in a foodborne illness outbreak can prevent additional people from becoming ill. So, the FDA has established an

serotype (or **strain**) A specific genetic variety of an organism.

incubation period The time elapsed between the consumption of a contaminated food and the emergence of sickness.

international network of laboratories that works to collectively produce detailed information about the genetics of the most common foodborne pathogens. This project, called GenomeTrakr, will undoubtedly contribute to better health for the U.S. population for years to come. Some of the most common infectious agents and their incubation periods are listed in Table 14.1. The methods by which these agents cause illness are described next.

14-1b Preformed Toxins

Some pathogenic organisms produce toxic substances while they are growing in food. Such a substance is called a **preformed toxin** because it already exists in the food before the food is eaten. Preformed toxins typically cause rapid reactions (occurring in 1 to 6 hours) such as nausea, vomiting, diarrhea, and sometimes, neurological damage.

Staphylococcus aureus (S. aureus) One bacterium that produces preformed toxins is *Staphylococcus aureus* (*S. aureus*). Although *S. aureus* is a common bacterium found on the skin and in the noses of up to 25 percent of healthy people and animals, it sometimes causes illness. In fact, this bacterium causes nearly 250,000 foodborne illnesses in the United States every year.[3] Foods commonly contaminated with *S. aureus* include raw and undercooked meat and poultry, cream-filled pastries, and unpasteurized dairy products. Symptoms of *S. aureus* infection include a sudden onset of severe nausea and vomiting, diarrhea, and abdominal cramps. These symptoms typically occur within 1 to 6 hours of the contaminated food's consumption.

An antibiotic-resistant strain of *S. aureus* called **methicillin-resistant *Staphylococcus aureus* (MRSA)** has received considerable attention from public health officials. MRSA was identified more than 40 years ago, and, at the time, it was believed to spread only through direct contact with an infected

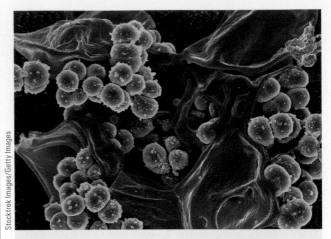

This scanning electron micrograph shows the antibiotic-resistant *Staphylococcus aureus*. In England, *S. aureus* is commonly known as Golden Staph, because *Staphylococcus aureus* translates literally to "golden cluster seed."

person—usually in a hospital environment. Scientists now know, however, that MRSA can spread through the sharing of towels and equipment in athletic and school facilities and that infection can occur via consumption of MRSA-contaminated foods. To date, there has only been one documented outbreak of foodborne illness caused by MRSA in the United States. In that case, researchers believe that an infected food preparer transmitted the bacterium to coleslaw. Three family members consumed the coleslaw and subsequently became ill.[4] Because MRSA cannot be treated effectively with antibiotics, public health officials are carefully monitoring for outbreaks of this bacterium—especially in regard to foodborne illness.

Clostridium botulinum (C. botulinum) Another bacterium that produces preformed toxins, *Clostridium botulinum* (*C. botulinum*), is found mainly in inadequately processed low-acid, home-canned foods such as green beans. A person can also become infected with this microbe by eating improperly canned commercial foods. *C. botulinum* infection causes an illness called **botulism**.

Improperly canned low-acid foods can contain live *Clostridium botulinum* bacteria, which can cause botulism.

preformed toxin A toxic substance that already exists in a food before the food is eaten.

methicillin-resistant *Staphylococcus aureus* (MRSA) An antibiotic-resistant strain of *S. aureus* that has received considerable attention from public health officials.

botulism The foodborne illness caused by *Clostridium botulinum*.

TABLE 14.1 INFECTIOUS AGENTS OF FOODBORNE ILLNESS, FOOD SOURCES, AND SYMPTOMS OF INFECTION

Organism	Incubation Period	Duration of Illness	Associated Foods	Signs and Symptoms
Bacteria				
Campylobacter jejuni	2–5 days	2–10 days	Raw and undercooked poultry, untreated water, and unpasteurized milk	Diarrhea (often bloody), abdominal cramping, nausea, vomiting, fever, and fatigue
Clostridium botulinum	12–72 hours	From days to months	Home-canned foods with low-acid contents, improperly canned commercial foods, and herb-infused oils	Vomiting, diarrhea, blurred vision, drooping eyelids, slurred speech, dry mouth, difficulty swallowing, and weak muscles
Clostridium perfringens	8–16 hours	24–48 hours	Raw and undercooked meats, gravy, and dried foods	Abdominal pain, watery diarrhea, vomiting, and nausea
Escherichia coli O157:H7	1–8 days	5–10 days	Raw and undercooked meats, raw fruits and vegetables, unpasteurized milk and juice, and contaminated water	Nausea, abdominal cramps, and severe diarrhea (often bloody)
Escherichia coli (enterotoxigenic)	1–3 days	Variable	Water and food contaminated with human feces	Diarrhea, abdominal cramps, and some vomiting
Listeria monocytogenes	9–48 hours for GI symptoms, 2–6 weeks for invasive disease	Variable	Raw and inadequately pasteurized dairy products and ready-to-eat luncheon meats and frankfurters	Fever, muscle aches, nausea, diarrhea, and premature delivery, miscarriage, or stillbirth
Salmonella	1–3 days	4–7 days	Raw poultry, eggs, and beef; fruit and alfalfa sprouts; and unpasteurized milk	Diarrhea, fever, abdominal pain, and severe headache
Shigella	24–48 hours	4–7 days	Raw and undercooked foods and water contaminated with human fecal material	Fever, fatigue, watery or bloody diarrhea, and abdominal pain
Staphylococcus aureus	1–6 hours	24–48 hours	Improperly refrigerated meats, potato and egg salads, and cream pastries	Severe nausea and vomiting, diarrhea, and abdominal pain
Vibrio cholerae	1–7 days	2–8 days	Contaminated water and many undercooked foods	Diarrhea and vomiting
Viruses				
Hepatitis A virus	15–50 days	2–12 weeks	Mollusks (oysters, clams, mussels, scallops, and cockles)	Jaundice, fatigue, abdominal pain, loss of appetite, nausea, diarrhea, and fever
Norovirus	12–48 hours	12–60 hours	Raw and undercooked shellfish and contaminated water	Nausea, vomiting, diarrhea, abdominal pain, headache, and fever
Parasites				
Trichinella (worm)	1–2 days for initial symptoms; others begin 2–8 weeks after infections	Months	Raw and undercooked pork and meats of carnivorous animals	Acute nausea, diarrhea, vomiting, fatigue, fever, and abdominal pain
Giardia intestinalis (protozoan)	1–2 weeks	Days to weeks	Contaminated water and many uncooked foods	Diarrhea, flatulence, and abdominal pain
Molds				
Aspergillus flavus	Days to weeks	Weeks to months	Wheat, flour, peanuts, and soybeans	Liver damage

Source: Adapted from Centers for Disease Control and Prevention. Diagnosis and management of foodborne illnesses: A primer for physicians and other health care professionals. Morbidity and Mortality Weekly Reports. 2004; 53:1–33. Available at: http://www.cdc.gov/mmwr/PDF/rr/rr5304.pdf. Murano P. Understanding food science and technology. Thompson/Wadsworth, 2003.

When improperly dried, the *Aspergillus* mold sometimes found in peanuts can produce aflatoxin, a dangerous poison.

14-1c Enteric (Intestinal) Toxins

In contrast to those that produce preformed toxins, some organisms produce harmful toxins after they enter the GI tract. Such a toxin is called an **enteric toxin** (or **intestinal toxin**). Enteric toxins draw water into the intestinal lumen, resulting in diarrhea. Although the symptoms of enteric toxin infections vary, incubation periods generally span one to five days—substantially longer than those of most preformed toxins.

Noroviruses **Norovirus**, previously called Norwalk or Norwalk-like virus, is an example of a pathogen that produces enteric toxins. Norovirus is the leading cause of disease outbreaks from contaminated food in the United States. Symptoms of norovirus infection

Mild cases of botulism cause vomiting and diarrhea, and severe cases can cause double vision, blurred vision, drooping eyelids, slurred speech, difficulty swallowing, dry mouth, and muscle weakness. In the most severe cases, botulism can cause paralysis, respiratory failure, and death. In 2007, the CDC reported an outbreak of eight cases of botulism in Indiana, Texas, and Ohio.[5] All infected persons had consumed a particular brand of hot dog chili sauce that was quickly recalled by the manufacturer. Although all of the infected individuals recovered, each endured several days of painful gastrointestinal (GI) symptoms. Because the botulism toxin is destroyed by high temperatures, some experts recommend that home-canned foods be boiled for 10 minutes before they are consumed.

Aspergillus and Aflatoxin Although most molds that grow on foods such as cheese and bread are not dangerous, some produce unsafe preformed toxins. An example is **aflatoxin**, which is produced by the *Aspergillus* mold found on some agricultural crops (such as peanuts, rice, and wheat). Aflatoxin is occasionally found in the milk of animals fed with contaminated feed. The consumption of aflatoxin is of great concern because this toxin can cause liver damage and cancer, and it is often fatal. Although many agricultural practices, such as sufficient drying, are used to minimize the risk of aflatoxin contamination in the United States, consumption of this toxin remains a significant public health issue throughout many regions of the world.

aflatoxin A toxin produced by the *Aspergillus* mold that is found on some agricultural crops.

enteric toxin (or **intestinal toxin**) A toxic substance produced by an organism after it enters the gastrointestinal tract.

norovirus An infectious pathogen that produces enteric toxins and often causes foodborne illness.

Using Technology to Track Food Safety

Are you interested in new food safety alerts and tips issued by the federal government? If so, you can download a free widget from the U.S. Food and Drug Administration (FDA) website (http://www.foodsafety.gov/widgets/). This low-maintenance application provides instant updates on the latest food safety recalls and offers useful food safety tips appropriate to the season. With the aid of this widget, you might learn that your favorite brand of chicken noodle soup has been recalled because of suspected contamination, or how to cook your turkey to the appropriate internal temperature for Thanksgiving. Keeping abreast of food safety tips will not only help keep you safe, but it will also provide a constant reminder that issues related to food safety can affect you every day.

Source: U. S. Department of Health and Human Services—http://www.foodsafety.gov/widgets/

typically include nausea, vomiting, diarrhea, and abdominal cramping. A person with a norovirus infection may also develop a low-grade fever, chills, headaches, muscle aches, and/or a general sense of fatigue. Symptoms usually begin one to two days after ingestion of the contaminated food. In 2008, a norovirus outbreak occurred on three college campuses. This outbreak resulted in approximately 1,000 cases of reported illness and 10 hospitalizations, prompting the closure of one of the three campuses. Although the cause of the related outbreaks (or even if they were traced to a common food) was never determined, the outbreaks eventually subsided. Still, because extensive opportunities for transmission created by shared living and dining areas put college campuses at particularly high risk for norovirus outbreaks, many nutritionists and administrators remain on high alert.[6] Foods that are commonly involved in outbreaks of norovirus illness are leafy greens (e.g., lettuce), fresh fruits, and shellfish (e.g., oysters). But, any food that is served raw or handled after being cooked can become contaminated.

Enterotoxigenic E. coli Some forms of *E. coli*, called **enterotoxigenic E. coli**, produce enteric toxins. The consumption of food or water contaminated with these bacteria can cause severe GI upset; symptoms including diarrhea, abdominal cramps, and nausea tend to occur within 6 to 48 hours after consumption. Foods and beverages typically contaminated with these forms of *E. coli* include uncooked vegetables, fruits, raw and undercooked meats and seafood, unpasteurized dairy products, and untreated tap water. In 2015, a foodborne *E. coli* outbreak associated with Chipotle Mexican Grill restaurants infected dozens of people from several states, mostly in Washington and Oregon. The responsible food (at the time this book was being revised) had not been identified.

Enterotoxigenic *E. coli* is the primary cause of a condition often referred to as *traveler's diarrhea*, a clinical syndrome resulting from the microbial contamination of food or water during or shortly after travel. If you have ever had severe diarrhea and/or abdominal cramps while traveling, you might have experienced this type of foodborne illness.

14-1d Enterohemorrhagic Pathogens

Some pathogens invade the cells of the intestine, seriously irritating the mucosal lining and causing fever, severe abdominal discomfort, and bloody diarrhea. These types of pathogens are called **enterohemorrhagic**, literally meaning that they cause bloody diarrhea and intestinal inflammation. Examples of enterohemorrhagic pathogens include *Salmonella* and an especially dangerous serotype of *E. coli* called *E. coli* O157:H7. Incubation periods for enterohemorrhagic pathogens generally span one to eight days.

Salmonella *Salmonella* is one of the most common causes of foodborne illness in the United States causing around 1 million illnesses, 19,000 hospitalizations and 380 deaths each year. Most persons infected with *Salmonella* develop diarrhea, fever, and abdominal cramps 12 to 72 hours after they are infected. It is typically found in raw poultry, eggs, beef, improperly washed fruit, alfalfa sprouts, and unpasteurized milk. The incubation period for this organism is one to three days, and symptoms can include severe GI upset

> **enterotoxigenic E. coli** A form of *E. coli* that produces enteric toxins.
>
> **enterohemorrhagic** Causing bloody diarrhea and intestinal inflammation.

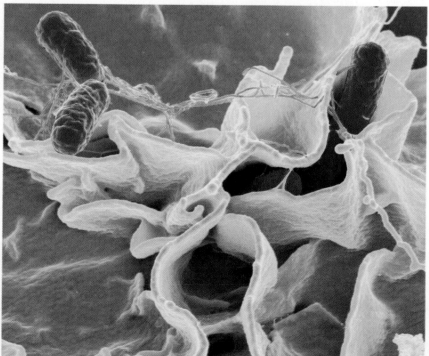

Salmonella (red), shown here invading cultured human cells, is one of the most common causes of foodborne illness.

© Rocky Mountain Laboratories/NIAID/NIH

and headaches. In 2015, close to 900 individuals became infected with *Salmonella* after consuming tainted cucumbers imported from Mexico.[7] This multistate outbreak resulted in over 100 hospitalizations and several deaths. *Salmonella* infections have also resulted from the consumption of raw alfalfa and mung bean sprouts. This is because raw sprouts are particularly prone to a variety of bacterial infections, and the FDA recommends that people with compromised immune systems (children, the elderly, and those with autoimmune conditions) avoid them.

E. coli O157:H7 and *E. coli* O104:H4
As mentioned, *E. coli* O157:H7 is an enterohemorrhagic serotype of *E. coli*. Unpasteurized milk, apple juice, and apple cider sometimes harbor this pathogenic organism, as does improperly prepared meat, including poultry. *E. coli* O157:H7 infection, which has an incubation period of one to eight days, results in nausea, abdominal cramps, and severe diarrhea that is often bloody. In 2006, 183 persons in 26 states were infected with *E. coli* O157:H7.[8] This outbreak rapidly gained national attention because numerous affected individuals were hospitalized and at least three died. Fresh spinach was identified as the source of contamination, and the FDA quickly advised consumers not to eat bagged fresh spinach or fresh spinach-containing products unless they were cooked at 71°C (160°F) for at least 15 seconds. Partly because of this outbreak, the FDA issued new guidance to the food industry to minimize microbial contamination of fresh-cut fruits and vegetables.[9] More recently (2015), 19 people in 7 states became infected after they consumed *E. coli* O157:H7-containing rotisserie chicken salad made and sold in Costco Wholesale stores. Although not confirmed, it is thought that the vegetables in the chicken salad were tainted, not the chicken itself.[10]

In 2011, an outbreak of a different pathogenic serotype of *E. coli*, *E. coli* O104:H4, resulted in hundreds—if not thousands—of illnesses and numerous deaths throughout the European Union. Several cases were also reported in the United States, but most of those affected had recently traveled to Germany. Authorities identified raw vegetable sprouts grown in Germany as the likely source of contamination.[11]

14-1e Parasites: Protozoa and Worms

A **parasite** is an organism that relies on another organism to survive. The consumption of parasite-infested foods can cause foodborne illness. An example of a parasite is the **protozoan**, a single-celled eukaryotic organism. Some protozoa can live as parasites in the intestinal tracts of animals and humans. As part of its reproductive cycle, a protozoan forms **cyst** sacs that are excreted in the feces. If cyst-containing feces come in contact with plants or animals used for food, the resulting food products can be contaminated as well. Thus, if you consume foods or beverages contaminated with protozoan cysts, you can develop a foodborne illness.

One parasitic protozoan, *Giardia intestinalis*, causes diarrhea, abdominal discomfort, and cramping. Symptoms typically begin one to two weeks after infection. *Giardia intestinalis* can be found in untreated swimming pools and hot tubs, and in rivers, ponds, and streams that have been contaminated with the feces of an infected animal or person. This is one reason why chlorine, which kills *Giardia intestinalis*, is typically added to water in public swimming and bathing areas, and why boiling, chemically treating, or filtering water from ponds, streams, and lakes is recommended before drinking.

Consuming foods that contain parasitic worms can also cause foodborne illness. Like protozoa, worms form cysts as part of their life cycles. Once they are ingested, cysts mature into worms that can cross the intestinal lining, travel through the blood, and eventually settle in various locations in the body, including muscles, eyes, and

parasite An organism that relies on another organism to survive.

protozoan A single-celled eukaryotic organism. Some protozoa are parasites.

cyst A stage of growth in the life cycle of a protozoan; excreted in the feces.

Consumption of untreated water is not advised because it may be contaminated with *Giardia intestinalis*.

© michaeljung/Shutterstock.com

the brain. An example of a parasitic worm is *Trichinella*, a roundworm that can invade a variety of animals, including pigs and some fish. Eating undercooked *Trichinella*-contaminated pork and seafood can result in the worm entering the body, causing muscle pain, swollen eyelids, and fever. Although, in some cases, the infection eradicates itself, severe infections can cause complications such as inflammation of the brain and cardiac arrhythmia (unsteady heart beat).

14-1f Prions

Although **prions** are not living organisms—in fact, they are not organisms at all—they may pose food safety concerns. A prion is an altered protein created when the secondary structure of a normal protein is disrupted. Prions can cause other normal proteins to unravel, setting off a cascade of similar reactions that converts hundreds of normal proteins into abnormal prions. When large numbers of prions build up in a cell, the cell can rupture and release its prions into the surrounding area, which in turn destroys other cells. Eventually, this process kills the surrounding cells and gives the infected tissue a spongy texture. Unfortunately, prions are extremely resilient. They retain the ability to damage other cells even after they have been exposed to extreme heat and/or acid. Therefore, merely cooking food does not destroy prions, and prions are not destroyed by the acidic conditions of the stomach. If a prion-containing food is consumed, the prions can be absorbed into the bloodstream. Though prions are inert (nonliving), some experts consider some prion-related diseases to be infectious.

Several diseases, such as **bovine spongiform encephalitis** (**BSE**, or **mad cow disease**), are known to be caused by prion ingestion.[12] Mad cow disease is characterized by loss of motor control, confusion, paralysis, wasting, and eventually death. **Creutzfeldt-Jakob disease**, another deadly disease caused by prions, is very rare: it occurs in only one in every million people each year.[13] Creutzfeldt-Jakob disease is usually attributed to direct infection by prions from contaminated medical equipment—during surgery, for example. Simultaneous to a BSE outbreak that occured in the United Kingdom in 1996, however, researchers discovered a new form of Creutzfeldt-Jakob disease called **variant Creutzfeldt-Jakob disease**. This disease has a relatively long incubation period (years) and is fatal. Considerable evidence links variant Creutzfeldt-Jakob disease to the consumption of BSE-contaminated products.[14] To date, more than 225 human cases of variant Creutzfeldt-Jakob disease have been reported worldwide, four of which were in the United States. It is believed, however, that the individuals diagnosed with this disease in the United States all were infected while traveling abroad. No treatment exists for either form of Creutzfeldt-Jakob disease, and nothing can slow either disease's progression.

Prions and Public Policy Because prions are found mainly in nerves and the brain, the World Health Organization (WHO) recommends that all governments prohibit the feeding of highly innervated tissue, such as the brain and spinal cord, from slaughtered cattle to other animals. Similarly, both Canada and the United States have banned the use of such products in human food, including dietary supplements, and in cosmetics. In 2006, Canada also banned the inclusion of cattle tissues capable of transmitting BSE in all animal feeds, pet foods, and fertilizers. Partly in response to concern about BSE, the U.S. Department of Agriculture (USDA) does not allow the slaughter of non-ambulatory (downer) cattle for food purposes.[15] These measures are part of an international effort to unify BSE policy to ensure the safety of human food.

14-2 WHAT NONINFECTIOUS SUBSTANCES CAUSE FOODBORNE ILLNESS?

Although the consumption of foods that contain infectious pathogens poses the greatest risk of foodborne illness, the consumption of inert (nonliving) noninfectious agents can also make you sick. Noninfectious agents include physical contaminants, such as glass and plastic, and other dangerous substances, such as toxins, heavy metals, and pesticides.

prions An altered protein that forms when the secondary structure of a normal protein is disrupted.

bovine spongiform encephalitis (**BSE**, or **mad cow disease**) A fatal disease caused by prion ingestion.

Creutzfeldt-Jakob disease A rare but fatal disease caused by a genetic mutation or exposure to prions during surgery.

variant Creutzfeldt-Jakob disease A form of Creutzfeldt-Jakob disease that may be caused by consumption of BSE-contaminated foods.

Red tides occur when marine algae produce brightly colored pigments and toxins that cause shellfish poisoning.

Don Paulson/PureStock/AGE Fotostock

14-2a Algal Toxins

One type of noninfectious foodborne illness, **shellfish poisoning**, can result from the consumption of particular types of contaminated fish and shellfish (e.g., clams and oysters). Certain marine animals consume large amounts of ocean algae that sometimes produce poisonous **marine toxin** compounds. Eating marine toxin–contaminated foods can cause tingling, burning, numbness, drowsiness, and difficulty breathing. People with shellfish poisoning may also experience a strange phenomenon called *hot–cold inversion*, whereby cold is perceived as hot and vice versa.

One variety of shellfish poisoning is associated with a phenomenon referred to as **red tide**. This phenomenon occurs when a particular marine alga begins to grow quickly. As it multiplies, it produces brightly colored pigments that are released and make the surrounding water appear red or brown. During this period of rapid growth, the algae also produce a potent toxin called **brevetoxin**, which, when consumed by humans, causes shellfish poisoning. Approximately 30 cases of shellfish poisoning are reported in the United States every year, with most cases occurring in the coastal regions of the Atlantic Northeast and Pacific Northwest.

14-2b Pesticides, Herbicides, Antibiotics, and Hormones

Because no food production system is foolproof, pesticides, herbicides, antibiotics, and hormones used in agriculture sometimes come in contact with food products. A number of national and international agencies work together to ensure that the presence of these contaminants in food is negligible or poses no known risk to consumers or the environment. These agencies include the Food and Agriculture Organization (FAO) of the United Nations, the U.S. Environmental Protection Agency (EPA), the FDA, and the USDA.

shellfish poisoning A type of noninfectious foodborne illness caused by the consumption of particular types of contaminated fish and shellfish.

marine toxin A poison produced by ocean algae.

red tide A phenomenon whereby certain ocean algae grow profusely and produce brightly colored pigments that make the surrounding water appear red or brown.

brevetoxin A potent toxin produced by red tide–causing algae that, when consumed, leads to shellfish poisoning.

An example of a dangerous compound addressed by a governmental agency is dichloro-diphenyltrichloro-ethane (DDT), a pesticide once used to kill mosquitoes and increase crop yields. The use of DDT was banned in the United States in 1972, when the compound was found to damage wildlife. **Bovine somatotropin (bST)**, otherwise known as bovine growth hormone, has attracted similar public attention in recent years. This hormone is produced naturally by cattle and is used in the dairy industry to increase milk production. Although some people are concerned about its safety, a substantial amount of research suggests that bST is safe for both cows and consumers, so there are no laws requiring milk produced by cows treated with bST to be labeled as such.[16] Similarly, although some people are concerned about it, glyphosate (the active agent in the herbicide Roundup) has been deemed perfectly safe for use. Even after substances such as bST and glyphosate have been approved, however, the FDA continues to evaluate their safety.

14-2c Food Allergies and Sensitivities

Certain noninfectious compounds can cause illness in small segments of the population, but not because they are toxic or poisonous. Instead, these compounds cause illness only for those who are sensitive or allergic to them. Because the percentage of individuals who have adverse reactions to these compounds is so small, however, their presence in food is not prohibited.

Monosodium glutamate (MSG), a flavor enhancer used in a multitude of commercially processed foods, causes severe headaches, facial flushing, and a generalized burning sensation in some people.[17] MSG must be listed on the label of any food to which it is added. Similarly, sulfite compounds are sometimes added to foods such as wine and dried fruits to enhance color and prevent spoilage. Unfortunately, the consumption of sulfites causes breathing difficulties in sulfite-sensitive individuals, especially those with asthma. The FDA requires food manufacturers to label all foods that

Monosodium glutamate (MSG), a compound that gives some people headaches and causes facial flushing, is commonly used in Asian cuisine.

contain at least 10 parts per million (ppm) sulfites. You can determine if a food has added sulfites by looking for the following terms on its food label.

- Sodium bisulfite
- Sodium metabisulfite
- Sodium sulfite
- Potassium bisulfite
- Potassium metabisulfite

Recall from Chapter 5 that exposure to proteins naturally present in some foods causes allergic reactions in certain individuals. People who are allergic to peanuts and other legumes, for example, can experience life-threatening allergic responses if they eat these foods. The most common food allergies are caused by proteins present in eggs, milk, peanuts, soy, and wheat. Researchers estimate that approximately 2 percent of adults and 5 percent of infants and young children have food allergies in the United States.

14-2d "Hot Topics" in Food Safety

Scientists and public health officials work to continually identify compounds in foods that may cause illness. Two examples are melamine and bisphenol A. In addition, segments of the public sometimes become interested in various food production methods and techniques, believing they may be harmful or particularly healthful. For instance, there is a growing trend for the consumption of unpasteurized dairy products, because some people consider them to be more "natural" and nutritious than their pasteurized counterparts. Another "hot topic" is whether genetically modified foods pose some sort of harm to the consumer or environment. These topics are briefly discussed here.

Melamine Melamine is a nitrogen-containing chemical that is typically used to make lightweight plastic objects such as dishes. In 2007, the consumption of pet food contaminated with **melamine** caused numerous dogs and cats to become ill and even die.[18] Gluten is a wheat protein commonly used to produce a number of foods, including animal foods. Because melamine is rich in nitrogen, some Chinese gluten manufacturers added melamine to their products to make them appear

bovine somatotropin (bST) A growth hormone produced by cattle and used in the dairy industry to increase milk production.

melamine A nitrogen-containing chemical typically used to make lightweight plastic objects.

higher in protein than they actually were. Investigations by the FDA confirmed that melamine-tainted gluten was used to make pet foods sold in the United States. After this story was released, public attention quickly shifted to the possibility that melamine could enter the human food supply as well. In response, the FDA and USDA issued a press release stating that consumption of meat and milk from animals fed melamine-tainted feed posed very low risk to human health.[19] This position was quickly revised, however, when melamine-contaminated milk products caused nearly 60,000 infants and children throughout China to become sick in 2008 and 2009.[20] The FDA now advises consumers worldwide to avoid using infant formula products, milk products, and products with milk-derived ingredients made in China.

Bisphenol A Some people believe that early exposure to **bisphenol A (BPA)**, a chemical found in some plastic food and beverage containers (including baby bottles), can pose a potential threat to health. Because BPA has been shown to increase the long-term risks of cancer and reproductive abnormalities in laboratory animals, scientists and public health officials are currently studying whether the consumption of foods and beverages from these types of containers poses a health risk to humans as well.[21] After reviewing the growing literature on BPA, the FDA announced in 2010 that it was concerned about the potential effects of BPA on the brain, behavior, and development in fetuses, infants, and young children. In response to its findings, the FDA has begun supporting efforts to decrease BPA exposure in the United States.[22] Individuals who want to decrease their own exposure to BPA should:

- Choose glass or metal containers over polycarbonate ones;
- Refrain from heating plastics in the microwave; and
- Wash plastic containers by hand instead of in the dishwasher.

Because of health concerns associated with bisphenol A (BPA), BPA-free plastic products are now widely available.

bisphenol A (BPA) A chemical found in some plastic food and beverage containers believed by some to pose a potential threat to health.

14-2e Is Raw Milk Better Than Pasteurized Milk?

As described previously, pasteurization has been used for over a century to kill potentially pathogenic organisms in foods and beverages—particularly milk and cheese. In some states, however, the sale of unpasteurized (raw) dairy products is allowed, and consumers have a choice as to which type to purchase. Some claim that raw milk is healthier because it is less processed, contains more of certain nutrients, or has some special properties that can solve health problems, and raw dairy products are experiencing a veritable boom in nationwide sales. Almost all of the nutritional benefits of drinking raw milk are available from pasteurized milk, however (without the risk of disease), and raw milk offers no known health benefits that cannot also be obtained by drinking pasteurized milk. Importantly, the risk posed by consuming raw milk is staggering.[23] Indeed, the chance of an outbreak caused by raw milk is at least 150 times higher than that caused by pasteurized milk. Between 2007 and 2012, 81 outbreaks in 26 states were linked to raw milk; most of these outbreaks involved children under the age of 5 years, and, not surprisingly, the vast majority of these outbreaks happened in states where selling raw milk was legal. Consequently, the CDC and many other health-related agencies and organizations highly recommend that we always choose pasteurized dairy products.

14-2f Genetically Modified Foods: Are They Safe?

Humans have always sought ways to produce abundant amounts of food. One method that has been used extensively is the careful selection of seeds and animals that naturally grow well in a particular environment due to their genetic makeup. For instance, Holstein cows (the black and white variety so common on dairy farms) have been selectively bred for generations to produce increased amounts of milk. Farmers and the agriculture industry also cross different varieties of seeds with genetic traits that complement each other, forming plant hybrids with optimal growing characteristics. More recently, researchers have developed ways to transfer specific genes into a plant or animal so that certain useful traits can be acquired—a technology referred to as genetic engineering. Plants and animals that have been modified using genetic engineering are referred to as **genetically modified organisms (GMO)**, and, when used in the food supply, they are called **genetically modified foods**. There are many reasons for developing genetically modified foods. For instance, some genetically modified plants better resist insects and other pests that have historically lowered crop yield. Other plants have been genetically modified so that they are not killed by commonly used herbicides, so that they do not bruise or brown after being sliced or peeled (as shown here with potatoes), so that they can grow in very wet (or extremely dry) soil, and so that they contain higher levels of essential nutrients or lower levels of compounds related to health risks. To date, the only genetically modified animal approved for human consumption is a variety of salmon that grows much more quickly than typical farmed salmon. Although some people are concerned about the safety of genetically modified foods, the U.S. Food and Drug Administration (FDA), which is tasked with overseeing their safety and use in agriculture, has concluded that genetically modified foods are indeed as safe and nutritious as their conventional counterparts. Because the FDA considers genetically modified foods identical to non-genetically modified varieties, they do not have to be labeled as having been produced from a genetically modified plant or animal. Although there are clearly demonstrated benefits regarding the use of GMOs, public concern about their safety continues. It is important that scientists maintain an open dialog to address consumers' concerns and to reassure the public that there is no evidence that the consumption of genetically modified foods negatively affects human health.

14-3 HOW DO FOOD MANUFACTURERS PREVENT CONTAMINATION?

Countless pathogens and inert compounds cause foodborne illness—it is simply beyond the scope of this book to describe each one in detail. However, it is important for you to understand how disease-causing agents of all types are transmitted from one food, surface, or utensil to another (a process known as **cross-contamination**), as well as from person to person. To do this, it is important to understand the techniques that food manufacturers use to keep your food safe.

14-3a Handling and Sanitation

To prevent foodborne illness, it is critical that those who handle food do so safely and sanitarily. Pathogens can be transmitted from an infected person to almost any food. Because pathogens spread so easily, food that is otherwise safe can be made unsafe in a matter of seconds. Therefore, food handlers (including those who harvest, process, and prepare the food) must avoid coughing or sneezing on the foods they work with. In addition, because many pathogens pass through the GI tract and are excreted in the feces, handlers must ensure that fecal matter does not come in contact with food. People who

FIGURE 14.1 LEGAL STATUS OF THE SALE OF RAW MILK, AND OUTBREAKS

Legal Status of the Sale of Raw Milk, and Outbreaks Linked to Raw Milk, by State, 2007–2012

○ # of outbreaks

Sale is legal
- Retail stores
- Restricted to farms
- Other regulations allow some sale

Sale is illegal
- Illegal
- Cow shares permitted

Source: http://www.cdc.gov/foodsafety/pdfs/legal-status-of-raw-milk-sales-and-outbreaks-map-508c.pdf.

genetically modified organisms (GMO) A plant or animal that has been altered using genetic engineering.

genetically modified foods A food that has been produced or manufactured using a genetically modified organism.

cross-contamination The process by which a disease-causing agent is transmitted from one food, surface, or utensil to another.

process and prepare foods are generally required to wear gloves while working and must thoroughly wash their hands after using the toilet.

14-3b Food Production, Preservation, and Packaging

Beyond safe food-handling practices, the food-processing industry adheres to numerous other regulations that help keep your food safe. For instance, the USDA requires that meat and poultry be inspected before they are sold and that guidelines for safe handling appear on packages. However, inspection and adherence to guidelines alone cannot guarantee that meat is pathogen free. Food manufacturers also use drying, salting, smoking, fermentation, heating, freezing, and irradiation to help keep food safe, and, next, we briefly consider these commonly used food preservation methods.

Drying, Salting, Smoking, and Fermentation
Meat has long been preserved by salting, smoking, and/or drying because these techniques—when done correctly—inhibit bacterial growth. Recall from Chapter 13 that fermentation, which involves the addition of yeast to a sugar-containing beverage, is used to produce alcohol. The addition of select yeasts and/or bacteria to food is also used to make products such as sauerkraut, cheese, yogurt, and pickles—many of which are considered probiotic foods. The process of fermentation promotes the growth of nonpathogenic organisms, which in turn minimizes the growth of pathogenic organisms and preserves the food.

Heat Treatment: Cooking, Canning, and Pasteurization
Because most microorganisms prefer living in an environment between 40 and 140°F (4 and 60°C), this temperature range is often referred to as the **danger zone**. Manufactures, restaurants, and individuals alike should avoid the danger zone when storing and serving food. Cooling a food to below 40°F or heating it to above 140°F inhibits microbial growth and can even kill existing microbes. For this reason, food

danger zone The temperature range between 40 and 140°F at which most microorganisms prefer to live.

Product	Minimum Internal Temperature & Rest Time
Beef, pork, veal, and lamb steaks, chops, roasts	145°F (63°C) and allow to rest for at least 3 minutes
Ground meats	160°F (72°C)
Ham, fresh or smoked (uncooked)	145°F (63°C) and allow to rest for at least 3 minutes
Fully cooked ham (to reheat)	Reheat cooked hams packaged in USDA-inspected plants to 140°F (60°C) and all others to 165°F (74°C)
All poultry (breasts, whole bird, legs, thighs, and wings, ground poultry, and stuffing)	165°F (74°C)
Eggs	160°F (72°C)
Fish & Shellfish	145°F (63°C)
Leftovers	165°F (74°C)
Casseroles	165°F (74°C)

TABLE 14.2 SAFE MINIMUM INTERNAL TEMPERATURES FOR VARIOUS TYPES OF MEATS, EGGS, LEFTOVERS, AND CASSEROLES

Source: U.S. Department of Agriculture. Safe minimum internal temperature chart. Available at: http://www.fsis.usda.gov/wps/portal/fsis/topics/food-safety-education/get-answers/food-safety-fact-sheets/safe-food-handling/safe-minimum-internal-temperature-chart/ct_index

manufacturers often heat-treat their products. Table 14.2 lists temperature guidelines for cooking, serving, and reheating foods. As you can see, different foods require different temperatures to maintain safety. Note that the USDA recommends cooking all whole cuts of meat (excluding poultry) to 145°F (63°C), as measured with a food thermometer placed in the thickest part of the meat, and then allowing the meat to rest for 3 minutes before carving or consuming.

Heating also preserves food for later consumption. Take the canning process, for example. After a food is packaged (or "canned") in a sanitized jar or can, the container—and

© marcociannarel/Shutterstock.com

Pasteurization is named for French microbiologist Louis Pasteur, who developed the process to prevent wine and beer from spoiling.

the food inside—is heated to a high temperature. This kills pathogens and creates a vacuum within the container. The oxygen-free environment within the container helps preserve the food because most food-borne illness–causing organisms require oxygen to grow. Although some organisms, such as *C. botulinum*, can survive in low-acid anaerobic conditions, ensuring that home-canned foods have the proper acidity and are heated sufficiently helps keep them safe. You can learn more about safe canning practices at the USDA website.[24]

Pasteurization, the process whereby food is partially sterilized through brief exposure to a high temperature, is perhaps the most common form of heat treatment used in food preservation. Foods that are typically pasteurized include milk, juice, spices, ice cream, and cheese. Pasteurization is especially effective for killing some forms of *E. coli*, *Salmonella*, *Campylobacter jejuni*, and *Listeria monocytogenes*. Several outbreaks of *Salmonella* poisoning stemming from unpasteurized apple cider prompted new laws regarding pasteurization to be enacted. Commercially available apple cider must be either pasteurized or labeled as nonpasteurized, and the FDA

recommends that homemade cider be heated for 30 minutes at 155°F (68°C) or 15 seconds at 180°F (82°C).

Cold Treatment: Cooling and Freezing As you have likely experienced, foods spoil less quickly if they are kept cold. To slow or halt the growth of microorganisms, foods should always be refrigerated at 40°F (4°C) or colder, or frozen soon after they are prepared. This helps prevent foods from staying in the danger zone for an extended period of time.

Irradiation In the 1950s, the National Aeronautics and Space Administration (NASA) first used **irradiation** to preserve food for space travel. Shortly thereafter, the FDA approved this food processing practice as a form of food preservation here on Earth. Today, irradiation is approved for meat, poultry, shellfish, eggs, fresh fruits, vegetables, and spices. During irradiation, foods are exposed to radiant energy that damages or destroys bacteria. Irradiation makes foods safer to eat and can dramatically increase shelf life. For example, compared to nonirradiated strawberries, which have a brief shelf life, irradiated strawberries last for several weeks without spoiling. Just as the use of x-rays to inspect luggage neither damages travelers' belongings nor makes them radioactive, irradiation does not make foods radioactive. Irradiation also does not compromise nutritional quality or noticeably change the taste, texture, or appearance of food. Still, irradiated foods must be labeled with the *radura symbol* (see Figure 14.2).

pasteurization A food preservation process whereby food is partially sterilized through brief exposure to a high temperature.

irradiation A food preservation process whereby a food is exposed to radiant energy that damages or destroys bacteria.

The difference in freshness between irradiated strawberries (left) and nonirradiated strawberries (right) picked the same day is quite amazing.

FIGURE 14.2 THE RADURA SYMBOL

Irradiated foods must include this symbol on their food labels.
Source: FDA.

14-4 # WHAT STEPS CAN YOU TAKE TO REDUCE FOODBORNE ILLNESS?

There are many precautions you can take to reduce your risk of foodborne illness. First, you should familiarize yourself with recommendations put forth by regulatory agencies and keep abreast of food safety alerts and recalls. Second, you should understand and utilize basic concepts related to safe food handling. In this section, you will learn about a variety of consumer advisory bulletins that are available from the FDA and USDA. You will also learn about various dates that are sometimes found on food labels, and a national campaign developed by the U.S. government that provides useful tips for consumers on how to avoid foodborne illness.

14-4a Interpreting Food Product Dating

You may have noticed that some food labels include various types of dates. Sometimes, this information helps the store determine how long to display the product for sale. This type of date is called a **sell-by date**, and although you should purchase the product before the date passes, the food is not necessarily unsafe to consume after that time. If the product has a sell-by date or no date, you should simply aim to cook or freeze it according to the times listed in Table 14.3. **Use-by dates** are simply manufacturers' suggestions for when you should consume the food for best quality or flavor. They do not imply anything about food safety. Because product dates are not really intended to be guides for the safe use of a product, you may wonder how long you should store the food in order to maintain best quality. The USDA recommends following these guidelines.

- Purchase the product before the sell-by date expires.
- If the food is perishable, take it home immediately after purchase and refrigerate it promptly. Freeze it if you can't use it within the time period recommended in Table 14.3.

sell-by date A date placed on a product's label to assist a grocer in knowing when the food should be removed from the sales display.

use-by dates A manufacturer's best prediction as to when the product will retain top quality and flavor.

| TABLE 14.3 | RECOMMENDED REFRIGERATOR STORAGE TIMES (AT OR BELOW 40°F) PRIOR TO COOKING OR FREEZING |

Type of Product	Maximal Storage Time Prior to Cooking/Freezing
Fresh and Uncooked Products	**After Purchase**
Poultry	1 or 2 days
Beef, veal, pork, and lamb	3 to 5 days
Ground meat and ground poultry	1 or 2 days
Fresh variety meats (e.g., liver, tongue, brain, kidney)	1 or 2 days
Cured ham, labeled "cook before eating"	5 to 7 days
Uncooked pork, beef, or turkey sausage	1 or 2 days
Eggs	3 to 5 weeks

Processed Products (Sealed When Purchased)	Unopened, After Purchase	After Opening
Cooked poultry	3 to 4 days	3 to 4 days
Cooked sausage	3 to 4 days	3 to 4 days
Sausage, hard/dry, shelf-stable	6 weeks	3 weeks
Corned beef, uncooked, in pouch with pickling juices	5 to 7 days	3 to 4 days
Vacuum-packed dinners	2 weeks	3 to 4 days
Bacon	2 weeks	7 days
Hot dogs	2 weeks	1 week
Luncheon meat	2 weeks	3 to 5 days
Ham, fully cooked	7 days	slices, 3 days; whole, 7 days
Ham, canned, labeled "keep refrigerated"	9 months	3 to 4 days
Ham, canned, shelf stable	2 years/pantry	3 to 5 days
Canned meat and poultry, shelf stable	2 to 5 years/pantry	3 to 4 days

Source: http://www.fsis.usda.gov/wps/portal/fsis/topics/food-safety-education/get-answers/food-safety-fact-sheets/food-labeling/food-product-dating/food-product-dating

- Once a perishable product is frozen, it doesn't matter if the date expires, because foods kept frozen are safe indefinitely.
- Follow the handling recommendations provided on the product.

14-4b Consumer Advisory Bulletins

Both the USDA and FDA maintain user-friendly websites and toll-free phone numbers that provide information about current food safety recommendations. Reliable USDA and FDA resources include the following.

- *General information*: www.foodsafety.gov
- *Recalls and alerts*: www.foodsafety.gov/recalls
- *To report a foodborne illness*: www.foodsafety.gov/report/index.html
- *USDA's meat and poultry hotline*: 1-888-MPHotline
- *FDA's food safety information hotline*: 1-888-SAFEFOOD

14-4c The Fight Bac!® Campaign

Although issues related to specific foodborne illnesses change over time, consumers are encouraged to follow some basic rules when handling, preparing, and storing food. To help consumers avoid the infectious agents that cause foodborne illness, the USDA and the Partnership for Food Safety Education developed a set of food safety guidelines called **Fight BAC!®** (see Figure 14.3). The four major components that comprise Fight BAC!® are clean, separate, cook, and chill.

Clean To prevent pathogens in fecal materials from contaminating your food, you should always wash your hands after using the bathroom, changing a diaper, or handling a pet. To do so effectively, wash your hands

Food Safety in an Emergency

Ensuring that food and water are safe to consume normally requires access to refrigeration, a stove or oven, and clean water. But what should you do when an emergency occurs or electricity is temporarily turned off? The USDA suggests that the best approach to an emergency is to be prepared in advance. A supply of bottled water is absolutely essential, and having three to four days' worth of nonperishable food at arm's reach can keep you both nourished and safe during an emergency situation. Consider what else you could do to prepare for a likely emergency. For example, if you live in a location that could be affected by a flood, store your emergency food on high shelves that would be safely out of the way of contaminated water. Coolers can keep food cold if the power will be out for more than 4 hours, so having a couple on hand (along with frozen gel packs) is advised. You can learn more about how to prepare for and how to keep food safe during an emergency at the USDA website (www.fsis.usda.gov/wps/portal/fsis/topics/food-safety-education/get-answers/food-safety-fact-sheets/emergency-preparedness). Remember the old saying, "An ounce of prevention is worth a pound of cure."

FIGURE 14.3 FIGHT BAC!

The Fight BAC! campaign was created to reduce the incidence of foodborne illnesses by educating Americans about safe food-handling practices at home and at work.

vigorously with soap for at least 20 seconds and rinse thoroughly under clean, warm running water. Dry your hands using clean paper or cloth towels. Using an antimicrobial gel can also help ensure that your hands are pathogen free, although gel sanitizers are not a substitute for proper hand washing. A clean cooking environment is also important; you should periodically sanitize counters, equipment, utensils, and cutting boards with a solution of 1 tablespoon of liquid chlorine bleach in 1 gallon of water.

Washing fresh fruits and vegetables (even those with skins and rinds that are not eaten) can also provide an important protection from foodborne pathogens and noninfectious agents (e.g., pesticides). Wash produce under clean, cool running water, and if possible, scrub with a clean

Fight BAC!® A public education program developed to reduce foodborne bacterial illness. The four major components that comprise FightBAC!® are clean, separate, cook, and chill.

brush. Fruits and vegetables should be dried with a clean paper or cloth towel. There is some evidence that commercial fruit and vegetable cleaners may help remove *E. coli* O157:H7 and *Salmonella* from produce, but the USDA has not taken a stance on the effectiveness of such products.[25] Note that, although you should wash produce, you should not wash raw meat, poultry, or fish, because doing so increases the likelihood of cross-contaminating otherwise noninfected foods and surfaces.

Separate To prevent cross-contamination, separate raw meat and seafood from other foods in your grocery cart, in your refrigerator, and while preparing a meal. For example, put raw meats in separate, sealed plastic bags in your grocery cart and use separate cutting boards when preparing fresh produce and raw meat. Do not return cooked meat to the same plate that was used to hold the raw meat unless you have thoroughly washed the plate with hot, soapy water.

© bikeriderlondon/Shutterstock.com

"Whereas washing fresh fruits and vegetables is recommended, you should not wash raw meat, poultry, or fish."

Cook Heat, which kills most dangerous microorganisms, can also alter the chemical compositions of some preformed toxins, making them less dangerous. The temperatures and cooking times required to kill pathogens depend on both the particular foods and the organisms or toxins present. To ensure that your cooked food is safe to eat, measure its *internal temperature* using a probe thermometer. (Also, thoroughly clean your thermometer between uses.) Before eating leftovers, reheat your food to its appropriate internal temperature. Just because a food has been previously cooked does not mean that it is pathogen free. In the late 1990s, a *Listeria monocytogenes* outbreak resulted in at least six deaths and two miscarriages. This outbreak prompted the recommendation that lunchmeats and frankfurters be heated before eating.[26] These recommendations may be especially important for pregnant women, infants, older adults, and people with impaired immunity (such as recipients of organ transplants or those with cancer), as they are more likely to become seriously ill when exposed to a pathogen.

Chill Because cold temperatures can slow the growth of microorganisms, you should always strive to keep perishable foods chilled. For instance, instead of marinating a piece of meat at room temperature, do so in a refrigerator. Similarly, thaw a frozen food in the refrigerator, submerged in an airtight bag in cold water, or in the microwave instead of at room temperature. After a meal, refrigerate foods as quickly as possible, and separate large amounts of foods into small, shallow containers to allow the food to cool quickly.

Even properly chilled foods can become sources of foodborne illness. For this reason, you should generally consume leftovers within three to four days. Indeed, per the old saying, "If in doubt, throw it out!"

14-4d Be Especially Careful When Eating Out

Making sure that your food is safe to eat can be especially difficult at a restaurant, picnic, potluck, or buffet. However, by considering a few simple questions, you can minimize your risk of foodborne illness in these environments. To gauge the safety of a food, ask yourself whether the four basic Fight Bac!® guidelines have likely been followed.

- Clean: How likely is it that the people handling and preparing this food used sanitary practices?
- Separate: Is there evidence that raw ingredients were kept separate from cooked ones?

© Yeko Photo Studio/Shutterstock.com

- Cook: Is it likely that this food was cooked properly and kept out of the thermic danger zone?
- Chill: Is there evidence that cold ingredients were kept cold?

Unfortunately, the basic rules of food safety often go neglected at picnics and social gatherings, resulting in foodborne illness. Use good judgment and eat only foods you know to be safe. It is better to be safe than sorry, after all!

14-5 WHAT STEPS CAN YOU TAKE TO REDUCE FOODBORNE ILLNESS WHILE TRAVELING?

There are added food safety concerns when traveling abroad or camping. You can keep up to date on current concerns by visiting a special website that the CDC maintains for this very purpose (www.cdc.gov/travel/). Although it is not always possible to prevent foodborne illness while traveling, you can substantially lower your risk by being vigilant in your food and beverage choices. Nonetheless, if you do experience traveler's diarrhea while abroad, it is important that you replace lost fluids and electrolytes as soon as symptoms begin to develop. Clear liquids are routinely recommended for adults, and those who develop three or more loose stools in an 8-hour period may benefit from antibiotic therapy.

14-5a Drink Only Purified or Treated Water

When traveling outside of the United States or camping in remote regions of the country, it is advisable to drink bottled water and avoid using ice. Before you drink water from a bottle, be sure that it has a fully sealed cap. If the seal is not intact, the container may have been refilled from an unknown source. If bottled water is not available, all water should be boiled for 1 minute before use. If at a high altitude (more than 2,000 meters above sea level), boil water for 3 minutes. If boiled or bottled water is not available, fresh water can be chemically treated to kill pathogens that may be present. Portable water filters can also be used to remove some pathogens.

14-5b Avoid or Carefully Wash Fresh Fruit and Vegetables

Although fresh produce that is contaminated with bacteria may not cause illness for local residents, it can cause serious illness for visitors. Thus, when traveling to a foreign country, avoid or carefully wash fresh fruit and vegetables that are eaten without peeling, such as grapes and peppers. Of course, only water that is known to be clean should be used to clean produce.

14-5c Traveling in Areas with Variant Creutzfeldt-Jakob Disease

The CDC recommends that when traveling in Europe or another area that has reported cases of BSE in cows, concerned travelers should consider either avoiding beef and beef products or selecting only beef products composed of solid pieces of muscle meat. These considerations, however, should be balanced with the knowledge that the risk of disease transmission is very low. Milk and milk products are not believed to pose any risk for transmission of BSE to consumers—even in areas of the world with an emerging incidence of variant Creutzfeldt-Jakob disease.

14-5d Emerging Issues of Food Biosecurity

The emergence of global terrorism has raised new concerns about food safety both at home and abroad. Indeed, a terrorist or militant faction could do serious widespread harm by contaminating America's food supply. As a result of this threat, **food biosecurity**—measures aimed at preventing the food supply from falling victim to planned contamination—has gained significant national attention in recent years. Of particular interest

> **food biosecurity** Measures aimed at preventing the food supply from falling victim to planned contamination.

Traveling abroad requires special precautions regarding food safety.

is the possibility that terrorists could use *C. botulinum* as a widespread disease-causing agent. A supply of botulism antitoxin is maintained by the CDC in the event that such an attack occurs. Congress authorized the **Public Health Security and Bioterrorism Preparedness and Response Act** (or **Bioterrorism Act**) in 2002 to ensure the continued safety of the U.S. food supply. Since its enactment, additional regulations have been approved to help ensure that foods grown and produced in the United States and abroad cannot be tampered with.[27] Clearly, food biosecurity will remain an important issue (for individuals and government agencies alike) for many years to come.

Outside of willful acts of terrorism, changes in food production and distribution systems may influence food safety and the risk of foodborne illness. For example, a contaminated food produced in a remote country might be distributed through American channels to dozens of states and sold at hundreds of grocery stores. From there, it might appear on your own table. Because production and distribution systems have become global enterprises, the origins of all foods must now be included on food labels. Still, as clearly illustrated by the outbreak of melamine contamination in pet food in 2007, the U.S. government has little power over whether optimal and honest food production policies are adhered to in other locations. Although the full impact of America's recent reliance on international food production and distribution is not known, some experts warn that fewer restrictions on food importation into the United States may increase the risk of foodborne illness. For such practices to continue, the health benefits imparted by the globalization of America's food supply must greatly outweigh its associated risks.

Public Health Security and Bioterrorism Preparedness and Response Act (or **Bioterrorism Act**) Federal legislation aimed at ensuring the continued safety of the U.S. food supply.

STUDY TOOLS 14

READY TO STUDY? IN THE BOOK, YOU CAN:

☐ Rip out the Chapter Review Card, which includes key terms and chapter summaries.

ONLINE AT WWW.CENGAGEBRAIN.COM, YOU CAN:

☐ Learn about eating meat from cloned animals in a short video.

☐ Recall key terms with matching activities.

☐ Interact with figures from the text to check your understanding.

☐ Prepare for tests with quizzes.

☐ Review the key terms with Flash Cards.

15 | Food Security, Hunger, and Malnutrition

LEARNING OUTCOMES

 15-1 List and explain factors associated with food security and food insecurity.

15-2 Understand the causes and consequences of food insecurity.

15-3 List and explain factors that contribute to global food insecurity.

15-4 Learn about ways you can take action against food insecurity.

After finishing this chapter go to **PAGE 364** for **STUDY TOOLS.**

 15-1 WHAT IS FOOD SECURITY?

You have surely experienced hunger—the physical drive to consume food. The word *hunger* is often used to describe the physical discomfort experienced by individuals who have consumed insufficient amounts of food. However, it is more commonly used on a global level to describe a chronic shortage of available food. While most people ease their hunger by eating, others do not always have this option. When sufficient food is not available or accessible, hunger can lead to serious physical, social, and psychological consequences. The prevalence of persistent hunger in the world is simply astonishing. Although nobody knows for sure the exact number of people who go hungry each day, the United Nations Food and Agriculture Organization (FAO) estimates that approximately 793 million people are undernourished globally.[1] This number continues to be unacceptably high. However, the good

food security A condition whereby a person is able to access sufficient amounts of nutritious food.

food insecurity A condition whereby a person does not have adequate physical, social, or economic access to food.

news is that in just one decade, the number of undernourished people in the world decreased by 167 million, and 216 million less than in 1990. Hunger affects people of all ages and in every country in the world—even wealthy countries such as the United States.

Food security is defined as the condition whereby a person is able to obtain sufficient amounts of nutritious food to support an active, healthy life. Conversely, **food insecurity** exists when a person does not have adequate physical, social, or economic access to food. Many people are surprised to learn that there is enough food produced in the world to provide every person with the calories he or she needs. In other words, food insecurity is not caused by insufficient worldwide food production. Rather, it is caused by an inability to obtain sufficient food to feed oneself or one's family. Other constraints, such as limited physical or mental function, can also contribute to food insecurity. In reality, the principal factors related to worldwide food insecurity are poverty, war, and natural disaster.

There are varying degrees of food insecurity. Although temporary hunger may not pose any serious health problems, chronic hunger and malnutrition can

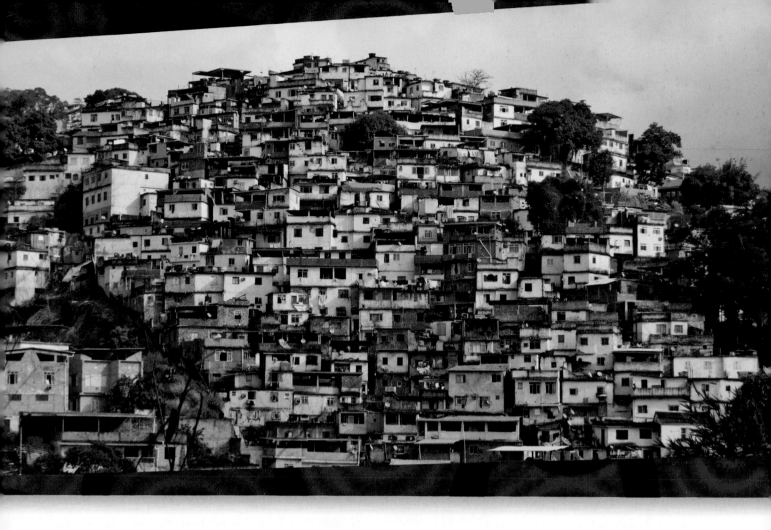

have severe consequences. Because many people live with uncertainty as to whether they will have enough to eat and/or the variety of food needed to attain all the essential nutrients, food insecurity and the hunger it causes are major social concerns in the world today.

15-1a Responses to Food Insecurity

People living in food-insecure households respond to the threat of hunger in different ways. Some take advantage of charitable organizations that assist people in need, while others resort to stealing, begging, and/or scavenging to obtain food.[2] For many people, food insecurity can cause feelings of alienation, deprivation, and distress. It can adversely affect family dynamics and social interactions within the larger context of community, and of course, it can result in hunger. Clearly, the short- and long-term consequences of food insecurity can be devastating for every person in a household.

15-1b Prevalence of Food Insecurity in the United States

How much money do you spend on food each week? According to the U.S. Department of Agriculture (USDA), American households spend about $50 per person on food purchases every week.[3] As you might expect, households experiencing food insecurity spend considerably less: about $38 per person each week. Determining the breadth and

> Food insecurity exists when a household does not have access to sufficient amounts of food.

depth of food insecurity in the United States is a difficult scientific challenge. For public health officials to develop effective strategies and target them to the appropriate populations, researchers must fully understand the scope and nature of the problem, however. Just as there is no single cause of food insecurity, there is no one solution to the problem.

Assessing the prevalence of food insecurity in prosperous countries can be particularly challenging because it is not usually associated with detectable signs of malnutrition. For this reason, clinical measurements of nutritional status (such as weight and height) are not always useful indicators of food insecurity. Instead, the prevalence of food insecurity in U.S. households is typically assessed using data regarding food availability and access. For example, the USDA issues a yearly survey that asks individuals questions about their behaviors related to food access and availability.[3] Depending on the responses to this survey, a person or household is classified as having either low food security or very low food security. Households classified as having **low food security** experience reduced food quality, variety, and/or desirability, although there is little indication of reduced food intake. This is not the case with households classified as having **very low food security**. These households are more likely to report disrupted eating patterns and reduced food intake.

Recent survey results indicate that 14 percent of American households (17.4 million households) were food insecure at least some time during the year in 2014, meaning they lacked access to enough food for an active, healthy lifestyle for all household members. The prevalence of very low food security was estimated to be about 6 percent, and the rates of food insecurity were substantially higher than the national average for households with incomes near or below the federal poverty cutoff, households with children headed by a single parent, Black- and Hispanic-headed households, and women living alone. In addition, food insecurity was highest in rural areas and lowest in suburban areas around large cities. Food-secure households spent 26 percent more for food than did food-insecure households.[3]

Poverty Is the Underlying Factor Associated with Food Insecurity

A person's risk of experiencing food insecurity in the United States is associated with income, ethnicity, family structure, and location of the home. Because many of these risk factors are interrelated, it can be difficult to determine the extent to which each one independently contributes to food insecurity. Although many factors may play a role, a link between income—specifically poverty—and food insecurity is indisputable. In fact, poverty is often the common thread that ties these factors together.

It is important to recognize that poverty is not just a problem for individuals without a source of income—many people who live in poverty maintain steady employment. Based on data provided by the U.S. Census Bureau, the poverty rate for children was 21 percent in 2014. Fourteen percent of people 18 to 64 years old live in poverty, and the poverty rate for people 65 years and older is 10 percent.[4]

When money is limited, people are often forced to reduce food-related expenses to pay for such things as housing, utilities, and health care.[5] Although one in three households living in poverty is food insecure, some households with incomes above the poverty line also experience food insecurity. An unexpected expense such as a medical problem or a repair bill can cause some people, at least temporarily, to not have the financial means to purchase sufficient amounts of food.

low food security A condition characterized by reduced food quality, variety, and/or desirability, but not reduced food intake.

very low food security A condition characterized by disrupted eating patterns and reduced food intake caused by a lack of food access and availability.

In the 48 contiguous states, the annual poverty threshold is $24,230 for a family of four.

Other Factors Associated with Food Insecurity

In addition to income level, many individual and socioeconomic factors can also predispose a person or family to food insecurity. For example, in the United States, food insecurity is more prevalent in certain regions of the country and among certain ethnic groups (see Figure 15.1).[3] Black and Latino households are at higher risk of food insecurity than most other racial and ethnic groups, and households headed by single women are at even higher risk. Individuals living in urban and rural areas are more likely to experience food insecurity than are those living in suburban regions. It is important to recognize, however, that all three of these factors (ethnicity, head of household, and living location) are strongly associated with income status. Again, poverty is the most telling risk factor of all.

Socioeconomically disadvantaged communities often lack access to affordable and/or nutritious foods. In such an environment, referred to as a **food desert**, residents rely on local food outlets that offer limited and expensive food choices that are low in nutritional value.[6] A lack of supermarkets within disadvantaged communities presents additional hardships. Individuals must either purchase groceries at local convenience stores or navigate public transportation to other communities that have more affordable market choices—a difficult task when transporting a week's worth (or more) of groceries. Socioeconomic disparities have adverse health outcomes, and disadvantaged populations tend to have high rates of obesity, type 2 diabetes, and cardiovascular disease. Marketplace incentives designed to attract food retailers into neighborhoods determined to be food deserts are currently underway. Nutritionists and urban developers alike hope that improved access to affordable and healthy foods such as fruits, vegetables, whole grains, and low-fat dairy products will

Millions of Americans struggle to get enough to eat. Although assistance programs provide food to those in need, many programs report increasing difficulty meeting demand.

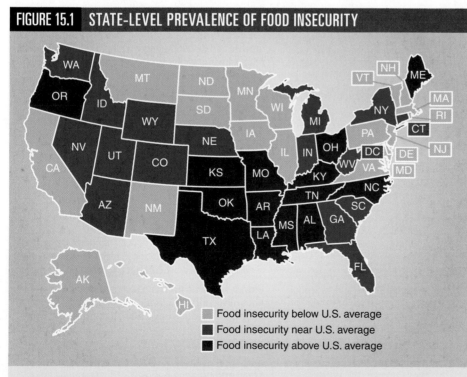

FIGURE 15.1 STATE-LEVEL PREVALENCE OF FOOD INSECURITY

Food insecurity below U.S. average
Food insecurity near U.S. average
Food insecurity above U.S. average

Rates of food insecurity vary considerably among U.S. states.

Source: Alisha Coleman-Jensen, Matthew P. Rabbitt, Christian Gregory, and Anita Singh. Household Food Security in the United States in 2014, ERR-194, U.S. Department of Agriculture, Economic Research Service, September 2015.

food desert An environment that lacks access to affordable and/or nutritious foods.

improve health and well-being among disadvantaged populations. You can learn more about where food deserts are located in the United States by visiting the USDA's Food Desert Locator (http://www.ers.usda .gov/data/fooddesert/).

 ## 15-2 WHAT ARE THE CONSEQUENCES OF FOOD INSECURITY?

Although food insecurity does not typically lead to starvation or nutrient deficiencies in the United States, with perhaps the exception of iron, it still represents a major public health concern. There are numerous consequences of food insecurity, many of which have been studied most extensively in women and children. For example, some studies have shown that mothers in food-insecure households shield their children from hunger by consuming less food themselves.[7] For this reason, women often experience the negative consequences of food insecurity before their children do. But despite parents' best efforts, it is children who most often experience the most significant long-term effects of living in a food-insecure household.[8] Although parents often try to protect their children from the realities of food insecurity, interviews with children reveal circumstantial awareness that is not always apparent to other family members.[9] Studies show that food-insecure children tend to have difficulties in school, earn lower scores on standardized tests, miss more days of school, exhibit more behavioral problems and depression, and be at increased risk of suicide.[10] Of course, these responses are not direct physiological consequences of food insecurity, but rather repercussions common to households experiencing this problem.

In addition to women and children, older adults are also at high risk for food insecurity.[11] However, they often experience food insecurity differently than children, young adults, and other population groups. Elders with limited mobility and poor health, for example, may have food available to them, but they may experience difficulty or anxiety associated with meal preparation and eating. Although older adults face unique challenges, poverty remains a significant indicator of food insecurity for this group. In fact, poverty rates are highest among older women and among older adults who live alone.[4] Many older people have relatively low fixed incomes, which they are unable to supplement with additional employment. Because income is often used to pay for basic living expenses (such as housing and utilities) rather than food, many elderly people do not have enough to eat. Perhaps not surprisingly, the prevalence of food insecurity among older adults is on the rise.[3]

15-2a Food-Based Assistance in the United States

Fortunately for many, there are a number of programs and services available in the United States to alleviate food insecurity. Some programs are federally funded, whereas others are community efforts staffed by volunteers. Select federally funded food assistance programs are listed in Table 15.1.

TABLE 15.1 SELECTED FEDERALLY FUNDED FOOD ASSISTANCE PROGRAMS	
Program	**Major Objective**
Child and Adult Care Food Program (CACFP)	Provides aid to child and adult care institutions and family or group daycare homes for the provision of nutritious foods that contribute to the wellness, healthy growth, and development of young children, as well as the health and wellness of older adults and chronically impaired disabled persons.
Expanded Food and Nutrition Education Program (EFNEP)	Assists low-income people in acquiring the knowledge, skills, attitudes, and behaviors necessary to maintain nutritionally balanced diets. Contributes to personal development and the improvement of the total family diet and nutritional well-being.
Supplemental Nutrition Assistance Program (SNAP)	Offers nutrition assistance to millions of eligible low-income individuals and families, while providing economic benefits to communities; previously referred to as the Food Stamp Program.
Head Start and Early Head Start Programs	Promotes school readiness and enhance the social and cognitive development of children by providing educational, health, nutritional, social, and other services to enrolled children and families.
National School Lunch, School Breakfast, Fresh Fruit and Vegetable, Special Milk, and Summer Food Service Programs	Provides children from low-income families with nutritious meals, including fresh fruits, vegetables, and milk, for free or at reduced cost, both during the school year and summer.
Special Supplemental Nutrition Program for Women, Infants, and Children (WIC)	Provides supplemental foods, health care referrals, and nutrition education to low-income pregnant, breastfeeding, and nonbreastfeeding postpartum women, and to infants and children (up to five years old) who are found to be at nutritional risk.

Supplemental Nutrition Assistance Program

The Supplemental Nutrition Assistance Program (SNAP), formerly known as the Food Stamp Program, is often the first line of defense against hunger for low-income households. Administered and operated by the USDA, SNAP currently helps more than 45 million people pay for food each month. Recent studies show that 44 percent of SNAP participants are children (18 years old or younger), with almost two-thirds of SNAP children living in single-parent households. In total, 76 percent of SNAP benefits go to households with children, 11.9 percent go to households with disabled persons, and 10 percent go to households with senior citizens.[12] Individuals who are eligible for SNAP are given *electronic benefit transfer (EBT)* cards that can be used like debit cards to make food purchases at grocery stores, convenience stores, and many farmers' markets. A household's monthly monetary allotment depends on the number of people in the household and household members' combined income. In 2015, the average SNAP client received a monthly benefit of $126, and the average household received $256 monthly.[12] SNAP participants can only use their EBT cards to purchase food items; cards cannot be used to buy beer, wine, liquor, cigarettes, or tobacco; any nonfood items; vitamins and medicines; foods that will be eaten in the store; or hot foods.[12] For some, applying for SNAP may be a daunting process, and for others, it may feel stigmatizing or demeaning.

Special Supplemental Nutrition Program for Women, Infants, and Children

Millions of women, infants, and children in the United States benefit from the Special Supplemental Nutrition Program for Women, Infants, and Children (WIC). This federally funded program, also administered by the USDA, assists pregnant women and families with young children in making nutritious food purchases. Individuals enrolled in WIC receive coupons that can be used to buy a variety of WIC-approved, nutrient-dense foods such as peanut butter, milk, rice, beans, cereal, and canned tuna. Many farmers' markets accept

The Special Supplemental Nutrition Program for Women, Infants, and Children (WIC) provides many important services to families in need.

WIC coupons, allowing parents to purchase a variety of fresh, locally grown fruits and vegetables. In addition to subsidizing food purchases, WIC provides health assistance and nutrition education to eligible women, young infants, and children. WIC's educational programs encourage exclusive breastfeeding and other optimal infant feeding guidelines set forth by the American Academy of Pediatrics.[13] Note that recent legislation will require that all agencies overseeing WIC benefits implement the use of EBT cards, rather than coupons, by 2020.[14]

Other Federally Funded Food-Based Assistance Programs

Other federally funded food-based assistance programs available in the Unites States include the **National School Lunch Program** and the **School Breakfast Program**. These federally funded programs provide nutritionally balanced meals (lunch and breakfast) either free of charge or at a reduced cost to school-age children. Administered by the

Supplemental Nutrition Assistance Program (SNAP) A federally funded program, formerly known as the Food Stamp Program, that helps low-income households pay for food.

National School Lunch Program A federally funded program that provides nutritionally balanced meals either free of charge or at a reduced cost to school-age children at lunch time.

School Breakfast Program A federally funded program that provides nutritionally balanced meals (lunch and breakfast) either free of charge or at a reduced cost to school-age children.

USDA, these programs are available in public schools, nonprofit private schools, residential child-care institutions, and after-school enrichment programs. The National School Lunch Program provided lunch for more than 31 million children each school day in 2012.[15] Since its enactment in 1946, the National School Lunch Program has served more than 224 billion lunches. In fact, because there is such a need, many schools and child-care programs also provide breakfast, lunch, and snacks to children throughout summer vacation. In fact, the USDA now supports a Summer Food Service Program to ensure that low-income children continue to receive nutritious meals when school is not in session. In 2010, the Institute of Medicine published a comprehensive document that provides guidance as to the optimal types and quantities of foods that should be served in the School Breakfast and National School Lunch Programs.[16] This document was followed by the passage of the Healthy Hunger-Free Kids Act of 2010, legislation that allows the USDA—for the first time in over 30 years—to make real reforms to the school lunch and breakfast programs and thus improve the critical nutritional safety net for millions of at-risk children.[17]

Privately Funded Food Assistance Programs

In addition to these and other government-funded programs, the private sector provides several services to make food more available to those in need. Private organizations include food recovery programs, food banks and pantries, and food kitchens—many of which are staffed by volunteers from the community. A **food bank** is an agency that collects donated foods and distributes them to local food pantries, shelters, and soup kitchens, while a **food pantry** is a program that provides canned, boxed, and sometimes fresh foods directly to individuals in need. Both food banks and food pantries rely on community donations to stock

The estimated value of wasted food in the United States is more than $161 billion/year, equaling 1,249 kcal/person/day. Food distribution programs (such as the one shown here) collect and distribute food that would otherwise be wasted.

their shelves with nonperishable and perishable items, which are then distributed to people who need them. A **food kitchen** is a program that serves prepared meals to members of the community, but mostly to those who are homeless or living in shelters.

A **food recovery program** such as that run by Feeding America® (formerly called America's Second Harvest) collects and redistributes discarded food that would have otherwise gone to waste. Recovered foods are donated to food pantries, emergency kitchens, and homeless shelters, providing millions of hungry people with food. The USDA estimates that, in the United States, 31 percent—or 133 billion pounds—of the 430 billion pounds of the available food supply at the retail and consumer levels goes to waste every year. The estimated value of this food loss is more than $161 billion

food bank An organization that collects donated foods and distributes them to local food pantries, shelters, and soup kitchens.

food pantry A program that provides canned, boxed, and sometimes fresh foods directly to individuals in need.

food kitchen A program that prepares and serves meals to members of the community who are in need.

food recovery program A program that collects and redistributes discarded food that would have otherwise gone to waste.

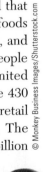

or 1,249 kcal/person/day.[18] There are many types of food recovery efforts, some of which include the following:

- *Field gleaning* programs gather and distribute agricultural crops that would otherwise not have been harvested.

- *Food rescue* initiatives collect unused perishable foods from retail grocery stores, gardens, restaurants, campus dining facilities, hotels, and/or caterers.

- *Nonperishable food collection* programs gather and donate damaged and/or dated canned and boxed foods from retail sources.

15-3 WHAT CAUSES WORLDWIDE HUNGER AND MALNUTRITION?

Because poverty is more prevalent in developing countries than in industrialized ones, food insecurity tends to be most ubiquitous and severe in nations with low *per capita* incomes. The Food and Agriculture Organization of the United Nations (FAO) estimates that about 795 million people are undernourished globally, down 167 million over the last decade, and 216 million less than in 1990–92. The decline is more pronounced in developing regions, despite significant population growth.[19] In recent years, progress has been hindered by slower and less inclusive economic growth, as well as political instability in some developing regions, such as Central Africa, the Middle East, and western Asia. You can see the prevalence of hunger around the world in Figure 15.2. For developing regions as a whole, the proportion of underweight children under five years of age has also declined. In summary, although there has been progress in decreasing global food insecurity, there is still much work to do in this regard.

15-3a Many Factors Contribute to Global Food Insecurity

The causes of global food insecurity are complex, and the factors that contribute to food insecurity in poor countries are often different from those that contribute to food insecurity in the United States. Most experts agree that global food insecurity is not caused by a lack

FIGURE 15.2 INTERNATIONAL PREVALENCE OF FOOD INSECURITY

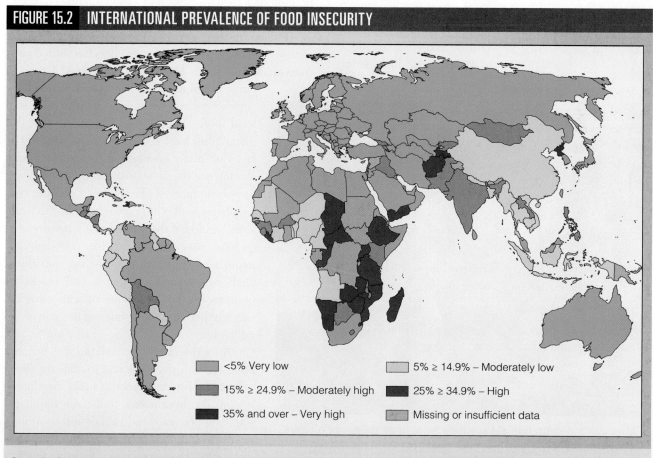

<5% Very low

5% ≥ 14.9% – Moderately low

15% ≥ 24.9% – Moderately high

25% ≥ 34.9% – High

35% and over – Very high

Missing or insufficient data

Source: Food and Agriculture Organization of the United Nations. State of food insecurity in the world. 2015. Available at: www.fao.org/hunger/en/

of available food on the international level. Rather, it is caused by diminished local food supplies, which are themselves caused by a variety of socioeconomic conditions. Political instability, a lack of available land for growing crops, population growth, and gender inequalities can all contribute to food instability.

Political Unrest The availability of and access to food are often limited by civil strife, war, and political unrest. Political turmoil can displace millions of people from their homes and force them to relocate to crude facilities set up for refugees. In countries with large refugee populations (such as Turkey, Pakistan, and Lebanon), food insecurity and malnutrition are rampant. According to the United Nations (UN), those living in refugee camps have the highest rates of disease and malnutrition of any group worldwide.[20] Because of the danger and logistical problems associated with political unrest, it can be difficult for relief agencies to provide much-needed aid to innocent civilians. Despite recent progress made by repatriation movements, the number of refugees is once again on the rise. Mainly attributable to violence taking place in the Middle East, the worldwide number of forcibly displaced persons is approaching 60 million. During 2014, an average of 42,500 individuals were

urbanization A population shift whereby large numbers of people move from rural to urban regions within a country.

forced to leave their homes and seek protection elsewhere, either within the borders of their own country or within other countries.[21]

Urbanization The use of land for reasons other than to grow crops for feeding a region's people and supporting local economies can contribute significantly to a food shortage.[22] Without the land to farm, people cannot produce adequate amounts and varieties of food for themselves and their families. As a result, many people relocate from rural to urban regions with the hope of finding employment opportunities. This shift in America's population, called **urbanization**, is both a consequence and cause of food shortages in many parts of the world. Urbanization and industrialization have had profound impacts on population demographics, transforming food systems and creating new nutritional challenges. For example, the expansion of large supermarket chains in urban areas has greatly impacted small food producers and retailers. Rather than buying food produced by local farmers, supermarket chains are more likely to utilize large consolidated food distribution centers. A shift in food production, procurement, and distribution systems has contributed to the displacement of workers, a decline in traditional food markets, and a fundamental change in local food culture.

Fueled by changes brought on by urbanization, the composition of diets among city-dwellers has shifted away from traditional foods and toward processed foods. In large, urban areas, fast-food restaurants have largely replaced street vendors who once sold local foods. Recall from Chapter 1 that this dietary shift has led to nutrition transition. The fact that many developing regions face food shortages at the same time that they experience increasing rates of obesity, heart disease, type 2 diabetes, and other diet-related health problems highlights the special challenges related to food insecurity worldwide. The simultaneous occurrence of two distinct nutritional challenges—food shortages and obesity—underscores the importance of addressing the needs of both the rural poor and those in urban migration.

Another challenge related to, but not exclusive to, urbanization is the lack of clean, affordable drinking water in many developing regions. Without access to a clean municipal water supply, many individuals and communities rely on nutrient-poor and calorie-rich sugar-sweetened beverages.[23] Aside from the fact that these are somewhat expensive, consuming

© kojoku/Shutterstock.com

Although humanitarian aid agencies try to meet the needs of those living in refugee camps, their efforts cannot match the immense needs for food, water, and medicine.

sugar-sweetened beverages instead of other more nutritious choices such as water and milk can cause health problems such as tooth decay and increased risk for obesity.

Some health experts argue that access to free (or at least affordable) drinkable water should be considered a basic human right. What do you think?

Population Growth   Population growth across some of the poorest regions of the world has significantly increased the challenge of providing adequate food and water to all those in need. Discouragingly, but perhaps not surprisingly, the countries with the fastest-growing populations tend to be those already burdened with staggering rates of hunger and malnutrition.[24] As a population proliferates, the ability of a government and assistance programs to satisfy the most basic needs—food and shelter—may be compromised even further.

Gender Inequality   In many developing nations, a gap exists between the opportunities that are made available to men and women. Some experts believe that promoting gender equality among underprivileged peoples holds the greatest promise for reversing the steady increase in global hunger.[25] For instance, providing equal educational and employment opportunities for girls and women might increase individual earning capacities, improve maternal nutrition and health, and bolster overall *per capita* income. Clearly, a lack of education and opportunity sustains the vicious cycle of poverty that is passed on from one generation to the next.

15-3b Global Food Insecurity and Malnutrition

Although there are many consequences of food insecurity in poor and developing countries, the most important and devastating consequence might be malnutrition. As you have learned, malnutrition is poor nutritional status that results from inadequate or excessive dietary intake. In the case of a food shortage, malnutrition takes the form of undernutrition, which has both short- and long-term effects on the health of individuals, families, and societies. In the case of the nutrition transition, malnutrition takes the form of overnutrition as urbanized individuals shift toward unhealthy, unbalanced diets.

Forms of Global Malnutrition   Some malnourished people consume sufficient calories but lack certain nutrients, whereas others lack both calories and nutrients. Iron, iodine, and vitamin A deficiencies, which are the three most common micronutrient deficiencies, affect billions of people worldwide. Countless infants and children suffer from iron and iodine deficiencies, both of which can impair growth and cognitive development. The lives of millions of preschool-age children are further compromised by vitamin A deficiency, which causes blindness and many other serious consequences.[26]

Simply put, the number of women, infants, and children in the world with micronutrient deficiencies is staggering. For instance, the World Health Organization (WHO) estimates that 2 billion people—over 30 percent of the world's population—are anemic, many due to iron deficiency.[27] Yet, the resources exist to rectify these problems: by distributing low-cost, nutrient-rich foods and/or nutrient supplements to those in need, world governments and agencies could eradicate micronutrient deficiencies throughout the world. Some efforts have been made to do just that, but meeting the needs of the hungry is an arduous, time-sensitive process. Although adequate nutrition is important throughout a child's life, the window of opportunity for an intervention to have the greatest impact is between conception and the child's second birthday—a period of time often referred to as "the first 1,000 days." The effects of persistent malnutrition on a child's health and development are largely irreversible after the age of 2.

Individual Consequences of Malnutrition   As with food insecurity, women, infants, and children are especially vulnerable to malnutrition. When experienced during

In many cultures around the world, women and girls are more likely than men and boys to be uneducated and experience food insecurity. Closing this gender gap is, therefore, an important global priority.

Saving the Lives of Children with Ready-to-Use Therapeutic Food

Malnutrition claims thousands of lives each day, but the burden of persistent hunger is particularly harmful to developing children. For example, almost one-half of the children in India are underweight and have impaired growth due to malnutrition by the time they are five years old.[1] Although providing children with food seems like the obvious solution, this process can be more complicated than you might think. For example, the children who need food the most often live in remote areas that are difficult to reach. A lack of refrigeration, clean water, and limited cooking amenities also complicates matters. For example, contaminated water and drought render powdered milk useless. Perishable foods spoil if not stored properly, and grains and cereals are not high-quality sources of protein or iron. Although it is difficult to provide adequate food to individuals living in these conditions, the availability of ready-to-use therapeutic foods (RUTF) has overcome many such geographic and logistic challenges. RUTF products are prepackaged, require no preparation, and are mainly composed of peanut butter, vegetable oil, milk powder, sugar, vitamins, and minerals. These nutrient-dense, high-quality protein products require no refrigeration, are easily distributed, and generally have long shelf lives. RUTFs are becoming the standard of care when it comes to refeeding malnourished children worldwide, and their use is enthusiastically supported by many organizations such as UNICEF.[29]

A ready-to-use therapeutic food is a nutrient-dense, high-quality protein product that has a long shelf life. The use of RUTFs has been credited with saving the lives of millions of children around the world.

Source: mananutrition

pregnancy, malnutrition can deplete a mother's nutritional stores and increase her risk of having a low-birth-weight (LBW) baby. Poor maternal nutritional status can also increase the risk of neonatal death. In fact, some studies estimate that nearly 60 percent of the deaths of infants and young children in the world are caused in part by malnutrition.[28] Because poor nutrition compromises the immune system, malnutrition can worsen the adverse effects of disease, leading to premature death. When a child is slightly underweight, risk of death increases to 2.5 times that of a child with a healthy weight.[28] Risk of death increases further when a child is severely underweight.

Even if it does not cause premature death, malnutrition can seriously affect a child's growth and development. Children with impaired growth tend to be in the lowest height-for-age percentiles on a growth reference curve. As a consequence of malnutrition, an estimated 24 percent of children younger than five years of age—149 million children—have impaired growth.[28] Africa has the highest percentage of undernourished infants and young children (28 percent), though it is followed closely by Asia (24 percent).[29] Even if a malnourished infant survives, impaired growth can compromise his or her health, well-being, and ability to function later in life.

Societal Consequences of Malnutrition

Beyond its effects on the health of the individual, malnutrition is harmful to societies. Extensive food insecurity and malnutrition can result in an entire nation of adults with reduced capacities for physical work and, therefore, lower work productivity overall. These consequences can have profound and long-term adverse effects on a country's economic growth and standard of living. Thus, not only is poverty a cause of hunger, but hunger is a cause of poverty. Addressing both food insecurity and malnutrition is critical to economic progress in underdeveloped nations.

Even if it does not cause premature death, malnutrition can seriously affect a child's growth and development. This Haitian toddler, showing signs of malnutrition, is being fed peanut butter bread given to her by missionaries.

15-3c International Organizations Provide Global Food-Based Assistance

Unlike most wealthy countries, impoverished countries often lack stable governments and have few programs in place to assist impoverished people. Developing countries typically depend on relief efforts provided by international organizations such as the WHO, the United Nations (UN), U.S. Peace Corps, and Heifer International®. Organizations such as these help individuals and entire communities make lasting changes that ultimately improve health and food security across the developing world. International interventions with the greatest impact include efforts to improve maternal nutrition during pregnancy and lactation, reduce societal inequities, provide access to health services, promote self-sufficiency, and reduce illiteracy.[30]

The United Nations Many international organizations are committed to the alleviation of world hunger. One such organization is the UN, a multinational organization first established in 1945 to promote peace through international cooperation and collective security. Today, the UN is comprised of 193 member states, and though the organization serves many purposes, combating international hunger is among its most important efforts. For example, the United Nations Children's Emergency Fund (UNICEF), a component of the UN, presented a conceptual framework in 2011 that provides communities with incentives for working to improve quality of life (such as by relieving hunger) in the world's poorest countries. Proposed incentives include financial reimbursement for families whose children attend school and for families who start their own small businesses.[31] This type of initiative is called a **conditional cash-transfer program**.

> **conditional cash-transfer program** An initiative that offers financial reimbursement for individuals or communities that work to improve quality of life.

The United Nations is a multinational organization that addresses a multitude of global problems, such as worldwide food insecurity.

FIGURE 15.3   UNITED NATION'S 2015 MILLENNIUM DEVELOPMENT GOALS

1 ERADICATE EXTREME POVERTY AND HUNGER

2 ACHIEVE UNIVERSAL PRIMARY EDUCATION

3 PROMOTE GENDER EQUALITY AND EMPOWER WOMEN

4 REDUCE CHILD MORTALITY

5 IMPROVE MATERNAL HEALTH

6 COMBAT HIV/AIDS, MALARIA AND OTHER DISEASES

7 ENSURE ENVIRONMENTAL SUSTAINABILITY

8 GLOBAL PARTNERSHIP FOR DEVELOPMENT

As the era of these Millennium Development goals comes to an end in 2016, the UN will launch its 2030 Agenda for Sustainable Development.

world, this unprecedented effort addressed the needs of the world's poorest countries. Its eight goals are illustrated in Figure 15.3. As you can see, many if not all of these goals are related to nutrition and food insecurity, particularly in resource-low populations. In 2015, the UN reported the good news that, overall, Goal 1 has almost been met at the global level. However, they noted marked differences in progress among individual countries. Clearly, there is more work to be done in this important arena. In response, the UN announced late in 2015 that it was launching a new initiative called the 2030 Agenda for Sustainable Development, which includes 17 goals aimed at realizing the human rights of all individuals and achieving gender equality and empowerment for all women and girls.[33] Particularly noteworthy is the UN's overarching goal "to end poverty and hunger, in all their forms and dimensions, and to ensure that all human beings can fulfill their potential in dignity and equality and in a healthy environment."

Peace Corps How can one person make a meaningful difference in the world, especially in regard to world hunger? One such example can be gleaned from President John F. Kennedy, who challenged students in 1961 to serve their country by working to improve quality of life for people in developing countries. This challenge served as the inspiration for a federally funded program called the **U.S. Peace Corps**. The mission of the Peace Corps was—and still is—to promote world peace and friendship by doing the following:

- Assisting interested countries in meeting their need for trained men and women;
- Bringing a better understanding of Americans to people in other countries; and
- Helping promote a better understanding of other peoples on the part of Americans.

Another UN effort, the Millennium Development Project, initiated in 2000, pledged to halve the proportion of people who live in hunger by the year 2015.[32] Endorsed by the majority of the countries in the

U.S. Peace Corps A federally funded volunteer program that promotes world peace and friendship worldwide.

Many Peace Corps volunteers find their work both challenging and rewarding.

Source: Peace Corps.

Since the Peace Corps was developed, more than 220,000 people have served in more than 140 countries. These volunteers work to improve developing nations by helping farmers grow crops, teaching mothers to better care for their children, and educating entire communities about health and disease prevention. Thus, the Peace Corps offers the opportunity to make a difference in the lives of others by addressing the problems of food insecurity and malnutrition throughout the world.

Heifer International Heifer International® is a humanitarian effort with a global commitment to foster environmentally sound farming methods that combat both hunger and environmental concerns. This organization recognizes that impoverished peoples often make decisions based on short-term needs rather than cultivating long-term solutions. Heifer International strives to teach families how to restore and manage land in ways that provide food and income for generations to come. Heifer's pragmatic problem-solving approach leads to novel long-term solutions that empower impoverished communities to provide for themselves. This focus on long-term development (rather than temporary relief) helps restore hope, health, and dignity among those with few resources. Most remarkable, however, is Heifer's *living loans* program, which ensures project sustainability. In a living loan, a community receives a gift of livestock that brings benefits such as food, wool, and nonmechanized power. To repay this loan, the community gifts the offspring of the livestock to another farmer or community. This

Need a quick gift idea? As an alternative to traditional gifts, consider making a contribution to a nutrition-related charitable organization in someone's name.

secondary donation repays the original debt, while bringing the hope of prosperity to others. This simple concept of passing on the gift is the founding philosophy of Heifer International, and the means by which it has fostered a living cycle of sustainability for over 65 years.

 15-4 **WHAT CAN YOU DO TO ALLEVIATE FOOD INSECURITY?**

Experts generally agree that there is enough food in the world to feed every living person.[34] Why then are food insecurity and malnutrition so pervasive and devastating? As you have learned, the causes of food insecurity and malnutrition differ by geographic region, gender, political climate, life stage, economic policy, ethnicity, and rate of population growth—among many other factors. Because the causes of food insecurity are so varied and intrinsically interrelated, it is important for nutritionists and policymakers to work both diligently and carefully in a culturally appropriate manner. Only by considering

the complexity of the issue, can we address the relative importance of each contributing factor. And it is only then that effective solutions can be developed.

Leading world health experts agree that improving food availability and access must be a global priority. Although malnutrition is a direct consequence of insufficient dietary intake, the ultimate causes of malnutrition often have more to do with economic and social circumstances. Even if all other factors are on the right track, a high prevalence of certain diseases (such as AIDS), violence, illiteracy, and political corruption can devastate food availability and access. Thus, to affect a genuine remedy for food insecurity and related malnutrition, the underlying societal problems must be addressed.

15-4a Taking Action against Hunger

Although the problem of food insecurity may seem staggering both at home and abroad, it is important to remember that an individual can do a lot to take action against hunger. Despite the scope of the problem, your actions can make a difference in the lives of others. Working alone or collectively toward the elimination of hunger and malnutrition is a worthwhile (and noble) personal and professional priority. For example, most universities and colleges offer alternative spring break opportunities that consist of faculty-led trips abroad. These types of volunteer experiences provide college students with exceptional experiential learning opportunities that are specifically designed to help students gain global perspectives. In addition, many such programs are centered around helping poverty-stricken populations, often times in a way that improves nutritional status. Many nutrition-related academic societies also focus some of their efforts on global nutrition. For example, the Academy of Nutrition and Dietetics (AND) is one of several professional organizations that work to alleviate world hunger by challenging its members to take action. For instance, the AND supports a Dietetic Practice Group to encourage dietitians to work with each other and with other health professionals to reduce poverty and hunger in their communities. The American Society for Nutrition (ASN) and the Nevin Scrimshaw International Nutrition Foundation (INF) are other organizations committed to the development of international strategies and policies that can help alleviate hunger and poverty around the world.

STUDY TOOLS 15

READY TO STUDY? IN THE BOOK, YOU CAN:

☐ Rip out the Chapter Review Card, which includes key terms and chapter summaries.

ONLINE AT WWW.CENGAGEBRAIN.COM, YOU CAN:

☐ Recall materials from the chapter in inline activities.

☐ Interact with figures from the text to check your understanding.

☐ Prepare for tests with quizzes.

☐ Review the key terms with Flash Cards.

NUTR
ONLINE

ACCESS TEXTBOOK CONTENT ONLINE—
INCLUDING ON SMARTPHONES!

Includes Videos & Other
Interactive Resources!

MANAGE MY COURSE ⌄   STUDENT

NUTR2

CHAPTER
1

Why Does Nutrition Matter?

CHAPTER
2

Choosing Foods Wisely

4LTR
PRESS

Access NUTR ONLINE at www.cengagebrain.com

ENDNOTES

1

1. Vicini J, Etherton T, Kris-Etherton P, Ballam J, Denham S, Staub R, Goldstein D, Cady R, McGrath M, Lucy M. Survey of retail milk composition as affected by label claims regarding farm-management practices. Journal of the American Dietetic Association. 2008;108:1198–203. Dangour AD, Dodhia SK, Hayter A, Allen E, Lock K, Uauy R. Nutritional quality of organic foods: A systematic review. American Journal of Clinical Nutrition. 2009;90:680–5. Drakou M, Birmpa A, Koutelidakis AE, Komaitis M, Panagou EZ, Kapsokefalou M. Total antioxidant capacity, total phenolic content and iron and zinc dialyzability in selected Greek varieties of table olives, tomatoes and legumes from conventional and organic farming. International Journal of Food Science and Nutrition. 2015;66:197–202.

2. Ward RE, German JB. Zoonutrients and health. Food Technology. 2003;57:30–36. Pappas E, Schaich KM, Gullett NP, Ruhul Amin AR, Bayraktar S, Pezzuto JM, Shin DM, Khuri FR, Aggarwal BB, Surh YJ, Kucuk O. Cancer prevention with natural compounds. Seminars in Oncology. 2010;37:258–81. Riccioni G, Speranza L, Pesce M, Cusenza S, D'Orazio N, Glad MJ. Novel phytonutrient contributors to antioxidant protection against cardiovascular disease. Nutrition. 2012;28:605–10.

3. Gobbetti M, Cagno RD, De Angelis M. Functional microorganisms for functional food quality. Critical Reviews in Food Science and Nutrition. 2010;50:716–27. Mattes RD, Dreher ML. Nuts and healthy body weight maintenance mechanisms. Asia Pacific Journal of Clinical Nutrition. 2010;19:137–41. Rao AV, Snyder DM. Raspberries and human health: A review. Journal of Agricultural Food Chemistry. 2010;58:3871–83. Tulipani S, Mezzetti B, Battino M. Impact of strawberries on human health: Insight into marginally discussed bioactive compounds for the Mediterranean diet. Public Health Nutrition. 2009;12:1656–62.

4. Lieberman HR. Cognitive methods for assessing mental energy. Nutrition and Neuroscience. 2007;10:229–42. Chelben J, Piccone-Sapir A, Ianco I, Shoenfeld N, Kotler M, Strous RD. Effects of amino acid energy drinks leading to hospitalization in individuals with mental illness. General Hospital Psychiatry. 2008;30:187–9. Clauson KA, Shields KM, McQueen CE, Persad N. Safety issues associated with commercially available energy drinks. Journal of the American Pharmacologic Association. 2008;48:e55–63.

5. Carey SS. A beginner's guide to scientific method, 4th ed. Belmont, CA: Wadsworth/Thomson Learning; 2011.

6. Ibid.

7. Health, United States, 2014: With Special Feature on Adults Aged 55–64. Hyattsville, MD. 2015. Available from: http://www.cdc.gov/nchs/data/hus/hus14.pdf. Xu JQ, Kochanek KD, Murphy SL, Tejada-Vera B. Deaths: Final data for 2007. National vital statistics reports. Hyattsville, MD:

National Center for Health Statistics. 2010; 58:19. Available from: http://www.cdc.gov/NCHS/data/nvsr/nvsr58/nvsr58_19.pdf.

8. Ibid.

9. Perls T, Terry D. Understanding the determinants of exceptional longevity. Annals of Internal Medicine. 2003; 139:445–9. Solon-Biet SM, Mitchell SJ, de Cabo R, Raubenheimer D, LeCouteur DG, Simpson SJ. Macronutrients and caloric intake in health and longevity. Journal of Endocrinology. 2015;226:R17–28. Pes GM, Tolu F, Dore MP, Sechi GP, Errigo A, Canelada A, Poulain M. Mail longevity in Sardinia, a review of historical sources supporting a causal link with dietary factors. European Journal of Clinical Nutrition. 2015;69:411–8.

10. Anderson RE, Smith RD, Benson ES. The accelerated graying of American pathology. Human Pathology. 1991;22:210–4.

11. Armstrong BL, Conn LA, Pinner RW. Trends in infectious disease mortality in the United States during the 20th century. Journal of the American Medical Association. 1999;281:61–6. Centers for Disease Control and Prevention. Achievements in public health, 1900–1999: Control of infectious diseases. Morbidity and Mortality Weekly Report. 1999;48:621–9.

12. U.S. Department of Health and Human Services. Office of Disease Prevention and Health Promotion. Healthy People 2020. ODPHP Publication No. B0132. November 2010. Available from: http://www.healthypeople.gov.

13. Health, United States, 2014: With Special Feature on Adults Aged 55–64. Hyattsville, MD. 2015. Available from: http://www.cdc.gov/nchs/data/hus/hus14.pdf.

2

1. Institute of Medicine. Dietary reference intakes for vitamin C, vitamin E, selenium, and carotenoids. Washington, DC: National Academies Press; 2000.

2. Institute of Medicine. Dietary reference intakes: Application in dietary assessment. Washington, DC: National Academies Press; 2000.

3. Institute of Medicine. Dietary Reference Intakes for energy, carbohydrate, fiber, fat, fatty acids, cholesterol, protein, and amino acids. Washington, DC: National Academies Press; 2005.

4. Welsh S, Davis C, Shaw A. A brief history of food guides in the United States—Food Guide Pyramid. Nutrition Today. 1992 (Nov.–Dec.); 6–11. Welsh S, Davis C, Shaw A. Development of the food guide pyramid—food guide pyramid. Nutrition Today. 1992 (Nov.–Dec.);12–15.

5. U.S. Department of Agriculture and U.S. Department of Health and Human Services. 2015–2020 Dietary Guidelines for Americans, 2015–2020. 8th ed. Washington, DC Government Printing Office. December 2015.

6. U.S. Department of Agriculture. (2015) FAQs—Country of Origin Labeling (Beef and Pork Repeal). Available at: http://www.ams.usda.gov/sites/default/files/media/FAQs%20-%20COOL%20Beef%20Pork%20Repeal.pdf.

3

1. Garneau NL, Nuessle TM, Sloan MM, Santorico SA, Coughlin BC, Hayes JE. Crowdsourcing taste research: Genetic and phenotypic predictors of bitter taste perception as a model. Frontiers in Integrative Neuroscience, Published online May 2014. Available from: http://www.ncbi.nlm.nih.gov/pmc/articles/PMC4035552/.

2. Spechler SJ. Clinical manifestations and esophageal complications of GERD. American Journal of Medical Sciences. 2003;326:279–84.

3. DeCross AJ, Marshall BJ. The role of Helicobacter pylori in acid-peptic disease. American Journal of Medical Sciences. 1993;306:381–92. Tan VP, Wong BC. Helicobacter pylori and gastritis: Untangling a complex relationship 27 years on. Journal of Gastroenterolgy and Hepatology. 2011;26:42–5. Leung FW. Risk factors for gastrointestinal complications in aspirin users: Review of clinical and experimental data. Digestive Diseases and Sciences. 2008;53:2604–15.

4. Meyer JH. Gastric emptying of ordinary food: Effect of antrum on particle size. American Journal of Physiology. 1980;239:G133–5. Hunt JN. Mechanisms and disorders of gastric emptying. Annual Review of Medicine. 1983; 34:219–29.

5. Dobrogosz WJ, Peacock TJ, Hassan HM. Evolution of the probiotic concept from conception to validation and acceptance in medical science. Advances in Applied Microbiology. 2010;72:1–41. Kolida S, Saulnier DM, Gibson GR. Gastrointestinal microflora: Probiotics. Advances in Applied Microbiology. 2006;59:187–219. Rastall RA. Bacteria in the gut: Friends and foes and how to alter the balance. Journal of Nutrition. 2004;134:2022S–6S. Pham TT, Shah NP. Biotransformation of isoflavone glycosides by bifidobacterium animalis in soymilk supplemented with skim milk powder. Journal of Food Science. 2007;72:316–24. Guslandi MJ. Probiotic agents in the treatment of irritable bowel syndrome. Journal of International Medical Research. 2007;35:583–9. Hedin C, Whelan K, Lindsay JO. Evidence for the use of probiotics and prebiotics in inflammatory bowel disease: A review of clinical trials. Proceedings of the Nutrition Society. 2007;66:307–15. DiBaise JK, Zhang H, Crowell MD, Krajmalnik-Brown R, Decker GA, Rittmann BE. Gut microbiota and its possible relationship with obesity. Mayo Clinic Proceedings. 2008;83:460–9. Kalliomäki M, Collado MC, Salminen S, Isolauri E. Early differences in fecal microbiota composition in children may predict overweight. American Journal of Clinical Nutrition. 2008;87:534–8. Cani PD, Delzenne NM. Gut microflora as a target for energy and metabolic homeostasis. Current Opinion in Clinical Nutrition and Metabolic Care. 2007;10:729–34. Crow JR, Davis SL, Chaykosky DM, Smith TT, Smith JM. Probiotics and fecal microbiota transplant for primary and secondary prevention of clostridium difficile infection. Pharmacotherapy. 2015;35:1016–25.

6. Miller FG, Colloca L, Kaptchuk TJ. The placebo effect: Illness and interpersonal health. Perspectives in Biology and Medicine.

2009;52:518–39. Wampold BE, Minami T, Tierney SC, Baskin TW, Bhati KS. The placebo is powerful: Estimating placebo effects in medicine and psychotherapy from randomized clinical trials. Journal of Clinical Psychology. 2005;6:835–54. Kaptchuk TJ, Kelley JM, Conboy LA, Davis RB, Kerr CE, Jacbson EE, Kirsch I, Schyner RN, Nam BH, Nguyen LT, Park M, Rivers AL, McManus C, Kokkotou E, Drossman DA, Goldman P, Lembo AJ. Components of placebo effect: Randomised controlled trial in patients with irritable bowel syndrome. British Medical Journal. 2008;336:1–8.

7. Xavier RJ, Podolsky DK. Unraveling the pathogenesis of inflammatory bowel disease. Nature. 2007;448:427–34. Lakatos PL. World recent trends in the epidemiology of inflammatory bowel diseases: Up or down? Journal of Gastroenterology. 2006;12:6102–8. Malik TA. Inflammatory Bowel Disease: Historical Perspective, Epidemiology, and Risk Factors. Surgical Clinics of North America. 2015;95:1105–22.

8. Fasano A, Berti I, Gerarduzzi T, Not T, Colletti RB, Drago S, Elitsur Y, Green P, Guandalini S, Hill ID, Pietzak M, Ventura A, Thorpe M, Kryszak D, Fornaroli F, Wasserman SS, Murray JA, Horvath K. Prevalence of Celiac Disease in at-risk and not-at-risk groups in the United States. A large multicenter study. Archives of Internal Medicine. 2003;163:286–92. Elli L, Branchi F, Tomba C, Villalta D, Norsa L, Ferretti F, Roncoroni L, Bardella MT. Diagnosis of gluten related disorders: Celiac disease, wheat allergy and non-celiac gluten sensitivity. World Journal of Gastroenterology. 2015;21:7110 9.

4

1. Stanhope KL, Havel PJ. Fructose consumption: Recent results and their potential implications. Annals of the New York Academy of Sciences. 2010;1190:15–24. Welsh JA, Sharma AJ, Grellinger L, Vos MB. Consumption of added sugars is decreasing in the United States. American Journal of Clinical Nutrition. 2011;94:726–34.

2. U.S. Department of Agriculture, Economic Research Service. 2012. Table 51—Refined cane and beet sugar: Estimated number of per capita calories consumed daily, by calendar year. Table 52—High fructose corn syrup: Estimated number of per capita calories consumed daily, by calendar year. Table 53-Other sweeteners estimated number of per capita calories consumed daily, by calendar year. Sugar and Sweeteners Yearbook 2012. Available from: http://www.ers.usda.gov/data-products/sugar-and-sweeteners-yearbook-tables.aspx.

3. Sanchez-Lozada LG, Mu W, Roncal C, Sautin YY, Abdelmalek M, Reungjui S, Le M, Nakagawa T, Lan HY, Yu X, Johnson RJ. Comparison of free fructose and glucose to sucrose in the ability to cause fatty liver. European Journal of Nutrition 2010;49:1–9. Vos MB, Lavine JE. Dietary fructose in nonalcoholic fatty liver disease. Hepatology. 2013;57:2424–31. Ma J, Karlsen C, Chung M, Jacques PF, Saltzman E, Smith CE, Fox CS, McKeown NM. Potential link between excess added sugar intake and ectopic fat: A systematic review of randomized controlled trials. Nutrition Reviews. 2015;[Epub ahead of print]. Dornas WC, de Lima WG, Pedrosa ML, Silva ME. Health Implications of High-Fructose Intake and Current Research. Advances in Nutrition. 2015;6:729–37.

4. Moeller, Fryhofer SA, Osbahr AJ 3rd, Robinowitz CB; Council on Science and Public Health, American Medical Association. The effects of high fructose corn syrup. Journal of the American College of Nutrition. 2009;28:619–26.

5. Available from: http://www.fda.gov/Food/GuidanceRegulation/GuidanceDocuments RegulatoryInformation/LabelingNutrition/ucm385663.htm.

6. Ervin RB, Ogden CL. U.S. Department of Health and Human Services, Centers for Disease Control and Prevention. 2013. NCHS Data Brief, No. 122: Consumption of Added Sugars Among U.S. Adults, 2005–2010.

7. National Cancer Institute. Sources of added sugars in the diets of the U.S. population ages 2 years and older, NHANES 2005–2006. Sources of Calories from Added Sugars among the US Population, 2005–06. Applied Research National Cancer Institute. 2014. Available from: http://healthcaredelivery.cancer.gov/.

8. U.S. Department of Agriculture and U.S. Department of Health and Human Services. Dietary Guidelines for Americans, 2010. 7th ed., Washington, DC: U.S. Government Printing Office, December 2010.

9. Johnson RK, Appel L, Brands M, Howard B, Lefevre M, Lustig R, Sacks F, Steffen L, Wyllie-Rosett J. Dietary sugars intake and cardiovascular health: A scientific statement from the American Heart Association. Circulation. 2009;120:1011–20.

10. Magnuson BA, Burdock GA, Doull J, Kroes RM, Marsh GM, Pariza MW, Spencer PS, Waddell WJ, Walker R, Williams GM. Aspartame: A safety evaluation based on current use levels, regulations, and toxicological and epidemiological studies. Critical Reviews in Toxicology. 2007;37:629–727.

11. Hackman DA1, Giese N, Markowitz JS, McLean A, Ottariano SG, Tonelli C, Weissner W, Welch S, Ulbricht C. Agave (Agave americana): An evidence-based systematic review by the natural standard research collaboration. Journal of Herbal Pharmacotherapy. 2006;6:101–22.

12. Lomax AR, Calder PC. Prebiotics, immune function, infection and inflammation: A review of the evidence. The British Journal of Nutrition. 2009;101:633–58.

13. Meyer D. Health benefits of prebiotic fibers. Advances in Food and Nutrition Research. 2015;74:47–91.

14. U.S. Department of Health and Human Services. Health Claims Meeting Significant Scientific Agreement. Available from: http://www.fda.gov/Food/GuidanceRegulation/GuidanceDocumentsRegulatoryInformation/LabelingNutrition/ucm064919.htm.

15. Meyer D, Stasse-Wolthuis M. The bifidogenic effect of inulin and oligofructose and its consequences for gut health. European Journal of Clinical Nutrition. 2009;63:1277–89.

16. Anderson JW, Baird P, Davis RH Jr, Ferreri S, Knudtson M, Koraym A, Waters V, Williams CL. Health benefits of dietary fiber. Nutrition Reviews. 2009;67:188–205.

17. Charalampopoulos D, Wang R, Pandiella SS, Webb C. Application of cereals and cereal components in functional foods: A review. International Journal of Food Microbiology. 2002;79:131–41.

18. Vuorisalo T, Arjamaa O, Vasemägi A, Taavitsainen JP, Tourunen A, Saloniemi I. High lactose tolerance in North Europeans: A result of migration, not in situ milk consumption. Perspectives in Biological Medicine. 2012;55:163–74.

19. Venn BJ, Green TJ. Glycemic index and glycemic load: Measurement issues and their effect on diet-disease relationships. European Journal of Clinical Nutrition. 2007;6:S122–31.

20. Astrup A, Meinert Larsen T, Harper A. Atkins and other low-carbohydrate diets: Hoax or an effective tool for weight loss? Lancet. 2004;364:897–9. Hu T, Bazzano LA. The low-carbohydrate diet and cardiovascular risk factors: Evidence from epidemiologic studies. Nutrition, Metabolism & Cardiovascular Diseases. 2014;24:337–43.

21. Ariza MA, Vimalananda VG, Rosenzweig JL. Reviews of endocrine and metabolic disorders. The economic consequences of diabetes and cardiovascular disease in the United States. 2010;11:1–10. Centers for Disease Control and Prevention. National Diabetes Statistics Report: Estimates of Diabetes and Its Burden in the United States, 2014. Atlanta, GA: U.S. Department of Health and Human Services; 2014. Available from: http://www.cdc.gov/diabetes/data/statistics/2014statisticsreport.html.

22. Han JC, Lawlor DA, Kimm SY. Childhood obesity. Lancet. 2010;375:1737–48. Gurnani M, Birken C, Hamilton J. Childhood Obesity: Causes, Consequences, and Management. Pediatric Clinics of North America. 2015;62:821–40.

23. Groop L, Lyssenko V. Genetics of type 2 diabetes. An overview. Endocrinology and Nutrition. 2009;56:S34–7.

24. Votruba SB, Jensen MD. Regional fat deposition as a factor in FFA metabolism. Annual Review of Nutrition. 2007;27:149–63. Gurnani M, Birken C, Hamilton J. Childhood Obesity: Causes, Consequences, and Management. Pediatric Clinics of North America. 2015;62:821–40.

25. Albright A. What is public health practice telling us about diabetes? Journal of the American Dietetic Association. 2008;108:S12–8. Huang ES, Basu A, O'Grady M, Capretta JC. Projecting the Future Diabetes Population Size and Related Costs for the U.S. Diabetes Care. 2009;32:2225–29.

26. Baier LJ, Hanson RL. Genetic studies of the etiology of type 2 diabetes in Pima Indians: Hunting for pieces to a complicated puzzle. Diabetes. 2004;53:1181–6.

27. Slavin JL, Jacobs D, Marquart L, Wiemer K. The role of whole grains in disease prevention. Journal of the American Dietetic Association. 2001;101:780–5. Hoy KM, Goldman JD. Fiber intake of the U.S. population. What We Eat in America, NHANES 2009–2010. Food Surveys Research Group, Dietary Data Brief No. 12 September 2014. Available from:http://www.ars.usda.gov/SP2UserFiles/Place/80400530/pdf/DBrief/12_fiber_intake_0910.pdf.

5

1. Vickery HB. The origin of the word protein. Yale Journal of Biology and Medicine. 1950;22:387-393.

2. Institute of Medicine. Dietary Reference Intakes for energy, carbohydrate, fiber, fat, fatty acids, cholesterol, protein, and amino acids. Washington, DC: National Academies Press; 2005.

3. Furst P, Stehle P. What are the essential elements needed for the determination of amino acid requirements in humans? Journal of Nutrition. 2004;134:1558S–65S.

4. Erlandsen H, Patch MG, Gamez A, Straub M, Stevens RC. Structural studies on phenylalanine hydroxylase and implications toward understanding and treating phenylketonuria.

Pediatrics. 2003;112:1557–65. van Spronsen FJ, Enns GM. Future treatment strategies in phenylketonuria. Molecular Genetics and Metabolism. 2010;99:S90–5.

5. American Academy of Pediatrics. Pediatric nutrition handbook, 6th ed. Kleinman RE, editor. Elk Grove Village, IL: American Academy of Pediatrics; 2008.

6. Reeds PJ, Garlick PJ. Protein and amino acid requirements and the composition of complementary foods. Journal of Nutrition. 2003;133:2953S–61S.

7. Schnog JB, Duits AJ, Muskeit FAJ, ten Cate H, Rojer RA, Brandjes DPM. Sickle cell disease: A general overview. Journal of Medicine. 2004;62:364–74. López C, Saravia C, Gomez A, Hoebeke J, Patarroyo MA. Mechanisms of genetically-based resistance to malaria. Gene. 2010;467:1–12.

8. Ibid.

9. Burdge GC, Hanson MA, Slater-Jefferies JL, Lillycrop KA. Epigenetic regulation of transcription: A mechanism for inducing variations in phenotype (fetal programming) by differences in nutrition during early life? British Journal of Nutrition. 2007;97:1036–46. Hanley B, Dijane J, Fewtrell M, Grynberg A, Hummel S, Junien C, Koletzko B, Lewis S, Renz H, Symonds M, Gros M, Harthoorn L, Mace K, Samuels F, van Der Beek EM. Metabolic imprinting, programming and epigenetics—A review of present priorities and future opportunities. British Journal of Nutrition. 2010;104:S1–25. Mathers JC. Early nutrition: Impact on epigenetics. Forum in Nutrition. 2007;60:42–8.

10. Taylor SL, Hefle SL. Food allergy. In: Present knowledge in nutrition, 9th ed. Bowman BA, Russell RM, editors. Washington, DC: ILSI Press; 2006.

11. National Institute of Allergy and Infectious Diseases. Food allergy. Report of the NIH expert panel on food allergy research. 2006. Available from: http://www.niaid.nih.gov /topics/foodallergy/ research/pages/reportfoodallergy.aspx.

12. Ibid.

13. Fuller MF, Reeds PJ. Nitrogen cycling in the gut. Annual Review of Nutrition. 1998;18:385–411. Rand WM, Pellet PL, Young VR. Meta-analysis of nitrogen balance studies for estimating protein requirements in healthy adults. American Journal of Clinical Nutrition. 2003;77:109–27.

14. Institute of Medicine. Dietary Reference Intakes for energy, carbohydrate, fiber, fat, fatty acids, cholesterol, protein, and amino acids. Washington, DC: National Academies Press; 2005.

15. Dewey KG. Energy and protein requirements during lactation. Annual Review of Nutrition. 1997;17:19–36.

16. American Dietetic Association, Dietitians of Canada, and the American College of Sports Medicine. Position of the American Dietetic Association, Dietitians of Canada, and the American College of Sports Medicine: Nutrition and athletic performance. Journal of the American Dietetic Association. 2009;109:509–27.

17. Campbell B, Kreider RB, Ziegenfuss T, LaBounty P, Roberts M, Burke D, Landis J, Lopez H, Antonio J. International Society of Sports Nutrition position stand: Protein and exercise. Journal of the International Society of Sports Nutrition. 2007;4:8.

18. Deldicque L, Francaux M. Functional food for exercise performance: Fact or foe? Current Opinions in Clinical Nutrition and Metabolic Care. 2008;11:774–81.Mero A. Leucine supplementation and intensive training. Sports Medicine. 1999;27:347–58.

19. Bedford JL, Barr SI. Diets and selected lifestyle practices of self-defined adult vegetarians from a population-based sample suggest they are more "health conscious." International Journal of Behavior, Nutrition, and Physical Activity. 2005;2:4. Haddad EH, Tanzman JS. What do vegetarians in the United States eat? American Journal of Clinical Nutrition. 2003;78:626S–32S.

20. Antony AC. Vegetarianism and vitamin B-12 (cobalamin) deficiency. American Journal of Clinical Nutrition. 2003;78:3– 6. Hunt JR. Bioavailability of iron, zinc, and other trace minerals from vegetarian diets. American Journal of Clinical Nutrition. 2003;78:633S–9S.

21. World Health Organization. Children: Reducing mortality. 2016. Available at: http://www.who.int/ mediacentre/factsheets/fs178/en/.

22. Jeejeebhoy KN. Protein nutrition in clinical practice. British Medical Bulletin. 1981;37:11–17. Waterlow JC. Classification and definition of protein-calorie malnutrition. British Medical Journal. 1972;3:566–9.

23. Golden M. The development of concepts of malnutrition. Journal of Nutrition. 2002;132:2117S–22S.

24. Kristensen KH, Wiese M, Rytter MJ, Ozcam J, Hansen LH, Namusoke H, Friis H, Nielsen DS. Gut microbiota in children hospitalized with oedematous and non-oedematous severe acute malnutrition in Uganda. PLoS Neglected Tropical Diseases. 2016; 10(1):e0004369. Smith MI, Yatsunenko T, Manary MJ, Trehan I, Mkakosya R, Cheng J, Kau AL, Rich SS, Concannon P, Mychaleckyj JC, Liu J, Houpt E, Li JV, Holmes E, Nicholson J, Knights D, Ursell LK, Knight R, Gordon JI. Gut microbiomes of Malawian twin pairs discordant for kwashiorkor. Science. 2013;339:548–54.

25. Hansen RD, Raja C, Allen BJ. Total body protein in chronic diseases and in aging. Annals of the New York Academy of Sciences. 2000;904:345–52. Chao A, Thun MJ, Connell CJ, McCullough ML, Miller PE, Lesko SM, Muscat JE, Lazarus P, Hartman TJ. Dietary patterns and colorectal adenoma and cancer risk: A review of the epidemiological evidence. Nutrition and Cancer. 2010;62:413–24.

26. World Cancer Research Fund/American Institute for Cancer Research. Food, nutrition, physical activity, and the prevention of cancer: A global perspective. Washington, DC: American Institute for Cancer Research; 2007.

27. Ibid.

6

1. Elias SL, Innis SM. Bakery foods are the major dietary source of trans-fatty acids among pregnant women with diets providing 30 percent energy from fat. Journal of the American Dietetic Association. 2002;102:46–51. Hayes KC, Pronczuk A. Replacing trans fat: The argument for palm oil with a cautionary note on interesterification. Journal of the American College of Nutrition. 2010;253S–84S.

2. Judd JT, Clevidence BA, Muesing RA, Wittes J, Sunkin ME, Podczasy JJ. Dietary trans fatty acids: Effects on plasma lipids and lipoproteins of healthy men and women. American Journal of Clinical Nutrition. 1994;59:861–8. Remig V, Franklin B, Margolis S, Kostas G, Nece T, Street JC. Trans fats in America: A review of their use, consumption, health implications, and regulation. Journal of the American Dietetic Association. 2010;110:585–92.

3. U.S. Food and Drug Administration. (2015). Final Determination Regarding Partially Hydrogenated Oils. Federal Register. Available at: https://www.federalregister.gov/ articles/2015/06/17/2015–14883/finaldetermination-regarding-partiallyhydrogenated-oils#h-13.

4. Hansen SN, Harris WS. New evidence for the cardiovascular benefits of long chain omega-3 fatty acids. Current Atherosclerosis Reports. 2007;9:434–40. Sala A, Folco G, Murphy RC. Transcellular biosynthesis of eicosanoids. Pharmacology Reports. 2010;62:503–10. Singh RK, Gupta S, Dastidar S, Ray A. Cysteinyl leukotrienes and their receptors: Molecular and functional characteristics. Pharmacology. 2010; 85:336–49.

5. Defilippis AP, Blaha MJ, Jacobson TA. Omega-3 fatty acids for cardiovascular disease prevention. Current Treatment Options in Cardiovascular Medicine. 2010;12:365–80. Wood DA, Kotseva K, Connolly S, Jennings C, Mead A, Jones J, Holden A, De Bacquer D, Collier T, De Backer G, Faergeman O, EUROACTION Study Group. Nurse-coordinated multidisciplinary, family-based cardiovascular disease prevention programme (EUROACTION) for patients with coronary heart disease and asymptomatic individuals at high risk of cardiovascular disease: A paired, cluster-randomised controlled trial. Lancet. 2008;371:1999–2012.

6. Mori TA, Beilin LJ. Omega-3 fatty acids and inflammation. Current Atherosclerosis Reports. 2004;6:461–7. Shahidi F, Miraliakbari H. Omega-3 (n-3) fatty acids in health and disease. Part 1: Cardiovascular disease and cancer. Journal of Medicinal Foods. 2004; 7:387–401. Wijendran V, Hayes KC. Dietary n-6 and n-3 fatty acid balance and cardiovascular health. Annual Review of Nutrition. 2004;24:597–615. Saltiel AR. Fishing out a sensor for anti-inflammatory oils. Cell. 2010;142:672–4.

7. Strumia R. Dermatologic signs in patients with eating disorders. American Journal of Clinical Dermatology. 2005;6:165–73.

8. Foster GD, Wyatt HR, Hill JO, Makris AP, Rosenbaum DL, Brill C, Stein RI, Mohammed BS, Miller B, Rader DJ, Zemel B, Wadden TA, Tenhave T, Newcomb CW, Klein S. Weight and metabolic outcomes after 2 years on a low-carbohydrate versus low-fat diet: A randomized trial. Annals of Internal Medicine. 2010;153:147–57. Vidon C, Boucher P, Cachefo A, Peroni O, Diraison F, Beylot M. Effects of isoenergetic high-carbohydrate compared with high-fat diets on human cholesterol synthesis and expression of key regulatory genes of cholesterol metabolism. American Journal of Clinical Nutrition. 2001;73:878–84.

9. Connor WE, Connor SL. Dietary treatment of familial hypercholesterolemia. Arteriosclerosis. 1989;9:91–105.

10. Allen RR, Carson L, Kwik-Uribe C, Evans EM, Erdman JW Jr. Daily consumption of a dark chocolate containing flavanols and added sterol esters affects cardiovascular risk factors in a normotensive population with elevated cholesterol. Journal of Nutrition. 2008;138:725–31. Klingberg S, Ellegård L, Johansson I, Hallmans G, Weinehall L, Andersson H, Winkvist A. Inverse relation between dietary intake of naturally occurring plant sterols and serum cholesterol in northern Sweden. American Journal of Clinical Nutrition. 2008;87:993–1001. Lin X, Racette SB, Lefevre M, Spearie CA, Most M, Ma L, Ostlund RE Jr. The effects of phytosterols present in natural food matrices on cholesterol metabolism and LDL-cholesterol: A controlled feeding trial. European Journal of Clinical Nutrition. 2010;64:1481–7.

11. Goldstein JL, Brown MS. A century of cholesterol and coronaries: From plaques to genes to statins. Cell. 2015 26;161:161–72.

12. Fernandez ML, Webb D. The LDL to HDL cholesterol ratio as a valuable tool to evaluate coronary heart disease risk. Journal of the American College of Nutrition. 2008;27:1–5.

13. Gerber PA, Berneis K. Regulation of low-density lipoprotein subfractions by carbohydrates. Current Opinion in Clinical Nutrition and Metabolic Care. 2012; 15:381–5; Desroches S, Lamarche B. Diet and low-density lipoprotein particle size. Current Atherosclerosis Reports 2004;6:453–60.

14. Esposito K, Ceriello A, Giugliano D. Diet and the metabolic syndrome. Metabolic Syndrome Related Disorders. 2007;5:291–6. Kritchevsky SB, Kritchevsky D, Kromhout D, de Lezenne Coulander C. Diet, prevalence and 10-year mortality from coronary heart disease in 871 middle-aged men. The Zutphen study. American Journal of Epidemiology. 1984;119:733–41. Mottillo S, Filion KB, Genest J, Joseph L, Pilote L, Poirier P, Rinfret S, Schiffrin EL, Eisenberg MJ. The metabolic syndrome and cardiovascular risk: A systematic review and meta-analysis. Journal of the American College of Cardiology. 2010;56:1113–32.

15. Lammert F, Wang DQ. New insights into the genetic regulation of intestinal cholesterol absorption. Gastroenterology. 2005;129:718–34. Yang Y, Ruiz-Narvaez E, Kraft P, Campos H. Effect of apolipoprotein E genotype and saturated fat intake on plasma lipids and myocardial infarction in the Central Valley of Costa Rica. Human Biology. 2007;79:637–47. Wu K, Bowman R, Welch AA, Luben RN, Wareham N, Khaw KT, Bingham SA. Apolipoprotein E polymorphisms, dietary fat and fibre, and serum lipids: The EPIC Norfolk study. European Heart Journal. 2007;28:2930–6.

16. Brown MS, Goldstein JL. How LDL receptors influence cholesterol and atherosclerosis. Scientific American. 1984;251:52–60. Dedoussis GV, Schmidt H, Genschel J. LDL-receptor mutations in Europe. Human Mutation. 2004;443–59.

17. Prentice RL. Women's Health Initiative studies of postmenopausal breast cancer. Advances in Experimental Medicine and Biology. 2008;617:151–60. Van Horn L, Manson JE. The Women's Health Initiative: Implications for clinicians. Cleveland Clinic Journal of Medicine. 2008;75:385–90. Wang J, John EM, Horn-Ross PL, Ingles SA. Dietary fat, cooking fat, and breast cancer risk in a multiethnic population. Nutrition and Cancer. 2008;60:492–504. Chan AT, Giovannucci EL. Primary prevention of colorectal cancer. Gastroenterology. 2010;138:2029–43.

18. Al-Serag HB. Obesity and disease of the esophagus and colon. Gastroenterology Clinics of North America. 2005;34:63–82. Key TJ, Schatzkin A, Willett WC, Allen NE, Spencer EA, Travis RC. Diet, nutrition and the prevention of cancer. Public Health Nutrition. 2004;7:187–200. McTiernan A. Obesity and cancer: The risks, science, and potential management strategies. Oncology. 2005;19:871–81. Vrieling A, Kampman E. The role of body mass index, physical activity, and diet in colorectal cancer recurrence and survival: A review of the literature. American Journal of Clinical Nutrition. 2010;92:471–90.

19. World Cancer Research Fund/American Institute for Cancer Research. Food, nutrition, physical activity, and the prevention of cancer: A global perspective. Washington, DC: AICR; 2007.

20. Food and Nutrition Board. Institute of Medicine. (2002/2005). Dietary reference intakes for energy, carbohydrate, fiber, fat, fatty acids, cholesterol, protein, and amino acids. The National Academies Press. Washington, DC.

21. Food and Nutrition Board. Institute of Medicine. (2002/2005). Dietary reference intakes for energy, carbohydrate, fiber, fat, fatty acids, cholesterol, protein, and amino acids. The National Academies Press. Washington, DC. U.S. Department of Agriculture and U.S. Department of Health and Human Services. 2015–2020 Dietary Guidelines for Americans, 2015–2020. 8th ed. Washington, DC. Government Printing Office. December 2015.

7

1. Carpenter KJ. A short history of nutritional science: Part 1 (1785–1885). Journal of Nutrition. 2003;133:638–45. Carpenter KJ. A short history of nutritional science: Part 2 (1885–1912). Journal of Nutrition. 2003;133:975–84. Carpenter KJ. A short history of nutritional science: Part 3 (1912–1944). Journal of Nutrition. 2003;133:3023–32. Carpenter KJ. A short history of nutritional science: Part 4 (1945–1985). Journal of Nutrition. 2003;133:3331–42.

2. Institute of Medicine. Dietary Reference Intakes for thiamin, riboflavin, niacin, vitamin B_6, folate, vitamin B_{12}, pantothenic acid, biotin, and choline. Washington, DC: National Academy Press; 1998.

3. Ibid.

4. West DW, Owen EC. The urinary excretion of metabolites of riboflavin in man. British Journal of Nutrition. 1963;23:889–98.

5. Cacciapuoti F. Hyper-homocysteinemia: A novel risk factor or a powerful marker for cardiovascular diseases? Pathogenetic and therapeutical uncertainties. Journal of Thrombosis and Thrombolysis. 2011 Jan 14. Scott JM. Homocysteine and cardiovascular risk. American Journal of Clinical Nutrition. 2000;72:33–4.

6. Moyers S, Bailey LB. Fetal malformation and folate metabolism: Review of recent evidence. Nutrition Reviews. 2001;7:215–4.

7. Mitchell LE, Adzick NS, Melchionne J, Pasquariello PS, Sutton LN, Whitehead AS. Spina bifida. Lancet. 2004;364: 1885–95.

8. Centers for Disease Control and Prevention. Spina Bifida and Anencephaly Before and After Folic Acid Mandate—United States, 1995–1996 and 1999–2000. MMWR 2004;53(17):362–5.

9. Pfeiffer CM, Caudill SP, Gunter EW, Osterloh J, Sampson EJ. Biochemical indicators of B vitamin status in the US population after folic acid fortification: Results from the National Health and Nutrition Examination Survey 1999–2000. American Journal of Clinical Nutrition. 2005;82:442–50. Pfeiffer CM, Johnson CL, Jain RB, Yetley EA, Picciano MF, Rader JI, Fisher KD, Mulinare J, Osterloh JD. Trends in blood folate and vitamin B-12 concentrations in the United States, 1988–2004. American Journal of Clinical Nutrition. 2007;86:718–27.

10. Toh BH, Alderuccio F. Pernicious anaemia. Autoimmunity. 2004;37:357–61.

11. Wojcik M, Burzynska-Pedziwiatr I, Wozniak LA. A review of natural and synthetic antioxidants important for health and longevity. Current Medical Chemistry. 2010;17:3262–88.

12. Jacob RA, Aiello GM, Stephensen CB, Blumberg JB, Milbury PE, Wallock LM, Ames BN. Moderate antioxidant supplementation has no effect on biomarkers of oxidant damage in healthy men with low fruit and vegetable intakes. Journal of Nutrition. 2003;133:740–3. Padayatty SJ, Katz A, Wang Y, Eck P, Kwon O, Lee J-H, Chen S, Corpe C, Dutta A, Dutta SK, Levine M. Vitamin C as an antioxidant: Evaluation of its role in disease prevention. Journal of the American College of Nutrition. 2003;22:18–35.

13. Stephen R, Utecht T. Scurvy identified in the emergency department: A case report. Journal of Emergency Medicine. 2001;21:235–7. Weinstein M, Babyn P, Zlotkin S. An orange a day keeps the doctor away: Scurvy in the year 2000. Pediatrics. 2001;108:E55.

14. Gerster H. Vitamin A—Functions, dietary requirements and safety in humans. International Journal of Vitamins and Nutrition Research. 1997;67:71–90. Hinds TS, West WL, Knight EM. Carotenoids and retinoids: A review of research, clinical, and public health applications. Journal of Clinical Pharmacology. 1997;37:551–8.

15. Campbell JK, Canene-Adams K, Lindshield BL, Boileau TWM, Clinton SK, Erdman JW. Tomato phytochemicals and prostate cancer risk. Journal of Nutrition. 2004;134:3486S–92S. Wertz K, Siler U, Goralczyk R. Lycopene: Modes of action to promote prostate health. Archives of Biochemistry and Biophysics. 2004;430:127–34. Beatty S, Nolan J, Kavanagh H, O'Donovan O. Macular pigment optical density and its relationship with serum and dietary levels of lutein and zeaxanthin. Archives of Biochemistry and Biophysics. 2004;430:70–6. Stringham JM, Hammond BR. Dietary lutein and zeaxanthin: Possible effects on visual function. Nutrition Reviews. 2005;63:59–64.

16. El Agamey A, Lowe GM, McGarvey DJ, Mortensen A, Phillip DM, Truscott G, Young AJ. Carotenoid radical chemistry and antioxidant/pro-oxidant properties. Archives of Biochemistry and Biophysics. 2004;430:37–48.

17. Stringham JM, Hammond BR. Dietary lutein and zeaxanthin: Possible effects on visual function. Nutrition Reviews. 2005;63:59–64.

18. The Alpha-Tocopherol, Beta Carotene Cancer Prevention Study Group. The effect of vitamin E and beta carotene on the incidence of lung cancer and other cancers in male smokers. New England Journal of Medicine. 1994;330:1029–35. Baron JA, Cole BF, Mott L, Haile R, Grau M, Church TR, Beck GJ, Greenberg ER. Neoplastic and antineoplastic effects of b-carotene on colorectal adenoma recurrence: Results of a randomized trial. Journal of the National Cancer Institute. 2003;95:717–22. Omenn GS, Goodman GE, Thornquist MD, Balmes J, Cullen MR, Glass A, Keogh JP, Meyskens FL, Valanis B, Williams JH, Barnhart S, Hammar S. Effects of a combination of beta carotene and vitamin A on lung cancer and cardiovascular disease. New England Journal of Medicine. 1996;334:1150–5.

19. American Academy of Dermatology. American Academy of Dermatology Association reconfirms need to boost vitamin D intake through diet and nutritional supplements rather than ultraviolet radiation. Available from: http://www.aad.org/aad/Newsroom/Vitamin1D1Consensus1Conf.htm. World Health Organization. Sunbeds, tanning and UV exposure. Available from: http://www.who.int/mediacentre/factsheets/fs287/en/print.html.

20. Yao Y, Zhu L, He L, Duan Y, Liang W, Nie Z, Jin Y, Wu X, Fang Y. A meta-analysis of the relationship between vitamin D deficiency and obesity. International Journal of Clinical and Experimental Medicine. 2015;8:14977–84.

21. Bikle DD. Vitamin D and skin cancer. Journal of Nutrition. 2004;134:3472S–8S. Gross MD. Vitamin D and calcium in the prevention

of prostate and colon cancer: New approaches for the identification of needs. Journal of Nutrition. 2005;135:326–31. Holick MF. Sunlight and vitamin D for bone health and prevention of autoimmune diseases, cancers, and cardiovascular disease. American Journal of Clinical Nutrition. 2004;6 Suppl:1678S–88S. Welsh J. Vitamin D and breast cancer: Insights from animal models. American Journal of Clinical Nutrition. 2004;80:1721S–4S.

22. Calvo MS, Whiting SJ, Barton CN. Vitamin D intake: A global perspective of current status. Journal of Nutrition. 2005;135:310–6.

23. Kreiter SR, Schwartz RP, Kirkman HN, Charlton PA, Calikoglu AS, Davenport ML. Nutrition rickets in African American breast-fed infants. Journal of Pediatrics. 2000;137:153–7. Pugliese MF, Blumberg DL, Hludzinski J, Kay S. Nutritional rickets in suburbia. Journal of the American College of Nutrition. 1998;17:637–41. Prentice A. Vitamin D deficiency: A global perspective. Nutrition Reviews, 2008;66:S153–64.

24. American Academy of Pediatrics. Pediatric nutrition handbook, 6th ed. Elk Grove Village, IL; 2008. Wagner CL, Greer FR, American Academy of Pediatrics Section on Breastfeeding, American Academy of Pediatrics Committee on Nutrition. Prevention of rickets and vitamin D deficiency in infants, children, and adolescents. Pediatrics. 2008;122:1142–52.

25. Institute of Medicine. Dietary Reference Intakes for calcium and vitamin D. Washington, DC: National Academies Press, 2011.

26. Bostick RM, Potter JD, McKenzie DR, Sellers TA, Kushi LH, Steinmetz KA, Folsom AR. Reduced risk of colon cancer with high intakes of vitamin E: The Iowa Women's Health Study. Cancer Research. 1992;15:4230–7. Kirsh VA, Hayes RB, Mayne ST, Chatterjee N, Subar AF, Dixon LB, Albanes D, Andriole GL, Urban DA, Peters U. PLCO Trial. Supplementation and dietary vitamin E, beta-carotene, and vitamin C intakes and prostate cancer risk. Journal of the National Cancer Institute. 2006;98:245–54. Kline K, Yu W, Sanders BG. Vitamin E and breast cancer. Journal of Nutrition. 2004;134:3458S–62S. Peters U, Littman AJ, Kristal AR, Patterson RE, Potter JD, White E. Vitamin E and selenium supplementation and risk of prostate cancer in the vitamins and lifestyle (VITAL) study cohort. Cancer Causes and Control. 2008;19:75–87. Slatore CG, Littman AJ, Au DH, Satia JA, White E. Long-term use of supplemental multivitamins, vitamin C, vitamin E, and folate does not reduce the risk of lung cancer. American Journal of Respiratory Critical Care Medicine. 2008;177:524–30.

27. Committee on Fetus and Newborn (American Academy of Pediatrics). Controversies concerning vitamin K and the newborn. Pediatrics. 2003;112:191–2.

28. U.S. Food and Drug Administration and Center for Food Safety and Applied Nutrition. Dietary supplement health and education act of 1994. Available from: http://www.cfsan.fda.gov/~dms/dietsupp.html.

8

1. Armstrong LE, Pumerantz AC, Roti MW, Judelson DA, Watson G, Dias JC, Sokmen B, Casa DJ, Maresh CM, Lieberman H, Kellogg M. Fluid, electrolyte, and renal indices of hydration during 11 days of controlled caffeine consumption. International Journal of Sport Nutrition and Exercise Metabolism. 2005;15:252–65. Zhang Y, Coca A, Casa DJ, Antonio J, Green JM, Bishop PA.

Caffeine and diuresis during rest and exercise: A meta-analysis. Journal of Science and Medicine in Sport. 2015;18:569–74.

2. Maughan RJ, Griffin J. Caffeine ingestion and fluid balance: A review. Journal of Human Nutrition and Dietetics. 2003;16:411–20.

3. Coris EE, Ramirez AM, Van Durme DJ. Heat illness in athletes: The dangerous combination of heat, humidity and exercise. Sports Medicine. 2004;34:9–16. Girard O, Brocherie F, Bishop DJ. Sprint performance under heat stress: A review. Scandinavian Journal of Medicine & Science in Sports. 2015;S1:79–89.

4. Bottled water: Pure drink or pure hype? Natural Resources Defense Council, 1999. Available from: http://www.nrdc.org /water/drinking/bw/bwinx.asp. Sloan AE. What, When and Where America Eats. Food Technology. 2014;68:20. Available from: http://www.ift.org/food-technology/past-issues/2010/january/features/america-eats/americaeats/.

5. Cook RJ, Barron JC, Papendick JI, Williams GJ. Impact on agriculture of the Mount St. Helens eruptions. Science. 1981;211:16–22.

6. Astrup A. Calcium for prevention of weight gain, cardiovascular disease, and cancer. American Journal of Clinical Nutrition. 2011;94:1159–60.

7. National Osteoporosis Foundation. Available from: http://nof.org/news/2948.

8. Syed FA, Ng AC. The pathology of the aging skeleton. Current Osteoporosis Reports. 2010;8:235–40.

9. Winsloe C, Earl S, Dennison EM, Cooper C, Harvey NC. Early life factors in the pathogenesis of osteoporosis. Current Osteoporosis Reports. 2009;7:140–4.

10. Straub DA. Calcium supplementation in clinical practice: A review of forms, doses, and indications. Nutrition in Clinical Practice. 2007;22:286–96.

11. Jackson RD, Shidham S. The role of hormone therapy and calcium plus vitamin D for reduction of bone loss and risk for fractures: Lessons learned from the Women's Health Initiative. Current Osteoporosis Reports. 2007;5:153–9.

12. Seeman E. Evidence that calcium supplements reduce fracture risk is lacking. Clinical Journal of American Society of Nephrology. 2010;1:S3–11.

13. Reid IR, Bristow SM, Bolland MJ. Calcium supplements: Benefits and risks. Journal of Internal Medicine. 2015;8:354–68.

14. Cashman KD. Calcium intake, calcium bioavailability and bone health. The British Journal of Nutrition. 2002;87:S169–77.

15. Davis JM, Burgess WA, Slentz CA, Bartoli WP. Fluid availability of sports drinks differing in carbohydrate type and concentration. American Journal of Clinical Nutrition. 1990;51:1054–7.

16. Evans GH, Shirreffs SM, Maughan RJ. Postexercise rehydration in man: The effects of osmolality and carbohydrate content of ingested drinks. Nutrition. 2009;25:905–13.

17. Kottke TE, Wu LA. Preventing heart disease and stroke: Messages from the United States. The Keio Journal of Medicine. 2001;50:274–9.

18. Dumler F. Dietary sodium intake and arterial blood pressure. Journal of Renal Nutrition. 2009;19:57–60.

19. Elkhalifa AM, Kinsara AJ, Almadani DA. Prevalence of hypertension in a population of healthy individuals. Medical Principles and Practice. 2011;20:152–5.

20. Reinhold JG, Faradji B, Abadi P, Ismail-Beigi F. Decreased absorption of calcium, magnesium, zinc and phosphorus by humans due

to increased fiber and phosphorus consumption as wheat bread. Nutrition Reviews. 1991;49:204–6.

21. Seiner R, Hesse A. Influence of a mixed and a vegetarian diet on urinary magnesium excretion and concentration. British Journal of Nutrition. 1995;73:783–90. Wisker E, Nagel R, Tanudjaja TK, Feldheim W. Calcium, magnesium, zinc, and iron balances in young women: Effects of a low-phytate barley-fiber concentration. American Journal of Clinical Nutrition. 1991;54:553–9.

22. Centers for Disease Control and Prevention. Recommendations to prevent and control iron deficiency in the United States. Mortality and Morbidity Weekly Report. 1998;47: 1–29. Stoltzfus RJ. Defining iron-deficiency anemia in public health terms: Reexamining the nature and magnitude of the public health problem. Journal of Nutrition. 2001;131: 565S–7S. Clark S.F. Iron deficiency anemia: Diagnosis and management. Current Opinions in Gastroenterology. 2009; 25:122–8.

23. Coad J, Conion C. Iron deficiency in women: Assessment, causes and consequences. Current Opinion in Clinical Nutrition and Metabolic Care. 2011;14:625–34.

24. Lopez MAA, Martos FC. Iron availability: An updated review. International Journal of Food Sciences and Nutrition. 2004;55:597–606. Zijp IM, Korver O, Tijburg LBM. Effect of tea and other dietary factors on iron absorption. Critical Reviews in Food Science and Nutrition. 2000;40:371–98.

25. Lopez A, Cacoub P, Macdougall IC, Peyrin-Biroulet L. Iron deficiency anemia. Lancet. 2015;15:SO140–6736.

26. Rosenzweig PH, Volpe SL. Iron, thermoregulation, and metabolic rate. Critical Reviews in Food Science and Nutrition. 1999;39:131–48. Cunningham-Rundles S, McNeeley DF, Moon A. Mechanisms of nutrient modulation of the immune response. Journal of Allergy and Clinical Immunology. 2005;115:1119–28. Failla ML. Trace elements and host defense: Recent advances and continuing challenges. Journal of Nutrition. 2003;133:1443S–7S.

27. Bryan J, Osendarp S, Hughes D, Calvaresi E, Baghurst K, van Klinken J-W. Nutrients for cognitive development in schoolaged children. Nutrition Reviews. 2004;62:295–306.

28. Gambling L, Danzeisen R, Fosset C, Andersen HS, Dunford S, Srai SKS, McArdle HJ. Iron and copper interactions in development and the effect on pregnancy outcome. Journal of Nutrition. 2003;133:1554S–5S.

29. Institute of Medicine. Dietary Reference Intakes for vitamin A, vitamin K, arsenic, boron, chromium, copper, iodine, iron, manganese, molybdenum, nickel, silicon, vanadium, and zinc. Washington, DC: National Academy Press; 2001.

30. Hoffman R, Benz E, Shattil S, Furie B, Cohen H, Silberstein L, McGlave P. Hematology: Basic principles and practice, 3rd ed. New York: Churchill Livingstone, Harcourt Brace; 2000.

31. Charles CV, Dewey CE, Hall A, Hak C, Channary S, Summerlee AJS. A Randomized Control Trial Using a Fish-Shaped Iron Ingot for the Amelioration of Iron Deficiency Anemia in Rural Cambodian Women. Tropical Medicine & Surgery. 2015;3:195–201.

Charles CV, Dewey CE, Daniell WE, Summerlee AJS. Iron-deficiency anaemia in rural Cambodia: Community trial of a novel iron supplementation technique. European Journal of Public Health. 2010;21:43–8.

Armstrong GR, Dewey CE, Summerlee AJS. Iron release from the Lucky Iron Fish™: Safety considerations. Asia Pacific Journal of Clinical Nutrition. 2015. Available from: http://www.apjcn.org/update/pdf/0000/0/2015-0188/2015-0188-online.pdf.

32. Armstrong GR, Summerlee AJS. The Etiology, Treatment and Effective Prevention of Iron Deficiency and Iron Deficiency Anemia in Women and Young Children Worldwide: A Review. Journal of Women's Health Care. 2014;4:1000213. Available from: http://www.omicsgroup.org/journals/the-etiology-treatment-and-effective-prevention-of-iron-deficiency-and-iron-deficiency-anemia-in-women-2167-0420.1000214.pdf.

33. Chuttani H, Gupta P, Gulati S, Gupta D. Acute copper sulfate poisoning. American Journal of Medicine. 1965;39:849–54. Bremner I. Manifestations of copper excess. American Journal of Clinical Nutrition. 1998;67:1069S–73S.

34. Zimmermann MB. Assessing iodine status and monitoring progress of iodized salt programs. Journal of Nutrition. 2004;134:1673–7.

35. Valko M, Izakovic M, Mazur M, Rhodes CJ, Telser J. Role of oxygen radicals in DNA damage and cancer incidence. Molecular and Cellular Biochemistry. 2004;266:37–56.

36. Balk EM, Tatsioni A, Lichtenstein AH, Lau J, Pittas AG. Effect of chromium supplementation on glucose metabolism and lipids: A systematic review of randomized controlled trials. Diabetes Care. 2007;30:2154–63.

37. Kozlovsky AS, Moser PB, Reiser S, Anderson RA. Effects of diets high in simple sugars on urinary chromium losses. Metabolism. 1986;35:515–8.

38. Cefalu WT, Hu FB. Role of chromium in human health and in diabetes. Diabetes Care. 2004;27:2741–51. Hopkins Jr. LL, Ransome-Kuti O, Majaj AS. Improvement of impaired carbohydrate metabolism by chromium (III) in malnourished infants. American Journal of Clinical Nutrition. 1968;21:203–11. Mertz W. Interaction of chromium with insulin: A progress report. Nutrition Reviews. 1998;56:174–7.

39. McNamara JP, Valdez F. Adipose tissue metabolism and production responses to calcium proprionate and chromium proprionate. Journal of Dairy Science. 2005;88:2498–507. Page TG, Southern LL, Ward TL, Thompson DLJ. Effect of chromium picolinate on growth and serum and carcass traits of growing-finishing pigs. Journal of Animal Science. 1993;71:656–62.

40. Costa M, Klein CB. Toxicity and carcinogenicity of chromium compounds in humans. Critical Reviews in Toxicology. 2006;36:155–63. Coyle YM, Minahjuddin AT, Hynan LS, Minna JD. An ecological study of the association of metal air pollutants with lung cancer incidence in Texas. Journal of Thoracic Oncology. 2006;1:654–61. Michaels D, Lurie P, Monforton C. Lung cancer mortality in the German chromate industry, 1958 to 1998. Journal of Occupational and Environmental Medicine. 2006;48:995–7.

41. Abumrad NN, Schneider AJ, Steel D, Rogers LS. Amino acid intolerance during total parenteral nutrition reversed by molybdate therapy. American Journal of Clinical Nutrition. 1981;34:2551–9.

42. Caruso TJ, Prober CG, GwaltneyJr JM. Treatment of naturally acquired common colds with zinc: A structured review. Clinical Infectious Diseases. 2007;45:569–74. Jackson JL, Lesho E, Peterson C. Zinc and the common cold: A metaanalysis revisited. Journal of Nutrition. 2000;130:1512S–5S. Kurugöl Z, Bayram N, Atik T. Effect of zinc sulfate on common cold in children: Randomized, double blind study. Pediatrics

International. 2007;49:842–7. Kurugöl Z, Akilli M, Bayram N, Koturoglu G. The prophylactic and therapeutic effectiveness of zinc sulphate on common cold in children. Acta Paediatrica. 2006;95:1175–81.

43. Quock RL, Chan JT. Fluoride content of bottled water and its implications for the general dentist. General Dentistry. 2009;57:29–33.

44. Dhar V, Bhatnagar M. Physiology and toxicity of fluoride. Indian Journal of Dental Research. 2009;20:350–5.

9

1. Liu J, Fox CS, Hickson, DA, May, WD, Hairston KG, Carr J, Taylor HA. Impact of abdominal visceral and subcutaneous adipose tissue on cardiometabolic risk factors: The Jackson heart study. Journal of Clinical Endocrinology and Metabolism. 2010; 95: 5419–26.

2. National Institutes of Health. Clinical guidelines on the identification, evaluation, and treatment of overweight and obesity in adults. National Institutes of Health, National Heart, Lung, and Blood Institute, Obesity Education Initiative. Available from: http://www.nhlbi.nih.gov/guidelines/obesity/practgde.htm.

3. Heinrich, KM, Jitnarin N, Suminski RR, Berkel L, Hunter CM, Alvarez L, Brundige AR, Peterson AL, Foreyt JP, Haddock CK, Poston WS. Obesity classification in military personnel: A comparison of body fat, waist circumference, and body mass index measurements. Military Medicine. 2008;17:67–73.

4. U.S. Department of Health and Human Services and U.S. Department of Agriculture. 2015–2020 Dietary Guidelines for Americans. 8th ed. December 2015. Available at: http://health.gov/dietaryguidelines/2015/guidelines/.

5. Betz MJ, Enerbäck S. Human Brown Adipose Tissue: What We Have Learned So Far. Diabetes. 2015;64:2352–60.

6. Geliebter A. Stomach capacity in obese individuals. Obesity Research. 2001;9:727–8. Khan S, Rock K, Baskara A, Qu W, Nazzal M, Ortiz J. Trends in bariatric surgery from 2008 to 2012. The American Journal of Surgery. 2015;15:S0002–9610.

7. French S, Robinson T. Fats and food intake. Current Opinion in Clinical Nutrition and Metabolic Care. 2003;6:629–34.

8. Inui A, Asakawa A, Bowers CY, Mantovani G, Laviano A, Meguid MM, Fujimiya M. Ghrelin, appetite, and gastric motility: The emerging role of the stomach as an endocrine organ. Federation of American Societies for Experimental Biology Journal. 2004;18:439–56.

9. Yanovski S. Sugar and fat: Cravings and aversions. Journal of Nutrition. 2003;133:835S–7S.

10. Hulbert AJ, Else PL. Basal metabolic rate: History, composition, regulation, and usefulness. Physiological and Biochemical Zoology. 2004;77:869–76.

11. Chahal HS, Drake WM. The endocrine system and ageing. The Journal of Pathology. 2007;211:173–80. McIntire KL, Hoffman AR. The endocrine system and sarcopenia: Potential therapeutic benefits. Current Aging Science. 2011;4:298–305. Ryall JG, Schertzer JD, Lynch GS. Cellular and molecular mechanisms underlying age-related skeletal muscle wasting and weakness. Biogerontology. 2008;9:213–28. Perrini S, Laviola L, Carreira MC, Cignarelli A, Natalicchio A, Giorgino F. The GH/IGF1 axis and signaling pathways in the muscle and bone:

Mechanisms underlying age-related skeletal muscle wasting and osteoporosis. Journal of Endocrinology. 2010;205:201–10. Nass R, Thorner MO. Impact of the GH-cortisol ratio on the age-dependent changes in body composition. Growth Hormone & IGF Research. 2002;12:147–61.

12. Luke A, Schoeller DA. Basal metabolic rate, fat-free mass, and body cell mass during energy restriction. Metabolism. 1992;41:450–6.

13. Brooks GA, Butte NF, Rand WM, Flatt JP, Caballero B. Chronicle of the Institute of Medicine physical activity recommendation: How a physical activity recommendation came to be among dietary recommendations. American Journal of Clinical Nutrition. 2004;79:921S–30S.

14. Nair KS, Halliday D, Garrow JS. Thermic response to isoenergetic protein, carbohydrate or fat meals in lean and obese subjects. Clinical Science. 1983;65:307–12.

15. Friedl KE. Can you be large and not obese? The distinction between body weight, body fat, and abdominal fat in occupational standards. Diabetes Technology and Therapeutics. 2004;6:732–49.

16. Katzmarzyk PT, Janssen I, Ardern CI. Physical inactivity, excess adiposity and premature mortality. Obesity Reviews. 2003;4:257–901.

17. National Center for Health Statistics. Health, United States, 2014: With Special Feature on Adults Aged 55–64. Hyattsville, MD. 2015. Available from: http://www.cdc.gov/nchs/data/hus/hus14.pdf#059.

18. Popkin BM, Gordon-Larsen P. The nutrition transition: Worldwide obesity dynamics and their determinants. International Journal of Obesity and Related Metabolic Disorders. 2004;3:S2–9.

19. Austin GL, Ogden LG, and O Hill J. Trends in carbohydrate, fat, and protein intakes and association with energy intake in normal-weight, overweight, and obese individuals: 1971– 2006. American Journal of Clinical Nutrition. 2011;93:836–43. U.S. Department of Agriculture. Food and Nutrition Service. Agriculture Fact Book. Profiling Food Consumption in America. Available from: http://www.usda.gov/factbook/chapter2.pdf.

20. Centers for Disease Control and Prevention and National Health and Nutrition Examination Survey, 2011–2012. Available from: http://www.cdc.gov/nchs/data/databriefs/db213.htm.

21. Ello-Martin JA, Ledikwe JH, Rolls BJ. The influence of food portion size and energy density on energy intake: Implications for weight management. American Journal of Clinical Nutrition. 2005;82:236S–41S. Wansink B, Kim J. Bad popcorn in big buckets: Portion size can influence intake as much as taste. Journal of Nutrition Education and Behavior. 2005;37:242–5.

22. Rolls BJ, Roe LS, Kral TVE, Meengs JS, Wall DE. Increasing the portion size of a packaged snack increases energy intake in men and women. Appetite. 2004;42:63–9. Do increased portion sizes affect how much we eat? Centers for Disease Control and Prevention. Available from: http://www.cdc.gov/nccdphp/dnpa/nutrition/pdf/portion_size_research.pdf.

23. Hetherington MM, Foster R, Newman T, Anderson AS, Norton G. Understanding variety: Tasting different foods delays satiation. Physiology and Behavior. 2006;87:263–71.

24. Chandon P, Wansink B. The biasing health halos of fast-food restaurant health claims: Lower calorie estimates and higher side-dish consumption intentions. Journal of Consumer Research. 2007;34:301–14.

25. Kim JY, Oh DJ, Yoon TY, Choi JM, Choe BK. The impacts of obesity on psychological well-being:

A cross-sectional study about depressive mood and quality of life. Preventive Medicine and Public Health. 2007;40:191–5.

26. Christakis NA and Fowler JH. The spread of obesity in a large social network over 32 years. New England Journal of Medicine. 2007;57:370–9.

27. National Center for Health Statistics (NCHS). Health, United States, 2010. Available from: http://www.cdc.gov/nchs/data/hus/hus10.pdf. Blackwell DL, Lucas JW, Clarke TC. Summary health statistics for U.S. adults: National health interview survey, 2012. Vital and health statistics. Series 10, Data from the National Health Survey. Feb 2014(260): 1–161. Available from: http://www.ncbi.nlm.nih.gov/pubmed/24819891. U.S. Department of Health and Human Services. 2008 Physical Activity Guidelines for Americans. Washington, DC: U.S. Department of Health and Human Services; 2008. Available from: http://www.health.gov/paguidelines/pdf/paguide.pdf.

28. American College of Sports Medicine (ACSM). ACSM's Guidelines for Exercise Testing and Prescription, 8th ed. Philadelphia: Lippincott Williams & Wilkins; 2008.

29. Wareham NJ, van Sluijs EM, Ekelund U. Physical activity and obesity prevention: A review of the current evidence. Proceedings of the Nutrition Society. 2005;64:229–47.

30. Hill JO, Wyatt HR, Reed GW, Peters JC. Obesity and the environment: Where do we go from here? Science. 2003;299:853–5.

31. Ingalls A, Dickie M, Snell GD. Obese, a new mutation in the house mouse. Journal of Heredity. 1950;41:317–8.

32. ColeZhang Y, Proenca R, Maffei M, Leopold L, Friedman JM. Positional cloning of the mouse obese gene and its human homologue. Nature. 1994;372:125–32.

33. FarooqiI S, O'RahillyI S. Monogenic obesity in humans. Annual Review of Medicine. 2005;56:443–58.

34. Marx J. Cellular warriors at the battle of the bulge. Science. 2003;299:846–9.

35. Kennedy AG. The role of the fat depot in the hypothalamic control of food intake in the rat. Proceedings of the Royal Society of London. 1953;140:578–92.

36. Banks WA. Blood-brain barrier as a regulatory interface. Forum in Nutrition. 2010;63:102–10. Tups A. Physiological models of leptin resistance. Journal of Neuroendocrinology. 2009;21:961–71.

37. Langhans W, Geary N (eds): Frontiers in Eating and Weight Regulation. Karger Medical and Scientific Publishers, 2010;63:123–32.

38. Wing RR, Phelan S. Long-term weight loss maintenance. American Journal of Clinical Nutrition. 2005;82:222S–5S.

39. National Institutes of Health, and National Heart, Lung and Blood Institute. Clinical guidelines on the identification, evaluation and treatment of overweight and obesity in adults—The evidence report. National Institutes of Health Publication Number 00–4084. Bethesda, MD: National Institutes of Health; October 2000.

40. Higgins JP. Smartphone Applications for Patients' Health and Fitness. The American Journal of Medicine. 2016;129:11–9.

41. Fogelholm M. Physical activity, fitness and fatness: Relations to mortality, morbidity, and disease risk factors. A systematic review. Obesity Reviews. 2010;11:202–21.

42. Barlow CE, Kohl HW III, Gibbons LW, Blair SN. Physical fitness, mortality and obesity. International Journal of Obesity and Related Metabolic Disorders. 1995;19:S41–4. McAuley PA, Kokkinos PF, Oliveira RB, Emerson BT, Myers JN. Obesity paradox and cardiorespiratory fitness in 12,417 male veterans aged 40 to 70 years. Mayo Clinic Proceedings. 2010;85:115–21.

43. U.S. Department of Agriculture and U.S. Department of Health and Human Services. Dietary Guidelines for Americans, 2010. 7th ed. Washington, DC: U.S. Government Printing Office, December 2010.

44. Catenacci VA, Ogden LG, Stuht J, Phelan S, Wing RR, Hill JO, Wyatt HR. Physical activity patterns in the National Weight Control Registry. Obesity. 2008;16:153–61. Butryn ML, Phelan S, Hill JO, Wing RR. Consistent self-monitoring of weight: A key component of successful weight loss maintenance. Obesity. 2007;15:3091–6. Phelan S, Wyatt H, Nassery S, Dibello J, Fava JL, Hill JO, Wing RR. Three-year weight change in successful weight losers who lost weight on a low-carbohydrate diet. Obesity. 2007;15:2470–7.

45. Westman EC, Yancy WS Jr, Vernon MC. Is a low-carb, low-fat diet optimal? Archives of Internal Medicine. 2005;165:1071–2.

46. Center for Nutrition Policy and Promotion and the US Department of Agriculture. Nutrition Insights. Is fat consumption really decreasing? Insight 5 April 1998. Available from: http://www.cnpp.usda.gov/Publications/NutritionInsights/insight5.pdf.

47. Willett WC. Dietary fat and body fat: Is there a relationship? Journal of Nutritional Biochemistry. 1998;9:522–4. Willett WC. Is dietary fat a major determinant of body fat? American Journal of Clinical Nutrition. 1998;67:556S–625S.

48. Sacks FM, Bray GA, Carey VJ, Smith SR, Ryan DH, Anton SC, McManus K, Champagne CM, Bishop LM, Laranjo N, Leboff MS, Rood JC, de Jonge L, Greenway FL, Loria CM, Obarzanek E, Willamson DA. Comparison of weight-loss diets with different compositions of fat, protein, and carbohydrates. New England Journal of Medicine. 2009;360:859–73. Shai I, Schwarzfuchs D, Henkin Y, Shahar DR, Witkow S, Greenberg I, Golan R, Fraser D, Bolotin A, Vardi H, Tangi-Rozental O, Zuk-Ramot R, Sarusi B, Brickner D, Schwartz Z, Sheiner E, Marko R, Katorza E, Thiery J, Fiedler GM, Blüher M, Stumvoll M, Stampfer MJ. Dietary Intervention Randomized Controlled Trial (DIRECT) Group. Weight loss with a low-carbohydrate, Mediterranean, or low-fat diet. The New England Journal of Medicine. 2008;359:229–41. Howard BV, Manson JE, Stefanick ML, Beresford SA, Frank G, Jones B, Rodabough RJ, Snetselaar L, Thomson C, Tinker L, Vitolins M, Prentice R. Low-fat dietary pattern and weight change over 7 years: The Women's Health Initiative Dietary Modification Trial. The Journal of the American Medical Association. 2006;295:39–49.

49. Bray GA, Paeratakul S, Popkin BM. Dietary fat and obesity: A review of animal, clinical and epidemiological studies. Physiology and Behavior. 2004; 83:549–55.

50. Taubes G. The soft science of dietary fat. Science. 2001;291:2536–45.

51. Foster GD, Wyatt HR, Hill JO, McGuckin BG, Brill C, Selma B, Szapary PO, Rader DJ, Edman JS, Klein S. A randomized trial of a low-carbohydrate diet for obesity. New England Journal of Medicine. 2003;248:2082–90. Halton TL, Hu FB. The effects of high protein diets on thermogenesis, satiety and weight loss: A critical review. The Journal of the American College of Nutrition. 2004;23:373–85.

52. Crowe TC. Safety of low-carbohydrate diets. Obesity Reviews. 2005;6:235–45. Manheimer EW, van Zuuren EJ, Fedorowicz Z, Pijl H. Paleolithic nutrition for metabolic syndrome: Systematic review and meta-analysis. The American Journal of Clinical Nutrition. 2015;102:922–32.

53. Bravata DM, Sanders L, Huang J, Krumholz HM, Olkin I, Gardner CD, Bravata DM. Efficacy and safety of low-carbohydrate diets: A systematic review. Journal of the American Medical Association. 2003;289:1838–49.

54. Schwenke DC. Insulin resistance, low-fat diets, and low-carbohydrate diets: Time to test new menus. Current Opinion in Lipidology. 2005;16:55–60.

10

1. Centers for Disease Control and Prevention, National Center for Health Statistics. Clinical growth charts. Available from: http://www.cdc.gov/growthcharts/.

2. Villar J, Merialdi M, Gulmezoglu AM, Abalos E, Carroli G, Kulier R, de Onis M. Characteristics of randomized controlled trials included in systematic reviews of nutritional interventions reporting maternal morbidity, mortality, preterm delivery, intrauterine growth restriction and small for gestational age and birth weight outcomes. Journal of Nutrition. 2003;133:1632S–9S. U.S. Department of Health and Human Services. Centers for Disease Control and Prevention. National Center for Health Statistics System. National Vital Statistics Reports. Infant mortality statistics from the 2006 period linked birth/infant death data set. 2010. Available from: http://www.cdc.gov/nchs/data/nvsr/nvsr58/nvsr58_17.pdf.

3. Gillman MW. Developmental origins of health and disease. New England Journal of Medicine. 2005;353:1848–50. Barker DJ. The foetal and infant origins of inequalities in health in Britain. Journal of Public Health Medicine. 1991;13:64–8. Barker DJ. Fetal programming of coronary heart disease. Trends in Endocrinology and Metabolism. 2002;13:364–8. Rasmussen KM. The "fetal origins" hypothesis: Challenges and opportunities for maternal and child nutrition. Annual Review of Nutrition. 2000;21:73–95.

4. Institute of Medicine. Weight gain during pregnancy: Reexamining the guidelines. Washington, DC: National Academies Press; 2009. Available from: http://www.iom.edu/pregnancyweightgain.

5. U.S. Department of Agriculture and US Department. ChooseMyPlate.gov. Eating Fish While you are Pregnant or Breastfeeding. December 2015. Available from: http://www.choosemyplate.gov/moms-food-safety-fish.

6. Beard JL. Effectiveness and strategies of iron supplementation during pregnancy. American Journal of Clinical Nutrition. 2000;71:1288S–94S.

7. Beard JL. Effectiveness and strategies of iron supplementation during pregnancy. American Journal of Clinical Nutrition. 2000;71:1288S–94S. Feinleib M, Beresford SA, Bowman BA, Mills JL, Rader JI, Selhub J, Yetley EA. Folate fortification for the prevention of birth defects: Case study. American Journal of Epidemiology. 2001;154:S60–9.

8. Centers for Disease Control and Prevention. Prevalence Estimates of Gestational Diabetes Mellitus in the United States, Pregnancy Risk Assessment Monitoring System (PRAMS), 2007–2010. Available from: http://www.cdc.gov/pcd/issues/2014/13_0415.htm.

9. Di Cianni G, Ghio A, Resi V, Volpe L. Gestational diabetes mellitus: An opportunity to prevent type 2 diabetes and cardiovascular disease in young women. Women's Health. 2010;6:97–105. Centers for Disease Control and Prevention.

Diabetes. Available from: http://www.cdc.gov/media/presskits/aahd/diabetes.pdf. Centers for Disease Control and Prevention. Diabetes and Pregnancy Gestational Diabetes. Available from: http://www.cdc.gov/pregnancy/documents/Diabetes_and_Pregnancy508.pdf.

10. Papageorghiou AT, Campbell S. First trimester screening for preeclampsia. Current Opinion in Obstetrics and Gynecology. 2006;18:594–600. Hutcheon JA, Lisonkova S, Joseph KS. Epidemiology of pre-eclampsia and the other hypertensive disorders of pregnancy. Best Practices & Research Clinical Obstetrics & Gynaecology. 2011;25:391–403. American Academy of Obstetrics and Gynecology. Hypertension in Pregnancy. Available from: http://www.acog.org/Resources-And-Publications/Task-Force-and-Work-Group-Reports/Hypertension-in-Pregnancy.

11. American Academy of Pediatrics. Pediatric nutrition handbook, 6th ed., Elk Grove Village, IL; 2008. American Academy of Pediatrics. 2005 AAP policy statement of breastfeeding and the use of human milk. Pediatrics. 2005;115:496–501.

12. Ryan AS, Wenjun Z, Acosta A. Breastfeeding continues to increase into the new millennium. Pediatrics. 2002;110:1103–9.

13. Grummer-Strawn LM, Scanlon KS, Fein SB. Infant feeding and feeding transitions during the first year of life. Pediatrics. 2008;122:S36–42.

14. McGuire MK, McGuire MA. Human milk: Mother Nature's prototypical probiotic food? 2015. Advances in Nutrition 6:112–23. Hunt KM, Foster JA, Forney LJ, Schütte UME, Beck DL, McGuire MK, McGuire MA. Characterization of the diversity and temporal stability of bacterial communities in human milk. PLoS ONE. 2011. 6(6): e21313.

15. Dewey KG. Impact of breastfeeding on maternal nutritional status. Advances in Experimental Medical Biology. 2004;554:91–100. Winkvist A, Rasmussen KM. Impact of lactation on maternal body weight and body composition. Journal of Mammary Gland Biology and Neoplasia. 1999;4:309–18.

16. Taylor JS, Kacmar JE, Nothnagle M, Lawrence RA. A systematic review of the literature associating breastfeeding with type 2 diabetes and gestational diabetes. Journal of the American College of Nutrition. 2005;24:320–6. Labbok MH. Effects of breasfeeding on the mother. Pediatric Clinics of North America. 2001;48:143–58.

17. American Academy of Pediatrics. New Mother's Guide to Breastfeeding 2e. 2011. Available from: https://www.healthychildren.org/English/ages-stages/baby/breastfeeding/Pages/Things-to-Avoid-When-Breastfeeding.aspx. U.S. Department of Agriculture and U.S. Department. ChooseMyPlate.gov. Nutritional Needs While Breastfeeding. Available from: http://www.choosemyplate.gov/moms-breastfeeding-nutritional-needs#sthash.iSZMmOMb.dpuf.

18. Lind T, Hernell O, Lonnerdal B, Stenlund H, Domellof M, Persson LA. Dietary iron intake is positively associated with hemoglobin concentration during infancy but not during the second year of life. Journal of Nutrition. 2004;134:1064–70.

19. Wagner CL, Greer FR. Prevention of rickets and vitamin D deficiency in infants, children, and adolescents. Pediatrics. 2008;122:1142–52.

20. American Academy of Pediatrics. Pediatric nutrition handbook, 6th ed., Elk Grove Village, IL; 2008. American Academy of Pediatrics. 2005. AAP policy statement of breastfeeding and the use of human milk. Pediatrics. 2005;115:496–501.

21. Centers for Disease Control and Prevention. Recommendations to prevent and control iron deficiency anemia in the United States. Morbidity Mortality Weekly Report. 2002;51:897–9. Baker RD, Greer FR. From the American Academy of Pediatrics. Clinical report—Diagnosis and prevention of iron deficiency and iron-deficiency anemia in infants and young children (0–3 years of age). Pediatrics. 2010;126:2010–576.

22. American Academy of Pediatrics Committee on Nutrition. The use and misuse of fruit juice in pediatrics. Pediatrics. 2001;107:1210–3. American Academy of Pediatrics. Pediatric nutrition handbook, 6th ed., Elk Grove Village, IL; 2008.

23. Grummer-Strawn LM, Scanlon KS, Fein SB. Infant feeding and feeding transitions during the first year of life. Pediatrics. 2008;122:S36–42.

24. Thygarajan A, Burks AW. American Academy of Pediatrics recommendations on the effects of early nutritional interventions on the development of atopic disease. Current Opinion in Pediatrics. 2008;20:698–702. Greer FR, Sicherer SH, Burks A, and the Committee on Nutrition and Section on Allergy and Immunology. Effects of early nutritional interventions on the development of atopic disease in infants and children: The role of maternal dietary restriction, breastfeeding, timing of introduction of complementary foods, and hydrolyzed formulas. Pediatrics. 2008;121:183–91.

25. American Academy of Pediatrics. healthychildren.org. Starting Health Foods. Available from: https://www.healthychildren.org/English/ages-stages/baby/feeding-nutrition/Pages/Switching-To-Solid-Foods.aspx.

26. Centers for Disease Control and Prevention. Trends in Allergic Conditions Among Children: United States, 1997–2011. Available from: http://www.cdc.gov/nchs/data/databriefs/db121.htm.

27. Boyce JA, Assa'ad A, Burks AW, Jones SM, Sampson HA, Wood RA, Plaut M, Cooper SF, Fenton MJ, Arshad SH, Bahna SL, Beck LA, Byrd-Bredbenner C, Camargo CA Jr, Eichenfield L, Furuta GT, Hanifin JM, Jones C, Kraft M, Levy BD, Lieberman P, Luccioli S, McCall KM, Schneider LC, Simon RA, Simons FE, Teach SJ, Yawn BP, Schwaninger JM. Guidelines for the Diagnosis and Management of Food Allergy in the United States: Summary of the NIAID-Sponsored Expert Panel Report. The Journal of Allergy and Clinical Immunology. 2010. 126:1105–18.

28. American Academy of Pediatrics. What you need to know about the new guidelines for the diagnosis and management of food allergy in the U.S. Available from: http://www2.aap.org/sections/allergy/allergy_guidelines_final_1.pdf.

29. American Academy of Pediatrics Committee on Nutrition. The use and misuse of fruit juice in pediatrics. Pediatrics. 2001;107:1210–3.

30. Koivisto Hursti UK. Factors influencing children's food choice. Annals of Medicine. 1999;31:26–32.

31. Patrick H, Nicklas TA. A review of family and social determinants of children's eating patterns and diet quality. Journal of American College of Nutrition. 2005;2:83–92.

32. U.S. Department of Agriculture. ChooseMyPlate.gov. Growth During the Preschool Years. Available from: http://www.choosemyplate.gov/preschoolers-growth.

33. National Center for Health Statistics. Health, United States, 2014: With Special Feature on Adults Aged 55–64. Hyattsville, MD. 2015. Available from: http://www.cdc.gov/nchs/data/hus/hus14.pdf.

34. Robinson TN. Television viewing and childhood obesity. Pediatric Clinics of North America. 2001;4:1017–25.

35. Crespo CJ, Smit E, Troiano RP, Bartlett SJ, Macera CA, Andersen RE. Television watching, energy intake, and obesity in US children: Results from the third National Health and Nutrition Examination Survey, 1988–1994. Archives of Pediatric and Adolescent Medicine. 2001;155:360–5. U.S. Department of Health and Human Services. Centers for Disease Control and Prevention. National Center for Health Statistics. Herrick KA, Fakhouri THI, Carlson SA, Fulton JE. TV Watching and Computer Use in U.S. Youth Aged 12–15, 2012. NCHS Data Brief. No 157, 2014. Available from: http://www.cdc.gov/nchs/data/databriefs/db157.pdf.

36. Soemantri AG, Pollitt E, Kim I. Iron deficiency anemia and educational achievement. American Journal of Clinical Nutrition. 1985;42:1221–8; Iannotti LL, Tielsch JM, Black MM, Black RE. Iron supplementation in early childhood: Health benefits and risks. American Journal of Clinical Nutrition. 2006;84:1261–76. Baker RD, Greer FR. The Committee on Nutrition. Diagnosis and prevention of iron deficiency and iron deficiency anemia in infants and young children (0–3 years of age). Pediatrics. 2010;126. Available from: http://pediatrics.aappublications.org/content/126/5/1040.

37. Salsberry PJ, Reagan PB, Pajer K. Growth differences by age of menarche in African American and White girls. Nursing Research. 2009;58:382–90.

38. Clemens, R, Kranz S, Mobley AR, Nicklas TA, Raimondi MP, Rodreguez JC, Slavin JL, Warshaw H. Filling America's fiber intake gap: Summary of a roundtable to probe realistic solutions with a focus on grain-based foods. The Journal of Nutrition. 2012;142:1390S–1401S.

39. United States Department of Agriculture. What we eat in America. NHANES 2011–2012. Available from: http://www.ars.usda.gov/SP2UserFiles/Place/80400530/pdf/1112/Table_1_NIN_GEN_11.pdf.

40. Nielsen SJ, Popkin BM. Changes in beverage intake between 1977 and 2000. American Journal of Preventive Medicine. 2004;27:205–10.

41. Centers for Disease Control and Prevention. Recommendations to prevent and control iron deficiency anemia in the United States. Morbidity Mortality Weekly Report. 2005;51:897–9. Baker RD, Greer FR. From the American Academy of Pediatrics. Clinical report—Diagnosis and prevention of iron deficiency and iron-deficiency anemia in infants and young children (0–3 years of age). 2010;126:2010–576.

42. Cavadini C, Siega-Riz AM, Popkin BM. US adolescent food intake trends from 1965 to 1996. Archives of Disease in Children. 2000;83:18–24.

43. Centers for Disease Control and Prevention. The state of aging and health in America. 2004. Available from: http://www.cdc.gov/aging/pdf/state_of_aging_and_health_in_america_2004.pdf.

44. Moretti C, Frajese GV, Guccione L, Wannenes F, De Martino MU, Fabbri A, Frajese G. Androgens and body composition in the aging male. Journal of Endocrinological Investigation. 2005;28:56–64. St-Onge MP, Gallagher D. Body composition changes with aging: The cause or the result of alterations in metabolic rate and macronutrient oxidation? Nutrition. 2010;26:152–55.

11

1. National Center for Health Statistics. Health, United States, 2012: With special feature on emergency care. Hyattsville, MD; 2012. Available from: http://www.cdc.gov/nchs/data/hus/hus12.pdf.

2. U.S. Department of Health and Human Services, Office of Disease Prevention and Health Promotion. 2008 physical activity guidelines for Americans. Washington, DC. Available from: http://health.gov/PAGuidelines/pdf/paguide.pdf.

3. Vuori IM. Health benefits of physical activity with special reference to interaction with diet. Public Health Nutrition. 2001;4:517–28. Williams MA, Haskell WL, Ades PA, Amsterdam EA, Bittner V, Franklin BA, Gulanick M, Laing ST, Stewart KJ, American Heart Association Council on Clinical Cardiology, American Heart Association Council on Nutrition, Physical Activity, and Metabolism. Resistance exercise in individuals with and without cardiovascular disease: 2007 update: A scientific statement from the American Heart Association Council on Clinical Cardiology and Council on Nutrition, Physical Activity, and Metabolism Circulation. 2007;116:572–84.

4. Garber CE, Blissmer B, Deschenes MR, Franklin BA, Lamonte MJ, Lee IM, Nieman DC, Swain DP. Quantity and quality for developing and maintaining cardiorespiratory, musculoskeletal, and neuromotor fitness in apparently healthy adults: guidance for prescribing exercise. Medicine & Science in Sports & Exercise. 2011;43:1334–59. Available from: http://journals.lww.com/acsm-msse/Fulltext/2011/07000/Quantity_and_Quality_of_Exercise_for_Developing.26.aspx.

5. Balady GJ, Arena R, Sietsema K, Myers J, Coke L, Fletcher GF, Forman D, Franklin B, Guazzi M, Gulati M, Keteyian SJ, Lavie CJ, Macko R, Mancini D, Milani RV. Clinician's guide to cardiopulmonary exercise testing in adults: A scientific statement from the American Heart Association. Circulation. 2010;122:191–225. Blair SN, Kohl HW, Barlow CE, Paffenbarger RS, Gibbons LW, Macera CA. Changes in physical fitness and all-cause mortality. A prospective study of health and unhealthy men. The Journal of the American Medical Association. 1995;273:1093–8. Booth FW, Gordon WE, Carlson CH, Hamilton MT. Waging war on modern chronic diseases: Primary prevention through exercise biology. Journal of Applied Physiology. 2000;88:774–87.

6. Bickel CS, Slade J, Mahoney E, Haddad F, Dudley GA, Adams GR. Time course of molecular responses of human skeletal muscle to acute bouts of resistance exercise. Journal of Applied Physiology. 2005;98:482–8.

7. Bemben DA, Bemben MG. Dose-response effect of 40 weeks of resistance training on bone mineral density in older adults. Osteoporosis International. 2011;22:179–86.

8. Bonnar BP, Deivert RG, Gould TE. The relationship between isometric contraction durations during hold-relax stretching and improvement of hamstring flexibility. Journal of Sports Medicine and Physical Fitness. 2004;44:258–61.

9. Harmer PA, Li F. Tai Chi and falls prevention in older people. Medicine and Sport Science. 2008;52:124–34. Day L, Hill KD, Stathakis VZ, Flicker L, Segal L, Cicuttini F, Jolley D. Impact of tai-chi on falls among preclinically disabled older people. A randomized controlled trial. Journal of the American Medical Directors Association. 2015 May 1;16(5):420–6. Mat S, Tan MP, Kamaruzzaman SB, Ng CT. Physical therapies for improving balance and reducing falls risk in osteoarthritis of the knee: A systematic review. Age and Ageing. 2015;44(1):16–24. Kendrick D, Kumar A, Carpenter H, Zijlstra GA, Skelton DA, Cook JR, Stevens Z, Belcher CM, Haworth D, Gawler SJ, Gage H, Masud T, Bowling A, Pearl M, Morris RW, Iliffe S, Delbaere K. Exercise for reducing fear of falling in older people living in the community. The Cochrane Database of Systematic Reviews. 2014. 28;11:CD009848.

10. Medbo, JI, Tabata, I. Relative importance of aerobic and anaerobic energy release during short-lasting exhausting bicycle exercise. Journal of Applied Physiology. 1989;67:1881–86.

11. Baker JS. McCormick MC, Robergs RA. Interaction among skeletal muscle metabolic energy systems during intense exercise. Journal of Nutrition and Metabolism. 2010;2010:905612. Gladden LB. Lactate metabolism: A new paradigm for the third millennium. The Journal of Physiology. 2004;558:5–30. Hall MM, Rajasekaran S, Thomsen TW, Peterson AR. Lactate: Friend or Foe. PM R.: The Journal of Function, Injury and Rehabilitation. 2016;8:S8–S15.

12. Connolly DA, Sayers SP, McHugh MP. Treatment and prevention of delayed onset muscle soreness. Journal of Strength and Conditioning Research. 2003;17:197–208. Lewis PB, Ruby D, Bush-Joseph CA. Muscle soreness and delayed-onset muscle soreness. Clinics in Sports Medicine. 2012;31:255–62. Volpe SL. Micronutrient requirements for athletes. Clinics in Sports Medicine. 2007;26:119–30.

13. Stellingwerff T. Contemporary nutrition approaches to optimize elite marathon performance. International Journal of Sports Physiology and Performance. 2013;8: 573–78.

14. Manore M, Thompson J. Energy requirements of the athlete: Assessment and evidence of energy efficiency. In: Burke L, Deakin V, eds. Clinical Sports Nutrition. 5th eds. Sydney, Australia: McGraw-Hill; 2015:114–39.

15. Institute of Medicine. Dietary Reference Intakes for energy, carbohydrate, fiber, fat, fatty acids, cholesterol, protein, and amino acids. Washington, DC: National Academies Press; 2005.

16. Barnes MJ. Alcohol: Impact on sports performance and recovery in male athletes. Sports Medicine. 2014; 44:909–19.

17. Burke LM, Hawley JA, Wong SH, Jeukendrup AE. Carbohydrates for training and competition. Journal of Sports Sciences. 2011;29:S17–27.

18. American College of Sports Medicine, Academy of Nutrition and Dietetics Dietitians of Canada. Joint Position Statement. Nutrition and Athletic Performance. Medicine & Science in Sports & Exercise. 2016;48:543–68. Available from: http://journals.lww.com/acsm-msse/Fulltext/2016/03000/Nutrition_and_Athletic_Performance.25.aspx.

19. Burke LM. Nutrition strategies for the marathon: Fuel for training and racing. Sports Medicine. 2007;37:344–7. Hawley JA, Schabort EJ, Noakes TD, Dennis SC. Carbohydrate-loading and exercise performance. An update. Sports Medicine. 1997; 24:73–81.

20. Burke LM, Kiens B, Ivy JL. Carbohydrates and fat for training and recovery. Journal of Sports Sciences. 2004;22:5–30.Thomas DE, Brotherhood JR, Brand JC. Carbohydrate feeding before exercise: Effect of glycemic index. International Journal of Sports Medicine. 1991;12180–6. Rehrer NJ, van Kemenade M, Meester W, Brouns F, Saris WH. Gastrointestinal complaints in relation to dietary intake in triathletes. International Journal of Sport Nutrition. 1992;2:48–59.

21. Tipton KD, Witard OC. Protein requirements and recommendations for athletes: Relevance of ivory tower arguments for practical recommendations. Clinics in Sports Medicine. 2007;26:17–36.

22. Institute of Medicine. Dietary Reference Intakes for energy, carbohydrate, fiber, fat, fatty acids, cholesterol, protein, and amino acids. Washington, DC: National Academies Press; 2005.

23. Mettler S, Mitchell N, Tipton KD. Increased protein intake reduces lean body mass loss during weight loss in athletes. Medicine and Science in Sports and Exercise. 2010;42:326–37. Phillips SM, Van Loon LJ. Dietary protein for athletes: From requirements to optimum adaptation. Journal of Sports Sciences. 2011;29:S29–38.

24. Craig WJ, Mangels AR. American Dietetic A. Position of the American Dietetic Association: Vegetarian diets. Journal of the American Dietetic Association. 2009;109:1266–82. Berning JR. The Vegetarian Athlete. In: Maughan RJ, ed. The Encyclopaedia of Sports Medicine: An IOC Medical Commission Publications, Sports Nutrition. West Sussux, UK: Wiley; 2014:382–91.

25. Simopoulos AP. Omega-3 fatty acids and athletics. Current Sports Medicine Reports. 2007;6:230–6.

26. Maughan RJ, King DS, Lea T. Dietary supplements. Journal of Sports Science. 2004;22:95–113.

27. Barr SI, Rideout CA. Nutritional considerations for vegetarian athletes. Nutrition. 2004;20:696–703. Fuhrman J, Ferreri DM. Fueling the vegetarian (vegan) athlete. Current Sports Medicine Reports. 2010;8:233–41.

28. Lukaski HC. Vitamin and mineral status: Effects on physical performance. Nutrition. 2004;20:632–44. Haymes E. Iron. In: Driskell J, Wolinsky I, eds. Sports Nutrition: Vitamins and Trace Elements. New York, NY: CRC/Taylor & Francis;2006:203–16.

29. Suedekum NA, Dimeff RJ. Iron and the athlete. Current Sports Medicine Reports. 2005;4:199–202.

30. Kunstel K. Calcium requirements for the athlete. Current Sports Medicine Reports. 2005;4:203–6.

31. Halliday TM, Peterson NJ, Thomas JJ, Kleppinger K, Hollis BW, Larson-Meyer DE. Vitamin D status relative to diet, lifestyle, injury, and illness in college athletes. Medicine and Science in Sports and Exercise. 2011;43:335–43. Cannell JJ, Hollis BW, Sorenson MB, Taft TN, Anderson JJ. Athletic performance and vitamin D. Medicine and Science in Sports and Exercise. 2009;415:1102–10. Larson-Meyer DE, Willis KS. Vitamin D and athletes. Current Sports Medicine Reports. 2010;9:220–6.

32. Peternelj TT, Coombes JS. Antioxidant supplementation during exercise training: Beneficial or detrimental? Sports Medicine. 2011; 41:1043–69. Watson TA, MacDonald-Wicks LK, Garg ML. Oxidative stress and antioxidants in athletes undertaking regular exercise training. International Journal of Sports Nutrition and Exercise Metabolism. 2005;15: 131–46.

33. Draeger CL, Naves A, Marques N, et al. Controversies of antioxidant vitamins supplementation in exercise: Ergogenic or ergolytic effects in humans? Journal of the International Society of Sports Nutrition. 2014;11: 4.

34. Sawka MN, Cheuvront SN, Carter R 3rd. Human water needs. Nutrition Reviews. 2005;63:S30–9.

35. Shirreffs SM. The importance of good hydration for work and exercise performance. Nutrition Reviews. 2005;63:S14–21.

36. Almond CS, Shin AY, Fortescue EB, Mannix RC, Wypij D, Binstadt BA, Duncan CN, Olson DP, Salerno AE, Newburger JW, Greenes DS. Hyponatremia among runners in the Boston Marathon. New England Journal of Medicine. 2005;352:1550–6.

37. Hsieh M. Recommendations for treatment of hyponatraemia at endurance events. Sports Medicine. 2004;34:231–8. Shirreffs SM, Sawka MN. Fluid and electrolyte needs for training, competition, and recovery. Journal of Sports Sciences. 2011;29: S39–46.

38. Sawka MN, Burke LM, Eichner ER, Maughan RJ, Montain SJ, Stachenfeld NS. American College of Sports Medicine position stand: Exercise and fluid replacement. American College of Sports Medicine. Medicine and Science in Sports and Exercise. 2007;39:377–90. American College of Sports M, Armstrong LE, Casa DJ, et al. American College of Sports Medicine position stand. Exertional heat illness during training and competition. Medicine and Science in Sports and Exercise. 2007;39:556–72.

39. Bergeron MF. Exertional heat cramps: Recovery and return to play. Journal of Sport Rehabilitation. 2007; 16:190–6.

40. Garth AK, Burke LM. What do athletes drink during competitive sporting activities? Sports Medicine. 2013;43:539–64.

41. Jeukendrup AE. Carbohydrate and exercise performance: The role of multiple transportable carbohydrates. Current Opinion in Clinical Nutrition and Metabolic Care. 2010;13:452–7.

42. Sherman WM, Doyle JA, Lamb DR, Strauss RH. Dietary carbohydrate, muscle glycogen, and exercise performance during 7 d of training. American Journal of Clinical Nutrition. 1993;57:27–31. Coyle EF. Timing and method of increased carbohydrate intake to cope with heavy training, competition and recovery. Journal of Sports Sciences. 1991;9:51–122. Ormsbee MJ, Bach CW, Baur DA. Pre-exercise nutrition: The role of macronutrients, modified starches and supplements on metabolism and endurance performance. *Nutrients*. 2014;6: 1782–1808.

43. Burke LM, Kiens B, Ivy JL. Carbohydrates and fat for training and recovery. Journal of Sports Science. 2004;22: 15–30.

44. Stevenson E, Williams C, Biscoe H. The metabolic responses to high carbohydrate meals with different glycemic indices consumed during recovery from prolonged strenuous exercise. International Journal of Sports Nutrition and Exercise Metabolism. 2005;15:291–307. Walton P, Rhodes EC. Glycaemic index and optimal performance. Sports Medicine. 1997;23:164–72.

45. Baty JJ, Hwang H, Ding Z, Bernard JR, Wang B, Kwon B, Ivy JL. The effect of a carbohydrate and protein supplement on resistance exercise performance, hormonal response, and muscle damage. Journal of Strength and Conditioning Research. 2007;21:321–9.

46. Howarth KR, Moreau NA, Phillips SM, Gibala MJ. Coingestion of protein with carbohydrate during recovery from endurance exercise stimulates skeletal muscle protein synthesis in humans. Journal of Applied Physiology. 2009;106:1394–402. Tang JE, Manolakos JJ, Kujbida GW, Lysecki PJ, Moore DR, Phillips SM. Minimal whey protein with carbohydrate stimulates muscle protein synthesis following resistance exercise in trained young men. Applied Physiology, Nutrition, and Metabolism. 2007;32:1132–8. Pennings B, Boirie Y, Senden JM, Gijsen AP, Kuipers H, van Loon LJ. Whey protein stimulates postprandial muscle protein accretion more

effectively than do casein and casein hydrolysate in older men. The American Journal of Clinical Nutrition. 2011;93:997–1005.

47. Speedy DB, Thompson JM, Rodgers I, Collins M, Sharwood K, Noakes TD. Oral salt supplementation during ultradistance exercise. Clinical Journal of Sport Medicine. 2002;12:279–84.

12

1. Paxton SJ. Body dissatisfaction and disordered eating. Journal of Psychosomatic Research. 2002;53:961–2. Voelker DK, Reel JJ, Greenleaf C. Weight status and body image perceptions in adolescents: Current perspectives. Adolescent Health, Medicine and Therapeutics. 2015;6:149–58.

2. Neumark-Sztainer D, Wall M, Larson NI, Eisenberg ME, Loth K. Dieting and disordered eating behaviors from adolescence to young adulthood: Findings from a 10-year longitudinal study. Journal of the American Dietetics Association. 2014;111:1004–11.

3. Wilson GT, Shafran R. Eating disorders guidelines from NICE. Lancet. 2005;365:79–81. Herpertz-Dahlmann B. Adolescent eating disorders: Update on definitions, symptomatology, epidemiology, and comorbidity. Child and Adolescent Psychiatric Clinics of North American. 2015;24:177–96.

4. National Association of Anorexia Nervosa and Associated Disorders. Eating Disorders Statistics. Available from: http://www.anad.org/get-information/about-eating-disorders/eating-disorders-statistics/.

5. Clarke LH. Older women's perceptions of ideal body weights: The tensions between health and appearance motivations for weight loss. Ageing and Society. 2002;22:751–3. Podfigurna-Stopa A, Czyzyk A, Katulski K, Smolarczyk R, Grymowicz M, Maciejewska-Jeske M, Meczekalski B. Eating disorders in older women. Maturitas. 2015;82:146–52.

6. Slevec JH, Tiggemann M. Predictors of body dissatisfaction and disordered eating in middle-aged women. Clinical Psychology Review. 2011;31:515–24.

7. Woodside BD, Garfinkel PE, Lin E, Goering P, Kaplan AS, Goldbloom DS, Kennedy SH. Comparisons of men with full or partial eating disorders, men without eating disorders, and women with eating disorders in the community. American Journal of Psychiatry. 2001;158:570–4. Raevuori A, Keski-Rahkonen A, Hoek HW. A review of eating disorders in males. Current Opinion in Psychiatry. 2014;27:426–30.

8. American Psychiatric Association. Diagnostic and statistical manual of mental disorders, 5th ed. (DSM-V). Washington, DC: American Psychiatric Association; 2013.

9. Eddy KT, Dorer DJ, Franko DL, Tahilani K, Thompson-Brenner H, Herzog DB. Diagnostic crossover in anorexia nervosa and bulimia nervosa: Implications for DSM-V. Williamson DA, Martin CK, Stewart T. Psychological aspects of eating disorders. Best Practice & Research. Clinical Gastroenterology. 2004;18:1073–88.

10. Heaner MK, Walsh BT. A history of the identification of the characteristic eating disturbances of Bulimia Nervosa, Binge Eating Disorder and Anorexia Nervosa. Appetite. 2013 Jun;65:185–8.

11. Peebles R, Wilson JL, Lock JD. Self-injury in adolescents with eating disorders: Correlates and provider bias. Journal of Adolescent Health. 2011;48:310–3.

12. Halmi KA, Tozzi F, Thornton LM, Crow S, Fichter MM, Kaplan AS, Keel P, Klump KL, Lilenfeld LR, Mitchell JE, Plotnicov KH, Pollice C, Rotondo A, Strober M, Woodside DB, Berrettini WH, Kaye WH, Bulik CM. The relation among perfectionism, obsessive-compulsive personality disorder and obsessive-compulsive disorder in individuals with eating disorders. International Journal of Eating Disorders. 2005;38:371–4.

13. Wolfe BE. Reproductive health in women with eating disorders. Journal of Obstetrics and Gynecology in Neonatal Nursing. 2005;34:255–63.

14. Teng K. Premenopausal osteoporosis, an overlooked consequence of anorexia nervosa. Cleveland Clinic Journal of Medicine. 2011;78:50–8.

15. Birmingham CL, Su J, Hlynsky JA, Goldner EM, Gao M. The mortality rate from anorexia nervosa. International Journal of Eating Disorders. 2005;38:143–6. Meczekalski B, Podfigurna-Stopa A, Katulski K. Long-term consequences of anorexia nervosa. Maturitas. 2013:75:215–20.

16. Patrick L. Eating disorders: A review of the literature with emphasis on medical complications and clinical nutrition. Alternative Medical Review. 2002;7:184–202. Westmoreland P, Krantz MJ, Mehler PS. Medical Complications of Anorexia Nervosa and Bulimia. The American Journal of Medicine. 2016;1:30–7.

17. Holm-Denoma JM, Witte TK, Gordon KH, Herzog DB, Franko DL, Fichter M, Quadflieg N, Joiner TE Jr. Deaths by suicide among individuals with anorexia as arbiters between competing explanations of the anorexia-suicide link. Journal of Affective Disorders. 2008;107:231–6.

18. Williams PM, Goodie J, Motsinger CD. Treating eating disorders in primary care. American Family Physician. 2008;77:187–95.

19. Hay P, Bacaltchuk J. Bulimia nervosa. Clinical Evidence. 2004;12:1326–47.

20. Cooper Z, Fairburn CG. Refining the definition of binge eating disorder and nonpurging bulimia nervosa. International Journal of Eating Disorders. 2003;34:S89–95.

21. Mitchell JE, Pyle RL, Eckert ED, Hatsukami D, Soll E. Bulimia nervosa in overweight individuals. Journal of Nervous and Mental Disease. 1990;178:324–7.

22. Thompson-Brenner H, Eddy KT, Franko DL, Dorer D, Vashchenko M, Herzog DB. Personality pathology and substance abuse in eating disorders: A longitudinal study. International Journal of Eating Disorders. 2008;41:203–8.

23. Forney KJ, Buchman-Schmitt JM, Keel PK, Frank GK. The medical complications associated with purging. The International Journal of Eating Disorders. 2016.

24. Smink FR, van Hoeken D, Hoek HW. Epidemiology of eating disorders: Incidence, prevalence and mortality rates. Current Psychiatry Reports. 2012;14:406–14.

25. Franko DL, Keshaviah A, Eddy KT, Krishna M, Davis MC, Keel PK. A longitudinal investigation of mortality in anorexia nervosa and bulimia nervosa. American Journal of Psychiatry. 2013;170:917–25.

26. American Psychiatric Association. Diagnostic and statistical manual of mental disorders, 5th ed. (DSM-V). Washington, DC: American Psychiatric Association; 2013.

27. Vanderlinden J, DalleGrave R, Fernandez F, Vandereycken W, Pieters G, Noorduin C. Which factors do provoke binge eating? An exploratory study in eating disorder patients. Eating and Weight Disorders. 2004;9:300–5.

28. Dansky BS, Brewerton TD, Kilpatrick DG. Comorbidity of bulimia nervosa and alcohol use disorders: Results from the National Women's Study. International Journal of Eating Disorders. 2000;27:180–90.

29. Striegel-Moore RH, Franko DL. Epidemiology of binge eating disorder. International Journal of Eating Disorders. 2003;34:S19–29.

30. Agh T, Kovács G, Pawaskar M, Supina D, Inotai A, Vokó Z.Epidemiology, health-related quality of life and economic burden of binge eating disorder: A systematic literature review. Eating and Weight Disorders. 2015;20:1–12.

31. Pagoto S, Bodenlos JS, Kantor L, Gitkind M, Curtin C, Ma Y. Association of major depression and binge eating disorder with weight loss in a clinical setting. Obesity. 2007;15:2557–9.

32. Niego SH, Kofman MD, Weiss JJ, Geliebter A. Binge eating in the bariatric surgery population: A review of the literature. International Journal of Eating Disorders. 2007;40:349–59.

33. Masheb RM, Grilo CM. On the relation of attempting to lose weight, restraint, and binge eating in outpatients with binge eating disorder. Obesity Research. 2000;8:638–45.

34. Schenck CH, Mahowald MW. Review of nocturnal sleep-related eating disorders. International Journal of Eating Disorders. 1994;15:343–56.

35. Aronoff NJ, Geliebter A, Zammit G. Gender and body mass index as related to the night-eating syndrome in obese outpatients. Journal of the American Dietetic Association. 2001;101:102–4.

36. Winkelman JW. Sleep-related eating disorder and night eating syndrome: Sleep disorders, eating disorders, or both? Sleep. 2006;29:949–54. Inoue Y. Sleep-related eating disorder and its associated conditions. Psychiatry and Clinical Neurosciences. 2015;69:309–20.

37. The Cleveland Clinic. Sleep-related eating disorders. Available from: http://my.clevelandclinic.org/disorders/sleep_disorders/hic_sleep-related_eating_disorders.aspx.

38. Winkelman JW. Sleep-related eating disorder and night eating syndrome: Sleep disorders. 2006;29:949–54.

39. Marcontell DK, Laster AE, Johnson J. Cognitive-behavioral treatment of food neophobia in adults. Journal of Anxiety Disorders. 2003;17:243–51.

40. Nicklaus S, Boggio V, Chabanet C, Issanchou S. A prospective study of food variety seeking in childhood, adolescence and early adult life. Appetite. 2005;44:289–97.

41. Pelchat ML. Of human bondage: Food craving, obsession, compulsion, and addiction. Physiology and Behavior. 2002;76:347–52.

42. Tuorila H, Mustonen S. Reluctant trying of an unfamiliar food induces negative affection for the food. Appetite. 2010;54:418–21.

43. Norris ML, Spettigue WJ, Katzman DK. Update on eating disorders: Current perspectives on avoidant/restrictive food intake disorder in children and youth. Neuropsychiatry Disease and Treatment. 2016;12:213–8.

44. Pope CG, Pope HG, Menard W, Fay C, Olivardia R, Phillips KA. Clinical features of muscle dysmorphia among males with body dysmorphic disorder. Body Image. 2005;2:395–400. AsHak WW,

Boton MA, Bensoussan JC, Dous GV, Nguyen TT, Powell-Hicks AL, Gardner JE, Ponton KM. Quality of life in body dysmorphic disorder. CNS Spectrums. 2012;17:167–75.

45. Wroblewska AM. Androgenic-anabolic steroids and body dysmorphia in young men. Journal of Psychosomatic Research. 1997;42:225–34.

46. Grieve FG. A conceptual model of factors contributing to the development of muscle dysmorphia. Eating Disorders. 2007;15:63–80.

47. Olivardia R, Pope HG Jr, Hudson JI. Muscle dysmorphia in male weightlifters: A case-control study. American Journal of Psychiatry. 2000;157:1291–6. Foster AC, Shorter GW, Griffiths MD. Muscle dysmorphia: Could it be classified as an addiction to body image? Journal of Behavioral Addictions. 2015;4:1–5.

48. Grieve FG. A conceptual model of factors contributing to the development of muscle dysmorphia. Eating Disorders. 2007;15:63–80.

49. Becker AE, Keel P, Anderson-Fye EP, Thomas JJ. Genes and/or jeans? Genetic and socio-cultural contributions to risk for eating disorders. Journal of Addictive Disorders. 2004;23:81–103.

50. Eddy KT, Hennessey M, Thompson-Brenner H. Eating pathology in East African women: The role of media exposure and globalization. Journal of Nervous and Mental Disorders. 2007;195:196–202.

51. George JB, Franko DL. Cultural issues in eating pathology and body image among children and adolescents. Journal of Pediatric Psychology. 2010;35:231–42.

52. National Eating Disorders Association. Statistics: Eating disorders and their precursors. Available from: http://www.nationaleatingdisorders.org.

53. Rubinstein S, Caballero B. Is Miss America an undernourished role model? Journal of the American Medical Association. 2000;283:1569.

54. Wonderlich AL, Ackard DM, Henderson JB. Childhood beauty pageant contestants: Associations with adult disordered eating and mental health. Eating Disorders. 2005;13:291–301.

55. Brown JD, Witherspoon EM. The mass media and American adolescents' health. Journal of Adolescent Health. 2002;31:153–70.

56. Fernández-Aranda F, Krug I, Granero R, Ramón JM, Badia A, Giménez L, Solano R, Collier D, Karwautz A, Treasure J. Individual and family eating patterns during childhood and early adolescence: An analysis of associated eating disorder factors. Appetite. 2007;49:476–85.

57. Coulthard H, Blissett J, Harris G. The relationship between parental eating problems and children's feeding behavior: A selective review of the literature. Eating Behavior. 2004;5:103–15.

58. Humphries LL, Wrobel S, Wiegert HT. Anorexia nervosa. American Family Physician. 1982;26:199–204.

59. Kluck AS. Family factors in the development of disordered eating: Integrating dynamic and behavioral explanations. Eating Behavior. 2008;9:471–83.

60. Mazzeo SE, Zucker NL, Gerke CK, Mitchell KS, Bulik CM. Parenting concerns of women with histories of eating disorders. International Journal of Eating Disorders. 2005;37:S77–9.

61. Agras S, Hammer L, McNicholas F. A prospective study of the influence of eating-disordered mothers on their children. International Journal of Eating Disorders. 1999;25:253–62.

62. Cassin SE, von Ranson KM. Personality and eating disorders: A decade in review. Clinical Psychology Review. 2005;25:895–916.

63. Bulik CM, Tozzi F, Anderson C, Mazzeo SE, Aggen S, Sullivan PF. The relation between eating disorders and components of perfectionism. American Journal of Psychiatry. 2003;160:366–8.

64. George JB, Franko DL. Cultural issues in eating pathology and body image among children and adolescents. Journal of Pediatriac Psychology. 2010;35:231–42.

65. Kaye WH, Bulik CM, Plotnicov K, Thornton L, Devlin B, Fichter MM, Treasure J, Kaplan A, Woodside DB, Johnson CL, Halmi K, Brandt HA, Crawford S, Mitchell JE, Strober M, Berrettini W, Jones I. The genetics of anorexia nervosa collaborative study: Methods and sample description. International Journal of Eating Disorders. 2008;41:289–300.

66. Bulik CM, Slof-Op't Landt MC, van Furth EF, Sullivan PF. The genetics of anorexia nervosa. Annual Review of Nutrition. 2007;27:263–75.

67. Bulik CM, Reba L, Siega-Riz AM, Reichborn-Kjennerud T. Anorexia nervosa: Definition, epidemiology, and cycle of risk. International Journal of Eating Disorders. 2005;37:S2–9.

68. Cox LM, Lantz CD, Mayhew JL. The role of social physique anxiety and other variables in predicting eating behaviors in college students. International Journal of Sport Nutrition. 1997;7:310–7. Reinking MF, Alexander LE. Prevalence of disordered-eating behaviors in undergraduate female collegiate athletes and nonathletes. Journal of Athletic Training. 2005;40:47–51.

69. Sudi K, Ottl K, Payerl D, Baumgartl P, Tauschmann K, Muller W. Anorexia athletica. Nutrition. 2004;20:657–61.

70. Salbach H, Klinkowski N, Pfeiffer E, Lehmkuhl U, Korte A. Body image and attitudinal aspects of eating disorders in rhythmic gymnasts. Psychopathology. 2007;40:388–93. Tan JO, Calitri R, Bloodworth A, McNamee MJ. Understanding eating disorders in elite gymnastics: Ethical and conceptual challenges. Clinics in Sports Medicine. 2016;35:275–92.

71. Johnson C, Powers PS, Dick R. Athletes and eating disorders: The National Collegiate Athletic Association study. International Journal of Eating Disorders. 1999;26:179–88. Nichols DL, Sanborn CF, Essery EV. Bone density and young athletic women. An update. Sports Medicine. 2007;37:1001–14.

72. Beals KA, Hill AK. The prevalence of disordered eating, menstrual dysfunction, and low bone mineral density among U.S. collegiate athletes. International Journal of Sport Nutrition and Exercise Metabolism. 2006;16:1–23. Unwin BK, Goodie J, Reamy BV, Quinlan J. Care of the college student. American Family Physician. 2013;88:596–604.

73. Barrow GW, Saha S. Menstrual irregularity and stress fractures in collegiate female distance runners. American Journal of Sports Medicine. 1988;3:209–16. Kelsey JL, Bachrach LK, Procter-Gray E, Nieves J, Greendale GA, Sowers M, Brown BW, Matheson KA, drawford LS, Cobb KL. Risk factors for stress fracture among young female cross-country runners. Medicine and Science in Sports and Exercise. 2007;39:1457–63.

74. Lock J, Agras WS, Bryson S, Kraemer HC. A comparison of short- and long-term family therapy for adolescent anorexia nervosa. Journal of the American Academy of Child and Adolescent Psychiatry. 2005;44:632–9.

13

1. French MT, Zavala SK. The health benefits of moderate drinking revisited: Alcohol use and self-reported health status. American Journal of Health Promotion. 2007;21:484–91. Nova E, Baccan GC, Veses A, Zapatera B, Marcos A. Potential health benefits of moderate alcohol consumption: Current perspectives in research. The Proceedings of the Nutrition Society. 2012;71:307–15.

2. U.S. Department of Health and Human Services and U.S. Department of Agriculture. 2015–2020 Dietary Guidelines for Americans. 8th ed. December 2015. Available at: http://health.gov/dietaryguidelines/2015/guidelines/.

3. Roberts C, Robinson SP. Alcohol concentration and carbonation of drinks: The effect on blood alcohol levels. Journal of Forensic and Legal Medicine. 2007;14:398–405.

4. Zintzaras E, Stefanidis I, Santos M, Vidal F. Do alcohol-metabolizing enzyme gene polymorphisms increase the risk of alcoholism and alcoholic liver disease? Hepatology. 2006;43:352–61.

5. Sumida KD, Hill JM, Matveyenko AV. Sex differences in hepatic gluconeogenic capacity after chronic alcohol consumption. Clinical Medicine and Research. 2007;5:193–202.

6. Nolen-Hoeksema S. Gender differences in risk factors and consequences for alcohol use and problems. Clinical Psychology Review. 2004;24:981–1010. Wilsnack SC, Wilsnack RW, Kantor LW. Focus on: Women and the costs of alcohol use. Alcohol Research. 2013;35:219–28.

7. Arranz S, Chiva-Blanch G, Valderas-Martínez P, Medina-Remón A, Lamuela-Raventós RM, Estruch R. Wine, beer, alcohol and polyphenols on cardiovascular disease and cancer. Nutrients. 2012;4:759–81.

8. Ellison RC. Balancing the risks and benefits of moderate drinking. Annals of the New York Academy of Sciences. 2002;957:1–6.

9. Djoussé L, Gaziano JM. Alcohol consumption and risk of heart failure in the Physicians' Health Study I. Circulation. 2007;115:34–9. Elkind MS, Sciacca R, Boden-Albala B, Rundek T, Paik MC, Sacco RL. Moderate alcohol consumption reduces risk of ischemic stroke: The Northern Manhattan Study. Stroke. 2006;37(1):13–9.

10. Wollin SD, Jones PJH. Alcohol, red wine and cardiovascular disease. Journal of Nutrition. 2001;131:1401–4. Das DK, Mukherjee S, Ray D. Resveratrol and red wine, healthy heart and longevity. Heart Failure Reviews. 2010;15:467–77.

11. Schmid B, Hohm E, Blomeyer D, Zimmermann US, Schmidt MH, Esser G, Laucht M. Concurrent alcohol and tobacco use during early adolescence characterizes a group at risk. Alcohol and Alcoholism. 2007;42:219–25. Hingson R, White A. New research findings since the 2007 Surgeon General's Call to Action to Prevent and Reduce Underage Drinking: A review. Journal of Studies on Alcohol and Drugs. 2014;75:158–69.

12. Mukamal KJ, Chiuve SE, Rimm EB. Alcohol consumption and risk for coronary heart disease in men with healthy lifestyles. Archives of Internal Medicine. 2006:166:2145–50.

13. Kabagambe EK, Baylin A, Ruiz-Narvaez E, Rimm EB, Campos H. Alcohol intake, drinking patterns, and risk of nonfatal acute myocardial infarction in Costa Rica. American Journal of Clinical Nutrition. 2005;82:1336–45.

14. Goldberg IJ, Mosca L, Piano MR, Fisher EA. AHA Science Advisory. Wine and your heart: A science advisory for healthcare professionals from the Nutrition Committee, Council on Epidemiology and Prevention, and Council on Cardiovascular Nursing of the American Heart Association. Stroke. 2001;32:591–4.

15. Kurth T, Everett BM, Buring JE, Kase CS, Ridker PM, Gaziano JM. Lipid levels and the risk of ischemic stroke in women. Neurology. 2007;68:556–62.

16. Agarwal DP. Cardioprotective effects of light-moderate consumption of alcohol: A review of putative mechanisms. Alcohol and Alcoholism. 2002;37:409–15.

17. Volpato S, Pahor M, Ferrucci L, Simonsick EM, Guralnik JM, Kritchevsky SB, Fellin R. Harris TB. Inhibitor-1 in Well-Functioning Older Adults. The Health, Aging, and Body Composition Study. Clinical Investigation and Reports. Circulation 2004; 109: 607–12. Collins MA, Neafsey EJ, Mukamal KJ, Gray MO, Parks DA, Das DK, Korthuis RJ. Alcohol in moderation, cardioprotection, and neuroprotection: Epidemiological considerations and mechanistic studies. Alcoholism: Clinical and Experimental Reseaarch. 2009;33:206–19.

18. Gresele P, Pignatelli P, Guglielmini G, Carnevale R, Mezzasoma AM, Ghiselli A, Momi S, Violi F. Resveratrol, at concentrations attainable with moderate wine consumption, stimulates human platelet nitric oxide production. Journal of Nutrition. 2008;138:1602–08. German JB, Walzem RL. The health benefits of wine. Annual Review of Nutrition. 2000;20:561–93. Das DK, Mukherjee S, Ray D. Resveratrol and red wine, healthy heart and longevity. Heart Failure Reviews. 2010;15:467–77.

19. Manari AP, Preedy VR, Peters TJ. Nutritional intake of hazardous drinkers and dependent alcoholics in the UK. Addiction Biology. 2003;8:201–10. Salaspuro M. Nutrient intake and nutritional status in alcoholics. Alcohol and Alcoholism. 1993;28:85–8.

20. Sayon-Orea C, Martinez-Gonzalez MA, Bes-Rastrollo M. Alcohol consumption and body weight: A systematic review. Nutrition Reviews. 2011;69:419–31.

21. Kesse E, Clavel-Chapelon F, Slimani N, van Liere M, E3N Group. Do eating habits differ according to alcohol consumption? Results of a study of the French cohort of the European Prospective Investigation into Cancer and Nutrition (E3N-EPIC). American Journal of Clinical Nutrition. 2001;74:322–7.

22. González-Reimers E, García-Valdecasas-Campelo E, Santolaria-Fernández F, Milena-Abril A, Rodríguez-Rodríguez E, Martínez-Riera A, Pérez-Ramírez A, Alemán-Valls MR. Rib fractures in chronic alcoholic men: Relationship with feeding habits, social problems, malnutrition, bone alterations, and liver dysfunction. Alcohol. 2005;37:113–7.

23. Baan R, Straif K, Grosse Y, Secretan B, El Ghissassi F, Bouvard V, Altieri A, Cogliano V. WHO International Agency for Research on Cancer Monograph Working Group. Carcinogenicity of alcoholic beverages. Lancet Oncology. 2007;8:292–3.

24. Allen NE, Beral V, Casabonne D, Kan SW, Reeves GK, Brown A, Green J. Moderate alcohol intake and cancer incidence in women. Million Women Study Collaborators. Journal of the National Cancer Institute. 2009;101:296–305. Wilsnack SC, Wilsnack RW, Kantor LW. Focus on women and the costs of alcohol use. Alcohol Research. 2013;35:219–28.

25. Blot WJ. Alcohol and cancer. Cancer Research. 1992;52:2119S–23S. Seitz HK, Meier P. The role of acetaldehyde in upper digestive tract cancer in alcoholics. Translational Research. 2007;149:293–7. Pelucchi C, Tramacere I, Boffetta P, Negri E, La Vecchia C. Alcohol consumption and cancer risk. Nutrition and Cancer. 2011;63:983–90.

26. Hamid A, Kaur J. Long-term alcohol ingestion alters the folate-binding kinetics in intestinal brush border membrane in experimental alcoholism. Alcohol. 2007;41:441–6.

27. U.S. Department of Agriculture, U.S. Department of Health and Human Services. Dietary Guidelines for Americans, 2015. 8th ed., Washington, DC: U.S. Government Printing Office, December 2015. Available from: http://health.gov/dietaryguidelines/2015/guidelines/appendix-9/.

28. Piano MR. Alcoholic cardiomyopathy: Incidence, clinical characteristics, and pathophysiology. Chest. 2002;121: 1638–50.

29. Sesso HD, Cook NR, Buring JE, Manson JE, Gaziano JM. Alcohol consumption and the risk of hypertension in women and men. Hypertension. 2008;168:884–90. Klatsky AL. Alcohol and cardiovascular diseases: Where do we stand today? Journal of Internal Medicine. 2015;278:238–50.

30. Djoussé L, Levy D, Benjamin EJ, Blease SJ, Russ A, Larson MG, Massaro JM, D'Agostino RB, Wolf PA, Ellison RC. Long-term alcohol consumption and the risk of atrial fibrillation in the Framingham study. American Journal of Cardiology. 2004;93:710–3.

31. Hanck C, Whitcomb DC. Alcoholic pancreatitis. Gastroenterology Clinics of North America. 2004;33:751–65.

32. Peters R, Peters J, Warner J, Beckett N, Bulpitt C. Alcohol, dementia and cognitive decline in the elderly: A systematic review. Age Ageing. 2008;37:505–12. Weyerer S, Schäufele M, Wiese B, Maier W, Tebarth F, van den Bussche H, Pentzek M, Bickel H, Luppa M, Riedel-Heller SG. German AgeCoDe Study group. Current alcohol consumption and its relationship to incident dementia: Results from a 3-year follow-up study among primary care attenders aged 75 years and older. Age and Ageing. 2011;40:456–63. Kabai P. Alcohol consumption and cognitive decline in early old age. Neurology. 2014;83:476.

33. Nordström P, Nordström A, Eriksson M, Wahlund LO, Gustafson Y. Risk factors in late adolescence for young-onset dementia in men: A nationwide cohort study. JAMA Internal Medicine. 2013;173:1612–18.

34. Siqueira L, Smith VC. Committee on Substance Abuse. Binge drinking. Pediatrics. 2015;136:e718–e726

35. National Alliance on Mental Illness. Alcohol and mental illness. Available from: http://www.nami.org/Template.cfm?Section=Smoking_Cessation&Template=/ContentManagement/ContentDisplay.cfm&ContentID=152818.

36. American Psychiatric Association. Diagnostic and statistical manual of mental disorders, 5th ed., DSM-V (Text Revision). Washington, DC: American Psychiatric Publishing, 2013.

37. Centers for Disease Control and Prevention. Excessive Drinking Is Draining the U.S. Economy. 2016. Available from: http://www.cdc.gov/features/costsofdrinking/Health, Department of Health and Human Services, 2000.

38. Hingson R, White A. New research findings since the 2007 Surgeon General's Call to Action to Prevent and Reduce Underage Drinking: A review. Journal of Studies on Alcohol and Drugs. 2014;75:158–69.

39. Hingson R, Heeren T, Winter M, Wechsler H. Magnitude of alcohol-related mortality and morbidity among U.S. college students ages 18–24: Changes from 1998 to 2001. Annual Review of Public Health. 2005;26:259–79. Saylor DK. Heavy drinking on college campuses: No reason to change minimum legal drinking age of 21. Journal of American College Health. 2011;59:330–3.

40. Turrisi R, Mallett KA, Mastroleo NR, Larimer ME. Heavy drinking in college students: Who is at risk and what is being done about it? Journal of General Psychology. 2006;133:401–20. Mastroleo NR, Logan DE. Response of colleges to risky drinking college students. Rhode Island Medical Journal. 2014;97:40–2.

41. National Institute on Alcohol Abuse and Alcoholism. College Drinking. Available from: http://www.niaaa.nih.gov/alcohol-health/special-populations-co-occurring-disorders/college-drinking.

42. Yi H, Chen CM, Williams GD. Surveillance report #76: Trends in alcohol-related fatal traffic crashes, United States, 1982–2004. Bethesda, MD: NIAAA, Division of Biometry and Epidemiology, Alcohol Epidemiologic Data System, August 2006. Available from: http://pubs.niaaa.nih.gov/publications/surveillance76/FARS04.pdf.

43. U.S. Department of Health and Human Services, National Institutes of Health, National Institute on Alcohol Abuse and Alcoholism. What colleges need to know now: An update on college drinking research. NIH Publication No. 07–5010, November 2007. Available from: http://www.collegedrinkingprevention.gov/1college_bulletin-508_361C4E.pdf.

44. Timberlake DS, Hopfer CJ, Rhee SH, Friedman NP, Haberstick BC, Lessem JM, Hewitt JK. College attendance and its effect on drinking behaviors in a longitudinal study of adolescents. Alcoholism, Clinical and Experimental Research. 2007;31:1020–30. National Institute on Alcohol Abuse and Alcoholism. College Drinking. December 2015. Available from: http://pubs.niaaa.nih.gov/publications/CollegeFactSheet/CollegeFactSheet.pdf.

45. Lange JE, Clapp JD, Turrisi R, Reavy R, Jaccard J, Johnson MB, Voas RB, Larimer M. College binge drinking: What is it? Who does it? Alcoholism, Clinical and Experimental Research. 2002;26:723–30. Howland J, Rohsenow DJ, Greece JA, Littlefield CA, Almeida A, Heeren T, Winter M, Bliss CA, Hunt S, Hermos J. The effects of binge drinking on college students' next-day academic test-taking performance and mood state. Addiction. 2010;105:655–65.

46. DeJong W, Larimer ME, Wood MD, Hartman R. NIAAA's Rapid Response to College Drinking problems initiative: Reinforcing the use of evidence-based approaches in college alcohol prevention. Journal of Studies on Alcohol and Drugs. 2009;16S:5–11. Dejong W. Finding common ground for effective campus-based prevention. Psychology of Addictive Behaviors. 2001;15:292–6.

47. Beck KH, Arria AM, Caldeira KM, Vincent KB, O'Grady KE, Wish ED. Social context of drinking and alcohol problems among college students. American Journal of Health Behavior. 2008;32:420–30.

48. U.S. Department of Health and Human Services, Office of Disease Prevention and Health Promotion, Healthy people 2020. Washington DC., 2011. Available from: http://healthypeople.gov/2020/about/default.aspx.

49. U.S. Department of Agriculture ChooseMyPlate.gov. Pregnancy & Breastfeeding. Available from: http://www.choosemyplate.gov/moms-breastfeeding-nutritional-needs#sthash.yNfM9eXG.dpuf.

14

1. U.S. Centers for Disease Control and Prevention. Estimates of foodborne illness in the United States. April 17, 2014. Available from: http://www.cdc.gov/foodborneburden/estimates-overview.html.

2. Trabulsi LR, Keller R, Tardelli Gomes TA. Typical and atypical enteropathogenic Escherichia coli. Emerging Infectious Diseases. 2002;8:508–13.

3. U.S. Food and Drug Administration. Bad Bug Book. 2012. Available from: http://www.fda.gov/downloads/Food/FoodborneIllnessContaminants/UCM297627.pdf.

4. Jones TF, Kellum ME, Porter SS, Bell M, Schaffner W. An outbreak of community-acquired foodborne illness caused by methicillin-resistant Staphylococcus aureus. Emerging Infectious Diseases. 2002;8:82–4.

5. U.S. Centers for Disease Control and Prevention. Botulism associated with canned chili sauce, July–August 2007. Updated August 24, 2007. Available from: http://www.cdc.gov/botulism/botulism.htm.

6. U.S. Centers for Disease Control and Prevention. Norovirus outbreaks on three college campuses—California, Michigan, and Wisconsin, 2008. Morbidity and Mortality Weekly Reports. 2009;58:1095–100. Available from: http://www.cdc.gov/mmwr/preview/mmwrhtml/mm5839a2.htm.

7. U.S. Centers for Disease Control and Prevention. Investigation update: Multistate outbreak of Salmonella Poona infections linked to imported cucumbers. November 19, 2015. Available from: http://www.cdc.gov/salmonella/poona-09–15/index.html.

8. U.S. Centers for Disease Control and Prevention. Ongoing multistate outbreak of Escherichia coli serotype O157:H7 infections associated with consumption of fresh spinach—United States, September 2006. Morbidity and Mortality Weekly Reports. 2006;55:1–2. Available from: http://www.cdc.gov/mmwr/preview/mmwrhtml/mm5538a4.htm.

9. U.S. Food and Drug Administration. Guidance for industry: Guide to minimize microbial food safety hazards of fresh-cut fruits and vegetables. February 2008. Available from: http://www.fda.gov/Food/GuidanceComplianceRegulatoryInformation/GuidanceDocuments/ProduceandPlanProducts/ucm064458.htm.

10. U.S. Centers for Disease Control and Prevention. Multistate outbreak of shiga toxin-producing Escherichia coli O157:H7 infections linked to Costco Rotisserie chicken salad. Available at: http://www.cdc.gov/ecoli/2015/o157h7-11-15/index.html.

11. Federal Institute for Risk Assessment, Federal Office of Consumer Protection and Food Safety, and Robert Koch Institute. Information update on EHEC outbreak. June 10, 2011. Available from: http://www.rki.de/cln_144/nn_217400/EN/Home/PM082011.html. U.S. Centers for Disease Control and Prevention. Investigation update: Outbreak of shiga toxin-producing E. coli O104 (STEC O104:H4) infections associated with travel to Germany. June 15, 2011. Available from: http://www.cdc.gov/print.do?url=http://www.cdc.gov/ecoli/2011/ecoliO104/.

12. Mostl K. Bovine spongiform encephalopathy (BSE): The importance of the food and feed chain. Forum in Nutrition. 2003;56:394–6.

13. Ryou C. Prions and prion diseases: Fundamentals and mechanistic details. Journal of Microbiology and Biotechnology. 2007;17:1059–70.

14. Belay ED, Schonberger LB. The public health impact of prion diseases. Annual Review of Public Health. 2005;26: 191–212. U.S. Centers for Disease Control and Prevention. vCJD (variant Creutzfeldt-Jakob disease). Available from: http://www.cdc.gov/ncidod/dvrd/vcjd/risk_travelers.htm. Roma AA, Prayson RA. Bovine spongiform encephalopathy and variant Creutzfeldt-Jakob disease: How safe is eating beef? Cleveland Clinic Journal of Medicine. 2005;72: 185–94.

15. U.S. Department of Agriculture. Available from: http://www.usda.gov.

16. Bovine somatotropin and the safety of cows' milk: National Institutes of Health technology assessment conference statement. Nutrition Reviews. 1991;49:227–32. Etherton TD, Kris-Etherton PM, Mills EW. Recombinant bovine and porcine somatotropin: Safety and benefits of these biotechnologies. Journal of the American Dietetic Association. 1993;93:177–80.

17. Walker R, Lupien JR. The safety evaluation of monosodium glutamate. Journal of Nutrition. 2000;130:1049S–52S.

18. U.S. Food and Drug Administration. Melamine pet food recall of 2007. Available from: http://www.fda.gov/animalveterinary/safetyhealth/recallswithdrawals /ucm129575.htm.

19. U.S. Food and Drug Administration. FDA/USDA joint news release: Scientists conclude very low risk to humans from food containing melamine. Available from: http://www.fda.gov/NewsEvents/Newsroom/PressAnnouncements/2007/default.htm.

20. U.S. Food and Drug Administration. Melamine contamination in China. Available from: http://www.fda.gov/NewsEvents/PublicHealthFocus/ucm179005.htm.

21. Durando M, Kass L, Piva J, Sonnenschein C, Soto AM, Luque EH, Muñoz-de-Toro M. Prenatal bisphenol A exposure induces preneoplastic lesions in the mammary gland in Wistar rats. Environmental Health Perspectives. 2007;115:80–6. Newbold RR, Jefferson WN, Padilla-Banks E. Long-term adverse effects of neonatal exposure to bisphenol A on the murine female reproductive tract. Reproductive Toxicology. 2007;24:253–8. U.S. Food and Drug Administration. Update on bisphenol A (BPA) for use in food. January 2010. Available from: http://www.fda.gov/NewsEvents/PublicHealthFocus/ucm064437.htm.

22. U.S. Food and Drug Administration. Update on bisphenol A (BPA) for use in food. January 2010. Available at: http://www.fda.gov/NewsEvents/PublicHealthFocus/ucm064437.htm.

23. Mungai EA, Behravesh CB, Gould LH. Increased outbreaks associated with nonpasteurized milk, United States, 2007–2012. Emerging Infectious Diseases. 21:1 doi:http://dx.doi.org/10.3201/eid2101.140447.

24. Kenney SJ, Beuchat LR. Comparison of aqueous commercial cleaners for effectiveness in removing Escherichia coli O157:H7 and Salmonella meunchen from the surface of apples. International Journal of Food Microbiology. 2002;25:47–55.

25. U.S. Centers for Disease Control and Prevention. Update: Multistate outbreak of listeriosis—United States, 1998–1999. Morbidity and Mortality Weekly Report. 1999;47:1117–8.

26. U.S. Congress. Public health security and bioterrorism preparedness and response act of 2002. June 12, 2002. Available from: http://www.fda.gov/oc/bioterrorism/Bioact.html.

15

1. FAO, IFAD, and WFP. 2015. The State of Food Insecurity in the world 2015. Meeting the 2015 international hunger targets: Taking stock of uneven progress. Rome, FAO.

2. Kempson KM, Palmer Keenan D, Sadani PS, Ridlen S, Scotto Rosato N. Food management practices used by people with limited resources to maintain food sufficiency as reported by nutrition educators. Journal of the American Dietetic Association. 2002;102:1795–9.

3. Alisha Coleman-Jensen, Matthew P. Rabbitt, Christian Gregory, and Anita Singh. Household Food Security in the United States in 2014, ERR-194, U.S. Department of Agriculture, Economic Research Service, September 2015.

4. DeNavas-Walt C,. Proctor DB. U.S. Census Bureau, Current Population Reports, P60–252, Income and Poverty in the United States: 2014, U.S. Government Printing Office, Washington, DC, 2015.

5. Olson CM. Nutrition and health outcomes associated with food insecurity and hunger. Journal of Nutrition. 1999;129:521S–4S.

6. Beaulac J, Kristjansson E, Cummins S. A systematic review of food deserts, 1966–2007. Preventing Chronic Disease. 2009;6:A105. Fuller D, Engler-Stringer R, Muhajarine N. Examining food purchasing patterns from sales data at a full-service grocery store intervention in a former food desert. Preventive Medicine Reports. 2015;2:164–9.

7. Kendall A, Olson CM, Frongillo EA Jr. Relationship of hunger and food insecurity to food availability and consumption. Journal of the American Dietetic Association. 1996;96:1019–24.

8. Cook JT, Black M, Chilton M, Cutts D, Ettinger de Cuba S, Heeren TC, Rose-Jacobs R, Sandel M, Casey PH, Coleman S, Weiss I, Frank DA. Are food insecurity's health impacts underestimated in the U.S. population? Marginal food security also predicts adverse health outcomes in young U.S. children and mothers. Advances in Nutrition. 2013;4:51–61. Rose-Jacobs R, Black MM, Casey PH, Cook JT, Cutts DB, Chilton M, Heeren T, Levenson SM, Meyers AF, Frank DA. Household food insecurity: Associations with at-risk infant and toddler development. Pediatrics. 2008;121:65–72.

9. Fram MS, Frongillo EA, Jones SJ, Williams RC, Burke MP, Deloach KP, Blake CE. Children are aware of food insecurity and take responsibility for managing food resources. Journal of Nutrition. 2011;141:1114–9.

10. Alaimo K, Olson CM, Frongillo EA. Family food insufficiency, but not low family income, is positively associated with dysthymia and suicide symptoms in adolescents. Journal of Nutrition. 2002;132:719–25.

11. Lee JS, Frongillo EA Jr. Nutritional and health consequences are associated with food insecurity among U.S. elderly persons. Journal of Nutrition. 2001;131:1503–9.

12. SNAP: Frequently Asked Questions, www.snaptohealth.org/snap/snap-frequently-asked-questions.

13. American Academy of Pediatrics. WIC program. Provisional section on breastfeeding. Position statement. Pediatrics. 2001;108:1216–7.

14. Food and Nutrition Service USDA. Special Supplemental Nutrition Program for Women, Infants and Children (WIC): Implementation of Electronic Benefit Transfer-Related Provisions. Federal Register. 2016;81:10433–10451.

15. Food and Nutrition Service (USDA). National school lunch program fact sheet. 2013. Available at: www.fns.usda.gov/sites/default/files/NSLPFactSheet.pdf.

16. Institute of Medicine. School meals: Building blocks for healthy children. Washington, DC: The National Academies Press, 2010.

17. U.S. Public law 111–296. Congress. Healthy hunger-free kids act of 2010. Available from: http://www.gpo.gov/fdsys/pkg/PLAW-111publ296/pdf/PLAW-111publ296.pdf.

18. Buzby JC., Wells HF, Hyman J. The estimated amount, value, and calories of postharvest food losses at the retail and consumer levels in the United States, EIB-121, U.S. Department of Agriculture, Economic Research Service, February 2014. Available at: www.ers.usda.gov/media/1282296/eib121.pdf.

19. FAO, IFAD, and WFP. 2015. The state of food insecurity in the world 2015. Meeting the 2015 international hunger targets: Taking stock of uneven progress. Rome, FAO. Available at: www.fao.org/3/a-i4646e.pdf.

20. International Food Policy Research Institute, Concern Worldwide, and Welthungerhilfe. 2010 global hunger index. Available from: http://www.ifpri.org/sites/default/files/publications/ghi10.pdf.

21. United Nations High Commissioner for Refugees (UNHCR). Statistical Yearbook, 2014. Geneva, Switzerland, 2015. Available at: www.unhcr.org/566584fc9.html.

22. Reardon T, Timmer P, Barrett C, Berdegué J. The rise of supermarkets in Africa, Asia and Latin America. American Journal of Agricultural Economics. 2003;85:1140–6.

23. Singh GM, Micha R, Khatibzadeh S, et al. Global, regional, and national consumption of sugar-sweetened beverages, fruit juices, and milk: A systematic assessment of beverage intake in 187 countries. PLoS One. 2015;10(8):e0124845.

24. El-Ghannam AR. The global problems of child malnutrition and mortality in different world regions. Journal of Health and Social Policy. 2003;16:1–26. Horton KD. Bringing attention to global hunger. Journal of the American Dietetic Association. 2008;108:435.

25. Task Force on Education and Gender Equality. United Nations Millennium Project 2005. Taking action: Achieving gender equality and empowering women. Available from: http://www.unmillennium-project.org/documents/Gender-complete.pdf.

26. Kennedy E, Meyers L. Dietary Reference Intakes: Development and uses for assessment of micronutrient status of women—a global perspective. American Journal of Clinical Nutrition. 2005;81:1194S–7S.

27. World Health Organization. Nutrition. Micronutrient deficiencies. Available at: www.who.int/nutrition/topics/ida/en/.

28. Milman A, Frongillo EA, de Onis M, Hwang JY. Differential improvement among countries in child stunting is associated with long-term development and specific interventions. Journal of Nutrition. 2005:135:1415–22.

29. Policy Advisory Unit in UNICEF's Division of Policy and Strategy. Position paper. Ready-to-use therapeutic food for children with severe acute malnutrition. Available at: www.unicef.org/media/files/Position_Paper_Ready-to-use_therapeutic_food_for_children_with_severe_acute_malnutrition_June_2013.pdf.

30. McCall E. Communication for development strengthening the effectiveness of the United Nations. United Nations Development Programme. 2011. Available from: http://www.unicef.org/cbsc/files/Inter-agency_C4D_Book_2011.pdf.

31. Behrman JR, Parker SW, Todd PE. Schooling impacts of conditional cash transfers on young children: Evidence from Mexico. Economic Development and Cultural Change. 2009;57:439–77.

32. United Nations Millennium Development Project. United Nations Millennium development goals. Millennium Development Project report 2010. Available from: http://www.un.org/millenniumgoals/pdf/MDG%20Report%202010%20En%20r15%20-low%20res%2020100615%20-.pdf.

33. United Nations. Resolution adopted by the General Assembly on 25 September 2015. 7-/1. Transforming our world: The 2030 Agenda for Sustainable Development. Available at: www.un.org/en/ga/search/view_doc.asp?symbol=A/RES/70/1&Lang=E.

34. Nord M, Coleman-Jensen A, Andrews M, and Carlson S. Household food security in the United States, 2009. ERR-108, U.S. Department of Agriculture, Economics Research Service. 2010. Available from: http://www.ers.usda.gov/Publications/ERR108/ERR108.pdf.

INDEX

A

AA. *See* Alcoholics Anonymous®, 327
AAP. *See* American Academy
    of Pediatrics
ABCD methods of nutritional
    assessment, 23
Absorption
    of alcohol, 318
    of amino acids, 110–111
    of carbohydrates, 81–87
    of chromium, 215
    of copper, 211
    defined, 55
    of iodine, 211–212
    of iron, 207–208
    of lipids, 142
    of monosaccharides, 86
    of nutrients, 67–68
    of protein, 108–111
    of selenium, 214
    of triglycerides, 139–144
ABV (alcohol by volume), 317
Academy of Nutrition and Dietetics
    (AND), 364
Acceptable Macronutrient
    Distribution Range (AMDR)
    for carbohydrates, 95
    for energy-yielding nutrients, 32
    lipids, consumption
        guidelines, 150
    for protein, 117
    for total lipid consumption, 148
    for weight loss, 242
Accessory organ, 55, 62
    gallbladder, 55
    liver, 55
    pancreas, 55
    salivary glands, 55
Acesulfame K, 78
Acetaldehyde dehydrogenase
    (ALDH), 321
Acrodermatitis enteropathica, 218
ACSM. *See* American College of
    Sports Medicine (ACSM)
Actin, 112
Active transport mechanism, 51, 52
Adaptive thermogenesis, 228

Added sugar, 76–77
Adenosine triphosphate (ATP), 52,
    53. *See also* Energy
    aerobic oxidative system, 288
    cell organelles and, 53
    and energy-yielding nutrients, 8
    generation during exercise, 286
    thiamin in production of, 154
Adequate intake (AI) level,
    27–31, 149
Adequate nutrient intake, 21–22
ADH. *See* Alcohol
    dehydrogenase (ADH)
Adipocyte, 135, 221
Adipokine, 222
Adipose tissue, 221, 232
    brown, 223–224
    subcutaneous, 222, 223
    visceral, 222
    white, 223
Adiposity, 223
Adolescents, or adolescence
    food choices of, 272
    growth and development
        during, 271
    micronutrients during, 272
    nutritional concerns and
        recommendations
        during, 272
    nutritional requirements change
        during, 271–272
    psychological issues associated
        with, 271–272
Adrenal glands, release of
    aldosterone, 190
Adulthood
    age-related changes in,
        nutrient and energy
        requirements, 272–274
    body composition and bone
        health during, 275–276
    changes in gastrointestinal
        tract, 276–277
    nutritional concerns and
        recommendations
        during, 275–278
    physical maturity and
        senescence, 274–275

Aerobic, 70
Aflatoxin, 334
AGA. *See* Appropriate for
    gestational age (AGA)
Agave, 78
Age-related changes in adult,
    nutrient and energy
    requirements, 272–274
AI. *See* Adequate intake (AI) levels
AIDS (acquired immunodeficiency
    syndrome), protein
    deficiency in, 120
Al-Anon®, 327
Alateen®, 327
Albumin, 113
    fluid balance, 113
    synthesis, 113
Alcohol, 317. *See also* Blood alcohol
    concentration (BAC)
    absorption of, 318
    acquaintance rape, alcohol use
        and, 327
    as beneficial to heart, 322
    central nervous system
        and, 320
    circulation of, 318–320
    excessive intake and nutritional
        status, 323
    fermentation, 317, 342
    health benefits of, 321–323
    impact of heavy consumption,
        nutritional status, 324
    metabolism, 321, 323–324
    production of, 317–318
    proof, 318
    recommendations for
        responsible use, 329
    serious health risks from
        heavy, 323–325
    stages of intoxication, 320
    tolerance, 321
    underage drinking on university
        campuses, 328
    use on college campuses,
        327–328
Alcohol abuse
    cancer risk from long-term,
        324–325

and cardiovascular system,
  325–326
  and cognition, 326
  and dementia, 326
  and pancreatic function, 326
  treating, 327
Alcohol by volume (ABV), 317
Alcohol dehydrogenase (ADH), 321
Alcoholic cardiomyopathy, 325
Alcoholic hepatitis, 324
Alcoholics Anonymous® (AA), 327
Alcohol use disorders (AUDs), 326
Aldosterone, 190
Algal toxins, 338
Alpha (α)-linolenic acid, 130
Alpha (α)-helix, 105
Alpha (α)-keto acid, 100
Alpha (α) end, 125
Alpha (α) naming system, 129
Altern®, 78
Alternative sweeteners, 77
Aluminum (Al), 219
Alveoli, 259
AMDR. *See* Acceptable
  Macronutrient Distribution
  Range (AMDR)
Amenorrhea, 303
American Academy of Pediatrics
  (AAP), 355
  on breastfeeding, 261
  on fluoride
    supplementation, 263
  on iron supplements, 265
  on nonmilk complementary
    foods, 265
  on vitamin D
    supplementation, 263
  on water for infants, 265
American Cancer Society, 14
American College of Sports
  Medicine (ACSM)
  exercise recommendations,
    healthy adults, 284–286
  physical activity
    recommendations, 282
American Psychiatric Association
  diagnostic criteria for BED, 306
  diagnostic criteria for feeding
    and eating disorders, 300
American Society for Nutrition
  (ASN), 364
America Second Harvest, 356
Amino acids
  absorption of, 110–111
  circulation of, 110–111

components of, 99–100
conditionally essential, 99–101
dietary reference intakes,
  115–116
essential, 99–101
glucose and energy, source of, 114
limiting, 101
nonessential, 99–101
and protein breakdown, 114
recycle and reuse of, 114–115
structure of, 99
Amino group, 99
Ammonia, 70, 114
Amphipathic, phospholipids, 136
Amylopectin, 78
Amylose, 78
Anabolic pathway, 70, 71
Anabolism, 70
Anaerobic, 70–71
Anaerobic capacity, 290
Anaerobic pathway, 71
Anal canal, 64
Anaphylaxis, 110
Anemia
  megaloblastic, macrocytic, 165
  microcytic, hypochromic, 161
  pernicious, 167
  sickle cell, 104
  sports, 294
Aneurysm, 147
Anorexia nervosa (AN), 299,
  301–304
  behaviors associated with, 302
  binge-eating/purging type, 301
  health concerns associated
    with, 302–303
  physical signs and symptoms, 303
  restricting type, 301
  thoughts and behaviors
    associated with, 303
  types, 301
Anthropometry, 23
Antibiotics, 338–339
Antibody, 112
Antidiuretic hormone (ADH), 190
Anti-obesity hormone, 237
Antioxidant, 168, 294–295
Apolipoprotein, 145
Apoprotein, 145
Appetite, 227
Appropriate for gestational age
  (AGA), 254
Arachidonic acid, 130
Ariboflavinosis, 157

Arsenic (As), 219
Ascites, 121
Ascorbic acid. *See* Vitamin C
Aspartame, 77
Aspergillus, 334
Atherosclerosis, 147
Athletes
  iron needs, 209
  protein requirement for, 117
  protein supplements for, 117
  risk of eating disorders in, 312–314
Atkins, Robert, 242
Alpha (α)-tocopherol, 178
Atom, 48
ATP. *See* Adenosine triphosphate
  (ATP)
Atrophy, 289
Autoimmune disease, 17
Avidin, 162
Avoidant/restrictive food intake
  disorder (AFRID), 308

B

Babies
  classification based on
    gestational age and birth
    weight, 253
  complementary foods, 4 and 6
    months of age, 265–267
  nutritional needs of, 263–267
Baby boomers, 274
Baby bottle tooth decay, 267
Bacon, Francis, 19
Balanced diet, 96
Bariatric surgery, 225
Basal metabolic rate (BMR)
  defined, 229
  factors influencing, 229–230
Basal metabolism, 229
B-complex vitamins, 152
BED. *See* Binge-eating disorder
  (BED)
Beriberi, 154
Beta (β)-oxidation, 134
Beta-carotene (β-carotene), 169
Beta (β)-folded sheet, 105
Bile, 62
Bile acid, 137
Binge drinking, 326
Binge-eating disorder (BED), 299,
  305–306
Bingeing, 301
Bioavailability, 67

oxytocin, 259–260
prolactin, 259–260
Lacteal, 68
Lactogenesis, 259
Lacto-ovo-vegetarian, 118
Lactose, 75
Lactose intolerance, 84
Lactose tolerance, evolution of, 85
Lactovegetarian, 118
Lanugo, 135
Large for gestational age (LGA), 254
Large intestine, 64–65
   undigested matter elimination, 64
LDL receptor, 145
Legume, 39
Leptin
   in communication of body's
      energy reserve to brain,
      238–239
   defects in, as role in obesity,
      238–239
   discovery of, as genetic clue
      to obesity, 237
Life cycle
   effect of on nutrient
      requirements, 246
   physiological changes during,
      246–248
   stages in, 248
Life expectancy
   around the globe, 16
   changes in, 15
   defined, 16
Lifespan, 274
Lingual lipase, 139
Linoleic acid, 130
Linolenic acid, 130
Lipids
   absorption of, 142
   alpha (α) naming system, 129
   energy provided by, 8
   and cardiovascular health,
      147–148
   circulation of, 142
   cis vs. trans, 127–129
   common names, 130
   consumption guidelines, 150
   defined, 125
   dietary recommendations for,
      148–150
   digestion of, 124
   double bonds, number and
      position, 126–127
   emulsification, 140
   fats and oils as, 124–125

fatty acids as, 125–126
and health effects, 146–148
as macronutrients, 6–7
functions of, 124
number of carbons (chain
   length), 126
omega (ω) naming system, 129–130
phospholipids and, 135–136
sterols and, 137–138
triglycerides and, 134
Lipolysis, 134
Lipoprotein, 143, 144–146
Lipoprotein lipase, 143
Listeria monocytogenes, 346
   and pasteurization, 343
Liver, 55
Long-chain fatty acid, 126
Long-chain ω-3 fatty acid, 149
Low birth weight (LBW), 253
Low-density lipoprotein
   (LDL), 145
Low food security, 352
Lucky Iron Fish™, 210
Lumen, 55
Lymphatic systems, nutrient
   circulation, 68

M

Macromolecule, 51
Macronutrient
   acceptable distribution ranges,
      (AMDR) 32
   classification, 6–7
   defined, 5
   during pregnancy, 255–257
   recommended intakes for,
      292–293
   and weight loss, 242
   weight loss, total calories
      vs., 244
Macular degeneration, 172
Mad cow disease, 337
Magnesium (Mg)
   content of selected foods, 205
   deficiency, 204
   dietary sources, 204
   and muscle contraction, 49
   recommended intake, 204
   toxicity, 204
Major mineral, 191
Malnutrition
   defined, 20
   forms of, 359

consequences of, 359–360
primary, 21, 24
secondary, 21, 24
societal consequences of, 360
Maltase, 82
Maltose, 76
Manganese (Mn)
   content of selected foods, 216
   deficiency, 216
   dietary sources, 216
   for muscle contraction, 216
   recommended intake, 216
   toxicity, 216
Marasmus, 120
Marine toxin, 338
Maturation, 248
Maximal oxygen consumption
   (VO₂ max), 290
Meals on Wheels®, 279
Meat factor, 207
Medical history, 24
Medium-chain fatty acid, 126
Megaloblastic, macrocytic
   anemia, 165
Melamine, 339
Menadione, 179
Menaquinone, 179
Menarche, 271
Menopause, 277
Messenger RNA (mRNA), 102
Metabolic pathway, 69–70
Metabolic syndrome, 222
Metabolic waste-products
   excretion, 71
Metabolism, 69
Metalloenzyme, 192
Metallothionine, 217
Methicillin-resistant Staphylococcus
   aureus (MRSA), 332
Micelle, 140
Microcytic, hypochromic
   anemia, 161
Micronutrient
   during adolescence, 272
   classification, 6–7
   defined, 5
   during pregnancy, 257
   recommended intakes for ,
      293–295
Microvillus, 62
Milk
   nutrients in, 4
   pasteurization, 340
Milk let down, 260

Mineral
  availability in body, 193
  calcium, 194–199
  chloride, 203
  defined, 191
  in food, 192–193
  magnesium, 204–205
  as micronutrients, 7
  phosphorus, 203–204
  potassium, 203
  roles of, 191–192
  sodium, 199
  sulfur, 205
  trace, 206–210
    arsenic, 219
    boron, 219
    chromium, 215–216
    copper, 211
    fluoride, 218–219
    iodine, 211
    iron, 206–209
    magnesium, 204–205
    molybdenum, 216
    nickel, 219
    selenium, 214–215
    silicon, 219
    vanadium, 219
    zinc, 217–218
Mitochondria, 52–53
Molecular formula, 51
Molecule, 50
Molybdenum (Mo)
  bioavailability, absorption, and
    functions, 216
  dietary sources, deficiency,
    toxicity, and recommended
    intake of, 216
Monoglyceride, 133
Monosaccharide
  absorption, 86
  defined, 72
  functions of, 72
  structures of, 74
Monosodium glutamate (MSG), 339
Monounsaturated fatty acid
  (MUFA), 126
Morbidity rate, 15
Mortality rate, 15
Mucosa, 57
Muscle
  involuntary, 54
  skeletal, 53
  smooth, 53
  soreness, exercise-related, 289
  tissue, 53

Muscle atrophy, 289
Muscle dysmorphia, 308–309
Muscle hypertrophy, 289
Muscle tissue, 53
Muscular endurance, 283
Muscular flexibility, 283
Muscular strength, 283
Mutations, 107
MyFoodapedia, 40
Myoglobin, 206
MyPlate food guidance system, 38,
    39, 40
  and alcohol use, pregnancy, 329
  food group recommendations
    by, 38

N

National Academy of Sciences, 26
National School Lunch
    Program, 355
Naturally occurring sugar, 76–77
Negative energy balance, 220
Negative nitrogen balance, 115
Neural tube defect, 164, 258
Neuromotor exercise, 286
Neurotransmitter, 225
Neutral energy balance, 220
Neutral nitrogen balance, 115
Nevin Scrimshaw International
    Nutrition Foundation
    (INF), 364
Niacin (vitamin B$_3$)
  bioavailability of, 159
  deficiency of, 158
  functions of, 157
  recommended intake, 159
  in reduction-oxidation
    reactions, 158
  sources of, 158
Niacin equivalent (NE), 157
Nickel (Ni), 219
Night blindness, 171
Night eating syndrome (NES), 307
Nitrogen
  balance and protein status, 115
  excretion, 114–115
Nocturnal sleep-related eating
    disorder, 307
Non-essential nutrient, 4
Nonexercise activity
    thermogenesis, 228
Nonheme iron, 207
Noninfectious agent, 330
Noninfectious disease, 17

Nonperishable food collection, 357
Nonprovitamin A carotenoid, 169
Norovirus, 334–335
Nucleus, 53
Nutraceutical, 66
Nutrient content claims, 41–43
Nutrient digestion and pancreas, 64
Nutrient
  absorption, 67–68
  adequacy of, 26–32
  adequate intake, 21–22
  circulation, 67–68
  classification of, 3
  conditionally essential, 3
  defined, 4
  density, 35
  energy-yielding, 7
  essential, 4
  inorganic, 4–5
  macronutrient, 5
  micronutrient, 5
  nonessential, 4
  organic, 4–5
  requirements of, 26–27
Nutrition
  claims, 12–14
  connection between health
    and, 14–19
  defined, 2
  and epigeneitc inheritance, 252
  facts panels, 149
  and health, 14–19
  reasons for studying, 19
  sources of information, 14
  transition, 18
Nutritional adequacy, 22
Nutritional deficiency, 20
Nutritional guidelines, for
    cardiovascular health,
    lipids, 148
Nutritional health, assessing, of
    nation, 18–19
Nutritional scientist, 2
  research conducted by, 9–12
Nutritional status, 21, 46
  assessment of, 23–24
Nutritional toxicity, 21
Nutritious snacking, 269

O

Obese, 230
Obesity
  central, 223
  and eating habits, 234–235

NUTR
ONLINE

PREPARE FOR TESTS ON
THE STUDYBOARD!

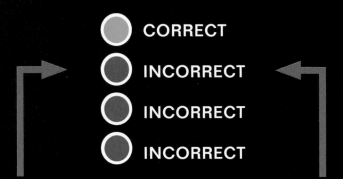

○ CORRECT
○ INCORRECT
○ INCORRECT
○ INCORRECT

**Personalize Quizzes
from Your StudyBits**

**Take Practice
Quizzes by Chapter**

CHAPTER QUIZZES
▶ Chapter 1
Chapter 2
Chapter 3
Chapter 4

4LTR
PRESS

Access NUTR ONLINE at www.cengagebrain.com

NUTR ONLINE—IT'S ALL IN THERE!

Working with NUTR Online

To help you take your reading outside the covers of NUTR, each new text comes with access to NUTR Online! You can read NUTR wherever and whenever you are online—NUTR Online will work on nearly any device, including smartphones!

But you can do more than just read!

- Create StudyBits from your highlights and take practice quizzes.
- Create StudyBits from photos and figures, and rate your understanding for later test prep.
- Use flashcards from key terms, and even create your own from StudyBits!
- Interact with a variety of digital study tools, including the following.

Videos, nutritional tutorials, animations, and digital walk-throughs of bodily processes

Interactive Figures and Matching Activities

Additional Images and Box Features

Key Terms and Definitions

Assessment

Log in to NUTR Online with the access code found in the front of your book at www.cengagebrain.com.

1-1

What Is Nutrition? · The term *nutrition* refers to how living organisms obtain and use food to support all the processes required for their existence. Because this process is complex, the study of nutrition incorporates a wide variety of scientific disciplines. · A dietitian has the credential *RD*, which stands for *registered dietitian*. Some dietitians are also involved in scientific research.

dietitian A nutrition professional who helps people make dietary changes and food choices to support a healthy lifestyle.

nutrition The science of how living organisms obtain and use food to support processes required for existence.

1-2

What Are Nutrients, and What Do They Do? · Essential nutrients must be consumed, whereas the body can make sufficient amounts of the nonessential nutrients when needed. When the body cannot make a typically nonessential nutrient in adequate amounts, it becomes conditionally essential. · The term *organic* is used by chemists to describe most substances that contain carbon and hydrogen atoms, whereas organic foods are those that are produced without using selected synthetic fertilizers, hormones, or other drugs. · Because a person must consume more than a gram of them every day, water, carbohydrates, proteins, and lipids are considered macronutrients. Because a person needs only very small amounts of vitamins and minerals, these substances are considered micronutrients. · Phytonutrients and zoonutrients are compounds found in plant- and animal-based foods, respectively. These substances are not considered traditional nutrients but may improve health. · Foods that contain enhanced amounts of essential nutrients, phytochemicals, or zoonutrients are called functional foods or super foods.

conditionally essential nutrient A normally nonessential nutrient that, under certain circumstances, becomes essential.

enrichment A type of fortification whereby specific amounts of selected nutrients are added to certain foods.

essential nutrient A substance that must be obtained from the diet to sustain life.

fortification The intentional addition of nutrients to a food.

functional food (or **super food**) A food that likely benefits human health by providing a high concentration of nutrients, phytochemicals, or zoochemicals.

inorganic compound A substance that does not contain carbon.

macronutrients A class of nutrients that humans need to consume in relatively large quantities (more than a gram per day).

micronutrients A class of nutrients that humans need to consume in relatively small quantities.

nonessential nutrient A substance that sustains life but does not necessarily need to be obtained from the diet.

nutrient A substance found in food that is used by the body for energy, maintenance of body structure, or regulation of chemical processes.

organic compound A substance that contains carbon and hydrogen atoms.

phytochemical (or **phytonutrient**) A compound found in plants that likely benefits human health beyond the provision of essential nutrients and energy.

zoochemical (or **zoonutrient**) A compound found in animal-based foods that likely benefits human health beyond the provision of essential nutrients and energy.

1-3

How Are Macronutrients and Micronutrients Classified? · Carbohydrates, proteins, and lipids all provide energy and are thus referred to as energy-yielding nutrients. · Carbohydrates, proteins, and lipids have many structural and regulatory functions in the body. · Water serves as the medium in which all chemical reactions occur, helps eliminate waste products, and regulates body temperature. · Vitamins, classified as either water or fat soluble, serve many purposes (mostly having to do with the regulation of chemical reactions). · At least 15 minerals, each of which serves a specific purpose, are considered to be essential nutrients. Many of these essential minerals have specific functions regarding structure, regulation, and energy use.

TABLE 1.1	GROUPING MACRONUTRIENTS AND MICRONUTRIENTS	
Macronutrients	**Micronutrients**
Carbohydrates	Vitamins
Proteins	Minerals
Lipids	
Water	

1-4

How Is the Energy in Food Measured? · Energy is not a nutrient, but in terms of nutrition, the body can use energy found in foods to grow, develop, move, and fuel the many chemical reactions required for life. The body transforms the chemical energy in foods into a usable form called ATP. · Carbohydrates and proteins contain 4 kilocalories of energy per gram of substance, while lipids contain 9 kilocalories per gram. · A kilocalorie is sometimes referred to as a Calorie (note the capital *C*) outside of scientific research, as on food labels. Therefore, 1 Calorie is equivalent to 1 kilocalorie, or 1,000 calories.

adenosine triphosphate (ATP) A chemical that provides energy to cells in the body.

calorie The unit of measurement used to express the amount of energy in a food.

energy The capacity to do work.

energy-yielding nutrient A nutrient that the body can use for energy.

1-5 **How Do Nutritional Scientists Conduct Their Research?** • Most research is conducted using a three-step process called the scientific method, which involves making an observation, generating an explanation (or hypothesis), and testing the explanation by conducting a study. • Epidemiologic studies are conducted to investigate correlations, whereas intervention studies can test causal relationships. • Scientists use techniques such as control groups, placebos, blinding, and random assignment to decrease study bias. • Sometimes it is not possible or practical to test a hypothesis using humans as participants. In these cases, researchers turn to animal models or cell cultures.

cause-and-effect relationship (or **causal relationship**) A relationship whereby an alteration to one variable causes a change in another variable.

confounding variable A coincidental factor related to a study's outcome but not of primary interest to the research; these should be accounted for in the study design or analysis.

control group Study participants who do not receive a treatment or intervention.

correlation (or **association**) A relationship whereby an alteration to one variable is related to a change in another variable.

dependent variable The outcome of interest in a research study.

double-blind study A human experiment in which neither the participants nor the scientists know to which group the participants have been assigned.

epidemiologic study A study in which data are collected from a group of people who are not asked to change their behaviors in any way.

hypothesis A prediction about the relationship between variables.

intervention study An experiment in which a variable is altered to determine its effect on another variable.

placebo An inert treatment given to the control group that cannot be distinguished from the actual treatment.

placebo effect A phenomenon whereby a study participant experiences an apparent effect of the treatment just because the participant believes that the treatment will work.

random assignment A research strategy whereby study participants have equal chance of being assigned to the treatment and the control group.

researcher bias A phenomenon whereby the researcher, usually inadvertently, influences the results of a study.

scientific method A series of steps used by scientists to objectively explain observations.

single-blind study A human experiment in which the participants do not know to which group they have been assigned.

1-6 **Are All Nutrition Claims Believable?** • Separating fact from fiction can be difficult, but it is largely possible to determine the validity of any nutrition claim. • Publication in a peer-reviewed journal indicates that the information is probably reliable. • Although most research is not likely to be influenced by the source of funding, it is possible for a funding agency to have biased the conclusions made by researchers. • It is important to consider whether the design of the study was appropriate and whether major public health groups support the study's conclusions.

peer-reviewed journal A publication that requires a group of scientists to read and approve a study before it is published.

1-7 **Nutrition and Health: What Is the Connection?** • Over the past century, the primary public health concerns have shifted from infectious diseases and nutritional deficiencies to chronic diseases and overnutrition. • These shifts are reflected by changes in morbidity and mortality rates and life expectancy: whereas infant mortality rates and mortality from infectious diseases have decreased, life expectancy and rates of chronic degenerative diseases have increased. • The shift from undernutrition to overnutrition or unbalanced nutrition as a society becomes more industrialized is called the nutrition transition. This phenomenon is strongly related to many of the chronic diseases facing humankind today. • Poor dietary practices are associated with a greater risk for chronic disease.

autoimmune disease An illness resulting from an inappropriate immune response that attacks the body's own cells.

chronic degenerative disease A noninfectious disease that develops slowly, persists over a long period of time, and tends to result in progressive breakdown of tissues and loss of function.

disease An abnormal condition of the body or mind that causes discomfort, dysfunction, or distress.

graying of America A phenomenon occurring in the United States characterized by an increasing proportion of the population over the age of 65.

infant mortality rate The number of infant deaths per 1,000 live births in a given year.

infectious disease An illness that is contagious, caused by a pathogen, and often times short-lived.

life expectancy A statistical prediction of the average number of years of life remaining for a person at a particular age.

morbidity rate The number of illnesses or diseases in a given period of time.

mortality rate The number of deaths that occur in a certain population group in a given period of time.

noninfectious disease An illness that is not contagious, does not involve an infectious agent, and often times long term and chronic.

nutrition transition A shift from undernutrition to overnutrition or unbalanced nutrition that often occurs as a society transitions to a more industrialized economy.

rate A measure of some event, disease, or condition within a specific time span.

risk factor A lifestyle, environmental, or genetic factor related to a person's chances of developing a disease. Just because something is a risk factor does not mean that it is causative in nature.

1-8 **Why Study Nutrition?** • Consuming a healthy balance of traditional nutrients, phytonutrients, and zoonutrients can decrease your risk of developing obesity, cardiovascular disease, high blood pressure, diabetes, and cancer. • As the occurrence of chronic diseases increases, it is ever more important to pay attention to what you eat throughout your entire life.

2-1 What Is Nutritional Status? • A person's nutritional status depends on whether sufficient amounts of nutrients and energy are available to support optimal bodily function. • Both undernutrition and overnutrition are examples of malnutrition, a state of poor nutrition caused by an imbalance between the body's nutrient requirements and nutrient availability. • Primary malnutrition is due to inadequate diet, whereas secondary malnutrition may be caused by other factors, such as illness. • Nutrient requirements vary greatly among individuals. They are influenced by genetic, lifestyle, and environmental factors.

malnutrition A state of poor nutritional status caused by an imbalance between the body's nutrient requirements and nutrient availability.

nutritional adequacy A condition whereby a person regularly consumes the required amount of a nutrient to meet his or her physiological needs.

nutritional deficiency A condition caused by inadequate intake of one or more essential nutrients.

nutritional status The extent to which a person's diet meets his or her individual nutrient requirements.

nutritional toxicity Overconsumption of a nutrient that results in dangerous toxic effects.

overnutrition A state of poor nutrition that occurs when the diet provides excess nutrients and/or energy.

primary malnutrition A condition whereby poor nutritional status is caused strictly by inadequate diet.

secondary malnutrition A condition whereby poor nutritional status is caused by factors other than diet, such as illness.

undernutrition The inadequate intake of one or more nutrients and/or energy.

2-2 How Is Nutritional Status Assessed? • Nutritional status can be assessed in several ways, including anthropometric measurements, biochemical measurements, clinical assessment, and dietary assessment (the *ABCD* methods). • Anthropometric measurements include measures of body dimensions and composition. • Biochemical analyses of blood and/or urine samples can provide detailed information about nutrient status. • Clinical assessment involves conducting a face-to-face medical history and physical examination to check for signs and symptoms of malnutrition. • Dietary assessment methods include diet recalls, food frequency questionnaires, and diet records.

anthropometric measurement A measurement of the body's physical dimensions or composition.

biochemical measurement Laboratory analysis of a biological sample, such as blood or urine, for a nutrient or other biomarker.

body composition The proportions of fat, water, lean tissue, and mineral (bone) mass that make up the body.

diet recall A retrospective dietary assessment method whereby a person

records and analyzes every food and drink consumed over a given time span.

diet record (or **food record**) A prospective dietary assessment method whereby a person records and analyzes every food and drink as it is consumed over a given time span.

dietary assessment The evaluation of adequacy of a person's dietary intake.

food frequency questionnaire A retrospective dietary assessment method

whereby food selection patterns are assessed over an extended period of time.

sign A physical indicator of disease, such as pale skin and brittle fingernails, that can be seen by others.

symptom A subjective manifestation of disease that generally cannot be observed by other people.

2-3 How Much of a Nutrient Is Adequate? • The Institute of Medicine has formulated a set of nutrient intake standards called the Dietary Reference Intakes (DRIs). These standards include the Estimated Average Requirements (EARs), Recommended Dietary Allowances (RDAs), Adequate Intake (AI) levels, and Tolerable Upper Intake Levels (ULs). • EARs estimate average nutrient requirements in various population groups; RDAs are based on EARs and can be utilized as nutrient-intake goals for individuals. • When the Institute of Medicine could not establish EARs and RDAs, AI levels were set as guidelines for intakes. • ULs are not dietary recommendations but are levels that should not be exceeded. • Estimated Energy Requirement (EER) equations and Acceptable Macronutrient Distribution Ranges (AMDRs) provide guidance as to total energy intake and distribution of energy intake from macronutrients.

Acceptable Macronutrient Distribution Range (AMDR) The recommended range of intake for a given energy-yielding nutrient, expressed as a percentage of total daily calorie intake.

Adequate Intake (AI) The daily intake of a nutrient that appears to support adequate nutritional status; established when RDAs cannot be determined.

Dietary Reference Intakes (DRIs) A set of four dietary reference standards used to assess and plan dietary intake: Estimated

Average Requirement, Recommended Dietary Allowance, Adequate Intake level, and Tolerable Upper Intake Level.

Estimated Average Requirement (EAR) The daily intake of a nutrient that meets the physiological requirements of half the healthy individuals in a given age, physiologic state, and sex.

Estimated Energy Requirement (EER) The average energy intake needed for a healthy person to maintain weight.

nutrient requirement The amount of a nutrient that a person must consume to promote optimal health.

Recommended Dietary Allowance (RDA) The daily intake of a nutrient that meets the physiological requirements of nearly all (roughly 97 percent) healthy individuals in a given life stage and sex.

Tolerable Upper Intake Level (UL) The highest level of usual daily nutrient intake likely to be safe.

2-4 **How Can You Assess and Plan Your Diet?** • The Dietary Guidelines for Americans provide science-based nutritional recommendations to promote optimal health and reduce the risk of chronic disease in the United States. • The Dietary Guidelines for Americans also incorporates the USDA Food Patterns, which outlines how much of each food group should be consumed to promote optimal health. • The Dietary Guidelines' Food Patterns forms the basis of MyPlate, a graphic representation of the food groups and recommended intakes. • The MyPlate graphic simply reminds Americans to eat healthful amounts of the five food groups using a familiar mealtime visual: a place setting. • The MyPlate graphic does not specify the numbers of servings recommended by the Food Patterns. The recommended amount of each food group depends on a person's age, sex, and physical activity level. • The MyPlate website provides in-depth information about the types and amounts of foods that fit into each food group, assistance with meal planning, and material for special subgroups (such as vegetarians). • MyPlate is a singular part of a much larger communications initiative developed to help U.S. consumers make better food choices. • The MyPlate website offers an excellent opportunity for you to conduct a self-assessment of your own dietary intake using the free Food Tracker.

Dietary Guidelines for Americans A series of recommendations that provide specific nutritional guidance and advice about physical activity, alcohol intake, and food safety.

eating patterns (food patterns) A combination of foods and beverages that constitute a person's overall dietary intake.

Food Tracker A component of the MyPlate website that allows individuals to conduct dietary self-assessments.

moderation A nutrition basic related to choosing overall serving sizes that fit within caloric needs.

MyPlate A visual food guide that illustrates the most important food intake pattern recommendations of the 2015 Dietary Guidelines for Americans.

nutrient density The relative ratio of a food's amount of nutrients to its total calories.

USDA Food Pattern A USDA publication that categorizes nutritionally similar foods into food groups and makes recommendations regarding the number of servings of each food group that should be consumed daily.

variety A nutrition basic related to consuming different types of foods and beverages within a single food group and across food groups.

2-5 **How Can You Use Food Labels to Plan a Healthy Diet?** • Food labels provide consumers with useful information they can use to make smart food choices. • Nutrition Facts panels provide information that can help us choose healthful foods. The FDA mandates that several critical elements be listed on every Nutrition Facts panel. • Daily Values (DVs) give consumers a benchmark as to whether a food is a good source of a particular nutrient. There are two basic types of DVs. The first type, used for select vitamins and minerals, represents a nutrient's recommended daily intake. The second type represents a nutrient's upper intake limit. • Nutrient content claims and health claims are valuable tools when planning a healthy diet.

Daily Value (DV) A benchmark as to whether a food is a good source of a particular nutrient. May represent a nutrient's recommended daily intake or upper limit.

health claim An FDA-approved statement that describes a specific health benefit of a food or food component.

nutrient content claim An FDA-regulated word or phrase that describes how much of a nutrient is in a food.

Nutrition Facts panel A required component of most food labels that provides information about the nutrient content of the food.

qualified health claim A health claim that has less scientific backing and must be accompanied by a disclaimer (or qualifier) statement.

regular health claim A health claim that is supported by considerable scientific research.

2-6 **Can You Put These Concepts into Action?** • Although dietary assessment is only one component of a complete nutritional assessment, it is an important first step toward a lifetime of health and nutritional awareness. • You now have the information needed to assess your own diet and begin to choose the right foods to improve it. • You might begin by conducting a dietary self-assessment using the MyPlate Food Tracker. • The food habits you establish now will not only affect your success in college but also influence your eating patterns and health for years to come.

FIGURE 2.8 THE MYPLATE GRAPHIC

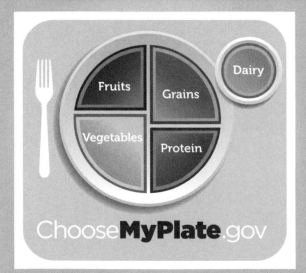

MyPlate illustrates the Dietary Guidelines for Americans recommended food consumption pattern as a consumer-friendly graphic.

Source: U. S. Department of Agriculture–www.choosemyplate.gov

3-1 Why Learn about Chemistry When Studying Nutrition?

• Chemistry is fundamental to the study of nutrition. Nutrients are chemicals, and the body's utilization of nutrients involves countless chemical reactions. • Many atoms have an equal number of positively and negatively charged particles, and therefore are neutral. Charged atoms, called ions, serve many vital functions in the body. • Identical atoms combine with each other to form elements. There are approximately 98 naturally occurring elements, 20 of which are essential to human health. • When chemical bonds join two or more atoms together, molecules are formed. Molecules can be very small or very large. • A molecular formula is a representation of the number and type of atoms present in a molecule. When a subscript number follows an element's symbol, it means that there are that many atoms of that type of element present. A number placed before the molecular formula means that there are many molecules of the substance present. For example, the molecular formula for three molecules of water is written $3H_2O$.

atom The basic unit of matter that makes up the world around us.

electrolyte A molecule that, when submerged in water, separates into individual ions.

element A pure substance made up of only one type of atom.

ion An atom that has a positive or negative electrical charge.

molecular formula A representation of the number and type of atoms present in a molecule.

molecule A unit of two or more atoms joined together by chemical bonds.

3-2 How Are Cells, Tissues, Organs, and Organ Systems Related?

• Molecules make up cells, which make up tissues, which in turn function as building blocks for organs, which work together as components of organ systems. • Transport mechanisms that do not require energy (such as simple diffusion, facilitated diffusion, and osmosis) are called passive transport systems, whereas mechanisms that require energy (such as carrier-mediated active transport) are called active transport systems. • Some organelles produce substances necessary to cellular activity, while others function as waste-disposal systems, assisting the recycling of worn-out cellular components. • There are four tissue types (epithelial, connective, muscle, and neural), which collectively carry out functions such as movement, communication, protection, and structure. • The nervous and endocrine systems work together to monitor our internal environment, respond to change, and restore balance when necessary. These mechanisms allow us to adapt in an ever-changing, complex environment so that we can maintain homeostasis.

active transport mechanism A transport mechanism that requires energy (ATP) to move substances across cell membranes.

carrier-mediated active transport An active transport mechanism whereby a substance moves from a region of lower concentration to a region of higher concentration with the assistance of a carrier molecule and energy.

cell A structural and functional unit that makes up body tissues.

connective tissue Tissue that supports, connects, and anchors structures in the body.

cytoplasm A gel-like fluid that fills every living cell.

epithelial tissue Tissue that helps protect the body.

facilitated diffusion A passive transport mechanism whereby a substance moves from a region of higher concentration to a region of lower concentration with the assistance of a carrier molecule.

homeostasis A state of balance or equilibrium.

hormones Chemical messengers released into the blood by the endocrine system that coordinate the activities of specific cells in the body.

muscle tissue Tissue that is used for movement.

neural tissue Tissue that facilitates communication throughout the body.

organ Two or more different types of tissues functioning together to perform a variety of related tasks.

organelles A structure that is responsible for a specific function within a cell.

osmosis The passive movement of water molecules across a cell membrane from a solution with low solute concentration to a solution with a higher solute concentration.

passive transport mechanism A transport mechanism that does not require energy (ATP) to move substances across cell membranes.

simple diffusion A passive transport mechanism whereby a substances moves from a region of higher concentration to a region of lower concentration without using energy or the assistance of a carrier molecule.

solutes A dissolved substance; a component of a solution.

tissue An aggregation of similarly structured and functioning cells that have grouped together to accomplish a common task.

3-3 What Happens during Digestion?

• The digestive system consists of the gastrointestinal (GI) tract and accessory organs, which release a variety of secretions needed for digestion. • The three functions of the digestive system are the chemical and physical breakdown of food (digestion), the transfer of nutrients into the blood or lymphatic circulatory systems (absorption), and the removal of undigested food residue (elimination). • After peristalsis propels a bolus down the esophagus, sphincters located throughout the GI tract regulate the flow of the luminal contents from one organ to the next. • The stomach is uniquely equipped to carry out two important functions: (1) mixing food with the gastric secretions that aid in chemical digestion and (2) temporarily storing food. • The hormone gastrin is released when food enters the stomach, stimulating the release of gastric juice. Food mixes with gastric juice and turns into chyme. • After leaving the stomach, the chyme passes into the small intestine. The small intestine is the primary site of chemical digestion and nutrient absorption, although some absorption occurs in the stomach and large intestine. • Secretions from the pancreas and gallbladder facilitate digestion in the small intestine.

absorption The movement of nutrients out of the GI tract and into the blood or lymph.

accessory organs Organs that are part of the digestive system, and assist with the process of digestion.

bile A fluid produced in the liver and released by the gallbladder that disperses large globules of fat into smaller droplets that are easier to digest.

bolus A soft, moist mass of chewed food.

chyme A semi-liquid paste resulting from the mixing of partially digested food with gastric juice in the stomach.

colon The first portion of the large intestine.

(continues)

(continued)

digestion The physical and chemical breakdown of food into a form that allows nutrients to be absorbed.

duodenum The first segment of the small intestine.

elimination The process whereby solid waste is removed from the body.

emulsification The breakdown of large fat globules into smaller droplets that aids in the overall process of digestion.

enterocytes Epithelial cells that make up villus.

enzyme A biological catalyst that accelerates a chemical reaction.

esophagus A narrow muscular tube that begins at the pharynx and ends at the stomach.

feces Solid waste consisting mainly of undigested and unabsorbed matter, dead cells, secretions from the GI tract, water, and bacteria.

gastric emptying The process by which food leaves the stomach and enters the small intestine.

gastric juice Digestive secretions that consist mainly of water, hydrochloric acid, digestive enzymes, mucus, and intrinsic factor.

gastric mucosal barrier A thick, gel-like substance that protects the stomach lining from acidic gastric juices.

gastrin A hormone released by endocrine cells in the stomach lining that stimulates the release of gastric juice and causes the muscular wall of the stomach to contract vigorously.

gastroesophageal reflux disease (GERD) A condition caused by chronic reflux of the stomach contents into the esophagus, irritating its lining.

gastroesophageal sphincter A circular muscle that regulates the flow of food from the esophagus to the stomach.

GI microbiota The natural microbial population that resides in the GI tract.

inflammatory bowel disease (IBD) Inflammatory conditions such as ulcerative colitis and Crohn's disease that affect the lining of the lower GI tract.

irritable bowel syndrome (IBS) A disorder that typically affects the lower GI tract, causing bouts of cramping, bloating, diarrhea, and constipation.

lumen The cavity that spans the entire length of the GI tract.

microvillus A tiny finger-like projection. Microvilli cover the lumenal surfaces of the enterocytes that line lumen of the small intestine.

mucosa The innermost lining of the gastrointestinal tract.

pancreatic juice A mixture of water, bicarbonate, and various enzymes released by the pancreas.

peristalsis A vigorous, wave-like muscular contraction that propels food from one region of the GI tract to the next.

pharynx A region at the back of the mouth that serves as the shared space between the oral and nasal cavities.

prebiotic food A typically fiber-rich food that may stimulate the growth of the microbial population in the large intestine.

probiotic food A food that contains live bacteria, some of which thrive in the colon.

pyloric sphincter A circular muscle that regulates the flow of chyme from the stomach into the small intestine.

rectum The segment of the large intestine that leads to the anal canal.

saliva A secretion released into the mouth by the salivary glands that moistens food and starts the chemical process of digestion.

sphincter A circular band of muscle that regulates the flow of food through the GI tract.

tight junctions Interlocking proteins between adjacent enterocytes that create an impermeable barrier.

transit time The amount of time it takes for food to travel the entire length of the GI tract.

villi Small finger-like projections that cover the inner lining of the small intestine (note that villus is the singular form of villi).

3-4 **What Happens after Digestion? Nutrient Absorption and Circulation** · Nutrient absorption is the process whereby nutrients are transported from the lumen of the GI tract into either the blood or lymph. The bioavailability of a particular nutrient can be influenced by physiological conditions, other dietary components, and certain medications. · Materials entering the large intestine consist mostly of undigested remains from plant-based foods, water, bile, and electrolytes. Muscles embedded within the intestinal wall squeeze the undigested food residue, and as material moves through the various regions of the colon, the water and electrolytes are absorbed and returned to the blood for reuse by the body. · Intestinal microbiota breaks down undigested food residue, produces nutrients, and inhibits the growth of other disease-causing bacteria. · Once the remaining material reaches the rectum (the last segment of the large intestine), it is ready to be eliminated from the body.

bioavailability The extent to which a nutrient is absorbed into the enterocyte.

lacteal A lymphatic vessel found in an intestinal villus into which nutrients circulate after they are absorbed.

3-5 **How Are Nutrients Metabolized and How Are Metabolic Waste Products Excreted from the Body?** · The delivery of nutrients and oxygen to cells is accomplished by the cardiovascular and lymphatic systems. · The cardiovascular system circulates nutrients and gases, while the lymphatic system circulates fat-soluble nutrients. · The circulatory systems, liver, kidneys, lungs, and skin work together to remove metabolic waste products and thus prevent the accumulation of toxins in the body. · Chemical energy is transferred into usable ATP through metabolic pathways. Each chemical reaction in a metabolic pathway requires the aid of at least one enzyme. · Catabolic pathways (both aerobic and anaerobic) break complex molecules into simpler ones, while anabolic pathways use energy to construct complex molecules from simpler ones.

aerobic A metabolic pathway that requires oxygen to function.

anabolism A series of metabolic reactions that uses energy to construct a complex molecule from simpler ones.

anaerobic A metabolic pathway that can function under conditions of low oxygen availability.

catabolism A series of metabolic reactions that breaks down a complex molecule into simpler ones, releasing energy in the process.

energy metabolism Chemical reactions that enable cells to use and store energy.

excretion The removal of metabolic waste products produced in cells.

metabolic pathway A series of interrelated chemical reactions that require enzymes.

metabolism The sum of chemical processes that occur within a living cell to maintain life.

4-1 What Are Simple Carbohydrates? · A carbohydrate consisting of a single sugar is called a monosaccharide, and a carbohydrate consisting of two sugars is called a disaccharide. Because of the small sizes of these molecules, monosaccharides and disaccharides are referred to as simple carbohydrates. · Monosaccharides include glucose (the most abundant monosaccharide in the human body), fructose (a naturally occurring monosaccharide found primarily in honey, fruits, and vegetables), and galactose (which occurs primarily as part of the disaccharide lactose). · The most common disaccharides are lactose (the most abundant carbohydrate in milk and other dairy products), maltose (a product of the enzymatic breakdown of starches), and sucrose (commonly known as refined table sugar). · Alternative low-calorie sweeteners such as saccharin, aspartame, and acesulfame K can be added to increase sweetness without increasing the caloric contents of foods.

carbohydrates An organic compound made up of one or more sugar molecules.

disaccharide A carbohydrate consisting of two monosaccharides bonded together.

fructose A naturally occurring monosaccharide found primarily in honey, fruits, and vegetables.

galactose A monosaccharide that exists primarily as part of a naturally occurring disaccharide found in dairy products.

glucose The most abundant monosaccharide in the human body; used extensively for energy.

high-fructose corn syrup (HFCS) A widely used sweetener consisting of glucose and fructose that is manufactured from cornstarch.

lactose A disaccharide comprised of galactose joined with glucose. It is the most abundant carbohydrate in milk and many other dairy products.

maltose A disaccharide composed of glucose joined with glucose. It is not found in many foods.

monosaccharide A carbohydrate consisting of a single sugar molecule.

photosynthesis A process whereby chlorophyll-containing plants produce glucose by combining carbon dioxide (CO_2) and water (H_2O) using energy harvested from sunlight.

simple carbohydrates (or **simple sugars**) A category of carbohydrates comprised of monosaccharides and disaccharides.

sucrose A disaccharide composed of fructose joined with glucose. It is naturally occurring and is most abundant in sugar cane and sugar beets.

4-2 What Are Complex Carbohydrates? · Complex carbohydrates are comprised of many monosaccharides bonded together. The types and arrangements of sugar molecules determine the shape and form of the polysaccharide. · Glucose molecules are arranged in starch in either an orderly unbranched linear chain (amylose) or a highly branched configuration (amylopection). Plants typically contain a mixture of these two types of starch. · Glycogen consists of glucose molecules bonded together in a highly branched arrangement and is found mainly in liver and skeletal muscle. · Dietary fibers are a diverse group of carbohydrates found in a variety of foods such as whole grains, legumes, vegetables, and fruits that are not digested or absorbed in the human small intestine. · Soluble dietary fiber tends to dissolve or swell in water, while insoluble dietary fiber remains relatively unchanged. Consumption of soluble fiber can help lower blood cholesterol levels, promote satiety, and lower blood glucose levels. · The nutritional value of grain is greatest when all three of its components—bran, germ, and endosperm—are present.

amylopectin A type of starch that consists solely of glucose molecules in a highly branched configuration.

amylose A type of starch that consists solely of glucose molecules arranged in a linear, unbranched chain.

bran The outer portion of a grain that contains most of the fiber.

complex carbohydrate (or **polysaccharide**) A category of carbohydrates comprised of many monosaccharides bonded together.

diverticular disease (or **diverticulosis**) A condition whereby pouches called *diverticula* form along the colon wall.

diverticulitis A condition whereby the diverticula become infected or inflamed.

endosperm The portion of a grain that contains mostly starch.

fiber (or **dietary fiber**) A diverse group of plant polysaccharides that are not digestible by human enzymes.

germ The portion of a grain that contains most of the vitamins and minerals.

glycogen A polysaccharide found primarily in liver and skeletal muscle that is comprised of glucose molecules.

insoluble fiber Dietary fiber that remains relatively unchanged in water.

soluble fiber Dietary fiber that tends to dissolve or swell in water.

whole-grain foods Cereal grains that contain bran, endosperm, and the germ in the same relative proportions as exist naturally.

4-3 How Are Carbohydrates Digested, Absorbed, and Circulated? · Carbohydrates undergo extensive chemical transformations as they move through the GI tract during digestion. Enzymes required for carbohydrate digestion are produced in the salivary glands, pancreas, and small intestine. · Starch digestion begins in the mouth, but most occurs in the small intestine via pancreatic amylase. Digestion is completed by the enzyme maltase, resulting in free (unbound) glucose molecules. · The digestion of disaccharides takes place entirely in the small intestine. Each disaccharide has its own specific digestive enzyme. · Glycemic response is a change in blood glucose following the ingestion of a carbohydrate-rich food. · Glycemic index (GI) and glycemic load (GL) are rating systems used to assess the glycemic response to specific foods. GL takes into account the typical portion of food consumed while GI does not.

glycemic index (GI) A rating system based on a scale of 0 to 100 used to compare the glycemic responses elicited by different foods.

glycemic load (GL) A rating system used to compare the glycemic responses associated with different foods that takes into account the typical portion of food consumed.

glycemic response The change in blood glucose following the ingestion of a food.

(continues)

Carbohydrates

(continued)

lactase An intestinal enzyme that digests lactose, releasing glucose and galactose molecules.

lactose intolerance A condition whereby the body does not produce enough of the enzyme lactase, making it difficult to digest lactose.

maltase An intestinal enzyme that digests maltose, releasing two glucose molecules.

pancreatic amylase An enzyme released from the pancreas that splits chemical bonds that hold glucose molecules together, eventually forming maltose.

salivary amylase An enzyme released from the salivary glands that breaks the chemical bonds in starch.

sucrase An intestinal enzyme that digests sucrose, releasing glucose and fructose molecules.

4-4 **How Does Your Body Regulate and Use Glucose?** • The pancreatic hormones insulin and glucagon play major roles in blood glucose regulation and energy storage. • Insulin, released when blood glucose is high, lowers blood glucose levels by (1) enabling cells to take up glucose from the blood; (2) increasing the rate at which glucose is used as an energy source; and (3) promoting the conversion of glucose to body fat, which is stored mostly in adipose tissue. • Glucagon, released when blood glucose levels are low, increases glucose availability by stimulating the breakdown of liver glycogen. As glycogen stores dwindle, glucagon stimulates gluconeogenesis. • When carbohydrate intake is limited and glycogen stores are depleted, the body minimizes protein loss by using ketones as an energy source.

epinephrine A hormone released from the adrenal glands that stimulates glycogenolysis in emergency situations.

glucagon A hormone secreted by the pancreas in response to low levels of blood glucose.

gluconeogenesis The synthesis of glucose from noncarbohydrate sources.

glycogenolysis The breakdown of glycogen into glucose.

glycolysis An anaerobic metabolic pathway made up of a series of chemical reactions that splits glucose into two three-carbon molecules.

hyperglycemia A condition characterized by high blood glucose.

hypoglycemia A condition characterized by low blood glucose.

insulin A hormone secreted by the pancreas in response to elevated levels of blood glucose.

insulin receptors Specialized proteins located on the outer membranes of certain types of cells that bind insulin and signal the cell to take up glucose.

ketones An organic compound used as an alternative energy source under conditions of limited glucose availability.

ketosis A condition characterized by excessive ketone accumulation in the blood.

pyruvate The end product of glycolysis; formed by the breakdown of glucose.

tricarboxylic acid (TCA) cycle (also called the Krebs cycle) An oxygen-requiring metabolic pathway consisting of a series of chemical reactions that ultimately generate energy (ATP).

4-5 **What Is Diabetes?** • Diabetes mellitus is a group of metabolic disorders characterized by elevated levels of glucose in the blood. The two main types of diabetes are referred to as type 1 diabetes and type 2 diabetes. • Type 1 diabetes, an autoimmune disease, occurs when the pancreas is no longer able to produce insulin. This causes blood glucose levels to become dangerously high. • Type 2 diabetes is caused by insulin resistance. Because the body's cells do not respond appropriately to insulin's signal, the amount of glucose taken up from the bloodstream is diminished.

diabetes mellitus A group of metabolic disorders characterized by elevated levels of glucose in the blood.

glucometer A medical device used to monitor the concentration of glucose in the blood.

insulin resistance A condition whereby insulin receptors throughout the body are less responsive to insulin.

type 1 diabetes A form of diabetes whereby the pancreas is no longer able to produce insulin, causing blood glucose levels to become dangerously high.

type 2 diabetes A form of diabetes whereby insulin resistance impairs the ability of certain cells to take up glucose from the blood.

4-6 **What Are the Recommendations for Carbohydrate Intake?** • The Institute of Medicine's Dietary Reference Intakes (DRIs) are based mainly on ensuring that the brain has adequate glucose for its energy needs. • Because some carbohydrate-rich foods are more nutrient-dense (and thus nutritional) than others, one must weigh a number of factors before deciding which carbohydrate-containing foods to consume. Following the national health guidelines can help eliminate much of the guesswork when it comes to determining which carbohydrate-rich foods to choose. • Health experts generally agree that a person should minimize his or her consumption of foods with high amounts of added sugar (such as cookies, soda, sugary cereals, and heavy syrups). • The Institute of Medicine's DRIs recommend that adults consume at least 14 grams of dietary fiber per 1,000 kilocalories, which is approximately 21–38 grams per day. • A sudden and/or large increase in fiber intake may cause a gastrointestinal problem such as diarrhea or constipation. Thus, a person should increase his or her fiber intake gradually and give the body time to adjust.

TABLE 4.5	ADDED SUGARS IN SELECTED FOOD ITEMS	
Food	**Serving Size**	**Added Sugar (g)[a]**
Soft drink	12 oz	43
Milkshake	10 oz	36
Fruit punch	8 oz	38
Chocolate candy	1.5 oz	24
Sweetened breakfast cereal	1 cup	15
Yogurt with fruit	1 cup	33
Ice cream	1 cup	28
Cake with frosting	1 slice	28
Cookies	2 (medium)	14
Jam or jelly	1 tbsp	2

[a]1 teaspoon equals 4.75 grams.

Source: U.S. Department of Agriculture database for the added sugars content of selected foods, Release 1 (2006). Available at: http://www.ars.usda.gov/Main/docs.htm?docid=12107

5-1 **What Are Proteins?** • Proteins are nitrogen-containing macronutrients made from amino acids linked together via peptide bonds. • Every amino acid contains a central carbon, a carboxylic acid group, an amino group, and an R-group. The body needs 20 amino acids; nine of these are essential nutrients.

α-keto acid A compound similar to an amino acid that does not have an amino group; used to synthesize nonessential amino acids.

amino acid A nitrogen-containing subunit that combines with other amino acids to form proteins.

amino group The nitrogen-containing component of an amino acid.

complete protein source A food that supplies an adequate and balanced amount of all the essential amino acids.

incomplete protein source A food that lacks or supplies low amounts of one or more of the essential amino acids.

limiting amino acid An essential amino acid that is insufficient or absent in an incomplete protein source.

peptide bond A chemical bond that joins amino acids.

phenylketonuria (PKU) A disorder caused by deficiency of an enzyme needed to convert the essential amino acid phenylalanine to the normally nonessential amino acid tyrosine; as a result, tyrosine cannot be made and becomes conditionally essential.

polypeptide A chain of amino acid subunits joined together through peptide bonds.

protein A nitrogen-containing macronutrient made from amino acids.

protein complementation The combining of diverse foods with different incomplete proteins to provide adequate amounts of all the essential amino acids.

R-group The side-chain component of an amino acid that distinguishes it from other amino acids.

transamination The process by which nonessential amino acids are synthesized.

5-2 **How Do Cells Make Proteins?** • Each gene in a DNA strand contains information about how a protein should be constructed. • Protein synthesis involves DNA, mRNA, ribosomes, tRNA, and amino acids. • A series of steps involving cell signaling, transcription, and translation produces hundreds of thousands of different proteins in the body.

cell signaling The process by which a cell is notified that it should make a particular protein.

chromosomes A substance comprised of coiled strands of DNA and special proteins; found in a cell's nucleus.

deoxyribonucleic acid (DNA) A chemical in the nucleus of a cell that provides the instructions for protein synthesis.

gene A subunit of a chromosome that tells a cell which amino acids are needed and in what order they must be arranged to synthesize a specific protein.

messenger RNA (mRNA) A chemical that carries the instructions contained in DNA outside of the nucleus.

ribosome A cellular component primarily involved in the assembly of proteins by involving mRNA and tRNA and the process of translation.

transcription The process by which mRNA is constructed using DNA as a template; occurs in the cell's nucleus.

transfer ribonucleic acid (tRNA) A chemical that carries amino acids to a ribosome to be assembled into a protein.

translation The process by which amino acids are joined via peptide bonds; occurs in the cell's cytoplasm.

5-3 **Why Is a Protein's Shape Critical to Its Function?** • The primary structure folds into secondary and tertiary structures. • Peptide chains can join together to form quaternary structures, and these can combine with prosthetic groups to form a protein's final shape. • Denaturation can negatively impact the ability of the protein to function.

α-helix A common secondary structure folding pattern that resembles the shape of a spiral staircase.

β-folded sheet A common secondary structure folding pattern that resembles the shape of a folded paper fan.

denaturation The process by which a protein's three-dimensional structure is altered.

primary structure (or **primary sequence**) The most basic level of protein structure; determined by the number and sequence of amino acids.

prosthetic group A nonprotein component of a protein that often contains minerals.

quaternary structure The most complex level of protein structure; occurs when two or more peptide chains join together.

secondary structure Organized and predictable folds that develop in portions of a peptide chain because charged portions of the amino acid backbone attract and repel each other.

sickle cell anemia (or **sickle cell disease**) A disease whereby an alteration in DNA results in the production of defective, misshapen molecules of the protein hemoglobin within red blood cells.

tertiary structure Additional folding of a peptide chain due to interactions between the amino acids' R-groups.

5-4 **What Is Meant by Genetics and Epigenetics?** • A person's genetic makeup (or genotype) is inherited from his or her parents. • Mutations in DNA can influence the cell's ability to produce a functional protein.

epigenetics Alterations in protein synthesis that do not involve changes in the DNA sequence.

epigenome A network of chemical variations around DNA that together determine which genes are active in a particular cell.

genotype The particular DNA inherited from one's parents.

mutation An alteration in a gene that occurs due to a chance of genetic modification.

phenotype The observable physical or biochemical characteristics of an organism.

Protein

5-5 **How Are Proteins Digested, Absorbed, and Circulated?** • The process of digestion disassembles food proteins into amino acids that are then absorbed and carried to cells where they are assembled into needed proteins. • Protein digestion involves hormones as well as a variety of proteases produced in the stomach, pancreas, and small intestine. • Protein-digesting enzymes are released as inactive proenzymes and then converted to their active forms (proteases) in the gastrointestinal tract. • Amino acids are absorbed primarily along the duodenum, where they enter the blood and circulate to the liver.

anaphylaxis A rapid immune response that causes a sudden drop in blood pressure, rapid pulse, dizziness, and a narrowing of the airways.

cholecystokinin (CCK) A hormone, secreted by intestinal cells, that signals the release of bile from the gallbladder and proenzyme proteases from the pancreas.

food allergy A condition whereby the body's immune system responds to a food derived peptide as if it were dangerous.

food intolerance (or **food sensitivity**) A condition whereby the body reacts negatively to a food or food component, but does not mount an immune response.

pepsin An enzyme needed for protein digestion.

pepsinogen The inactive form of pepsin.

proenzyme (also called a **zymogen**) An inactive precursor of an enzyme.

protease An enzyme that breaks peptide bonds between amino acids.

secretin A hormone, secreted by intestinal cells, that signals the release of sodium bicarbonate and proteases from the pancreas.

5-6 **Why Do We Need Proteins and Amino Acids?** • Amino acids are used to synthesize proteins needed for structure, catalysis, movement, transport, communication, and protection. • Proteins can be broken down and used for ATP production; some amino acids can be converted to glucose. • Amino acids themselves serve many diverse roles. Some regulate protein synthesis or breakdown, others are involved in cell communication, and still others are converted to neurotransmitters and other signaling molecules. • During times of energy abundance, amino acids are transformed into fat and stored in adipose tissue.

albumin A protein present in the blood that plays an important role in regulating fluid balance.

antibody (or **immunoglobulin**) A protein that helps fight infection.

edema A condition whereby low levels of albumin in the blood cause fluid to accumulate in body tissues or cavities.

proteolysis The breakdown of proteins.

5-7 **How Does the Body Recycle and Reuse Amino Acids?** • One's protein status can be assessed by comparing protein intake to the amount of nitrogen lost in body secretions. • Knowing whether a person is in neutral, positive, or negative nitrogen balance can help a clinician diagnose and treat certain disease states and physiologic conditions.

negative nitrogen balance When nitrogen intake is less than nitrogen loss.

neutral nitrogen balance When nitrogen intake equals nitrogen loss.

positive nitrogen balance When nitrogen intake is greater than nitrogen loss.

protein turnover The continual coordinated process of protein breakdown and synthesis.

urea A nitrogen-containing substance produced by the body when it breaks down amino acids; excreted in the urine.

5-8 **How Much Protein Do You Need?** • Protein must be consumed to supply the essential amino acids and the nitrogen needed to synthesize nonessential amino acids and other nonprotein, nitrogen-containing compounds such as DNA. • In general, protein requirements are greater for males than females and are highest during infancy, pregnancy, and lactation. A typical college-age male needs 56 g/day of protein, whereas a comparable female needs 46 g/day. • It is recommended that healthy adults consume approximately 0.8 grams of protein daily for each kilogram of body weight. Protein requirements may be higher for athletes. • The Institute of Medicine's Acceptable Macronutrient Distribution Range (AMDR) for protein is 10 to 35 percent of energy.

5-9 **Can Vegetarian Diets Be Healthy?** • Consuming adequate amounts of protein is typically not difficult for vegetarians, although they may be at greater risk for calcium, zinc, iron, and vitamin B_{12} deficiencies.

lacto-ovo-vegetarian A vegetarian who consumes dairy products and eggs in an otherwise plant-based diet.

lactovegetarian A vegetarian who consumes dairy products (but not eggs) in an otherwise plant-based diet.

vegan A vegetarian who consumes no animal products.

vegetarian A person who does not consume or consumes only some foods and beverages made from animal products.

5-10 **What Are the Consequences of Protein Deficiency and Excess?** • Treatment of children and adults with PEM is multifaceted, and usually involves a variety of factors including the provision of adequate nutrition. • For most people, excess protein intake does not result in health complications. • There is growing evidence that excessive consumption of red or processed meat is associated with increased risk of colorectal cancer.

ascites A condition characterized by fluid accumulation in the abdominal cavity.

kwashiorkor A form of PEM characterized by severe edema in the extremities and sometimes the abdomen.

marasmus A form of PEM characterized by extreme wasting of muscle and loss of adipose tissue.

protein-energy malnutrition (PEM) A condition whereby protein deficiency is accompanied insufficient energy, and usually, one or more micronutrients.

CHAPTER REVIEW

6-1

What Are Lipids? · A lipid that is liquid at room temperature is called an oil, whereas one that is solid is called a fat. · Fatty acids, made of carbon, hydrogen, and oxygen atoms, are the most abundant type of lipid in the body. A chain of carbon atoms with alpha (α) and omega (ω) ends forms the fatty acid's backbone. The number of carbon atoms in the backbone determines its chain length. · Double bonds can be in either *cis* or *trans* configuration, depending on the placement of the hydrogens. · Alpha (α) nomenclature designates the number of carbon atoms, the number and placement of double bonds, and the type of double bonds relative to the alpha end. · Omega (ω) nomenclature is similar, although the placement of double bonds is determined from the omega end.

alpha (α) end The end of a fatty acid that contains the carboxylic acid (–COOH) group.

chain length The number of carbon atoms in a fatty acid.

cis double bond A carbon–carbon double bond in which the hydrogen atoms are positioned on the same side of the double bond.

fat A lipid that is solid at room temperature.

fatty acids The most abundant type of lipid; comprised of a chain of carbons with a methyl (–CH₃) group on one end and a carboxylic acid (–COOH) group on the other.

hydrophobic A substance that does not easily mix with water.

lipid An organic macronutrient that is relatively insoluble in water and relatively soluble in organic solvents.

long-chain fatty acid A fatty acid with more than 12 carbon atoms.

medium-chain fatty acid A fatty acid with 8 to 12 carbon atoms.

monounsaturated fatty acid (MUFA) A fatty acid with one carbon–carbon double bond.

oil A lipid that is liquid at room temperature.

omega (ω) end The end of a fatty acid that contains the methyl (–CH₃) group.

omega-3 (ω-3) fatty acid A fatty acid in which the first double bond is located between the third and fourth carbons from the omega (ω) end.

omega-6 (ω-6) fatty acid A fatty acid in which the first double bond is located between the sixth and seventh carbons from the omega (ω) end.

partial hydrogenation A process whereby oils are converted into fats by changing many of the carbon–carbon double bonds into carbon–carbon single bonds; some *trans* fatty acids are also produced.

polyunsaturated fatty acid (PUFA) A fatty acid with more than one carbon–carbon double bond.

saturated fatty acid (SFA) A fatty acid with only carbon–carbon single bonds.

short-chain fatty acid A fatty acid with fewer than eight carbon atoms.

trans double bond A carbon–carbon double bond in which the hydrogen atoms are on opposite sides of the double bond.

trans fatty acid A fatty acid containing at least one *trans* double bond.

unsaturated fatty acid A fatty acid with at least one carbon–carbon double bond.

6-2

What Are Essential, Conditionally Essential, and Nonessential Fatty Acids? · The two essential fatty acids are linoleic acid and linolenic acid. Linoleic and linolenic acids are metabolized to longer-chain fatty acids and other compounds such as eicosanoids. · Because of the extensive amounts of linoleic and linolenic acids stored in adipose tissue, essential fatty acid deficiencies are rare.

desaturation The process whereby a fatty acid's carbon–carbon single bonds are converted to double bonds.

eicosanoids A diverse group of hormone-like compounds that help regulate the immune and cardiovascular systems and act as chemical messengers.

elongation The process whereby fatty acid chain length is increased by the addition of carbon atoms.

linoleic acid An essential ω-6 fatty acid with 18 carbons and two double bonds.

linolenic acid (or **α-linolenic acid**) An essential ω-3 fatty acid with 18 carbons and three double bonds.

6-3

What Is the Difference between Mono-, Di-, and Triglycerides? · Monoglycerides consist of one fatty acid attached to a glycerol molecule, while diglycerides have two fatty acids, and triglycerides have three. · The metabolic process by which fatty acids combine with glycerol to form triglycerides is called lipogenesis, and the metabolic process by which fatty acids in a triglyceride are separated from their glycerol backbone is called lipolysis. · Fatty acids not required for energy or other functions are stored as triglycerides in adipose tissue.

ß-oxidation The metabolic breakdown of fatty acids into 2-carbon units that are used to produce ATP.

adipocyte A specialized cell, found in adipose tissue, that can accumulate large amounts of triglycerides.

diglyceride A lipid comprised of two fatty acids attached to a glycerol backbone.

lanugo Very fine hair that grows as a physiological response to insufficient body fat.

lipogenesis The metabolic process by which fatty acids combine with glycerol to form triglycerides.

lipolysis The metabolic process by which a triglyceride's fatty acids are removed from the glycerol backbone.

monoglyceride A lipid comprised of one fatty acid attached to a glycerol backbone.

subcutaneous adipose tissue Adipose tissue located directly under the skin.

triglyceride A lipid comprised of three fatty acids attached to a glycerol backbone.

visceral adipose tissue Adipose tissue located around the vital organs in the abdomen.

6-4 **What Are Phospholipids and Sterols?** · A phospholipid consists of a glycerol, two fatty acids, and a phosphate-containing head group. Phospholipids make up cell membranes and circulate lipids throughout the body. · Sterols are multi-ring compounds that serve many roles in the body. Phytosterols are sterol-like compounds found in plants. · Cholesterol is an important component of cell membranes and a precursor for the synthesis of steroid hormones such as estrogen and testosterone.

amphipathic A characteristic of a substance that contains both hydrophilic and hydrophobic portions.

bile acid An amphipathic substance important for digestion and absorption of lipids.

cholesterol A sterol used to synthesize bile acids and a variety of hormones such as testosterone.

head group A phosphate-containing, hydrophilic component of a phospholipid.

hydrophilic A substance that mixes easily with water.

phospholipid A lipid composed of a glycerol molecule bonded to two hydrophobic fatty acids and a hydrophilic head group.

phytosterol A sterol-like compound made by plants.

sterol A type of lipid with a multiple-ring chemical structure.

6-5 **How Are Triglycerides Digested, Absorbed, and Circulated?** · Lipid digestion occurs in the mouth (lingual lipase), stomach (gastric lipase), and small intestine (pancreatic lipase). Bile emulsifies large lipid globules into smaller droplets in the small intestine. · Short- and medium-chain fatty acids are transported into intestinal cells unassisted because they are relatively water soluble. · More hydrophobic compounds are first repackaged into micelles within the intestinal lumen before moving into the intestinal cells. These lipids are incorporated into chylomicra and circulated in the lymph. · Chylomicra deliver dietary fatty acids to cells via lipoprotein lipase.

chylomicron A particle, composed of relatively hydrophobic lipids, cholesterol, and phospholipids, that transports dietary lipids in the lymph for circulation.

chylomicron remnant A residual fragment of a chylomicron that is taken up by the liver.

emulsification The process whereby large lipid globules are broken down and stabilized into smaller lipid droplets.

gastric lipase An enzyme, produced in the stomach, that cleaves fatty acids from glycerol molecules.

lingual lipase An enzyme, produced by the salivary glands, that cleaves fatty acids from glycerol molecules.

lipoprotein A particle that transports lipids throughout the body.

lipoprotein lipase An enzyme that enables chylomicra and other lipoproteins to deliver fatty acids to cells.

micelle A small droplet of fat formed, via emulsification, in the small intestine.

pancreatic lipase An enzyme, produced in the pancreas, that completes triglyceride digestion by cleaving fatty acids from glycerol molecules.

6-6 **What Are the Types and Functions of Various Lipoproteins?** · The liver produces very-low-density lipoproteins (VLDLs) that deliver dietary and endogenously produced fatty acids to cells. · The loss of fatty acids from a VLDL results in its conversion to an intermediate-density lipoprotein (IDL) and ultimately a low-density lipoprotein (LDL). · LDLs deliver cholesterol to cells. The liver produces high-density lipoproteins (HDLs), which pick up excess cholesterol and return it to the liver. · Too much LDL can result in the buildup of plaque. High levels of LDL and low levels of HDL are associated with increased risk of cardiovascular disease.

apoprotein (or **apolipoprotein**) A protein embedded within the outer shell of a lipoprotein that enables it to interact with cells.

high-density lipoprotein (HDL) A lipoprotein that collects excess cholesterol from cells and transports it back to the liver.

intermediate-density lipoprotein (IDL) A lipoprotein that delivers fatty acids to cells that need them.

LDL receptor A specialized type of protein, located on cell membranes, that binds to the apoproteins embedded in the surface of an LDL, allowing it to be taken up and broken down by the cell.

low-density lipoprotein (LDL) A lipoprotein that delivers cholesterol to cells.

plaque A fatty substance that builds up within the walls of blood vessels.

reverse cholesterol transport A process whereby high-density lipoproteins remove excess cholesterol from cells and carry it back to the liver.

very-low-density lipoprotein (VLDL) A lipoprotein, made by the liver, that delivers fatty acids to cells.

6-7 **How Are Dietary Lipids Related to Health?** · Excess energy intake is a major factor in the development of obesity and related complications such as cardiovascular disease. · Dietary fat intake may play a role in the development of cancer. However, this effect is likely primarily due to the influence of dietary fat on obesity.

aneurysm An outward bulging of a blood vessel.

atherosclerosis A narrowing and stiffening of the blood vessels due to plaque buildup, causing the restriction of blood flow.

blood clot A small, insoluble particle made of clotted blood and clotting factors.

cardiovascular disease A disease of the heart or vascular system.

heart disease A slowing or complete obstruction of blood flow to the heart.

stroke A slowing or complete obstruction of blood flow to the brain.

6-8 **What Are Some Overall Dietary Recommendations for Lipids?** · Consuming 20–35 percent of one's calories from lipid, limiting intake of SFA to less than 7 percent of calories, choosing foods low in *trans* fatty acids and emphasizing foods high in PUFAs and MUFAs are common lipid-specific dietary recommendations.

CHAPTER 7 LEARNING OBJECTIVES / KEY TERMS

7-1 **How Did Scientist First Discover Vitamins?** • The essential vitamins are classified as water soluble or fat soluble, depending on their chemical solubilities in water and lipids. • Water- and fat-soluble vitamins are often added to foods—this is called fortification. When certain nutrients are added in certain amounts, a fortified food can be labeled as being "enriched."

7-2 **Water-Soluble Vitamins** • Water-soluble vitamins dissolve in water (as opposed to in lipids). • You can prevent excessive nutrient loss in your foods by properly preparing and storing them.

enrichment The fortification of a select group of foods with FDA-specified levels of thiamin, niacin, riboflavin, folate, and iron.

7-3 **B Vitamins** • Thiamin is involved in energy metabolism and the synthesis of DNA and RNA. • Riboflavin is involved in energy metabolism, the synthesis of a variety of vitamins, nerve function, and protection of biological membranes. • Niacin is involved in energy and vitamin C metabolism and the synthesis of fatty acids and proteins. Niacin deficiency causes pellagra. • Pantothenic acid is a nitrogen-containing vitamin involved in energy metabolism, hemoglobin synthesis, and phospholipid synthesis. • Vitamin B_6 is involved in glycogenolysis, metabolism of proteins and amino acids, synthesis of neurotransmitters and hemoglobin, and regulation of steroid hormone function. The vitamin exists in several forms, the most common of which are pyridoxine, pyridoxal, pyridoxamine, and pyridoxal phosphate. • Biotin is involved in energy metabolism. • Folate refers to a group of related vitamins involved in single-carbon transfers, amino acid metabolism, and DNA synthesis. The active form of folate in the body is tetrahydrofolate acid (THF). • Vitamin B_{12} is involved in energy metabolism and methionine production.

ariboflavinosis A condition caused by riboflavin deficiency whereby cheilosis, stomatitis, glossitis, muscle weakness, and confusion occur.

beriberi A life-threatening condition caused by thiamin deficiency. There are four forms: wet, dry, infantile, and cerebral beriberi.

biotin (vitamin B_7) A water-soluble vitamin involved in energy metabolism that is obtained both from the diet and biotin-producing bacteria in the large intestine.

burning feet syndrome A condition believed to be caused by pantothenic acid deficiency whereby tingling in the feet and legs, fatigue, weakness, and nausea may occur.

cheilosis A condition characterized by sores on the outside and corners of the lips.

coenzyme A vitamin that facilitates the function of an associated enzyme.

dietary folate equivalent (DFE) A unit of measure for the approximate amount of folate in a food that is absorbed by the body.

folate (vitamin B_9) A group of related water-soluble vitamins involved in single-carbon transfers, amino acid metabolism, and DNA synthesis.

glossitis Inflammation of the tongue.

intrinsic factor A protein produced by the stomach that is needed for vitamin B_{12} absorption.

megaloblastic, macrocytic anemia A condition caused by folate deficiency whereby cells (including red blood cells) remain large and immature.

microcytic, hypochromic anemia A condition characterized by low concentrations of hemoglobin, which causes red blood cells to be small and light in color.

neural tube defect A malformation whereby neural tissue does not form properly during fetal development.

niacin (vitamin B_3) An essential water-soluble vitamin involved in energy and vitamin C metabolism and the synthesis of fatty acids and proteins.

niacin equivalent (NE) A unit of measure for the combined amounts of niacin and tryptophan in food.

pantothenic acid (vitamin B_5) A nitrogen-containing water-soluble vitamin involved in energy metabolism, hemoglobin synthesis, and phospholipid synthesis.

pellagra A condition caused by niacin deficiency whereby dermatitis, dementia, diarrhea, and/or death may occur.

pernicious anemia An autoimmune disease caused by vitamin B_{12} deficiency, whereby antibodies destroy the stomach cells that produce intrinsic factor.

reduction-oxidation (redox) reaction A transfer of oxygen or electrons from one molecule to another.

riboflavin (vitamin B_2) An essential water-soluble vitamin involved in energy metabolism, the synthesis of a variety of vitamins, nerve function, and protection of biological membranes.

spina bifida A failure of the neural tube to close properly during the first months of fetal life.

stomatitis Inflammation of the mouth, often caused by dietary deficiencies.

tetrahydrofolate (THF) The active form of folate.

thiamin (vitamin B_1) An essential water-soluble vitamin involved in energy metabolism and the synthesis of DNA and RNA.

vitamin B_6 A water-soluble vitamin involved in the metabolism of proteins and amino acids, the synthesis of neurotransmitters and hemoglobin, glycogenolysis, and regulation of steroid hormone function.

vitamin B_{12} (cobalamin) A water-soluble vitamin involved in energy metabolism and methionine production.

7-4 **Vitamin C** • Vitamin C (ascorbic acid) is a water-soluble vitamin that serves antioxidant functions within the body. • Vitamin C is found in many foods, including a variety of fruits and vegetables. Its deficiency causes scurvy, characterized by bleeding gums and poor wound healing.

antioxidant A compound that donates electrons or hydrogen ions to other substances, inhibiting oxidation.

free radicals A highly reactive molecule with one or more unpaired electrons; destructive to cell membranes, DNA, and proteins.

scurvy A condition caused by vitamin C deficiency, whereby bleeding gums, bruising, poor wound healing, and skin irritations occur.

vitamin C (ascorbic acid) A water-soluble vitamin that serves as an antioxidant within the body.

7-5 **Fat-Soluble Vitamins** • Fat-soluble vitamins are absorbed mostly in the small intestine, requiring the presence of dietary lipids as well as the action of bile. • Unlike water-soluble vitamins, your body can store most of the fat-soluble vitamins.

7-6 **Vitamin A and the Carotenoids** · Vitamin A refers to a group of compounds called retinoids, which includes retinol, retinoic acid, and retinal. · Vitamin A and their cousins, the carotenoids, are important in regulation of growth, reproduction, vision, immune function, gene expression, and bone formation. The carotenoids are also potent antioxidants and protect proteins, DNA, and cell membranes from free radical damage. · Good sources of preformed vitamin A are liver, fish, whole milk, and fortified foods. Vitamin A deficiency can result in blindness, infection, and death.

beta-carotene (ß-carotene) A provitamin A carotenoid made from two molecules of retinal.

carotenoid A dietary compound with a similar structure to those of the retinoids. Some, but not all, can be converted to vitamin A.

cell differentiation The process by which a nonspecialized, immature cell type becomes a specialized, mature cell type.

hypercarotenodermia A condition whereby carotenoids accumulate in the skin, causing it to become yellow-orange.

hyperkeratosis A condition caused by vitamin A deficiency, whereby skin and nail cells overproduce the protein keratin, causing them to become rough and scaly.

hypervitaminosis A condition caused by vitamin A toxicity, whereby blurred vision, liver abnormalities, and reduced bone strength occur.

macular degeneration A chronic disease that causes deterioration of the retina.

night blindness A condition characterized by an impaired ability to see in low-light environments.

nonprovitamin A carotenoid A carotenoid that cannot be converted to vitamin A.

provitamin A carotenoid A carotenoid that can be converted to vitamin A.

retinoid (or **preformed vitamin A**) A vitamin A compound.

retinol activity equivalent (RAE) A unit of measure for the combined amounts of preformed vitamin A and provitamin A carotenoids in a food.

vitamin A deficiency disorder (VADD) A spectrum of health-related consequences caused by vitamin A deficiency.

xerophthalmia A condition caused by vitamin A deficiency, whereby the cornea and other portions of the eye are damaged, leading to dry eyes, scarring, and even blindness.

7-7 **Vitamin D** · There are two forms of vitamin D: ergocalciferol (vitamin D_2) is found in plant sources and cholecalciferol (vitamin D_3) is found in animal foods and synthesized in the body. · In the presence of sunlight, the skin can produce vitamin D from a cholesterol metabolite. · Vitamin D deficiency can cause rickets in children and osteomalacia and osteoporosis in adults. · α-Tocopherol, the most biologically active form of vitamin E, is found in oils, nuts, seeds, and some fruits and vegetables. Vitamin E functions mainly as an antioxidant, protecting biological membranes from free radical damage. Vitamin E may also protect the eyes from cataract formation and influence cancer risk by decreasing DNA damage.

calcitriol The active form of vitamin D produced in the kidneys. Also known as 1,25-dihydroxyvitamin D_3 (1,25-[OH]$_2D_3$) .

cholecalciferol (vitamin D_3) A form of vitamin D that is found in animal-based foods, fortified foods, and supplements and also synthesized in the body.

ergocalciferol (vitamin D_2) A form of vitamin D found in plant-based foods, fortified foods, and supplements.

hypercalciuria A condition characterized by elevated urine calcium levels.

osteomalacia An adult condition, caused by vitamin D deficiency, whereby bones become soft and weak.

osteoporosis A serious disease, caused by vitamin D deficiency, whereby bones become weak and porous.

parathyroid hormone (PTH) A hormone released from the parathyroid gland that stimulates the conversion of 25-(OH) D_3 to calcitriol in the kidneys.

previtamin D_3 (or precalciferol) An intermediate product made in the skin during the conversion of a cholesterol derivative to cholecalciferol.

prohormone A compound converted to an active hormone in the body.

rickets A childhood condition caused by vitamin D deficiency, whereby slow growth and bone deformation occur.

vitamin D_3 (or cholecalciferol) The form of vitamin D that is made in the skin and diffuses into the blood and circulates to the liver.

7-8 **Vitamins E and K** · Vitamin K refers to three related compounds: phylloquinone, menaquinone, and menadione. Vitamin K is essential to blood clotting and proper bone mineralization. · Dark green vegetables are often good sources of vitamin K, and light and heat can destroy this vitamin in foods. In infants, severe vitamin K deficiency can cause a fatal condition called vitamin K deficiency bleeding.

α-tocopherol The most biologically active form of vitamin E.

coagulation The process by which blood clots are formed.

menadione A form of vitamin K produced commercially.

menaquinone A form of vitamin K produced by bacteria in the large intestine.

phylloquinone A form of vitamin K found naturally in plant-based foods.

vitamin K deficiency bleeding A disease caused by vitamin K deficiency, whereby uncontrollable internal bleeding occurs.

7-9 **Taking Dietary Supplements** · If you have difficulty consuming a good variety and balance of healthy foods in adequate amounts, taking a dietary supplement may help you obtain appropriate amounts of the essential nutrients. · It is important to make informed decisions about which supplements to take and which to avoid.

dietary supplement Product intended to supplement the diet that contains vitamins, minerals, amino acids, herbs or other plant-derived substances, or a multitude of other compounds.

8-1 **Water** • Water is the most abundant substance in the body. It is critical to many vital bodily functions, such as energy metabolism and temperature regulation. • Water crosses cell membranes via osmosis. The driving force behind osmosis is the difference in solute concentrations across these membranes. • Hydrolysis and condensation reactions break and create chemical bonds via the release and addition of water, respectively. • Dehydration influences cognitive function, ability to engage in aerobic activity, temperature regulation, and risk of urinary tract infections. The body responds to dehydration by releasing antidiuretic hormone (ADH), which decreases the amount of water excreted in the urine. • Individuals who maintain high levels of physical activity and/or live in hot environments are advised to consume additional water.

aldosterone A hormone produced by the adrenal glands in response to low blood pressure.

antidiuretic hormone (ADH, or **vasopressin)** A hormone released by the pituitary gland during periods of low blood volume that decreases the amount of water excreted in the urine.

condensation reaction A chemical reaction whereby a chemical bond joins two molecules together, releasing water in the process.

dehydration A condition characterized by an insufficient amount of water in the body.

evaporative cooling The process whereby sweat evaporates from the skin, taking heat with it.

extracellular fluid Fluid located outside of a cell.

hydrolysis reaction A chemical reaction whereby a chemical bond is broken by the addition of a water molecule.

intercellular fluid Extracellular fluid that fills spaces between or surrounding cells.

intracellular fluid Fluid located inside of a cell.

intravascular fluid Extracellular fluid located in blood and lymph.

solute A substance dissolved in a fluid.

solution A liquid in which a solute is uniformly distributed in a solvent.

solvent A substance that can dissolve other substances to form a solution.

8-2 **Minerals** • In nutrition, a mineral is an inorganic substance other than water. Because your body requires them in very small amounts, dietary minerals are considered micronutrients. • A major mineral is one required in an amount greater than 100 mg/day. A trace mineral is one required only in a minute amount (less than 100 mg/day). • Minerals are found in both plant and animal sources, but animal-based foods tend to have higher mineral contents than do plant-based foods. • In general, major minerals are absorbed in the small intestine and circulate throughout the body in the blood; excess amounts are excreted in the urine, and toxicities are rare.

cofactor A mineral that activates an enzyme by combining with it.

major mineral A mineral required in an amount greater than 100 mg/day.

metalloenzyme An enzyme that is activated when it combines with a mineral.

mineral An inorganic substance other than water that is required by the body in small amounts.

trace mineral A mineral required in an amount less than 100 mg/day.

8-3 **Calcium** • Calcium (Ca) is a major mineral located throughout the skeleton that is plays a critical structural role in bones and teeth. In the blood, calcium regulates vital body functions such as muscle contractions, blood clot formation, neural signaling, energy metabolism, and blood pressure regulation.

bone density The amount of bone tissue in a segment of bone.

bone mass The amount of minerals contained in bone.

bone remodeling (or **bone turnover**) The continuous process by which older and damaged bone is replaced by new bone.

bone resorption The process whereby osteoclasts break down bone, releasing minerals into the blood.

calcitonin A hormone released by the thyroid gland that decreases calcium loss from bone, decreases calcium absorption

in the small intestine, and increases calcium loss in the urine.

calcium (Ca) A major mineral that has an important structural role in bone and teeth; as well as regulating vital body functions such as blood clot formation, muscle contraction, neural signaling, energy metabolism, and blood pressure regulation, muscle and nerve function, and energy metabolism.

hydroxyapatite A large crystal-like molecule that combines with other minerals to form the structural matrix of bones and teeth.

kyphosis (or **dowager's hump**) A curvature of the upper spine.

osteoblast A bone cell that promotes bone formation.

osteoclast A bone cell that promotes the breakdown of older bone.

osteopenia A condition characterized by moderate bone loss in adults.

osteoporosis A condition characterized by brittle and fragile bones caused by a progressive loss of calcium and other bone minerals.

peak bone mass The point whereby bones have reached their maximum strength and density.

8-4 **Electrolytes: Sodium, Chloride, and Potassium** • An electrolyte is a molecule that produces charged ions when dissolved in water. The three most abundant ions in the body are sodium, chloride, and potassium. • Electrolytes and their resultant ions are critical to nerve and muscle function, and help regulate blood volume and blood pressure. • Although a high intake of salt has long been associated with increased risk for hypertension, this is true only in a segment of the population.

chloride (Cl⁻) A negatively charged ion (anion) found primarily in fluids surrounding cells in the body.

diuretic A substance or drug that helps the body eliminate water.

hypokalemia A condition characterized by low blood potassium concentration.

hyponatremia A condition characterized by low blood sodium concentration.

potassium (K⁺) The major positive ion (cation) found inside cells.

renin An enzyme, secreted by the kidneys, that converts angiotensinogen to angiotensin.

sodium (Na⁺) A positive ion (cation) found primarily in fluids surrounding cells in the body.

8-5 **Electrolytes: Phosphorus and Magnesium** · Phosphorus (P) is a major mineral that is essential to cell membranes, bone and tooth structure, DNA, RNA, ATP, lipid transport, and a variety of other processes in the body. · Magnesium (Mg) is a major mineral that is important to many physiological processes, such as energy metabolism and enzyme function.

magnesium (Mg) A major mineral that is important to many physiological processes, such as energy metabolism and enzyme function.

phosphorus (P) A major mineral that is essential to cell membranes, bone and tooth structure, DNA, RNA, ATP, lipid transport, and a variety of processes in the body.

8-6 **Sulfur** · The body needs sulfur to synthesize compounds required for healthy connective tissue and nerve function. Sulfur is also an important component of the B vitamins thiamin and biotin, and is therefore essential to energy metabolism.

sulfur (S) A major mineral that is a component of certain amino acids (e.g., cysteine and methionine).

8-7 **Trace Minerals** · Although they are only needed in minute amounts, the trace minerals are vital to your health. · The bioavailability of trace minerals is influenced by factors such as nutritional status, genetics, the aging process, and interactions with other food components. Trace minerals are absorbed primarily in the small intestine. The processes by which blood mineral levels are regulated vary from mineral to mineral. · Iron (Fe) is a trace mineral that is necessary to oxygen and carbon dioxide transport, energy metabolism, the stabilization of free radicals, and the synthesis of DNA. Two general forms of iron are found in foods: heme iron and nonheme iron. Heme iron is readily absorbed, whereas nonheme iron is not. · Copper (Cu) is an essential trace mineral that acts as a cofactor for nine enzymes involved in reduction-oxidation (redox) reactions. · Iodine (I) is an essential trace mineral that serves as a component of the thyroid hormones. · Selenium (Se) is an essential trace mineral that is critical to redox reactions, thyroid function, and the activation of vitamin C. Selenium is a component of selenoproteins. · Chromium (Cr) is an essential trace mineral that may be critical to proper insulin function. · Manganese (Mn) is an essential trace mineral that is a cofactor for metalloenzymes needed for bone formation, glucose production, and energy metabolism. · Molybdenum (Mo) is an essential trace mineral that is a cofactor for several important metalloenzymes needed for amino acid and purine metabolism. · Zinc (Zn) is an essential trace mineral involved in gene expression, immune function, and cell growth.

acrodermatitis enteropathica A genetic abnormality that causes secondary zinc deficiency.

ceruloplasmin A transport protein that binds to copper and transports it in the blood.

chelator A compound that binds to nonheme iron in the intestine, making it unavailable for absorption.

chromium (Cr) A trace mineral that may be critical to proper insulin function.

chromium picolinate A form of chromium taken as an ergogenic aid by some athletes.

copper (Cu) An essential trace mineral that acts as a cofactor for nine enzymes involved in redox reactions.

cretinism A severe form of iodine deficiency that affects babies born to iodine-deficient mothers.

dental fluorosis Pitting and mottling of teeth caused by excessive fluoride intake.

ferritin A protein complex that stores iron and releases it only when it is needed.

fluoride (F⁻) A trace mineral that strengthens bones and teeth.

goiter A sign of iodine deficiency characterized by an enlarged thyroid gland.

goitrogen substances that interfere with iodine uptake by the thyroid gland, which subsequently impairs the production of thyroid hormones.

hematocrit The percentage of blood volume composed of red blood cells.

heme iron Iron that is part of hemoglobin and myoglobin and is found only in meat.

hemoglobin A complex protein that transports oxygen from the lungs to cells and carbon dioxide from the cells to the lungs.

hepcidin A protein released by the liver that regulates the amount of iron circulating in the blood.

hereditary hemochromatosis A genetic abnormality whereby too much iron is absorbed, causing it to accumulate in the body.

iodine (I) An essential trace mineral that serves as a component of the thyroid hormones.

iodine deficiency disorders (IDDs) A broad spectrum of conditions caused by iodine deficiency.

iron (Fe) A trace mineral needed for oxygen and carbon dioxide transport, energy metabolism, stabilization of free radicals, and synthesis of DNA.

Keshan disease A disease caused by severe selenium deficiency that mostly affects children and causes serious heart problems.

manganese (Mn) An essential trace mineral that is a cofactor for metalloenzymes needed for bone formation, glucose synthesis, and energy metabolism.

meat factor A compound found in meat that increases the bioavailability of nonheme iron.

metallothionine An intestinal protein that regulates the amount of zinc released into the blood.

molybdenum (Mo) An essential trace mineral that is a cofactor for several important metalloenzymes needed for amino acid and purine metabolism.

myoglobin An oxygen-storage molecule located within muscles.

nonheme iron Iron that is not part of hemoglobin and myoglobin and is found primarily in plant-based foods.

selenium (Se) An essential trace mineral that is critical to redox reactions, thyroid function, and the activation of vitamin C.

selenoprotein A protein composed of selenium-containing amino acids.

skeletal fluorosis Weakening of the skeleton caused by excessive fluoride intake.

superoxide dismutase A copper-containing enzyme that helps stabilize highly reactive free radical molecules.

thyroid-stimulating hormone (TSH) A hormone, produced by the pituitary gland, that regulates iodine uptake by the thyroid gland.

transferrin A protein in the blood that binds to dietary iron once it is released from an intestinal cell.

zinc (Zn) An essential trace mineral involved in gene expression, immune function, and cell growth.

8-8 **Other Important Trace Minerals** · Fluoride (F⁻) is a trace mineral that strengthens bones and teeth. · In addition to the trace minerals known to be essential to human life, many others may influence health. These include nickel, aluminum, silicon, vanadium, arsenic, and boron. Although scientists do not know whether these minerals are important to human health, there is evidence that they influence the health of other animals.

CHAPTER 9 LEARNING OBJECTIVES / KEY TERMS

9-1 What Is Energy Balance? • The term *energy balance* describes the relationship between energy intake and energy expenditure. • A person is in positive energy balance when energy intake exceeds energy expenditure, negative energy balance when energy intake is less than energy expenditure, and neutral energy balance when energy intake equals energy expenditure. • The number and size of adipocytes determine the body's fat mass. Adipose tissue serves as the body's primary energy reserve. • When body fat increases, adipocytes can increase in size (hypertrophic growth), number (hyperplasic growth), or both. When body fat decreases, adipocytes decrease in size but not in number. • The distribution of visceral adipose tissue (VAT) and subcutaneous adipose tissue (SCAT) holds important information about a person's health. Central obesity, a predominant accumulation of VAT, puts a person at greater risk for weight-related health problems. Brown adipose tissue plays an important role in regulating body temperatures, whereas the prime function of white adipose tissue is energy storage.

adipokines Signaling proteins secreted by adipocytes that cause a cellular response in another organ.

brown adipose tissue (BAT) A type of connective tissue especially abundant in newborns and in hibernating mammals that generates body heat.

central obesity (or **central adiposity**) The predominant accumulation of adipose tissue that surrounds internal organs in the abdominal cavity.

metabolic syndrome A cluster of conditions that occur together and increase a person's risk of heart disease, stroke, and type 2 diabetes.

negative energy balance A condition whereby energy intake is less than energy expenditure.

neutral energy balance A condition whereby energy intake equals energy expenditure.

positive energy balance A condition whereby energy intake is greater than energy expenditure.

waist circumference A measurement used as an indicator of central adiposity.

white adipose tissue (WAT) A type of connective tissue that not only produces hormones, but also stores energy and regulates body temperature.

9-2 What Factors Influence Energy Intake? • *Hunger* is the basic physiological need for food, whereas satiety is the sensation of having eaten enough. • A complex region of the brain called the hypothalamus regulates energy intake and thus body weight by balancing feelings of hunger and satiety. • Signals from the gastrointestinal (GI) tract such as gastric stretching and GI hormones play a role in the regulation of short-term food intake. Circulating concentrations of glucose, fatty acids, and amino acids also influence hunger and satiety. • The majority of known GI hormones inhibit food intake, with the exception of ghrelin, which stimulates hunger. • *Appetite* reflects psychological factors that influence hunger and satiety. • A *food aversion* is a strong psychological dislike of a particular food and a food craving (in most cases) is a strong psychological desire for a particular food.

appetite A psychological longing or desire for food.

bariatric surgery Surgical procedure performed to promote weight loss.

food aversion A strong psychological dislike of a particular food.

food craving A strong desire for a particular food.

gastric banding A type of bariatric surgery whereby an adjustable, fluid-filled band is wrapped around the upper portion of the stomach, dividing it into a small upper pouch and a larger lower pouch.

gastric bypass A surgical procedure that involves reducing the size of the stomach and bypassing a segment of the small intestine so that overall nutrient absorption is decreased.

ghrelin A hormone secreted by cells in the stomach lining that stimulates hunger.

hunger A basic physiological drive to consume food.

hypothalamus A region of the brain that regulates energy intake by balancing hunger and satiety.

neurotransmitters Hormone-like substances released by nerve cells; transmit electrical impulses from one nerve cell to the next.

satiety A physiological response whereby a person feels he or she has consumed enough food.

sleeve gastrectomy A surgical weight-loss procedure that involves the removal of a large part of the stomach, creating a smaller sleeve- or tubular-shaped stomach pouch.

9-3 What Determines Energy Expenditure? • The body expends energy to maintain basal metabolism, engage in physical activity, and process food. These three components make up most of a person's total energy expenditure (TEE). • Smaller components of the TEE include adaptive thermogenesis and nonexercise activity thermogenesis. • The major factors influencing basal metabolic rate include body shape, body composition, age, sex, nutritional status, and genetics.

adaptive thermogenesis A temporary expenditure of energy that enables the body to adapt to temperature changes in the environment and physiological conditions.

basal metabolic rate (BMR) The amount of energy expended per hour (kcal/hour) so that the body can carry out basic, involuntary physiological functions.

basal metabolism An expenditure of energy to sustain basic, involuntary life functions such as respiration, beating of the heart, nerve function, and muscle tone.

indirect calorimetry A technique that provides an estimate of energy expenditure based on oxygen consumption and carbon dioxide production.

nonexercise activity thermogenesis An expenditure of energy associated with spontaneous movement such as fidgeting and maintenance of posture.

thermic effect of food (TEF, or **diet-induced thermogenesis)** The energy needed to digest, absorb, and metabolize nutrients following a meal.

total energy expenditure (TEE) The collective sum of energy used by the body.

9-4 How Are Body Weight and Composition Assessed?
• Overweight is excess weight for a given height, whereas obesity is an abundance of body fat in relation to lean tissue. • Indices such as height–weight tables and body mass index (BMI) are used to assess body weight. • BMI, based on the ratio of weight to height squared, is also considered a good indicator of body fat. • BMI is a useful method to determine if a person is at a healthy weight-for-height, but health professionals sometimes want to know a person's individual body composition. • Measures of body composition include hydrostatic weighing, dual-energy x-ray absorptiometry (DEXA), bioelectrical impedance, and the skinfold thickness method.

bioelectrical impedance A method of estimating body composition in which a weak electric current is passed through the body.

body mass index (BMI) An indirect measure of body fat calculated by dividing a person's body weight by their height squared.

dual-energy x-ray absorptiometry (DEXA) A method of estimating body composition in which low-dose x-rays are used to visualize fat and fat-free compartments of the body.

hydrostatic weighing A method of estimating body composition in which a person's body weight is measured in and out of water.

hyperplastic growth A process whereby new adipocytes are formed.

hypertrophic growth A process whereby adipocytes fill with lipid causing them to enlarge.

obese Excess body fat.

overweight Excess weight for a given height.

skinfold caliper An instrument used to measure the thickness of skin and subcutaneous fat.

skinfold thickness method A method of estimating body composition whereby a skinfold caliper is used to measure the thickness of skin and subcutaneous fat at various locations on the body.

9-5 Genetics versus Environment: What Causes Obesity?
• Many factors resulting in increased energy intake and/or decreased energy expenditure have contributed to the increasing prevalence of obesity in the United States. These include lifestyle, socioeconomic, and cultural factors. • One of the most important factors that influence body weight is the amount and energy density of the food we consume. • Although the hormone leptin has not proved effective for treating human obesity in most cases, its discovery has led to important insights into body weight regulation.

db gene The gene that codes for the leptin receptor.

db/db (diabetic) mouse A mouse with a gene mutation for the leptin receptor, which leads to obesity.

exercise Planned, structured, and repetitive bodily movement done to improve or maintain physical fitness.

hyperphagic Exhibiting excessive hunger and food consumption.

leptin A potent hormone (adipokine) produced primarily by adipose tissue that signals satiety.

ob gene The gene that codes for the hormone leptin.

ob/ob (obese) mouse A mouse with a gene mutation that impairs leptin production, which leads to obesity.

physical activity Any bodily movement that results in energy expenditure.

9-6 How Are Energy Balance and Body Weight Regulated?
• The set point theory of body weight regulation suggests that body weight is generally stable because of the body's ability to adjust energy intake and expenditure. • Leptin plays a role in regulating body weight by communicating the body's energy reserve to the brain. • When body fat increases, so does the concentration of leptin circulating in the blood, typically resulting in overall decreased energy intake and increased energy expenditure. When body fat decreases, leptin production also decreases, resulting in overall increased energy intake and decreased energy expenditure. • Most obese people produce adequate amounts of leptin, but defects in leptin responsiveness may lead to obesity.

set point theory A scientific concept whereby hormones circulating in the blood are theorized to regulate body weight by communicating the amount of adipose tissue in the body to the brain, which adjusts energy intake and expenditure to maintain neutral energy balance.

9-7 What Is the Best Approach to Weight Loss? • Maintaining weight loss requires lasting lifestyle changes such as eating moderate amounts of nutrient-dense foods and engaging in regular exercise. • The majority of individuals who lose weight and maintain weight loss do so through a combination of adhering to a healthy diet and engaging in regular physical activity. A healthy weight-loss goal might be to reduce one's current weight by 5–10 percent. • Proponents of low-fat diets believe that reducing one's fat intake leads to the consumption of fewer calories and therefore to greater weight loss, whereas advocates of low-carbohydrate diets believe that increased carbohydrate intake can cause insulin levels to rise, leading to weight gain. • Limiting carbohydrates can help prevent rises in insulin. Because increases in blood insulin levels may contribute to weight gain, limiting carbohydrates may help a person lose weight (or at least not gain weight).

TABLE 9.4 BEHAVIORS ASSOCIATED WITH HEALTHY WEIGHT MANAGEMENT

- Focusing attention on an overall healthy eating pattern that emphasizes nutrient-dense foods.
- Monitoring food intake, body weight, and physical activity to help make oneself more aware of what and how much is eaten and drunk.
- Selecting small-size and lower-calorie options when eating out.
- Preparing and serving smaller food portions, especially foods and beverages that are high in calories.
- Eating nutrient-dense foods to improve nutrient intake and healthy body weight.
- Reducing screen time to 1–2 hours each day; getting adequate, restful sleep (7–8 hours); and increasing physical activity to help with weight loss and weight maintenance.

10-1 **What Physiological Changes Take Place during the Human Life Cycle?** • Periods of growth, development, maturation, and senescence coincide with specific life stages: infancy, childhood, adolescence, adulthood, and for women, the special life stages of pregnancy and lactation. • Changes in body size and composition influence nutrient requirements. • When physical maturity is reached, cell turnover achieves equilibrium. As a person ages, the rate of new cell formation slows, resulting in senescence.

cell turnover The cyclical process by which cells form and break down.

development A change in the attainment or complexity of a skill or function.

growth A physical change that results from an increase in either cell size or number.

senescence The gradual physiological deterioration that occurs with age.

10-2 **What Are the Major Stages of Prenatal Development?** • The two stages of prenatal development are the embryonic period and the fetal period. • The placenta transfers nutrients, oxygen, and other substances from the mother's blood to the fetus; it also transfers waste products from the fetus to the mother. • Babies born with gestational ages between 37 and 42 weeks are considered full-term infants, whereas those born with gestational ages less than 37 weeks are considered preterm (or premature) and those born with gestational ages greater than 42 weeks are considered post-term. • Many preterm infants are born with low birth weight (LBW), which can also be caused by intrauterine growth restriction (IUGR). Babies who experience IUGR are often small for gestational age (SGA). • Gestational age is determined by counting the number of weeks between the first day of a woman's last normal menstrual period and birth.

appropriate for gestational age (AGA) A baby that has a birth weight between the tenth and ninetieth percentiles for gestational age.

blastocyst A dense sphere of cells that implants itself into the lining of the uterus.

critical period A period in prenatal development during which adverse effects on growth and development are irreversible.

developmental origins of health and disease hypothesis A hypothesis that states that less than optimal conditions in the uterus or during early infancy may cause permanent changes in the structure and function of organs and tissues, predisposing individuals to certain chronic diseases later in life.

embryo A developing human as it exists from the start of the third week to the end of the eighth week after fertilization.

embryonic period The period of prenatal development that spans from conception through the eighth week of gestation.

embryonic phase The stage of the embryonic period (2–8 weeks) during which organs and organ systems form.

fetal alcohol spectrum disorder A range of physical and behavioral problems that can arise in a child as a consequence of alcohol consumption by a woman during pregnancy.

fetal period The stage of prenatal development that begins at the ninth week of pregnancy and ends at birth.

fetus A developing human as it exists from the start of the ninth week of pregnancy until birth.

full-term A baby born with gestational age between 37 and 42 weeks.

gestation length The period of time between conception and birth.

gestational age The number of weeks from the first day of a woman's last normal menstrual cycle.

intrauterine growth restriction (IUGR) Slow growth while in the uterus.

large for gestational age (LGA) A baby that has a birth weight above the ninetieth percentile for gestational age.

low birth weight (LBW) A baby that weighs less than 5 lb, 8 oz (2,500 g) at birth.

placenta An organ, made of embryonic and maternal tissues, that supplies nutrients and oxygen to the developing child.

placental insufficiency A complication of pregnancy that can arise when the placenta is unable to deliver adequate nutrients and oxygen to the developing baby.

postterm A baby born with a gestational age greater than 42 weeks.

pre-embryonic phase The early phase of the embryonic period (0–2 weeks) that begins with fertilization and continues through the formation of the blastocyst.

preterm (or **premature**) A baby born with a gestational age less than 37 weeks.

small for gestational age (SGA) A baby that has a birth weight below the tenth percentile for gestational age.

teratogens An agent or condition that can disrupt prenatal growth and development.

zygote An ovum that has been fertilized by a sperm.

10-3 **What Are the Nutrition Recommendations for a Healthy Pregnancy?** • A pregnant woman who gains an appropriate amount of weight, eats a healthy diet, and refrains from smoking can increase the likelihood of having a full-term baby born with a healthy birth weight. • During pregnancy, additional energy is needed to support the growth of the fetus, placenta, and maternal tissues. Carbohydrates should remain the primary energy source during this time. • Additional protein is needed for the formation of fetal and maternal tissues. Dietary fat should provide 20–35 percent of total calories during pregnancy. • Extra calcium is necessary for a fetus to grow and develop, although changes in maternal physiology can accommodate these needs without increasing dietary intake. • Iron is essential for the formation of hemoglobin and the growth and development of the fetus and the placenta. • Folate is critical to cell division and development of the nervous system. A woman with poor folate status is at increased risk of having a baby with a neural tube defect. • Smoking during pregnancy increases the risk of bearing a preterm or LBW baby. • Two complications that can arise during the third trimester of pregnancy are gestational diabetes and pre-eclampsia. • Gestational diabetes is brought on by hormonal changes during pregnancy that cause maternal tissues to become less responsive to insulin. • Pre-eclampsia is characterized by high blood pressure and the presence of protein in the urine. Sometimes a serious complication called eclampsia can arise, which can compromise the health of the mother and the fetus.

eclampsia A serious complication of pregnancy that is typically preceded by a condition called pre-eclampsia, and is characterized by the onset of seizures.

gestational diabetes A form of diabetes that develops when pregnancy-related hormonal changes cause cells to become less responsive to insulin.

pica The urge to consume nonfood items.

pre-eclampsia Pregnancy-related condition characterized by high blood pressure, a sudden increase in weight, swelling due to fluid retention, and protein in the urine.

10-4 Why Is Breastfeeding Recommended during Infancy?
· Milk production and the release of milk from the alveoli into the milk ducts are regulated in part by the hormones prolactin and oxytocin. · The process by which milk is released from the alveoli into the milk ducts is called milk let-down. · The first secretion from the mammary glands is called colostrum, which is rich in both nutrients and substances that help prevent disease in the newborn infant. · Breastfeeding is the preferred method of nourishing infants because human milk provides many nutritional and immunological benefits.

alveoli Structures made up of milk-producing cells.

colostrum The first secretion from the mammary glands released after giving birth that provides nutrients and substances that help prevent disease.

lactation The production and release of milk.

lactogenesis Structural and functional changes in the mammary glands that begin soon after conception and continue until after the baby is born, enabling mammary glands to produce sufficient amounts of milk.

let-down The active process whereby milk is forced out of the alveoli and into the milk ducts.

oxytocin A hormone that causes the muscles around the alveoli to contract.

prolactin A hormone that stimulates alveolar cells to produce milk.

10-5 What Are the Nutritional Needs of Infants?
· During the first year of life, a healthy baby's weight almost triples and his or her length increases by up to 50 percent. · In communities without fluoridated water, fluoride supplements may be necessary after six months of age. Iron supplements are recommended for infants who exclusively breastfeed during the second six months of life. Vitamin D supplements are recommended for breastfed infants beginning in the first few days of life. · The American Academy of Pediatrics recommends exclusively breastfeeding throughout the first four to six months of life. Human milk and/or infant formula should be the primary source of nutrients and energy throughout the first year of life. · Nonmilk complementary foods can be added sometime between four and six months of age, depending on an infant's readiness. · Older infants should be given a wider variety of foods, although they should be chosen carefully to pose minimal risk for choking.

baby bottle tooth decay A condition whereby dental caries occur in an infant who habitually sleeps with bottles filled with milk, formula, juice, or any other carbohydrate-containing beverage.

10-6 What Are the Nutritional Needs of Toddlers and Young Children?
· Growth is monitored using growth charts, and BMI-for-age is used to assess adiposity in children over 2 years of age. · Following six general guidelines regarding common childhood feeding problems can encourage healthy eating. · Nutrients needed for bone health (such as calcium) and to support growth (such as iron) are particularly important during childhood.

10-7 How Do Nutritional Requirements Change during Adolescence?
· Hormonal changes begin the physical transformation from childhood into adolescence, causing shifts in height, weight, and body composition. · Because the timing of these changes varies, adolescents of the same age can differ in terms of physical maturation and nutritional requirements. · Linear growth (height) is largely completed at the end of the adolescent growth spurt, although bone mass continues to increase into early adulthood. · During adolescence, males experience an increase in percentage of lean body mass, whereas the opposite occurs in females.

menarche The first occurrence of menstruation.

puberty Maturation of the reproductive system.

10-8 How Do Age-Related Changes in Adults Influence Nutrient and Energy Requirements?
· As individuals grow older, they experience a relative loss of lean mass and increase in fat mass. For this reason, energy requirements decrease. · Many factors can contribute to inadequate food intake in older adults, including poor oral health, altered taste, and decreased ability to smell. · Age-related changes in the gastrointestinal tract can also affect nutritional status. · A decrease in production of gastric secretions can impair absorption of iron, calcium, biotin, folate, cobalamin, and zinc. · Menopause marks the cessation of menstrual cycles. · Services that provide food to older adults include congregate meal programs and meal delivery programs such as Meals on Wheels.

lifespan The maximum number of years of life attainable by a member of a particular species.

menopause A life stage characterized by very little estrogen production, which causes the menstrual cycle to stop.

perimenopausal A life stage characterized by a natural decline in estrogen production.

11-1

What Are the Health Benefits of Physical Activity?

• Physical activity includes day-to-day activities such as gardening, household chores, walking, and leisure-time activities. Exercise is a subcategory of physical activity defined as planned, structured, and repetitive bodily movement done to improve or maintain physical fitness. • While the importance of physical activity may be common knowledge, nearly one-half of all Americans lead sedentary lifestyles. • Participation in at least 2½ hours of moderate-intensity aerobic activity and two sessions of muscle-strengthening activity per week is recommended for optimal health. • Physical fitness is measured in terms of health parameters which as cardiorespiratory fitness, muscular strength, endurance, body composition, flexibility, balance, agility, reaction time, and power. A physical fit person is able to carry out daily tasks with vigor and alertness, without undue fatigue. • A well-balanced physical fitness program includes movement that improves cardiorespiratory fitness, muscular strength, muscular endurance, and muscular flexibility.

cardiorespiratory fitness A measure of the circulatory and respiratory systems' ability to supply oxygen and nutrients to working muscles during sustained physical activity.

muscular endurance The ability of a muscle or a muscle group to repeatedly exert resistance without becoming fatigued.

muscular flexibility A measure of the range of motion around a joint.

muscular strength The maximal force exerted by muscles during physical activity.

physical fitness Measurable health parameters (cardiorespiratory fitness, muscular strength, endurance, body composition, flexibility, balance, agility, reaction time, and power) that enable a person to carry out daily tasks with vigor and alertness, without undue fatigue.

proprioceptive neuromuscular facilitation stretching Exercise that improves equilibrium, agility, and balance by improving spatial and positional body awareness.

resistance exercise Strength-building activities that challenge specific groups of muscles by making them work against an opposing force.

sarcopenia Age-related loss of muscle tissue.

11-2

How Does Energy Metabolism Change during Physical Activity?

• At rest, the body expends approximately 1.0–1.5 kcal/minute. During physical exertion, however, energy expenditure can increase substantially. • ATP can be generated in both the presence and relative absence of oxygen. These conditions are called aerobic and anaerobic, respectively. • Aerobic pathways generate large amounts of ATP over an extended period of time, whereas anaerobic pathways generate smaller amounts of ATP for shorter periods of time. • Until the cardiovascular system is able to increase oxygen delivery to active muscles, the high-energy compound creatine phosphate can be broken down and combined with ADP to produce ATP. This is the simplest and most rapid means by which active muscles generate ATP, although overall yield is low. • If physical activity continues beyond the capacity of the glycolytic and phosphagen systems, muscles must utilize the aerobic (oxidative) system for ATP production. Although aerobic pathways are slower than anaerobic pathways in regard to the rate of ATP formation, the energy yield is rich.

creatine phosphate A high-energy compound that is broken down to produce ATP.

exercise duration The number of minutes or hours an activity is performed in one session.

exercise intensity The magnitude of the effort required to perform an activity or exercise.

glycolytic system An energy system characterized by the anaerobic metabolism of glucose (glycolysis).

oxidative system An aerobic energy system characterized by steady ATP production in an oxygen-rich environment.

phosphagen system The simplest and most rapid energy system; characterized by the use of creatine phosphate to produce ATP.

11-3

What Physiologic Adaptations Occur in Response to Athletic Training?

• As your body acclimates to the rigors of frequent exercise, physiological changes called adaptation responses enable it to perform and recover more efficiently and effectively. • Strength training (such as weight lifting) challenges muscles and helps increase muscle strength, power, endurance, and mass. Training-induced adaptations cause muscles to undergo hypertrophy, meaning that they become larger. Muscles that are not frequently challenged undergo atrophy, meaning that they become smaller. • Endurance training (such as running and swimming) is beneficial to pulmonary and cardiovascular function, and can help improve aerobic capacity (the ability to produce ATP in an oxygen-rich environment). These adaptive responses increase maximal oxygen consumption (or VO_2 max).

adaptation responses A series of physiological changes that enables the body to perform and recover more efficiently and effectively in response to training.

anaerobic capacity The maximum amount of work that can be performed under anaerobic conditions.

endurance training Exercise that entails steady, low- to moderate-intensity exercise that persists for an extended duration.

ergogenic aids A substance taken to enhance athletic performance.

heart rate The number of heartbeats (contractions) per unit of time.

interval training Exercise that entails alternating short, fast bursts of intense exercise with slower, less demanding activity.

maximal oxygen consumption (or VO_2 **max**) A measure of the cardiovascular system's capacity to deliver oxygen to muscles.

muscle atrophy A process whereby muscles decrease in size.

muscle hypertrophy An adaptation response to exercise characterized by the enlargement of muscles.

strength training Exercise that increases skeletal muscle strength, power, endurance, and mass.

stroke volume The amount of blood pumped out of the heart (left ventricle) per heartbeat.

ventilation rate Breaths expelled per minute.

How Does Physical Activity Influence Dietary Requirements? · Athletic individuals must consume enough energy to support normal daily activities as well as those performed as exercise. · Carbohydrates maintain glycogen stores, whereas protein builds, maintains, and repairs muscles. · High-quality protein sources are necessary to maintain, build, and repair tissue. · Dietary fat provides essential fatty acids and is an important source of energy. · Iron is essential to oxygen and carbon dioxide transport, and calcium is essential to the building, maintenance, and repair of bone tissue. · Antioxidant nutrients help to counteract the oxidative damage of free radicals. · Excess sweating can disrupt fluid and electrolyte balance, which impairs body temperature and fluid regulation. · Athletes can prevent dehydration by consuming adequate amounts of fluids before, during, and after exercise. · To stay fully hydrated, athletes require greater amounts of water than generally recommended for adult men and women (11 and 16 cups, respectively). Enough water is needed to replace water loss associated with sweating. · When an excess of water is consumed, sodium concentrations in the blood can decrease. Therefore, in some cases, electrolyte repletion is equally important and fluid intake. · Without adequate fluid intake, athletes are at increased risk of dehydration. This can impair sweating, which in turn can result in heat exhaustion, and in extreme cases heat stroke.

carbohydrate loading A technique used by some athletes to increase the amount of glycogen stored in the body (liver and skeletal muscle) so that they can sustain activity for long periods of time.

heat exhaustion A rise in body temperature that occurs when the body has difficulty dissipating heat.

heat stroke A serious condition that can develop if body temperature continues to rise beyond heat exhaustion.

hemodilution A disproportionate increase in plasma volume as compared to the synthesis of new red blood cells.

hyponatremia Diminished blood sodium concentration.

oxidative stress An imbalance between the body's natural oxidative stress defense system (antioxidants) and free radicals, which results in an impaired ability to prevent cellular oxidative damage.

sports anemia A temporary type of anemia that occurs at the onset of a training program due to hemodilution.

TABLE 11.1 | 2008 PHYSICAL ACTIVITY GUIDELINES FOR AMERICANS

Population Group	Physical Activity Recommendations
Children and adolescents (ages 6–17)	· Children and adolescents should engage, on a daily basis, in at least 1 hour of physical activity primarily consisting of either moderate- or vigorous-intensity aerobic physical activity. · Children and adolescents should participate in both muscle- and bone-strengthening activities at least 3 days per week.
Adults (ages 18–64)	· Adults should strive for 2.5 hours per week of moderate-intensity, or 1.25 hours per week of vigorous-intensity aerobic physical activity. Or, they should participate in an equivalent combination of moderate- and vigorous-intensity aerobic physical activity. · Aerobic activity should be performed in episodes of at least 10 minutes, spread throughout the week. · Additional health benefits can be obtained by increasing exercise to 5 hours per week of moderate-intensity aerobic physical activity, 2.5 hours per week of vigorous-intensity physical activity, or an equivalent combination of both. · In addition to aerobic activities, muscle-strengthening activities that involve all major muscle groups should be performed 2 or more days per week. · If necessary, adults should gradually increase the time spent doing aerobic physical activity and decrease caloric intake in order to achieve neutral energy balance and a healthy weight.
Older adults (ages 65+)	· Older adults should follow the adult guidelines if possible, but individuals with chronic conditions should be as physically active as their abilities allow. · If they are at risk of falling, older adults should also do exercises that maintain or improve balance.
Pregnant and postpartum women	· Pregnant and postpartum women should strive for 2.5 hours of moderate-intensity aerobic activity throughout the week. · Women who regularly engage in vigorous-intensity aerobic activity or high amounts of activity can continue their activity, provided that their condition remains unchanged and they have consulted with their health care providers. · After the first trimester, pregnant women should avoid exercises that involve lying on the back. They should also avoid activities that increase the risk of falling or abdominal trauma, such as contact and/or collision sports.
Children and adolescents with disabilities	· When possible, children and adolescents with disabilities should meet the guidelines for all children—or they should engage in as much activity as conditions allow.
Adults with disabilities	· Adults with disabilities should follow the adult guidelines if possible and be as physically active as their abilities allow.

Source: Adapted from U.S. Department of Health and Human Services. 2008 physical activity guidelines for Americans. Washington, DC: U.S. Department of Health and Human Services, 2008.

12-1 **What Is the Difference between Disordered Eating and Eating Disorders** • Disordered eating is characterized by unhealthy eating patterns such as irregular eating, consistent undereating, and/or consistent overeating. • Although disordered eating patterns can be disturbing to others, they typically do not persist long enough to cause serious physical harm. For some people, however, disordered eating develops into a full-blown eating disorder. • The three main types of eating disorders are anorexia nervosa (AN), bulimia nervosa (BN), and binge eating disorder (BED). • A category called other specified feeding or eating disorders (OSFED) includes some, but not all, of the diagnostic criteria for AN, BN, and/or BED. • AN is characterized by a fear of weight gain, distorted body image, and food restriction. People with AN typically limit their food intakes as well as the varieties of foods consumed. Some people with AN also exercise obsessively. • BN involves cycles of bingeing and purging. Individuals with BN often experience regret and loss of control after bingeing. These feelings can lead to depression, which increases the likelihood of future binge–purge cycles. • BED is characterized by repeated episodes of binge eating that are not followed by purging behaviors. A person with BED typically consumes large amounts of food within a relatively short period of time.

anorexia nervosa (AN) An eating disorder characterized by an irrational fear of gaining weight or becoming obese.

anorexia nervosa, binge-eating/purging type A subcategory of anorexia nervosa characterized by both food restriction and periods of bingeing and purging.

anorexia nervosa, restricting type A subcategory of anorexia nervosa characterized by food restriction and/or excessive exercise.

binge-eating disorder Repeated episodes of binge eating; characterized by out of control consumption of unusually large amounts of food in a short period of time.

bingeing Uncontrolled consumption of large quantities of food in a relatively short period of time.

bulimia nervosa (BN) An eating disorder characterized by repeated cycles of bingeing and purging.

disordered eating An eating pattern characterized by unhealthy eating behaviors.

eating disorder An extreme disturbance in eating behaviors that can be both physically and psychologically harmful.

Other Specified Feeding or Eating Disorders (OSFED) A category of feeding or eating behaviors that include some, but not all, of the diagnostic criteria for anorexia nervosa, binge eating, and/or bulimia nervosa.

purging Self-induced vomiting and/or excessive exercise, misuse of laxatives, diuretics, and/or enemas.

12-2 **Are There Other Disordered Eating Behaviors?** • Many disordered eating behaviors exist in addition to the eating disorders formally recognized by the American Psychiatric Association. • Nocturnal sleep-related eating disorder (SRED) is characterized by eating at night without recollection of having done so. • Night eating syndrome (NES) is characterized by a cycle of daytime food restriction, excessive food intake in the evening, and nighttime insomnia. • Food neophobia is an eating disturbance characterized by an irrational fear or avoidance of trying new foods as well as unusual food rituals and practices. • Avoidant/restrictive food intake disorder (AFRID) is an eating disturbance characterized by food aversions. The failure to consume adequate nutrients and energy often results in severe undernutrition. However, weight loss is not associated with a desire to become thin, but rather a consequence of nutritional problems. • Muscle dysmorphia is observed primarily in men who work out excessively to increase muscularity and have intense fears of being too small, too weak, and/or too skinny.

avoidant/restrictive food intake disorder (AFRID) An eating disturbance characterized by persistent failure to satisfy nutrition and/or energy requirements due to an aversion to food.

food neophobia A disordered eating pattern characterized by an irrational fear or avoidance of new foods.

muscle dysmorphia A disorder characterized by a preoccupation with increasing muscularity.

night eating syndrome (NES) A disordered eating pattern characterized by a cycle of daytime food restriction, excessive food intake in the evening, and nighttime insomnia.

nocturnal sleep-related eating disorder (SRED) A disordered eating pattern characterized by eating while asleep without any recollection of having done so.

restrained eater Individuals who suppress their desire for food and avoid eating for long periods of time between binges.

12-3 **What Causes Eating Disorders?** • There are many theories as to the factors that contribute to the development of eating disorders, but there are no simple answers. • Eating disorders are more prevalent in cultures where food is abundant and slimness is valued. However, as multicultural youth acculturate to mainstream American values, distinct cultural norms that once protected them from eating disorders may be eroding. • Major media outlets that glamorize unrealistically thin and overly muscular bodies have long been criticized for evoking a sense of inadequacy in impressionable children and young adults. • In addition to the media, a person's circle of peers may also contribute to the development of eating disorders. • Enmeshed family dynamics promote dependency, which may lay the foundation for the emergence of eating disorders later in life. Children raised in a chaotic environment may later develop eating disorders as a way to fill emotional emptiness, to gain attention, or to suppress emotional conflict. • Children of women with eating disorders are at increased risk of developing eating disorders themselves. • Personality traits associated with eating disorders include low self-esteem, lack of self-confidence, obsessiveness, and feelings of helplessness, anxiety, and depression. • Brain chemicals and other biological factors might play a role in the development of an eating disorder.

chaotic family (or **disengaged family**) A style of family interaction characterized by a lack of cohesiveness and little parental involvement.

enmeshment A style of family interaction whereby family members are overly involved with one another and have little autonomy.

food preoccupation Spending an inordinate amount of time thinking about food.

12-4

Are Athletes at Increased Risk for Eating Disorders?

• While some studies indicate that the prevalence of eating disorders among female student-athletes and nonathletes does not differ, other studies indicate otherwise. • The prevalence of disordered eating and eating disorders among collegiate athletes is estimated to be somewhere between 15 and 60 percent. • Sports that favor a thin physical appearance are likely to have more athletes with eating disorders than those for which size is not as important. • Because athletes tend to be competitive people and may equate their self-worth with athletic success, they may be especially willing to engage in risky weight-loss practices to achieve that success. • Female athletes are at risk for developing the female athlete triad: disordered eating/eating disorder, menstrual irregularities such as amenorrhea, and osteopenia.

female athlete triad A combination of three interrelated conditions: disordered eating (or eating disorder), menstrual dysfunction, and osteopenia.

12-5

How Can Eating Disorders Be Prevented and Treated?

• People with eating disorders are at risk for serious medical and/or emotional problems. Thus, they need treatment from qualified health professionals. • Eating disorder prevention involves educational strategies that focus on self-esteem and encourage healthy behaviors instead of simply focusing on the dangers of eating disorders. • Once an eating disorder has developed, it is important to seek treatment from a qualified team of professionals, which include mental health specialists who can help address and treat underlying psychological issues, medical doctors who can treat physiological complications, and a dietitian who can recommend healthy food choices. • A person with an eating disorder may not recognize or be unable to admit that he or she has a problem. Concerns expressed by friends and family members often go ignored or dismissed, making loved ones feel confused and frustrated by their inability to help. • Treatment goals for people with AN include achieving a healthy body weight, resolving psychological issues (such as low self-esteem and distorted body image), and establishing healthy eating patterns. • Treatment goals for people with BN include the reduction and eventual elimination of bingeing and purging. • The sooner a person with an eating disorder gets help, the better the chance of full recovery is. Regardless of when that happens, recovery can be a long, trying process.

TABLE 12.4	PROMOTING A HEALTHY BODY IMAGE AMONG CHILDREN AND ADOLESCENTS

- Encourage children to focus on positive body features instead of negative ones.
- Help children understand that everyone has a unique body size and shape.
- Be a good role model for children by demonstrating healthy eating behaviors.
- Resist making negative comments about your own weight or body shape.
- Focus on positive, nonphysical traits such as generosity, kindness, and a friendly laugh.
- Do not criticize a child's appearance.
- Never associate self-worth with physical attributes.
- Prepare a child for puberty in advance by discussing physical and emotional changes.
- Enjoy meals together as a family.
- Discuss how the media can negatively affect body image.
- Avoid using food as a way to reward or punish.

Source: Adapted from Story M, Holt K, Sofka D. Bright futures in practice: Nutrition, 2nd ed. Arlington, VA: National Center for Education in Maternal and Child Health, 2002.

13-1 What Is Alcohol and How Is It Produced?

Although alcohol has a rich history of pageantry and ritual, adults who consume alcohol should do so safely and in moderation. • The type of alcohol in beer, wine, and distilled liquor is called ethanol, which is produced through the process of fermentation. • During fermentation, yeast produces ethanol by metabolizing sugar. To increase the alcohol concentration, fermented beverages can be distilled. • Alcohol does not require digestion prior to its absorption in the small intestine and stomach. • Once absorbed, alcohol circulates in the blood. Its concentration is expressed as blood alcohol concentration (BAC), which is affected by a person's body size and composition. • Alcohol is a central nervous system depressant and acts as a sedative in the brain. • Alcohol can cause a temporary loss of inhibition (disinhibition), making a person feel relaxed and more outgoing while at the same time impairing judgment and reasoning.

alcohol An organic compound that has one or more hydroxyl (–OH) groups attached to carbon atoms.

alcohol by volume (ABV) The percentage of ethanol in a given volume of liquid.

blood alcohol concentration (BAC) A unit of measure for the amount of alcohol in the blood, measured in grams per deciliter (g/dL).

disinhibition A loss of inhibition.

distillation The process whereby alcohol vapors are condensed and collected to increase alcohol content.

ethanol The type of alcohol found in alcoholic beverages.

fermentation The process whereby alcoholic beverages are produced via the addition of yeast to grains or fruit; results in the conversion of sugars into ethanol and carbon dioxide.

proof A measure of the alcohol content of distilled liquor; proof is twice the percent of alcohol by volume.

13-2 How Is Alcohol Metabolized?

Most alcohol is metabolized through a two-step metabolic pathway. • The first step in this metabolic pathway requires the enzyme alcohol dehydrogenase (ADH), which is found primarily in the liver. The second step in the metabolic process, catalyzed by the enzyme acetaldehyde dehydrogenase (ALDH), converts acetaldehyde to acetate. • Genetic differences in ADH and ALDH activity can affect a person's ability to metabolize alcohol. • Chronic alcohol consumption can activate a group of liver enzymes that assist in alcohol metabolism, reducing the amount of time a person remains intoxicated, and often leading to alcohol tolerance. • If a heavy drinker develops a cross-tolerance to a drug, its actions in the body can decrease even though its concentration in the blood may reach dangerously high levels.

acetaldehyde dehydrogenase (ALDH) A liver enzyme that converts acetaldehyde to acetate.

alcohol dehydrogenase (ADH) A liver enzyme that converts ethanol to acetaldehyde.

tolerance A response to chronic drug or alcohol exposure that results in the body being able to metabolize increasingly larger quantities.

13-3 Does Alcohol Have Any Health Benefits?

The 2015 Dietary Guidelines for Americans states that if alcohol is to be consumed, consumption should not exceed one drink per day for women or two drinks per day for men, and that alcohol should only be consumed by nonpregnant adults of legal drinking age. • When consumed in moderation, alcohol is associated with decreased risks of cardiovascular disease, gallstones, age-related memory loss, and even type 2 diabetes. • Adults who consume an average of one to two alcoholic drinks a day have a 30-35 percent lower risk of cardiovascular disease than adults who do not consume alcohol; these benefits disappear when alcohol intake becomes excessive. • There is considerable evidence that the antioxidant resveratrol, found in the skin of red grapes, helps reduce inflammation and atherosclerosis.

resveratrol An antioxidant found in the skin of red grapes.

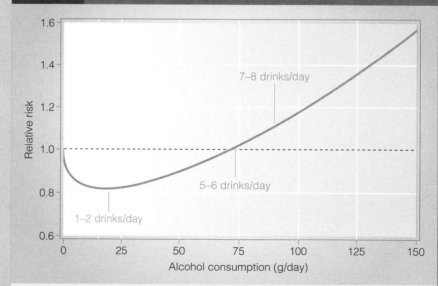

FIGURE 13.2 RELATIONSHIP BETWEEN AVERAGE ALCOHOL CONSUMPTION AND CORONARY HEART DISEASE

Light-to-moderate alcohol consumption is associated with reduced risk of coronary heart disease. At high levels (7–8 drinks/day), the risk of coronary heart disease increases.

Source: Corrao G, Rubbiati L, Bagnardi V, Zambon A, Poikolainen K. Alcohol and coronary heart disease: A meta-analysis. Addiction. 2000, 95:1505–23. Reprinted with permission of Blackwell Publishing Ltd.

13-4

What Serious Health Risks Does Heavy Alcohol Consumption Pose? • While moderate alcohol intake may provide health benefits for middle-aged adults, alcohol is clearly hazardous to one's health when consumed in excess. • As a result of chronic, heavy consumption of alcohol, an array of health problems can develop, collectively referred to as alcohol use disorders (AUDs). • Habitual drinking can adversely affect nutritional status by causing both primary and secondary malnutrition. • Long-term alcohol abuse can impair liver function, leading to fatty liver, alcoholic hepatitis, and cirrhosis. • Heavy drinking increases a person's risk of developing certain types of cancer, especially those of the mouth, esophagus, colon, liver, and breast; alcohol may act as both a carcinogen and a co-carcinogen. • Heavy drinking can cause alcoholic cardiomyopathy, high blood pressure, dementia, and cardiac arrhythmia. • A chronically high alcohol intake (consuming at least seven drinks a day for more than five years) increases a person's risk of developing pancreatitis.

alcoholic cardiomyopathy A serious condition whereby the heart muscle weakens in response to chronic alcohol consumption.

alcoholic hepatitis Inflammation of the liver caused by obstructed blood flow.

binge drinking The consumption of four or more drinks for women and five or more drinks for men over a 2-hour period.

cardiac arrhythmia An irregular heartbeat that can be caused by a high intake of alcohol.

cirrhosis A condition characterized by the presence of scar tissue in the liver.

fatty liver A condition caused by chronic alcohol consumption characterized by the accumulation of triglycerides in the liver.

pancreatitis A painful condition characterized by inflammation of the pancreas.

13-5

How Does Alcohol Abuse Contribute to Individual and Societal Problems? • When alcohol is the cause of significant difficulties in a person's life, drinking is problematic. • Organizations such as Alcoholics Anonymous®, Al-Anon®, and Alateen® provide support and fellowship for people suffering from the effects of alcohol abuse. • Alcohol abuse, rampant on many college campuses, is associated with a wide range of risky behaviors and negative consequences such as vandalism, violence, acquaintance rape, unprotected sex, and death. • Students who binge drink are more likely than those who do not to miss class, have lower academic rankings, experience trouble with campus law enforcement, and drive while intoxicated. • Colleges and universities must work together with communities and students to create a culture that discourages high-risk drinking.

TABLE 13.1	SERVING SIZES, ENERGY CONTENTS, AND ALCOHOL CONTENTS OF SELECTED ALCOHOLIC BEVERAGES		
Beverage	**Serving Size (oz [mL])**	**Energy (kcal/serving)[a]**	**Alcohol (g/serving)**
Light beer	12 (355)	103	11
Regular beer	12 (355)	153	13
White wine	5 (148)	121	14
Red wine	5 (148)	125	14
Distilled beverages[b]			
80 proof	1.5 (44)	96	14
90 proof	1.5 (44)	110	16
100 proof	1.5 (44)	123	18
Crème de menthe	1.5 (44)	186	15
Daiquiri	4 (118)	225	14
Whiskey sour	4.5 (133)	217	19
Piña colada	4.5 (133)	245	14

[a] Note that some alcoholic beverages contain energy-yielding nutrients other than alcohol. Therefore, caloric content cannot always be calculated simply by multiplying grams of alcohol by 7 kcal/g.
[b] Distilled beverages include gin, rum, vodka, whiskey, and other liquors.
Source: USDA National Nutrient Database for Standard Reference, Release 23. Nutrient Data Laboratory. Available at: https://ndb.nal.usda.gov/

Al-Anon® An organization that provides support for family members and friends of alcoholics.

Alateen® An organization that provides support for teenage children of alcoholics.

alcohol use disorders (AUDs) A spectrum of disorders that encompasses both recurrent excessive drinking and alcoholism.

Alcoholics Anonymous® (AA) An organization that offers a 12-step program that provides fellowship and support for individuals who want to achieve sobriety and stay sober.

CHAPTER 14 LEARNING OBJECTIVES / KEY TERMS

14-1

What Causes Foodborne Illness? • You are exposed to thousands of microscopic organisms (microbes) every day. Microbes populate the world you live in, and many even serve useful purposes for you and for your health. • Other microbes, however, are pathogenic (disease causing). • Foodborne illnesses are sometimes referred to as food poisoning. You can contract a foodborne illness by ingesting an infectious agent such as a bacterium, virus, mold, fungus, or parasite. • You can also contract a foodborne illness by ingesting a noninfectious agent such as a nonbacterial toxin; a chemical residue from processing, pesticides, or antibiotics; or a physical hazard such as glass or plastic. • Some bacterial serotypes are harmless, whereas others cause disease. • Because different pathogens have different incubation periods, this information is useful in determining the cause of a foodborne illness. • Some pathogenic organisms produce toxic substances while they are growing in food. These include methicillin-resistant *S. aureus* (MRSA), *C. botulinum*, and *Aspergillus*. • Some organisms produce harmful toxins after they enter the gastrointestinal (GI) tract. These include noroviruses and enterotoxigenic *E. coli*. • Some pathogens invade the cells of the intestine, seriously irritating the mucosal lining and causing fever, severe abdominal discomfort, and bloody diarrhea. These include *Salmonella* and *E. coli* O157:H7. • A parasite is an organism that relies on another organism to survive. The consumption of parasite-infested foods can cause foodborne illness. An example of a parasite is a protozoan. • Although prions are not living organisms, they may pose food safety concerns. Several diseases are known to be caused by prion ingestion.

aflatoxin A toxin produced by the *Aspergillus* mold that is found on some agricultural crops.

botulism The foodborne illness caused by *Clostridium botulinum*.

bovine spongiform encephalopathy (**BSE**, or **mad cow disease**) A fatal disease caused by prion ingestion.

Creutzfeldt-Jakob disease A rare but fatal disease caused by a genetic mutation or exposure to prions during surgery.

cyst A stage of growth in the life cycle of a protozoan; excreted in the feces.

enteric toxin (or **intestinal toxin**) A toxic substance produced by an organism after it enters the gastrointestinal tract.

enterohemorrhagic Causing bloody diarrhea and intestinal inflammation.

enterotoxigenic *E. coli* A form of *E. coli* that produces enteric toxins.

foodborne illness A disease caused by the ingestion of unsafe food.

incubation period The time elapsed between the consumption of a contaminated food and the emergence of sickness.

infectious agent (or **pathogen**) A living microorganism that can cause illness.

methicillin-resistant *Staphylococcus aureus* (MRSA) An antibiotic-resistant strain of *S. aureus* that has received considerable attention from public health officials.

noninfectious agent An inert (nonliving) substance that can cause illness.

norovirus An infectious pathogen that produces enteric toxins and often causes foodborne illness.

parasite An organism that relies on another organism to survive.

preformed toxin A toxic substance that already exists in a food before the food is eaten.

prion An altered protein that forms when the secondary structure of a normal protein is disrupted.

protozoan A single-celled eukaryotic organism. Some protozoa are parasites.

serotype (or **strain**) A specific genetic variety of an organism.

variant Creutzfeldt-Jakob disease A form of Creutzfeldt-Jakob disease that may be caused by consumption of prion-contaminated foods.

14-2

What Noninfectious Substances Cause Foodborne Illness? • Inert (noninfectious) compounds in foods can also cause foodborne illness. Noninfectious agents include physical contaminants (such as glass and plastic) and other dangerous substances such as toxins, heavy metals, and pesticides. • Shellfish poisoning can result from the consumption of particular types of contaminated fish and shellfish (e.g., clams and oysters). Eating marine toxin-contaminated foods can cause tingling, burning, numbness, drowsiness, and difficulty breathing. • Because no food production system is foolproof, illness-causing pesticides, herbicides, antibiotics, and hormones sometimes come in contact with food products. • Three foodborne substances that have garnered significant public health attention in recent years are acrylamide, melamine, and bisphenol A (BPA).

bisphenol A (BPA) A chemical found in some plastic food and beverage containers believed by some to pose a potential threat to health.

bovine somatotropin (bST) A growth hormone produced by cattle and used in the dairy industry to increase milk production.

brevetoxin A potent toxin produced by red tide–causing algae that, when consumed, leads to shellfish poisoning.

genetically modified foods A food that has been produced or manufactured using a genetically modified organism; not related to foodborne illness.

genetically modified organisms (GMO) A plant or animal that has been altered using genetic engineering; not related to foodborne illness.

marine toxin A poison produced by ocean algae.

melamine A nitrogen-containing chemical typically used to make lightweight plastic objects.

red tide A phenomenon whereby certain ocean algae grow profusely and produce brightly colored pigments that bloom outward and make the surrounding water appear red or brown.

shellfish poisoning A type of noninfectious foodborne illness caused by the consumption of particular types of contaminated fish and shellfish.

14-3

How Do Food Manufacturers Prevent Contamination?
• Disease-causing agents of all types can be transmitted from one food, surface, or utensil to another through cross-contamination. • To prevent foodborne illness, it is critical that those who handle food do so safely and sanitarily. The food-processing industry adheres to additional regulations to keep food safe, but inspection and adherence to guidelines alone cannot guarantee that foods are pathogen free. • Salting, drying, and smoking remove water from foods, inhibiting pathogenic growth; fermentation involves the addition of nonpathogenic organisms that inhibit the growth of dangerous ones. • Manufactures, restaurants, and individuals alike should avoid the danger zone (40°F–140°F) when storing and serving food. • Methods that slow the growth of or kill pathogens include heating, pasteurization, freezing, and irradiation.

cross-contamination The process by which a disease-causing agent is transmitted from one food, surface, or utensil to another.

danger zone The temperature range between 40°F and 140°F, in which most microorganisms prefer to live.

irradiation A food preservation process whereby a food is exposed to radiant energy that damages or destroys bacteria.

pasteurization A food preservation process whereby food is partially sterilized through brief exposure to a high temperature.

14-4

What Steps Can You Take to Reduce Foodborne Illness?
• There are many precautions you can take to reduce your risk of foodborne illness. • Both the USDA and FDA maintain user-friendly websites and toll-free phone numbers that provide information about current food safety recommendations. • The four major components that comprise Fight BAC!® are clean, separate, cook, and chill. • Making sure that your food is safe to eat can be especially difficult at a restaurant, picnic, potluck, or buffet. Food safety recommendations are even more important in these environments.

Fight BAC!® A public education program developed to reduce foodborne bacterial illness. The four major components that comprise Fight BAC!® are clean, separate, cook, and chill.

sell-by date A date placed on a product's label to assist a grocer in knowing when the food should be removed from the sales display.

use-by dates A manufacturer's best prediction as to when the product will retain top quality and flavor.

14-5

What Steps Can You Take to Reduce Foodborne Illness while Traveling?
• While it is not always possible to prevent foodborne illness while traveling, you can substantially lower your risk by being vigilant in your food and beverage choices. • If you do experience traveler's diarrhea while abroad, it is important that you replace lost fluids and electrolytes as soon as symptoms begin to develop. • Drink bottled water and avoid using ice. Before you drink water from a bottle, be sure that it has a fully sealed cap. • Although fresh produce that is contaminated with bacteria may not cause illness for local residents, it can cause serious illness for visitors. • Concerned travelers visiting regions where variant Creutzfeldt-Jakob disease has been reported might want to consider either avoiding beef and beef products or selecting only beef products composed of solid pieces of muscle meat. • Food biosecurity and potential bioterrorism have gained significant national attention in recent years. • Outside of willful acts of terrorism, changes in food production and distribution systems may influence food safety and the risk of foodborne illness.

food biosecurity Measures aimed at preventing the food supply from falling victim to planned contamination.

Public Health Security and Bioterrorism Preparedness and Response Act (or **Bioterrorism Act**) Federal legislation aimed at ensuring the continued safety of the U.S. food supply.

FIGURE 14.3   FIGHT BAC!

Used with permission of Partnership for Food Safety Education. www.fightbac.org

The Fight BAC! campaign was created to reduce the incidence of foodborne illnesses by educating Americans about safe food-handling practices at home and at work.

15-1 What Is Food Security?

· Most people are able to ease their hunger, as they have access to sufficient amounts of nutritious food. Others, however, are not. When sufficient food is not available or accessible, hunger can lead to serious physical, social, and psychological consequences. · The United Nations Food and Agriculture Organization (FAO) estimates that approximately 925 million people worldwide experience persistent hunger. · Food insecurity exists when people do not have adequate physical, social, or economic access to food. · Households classified as having low food security experience reduced food quality, variety, and/or desirability, whereas households classified as having very low food security are also likely to report disrupted eating patterns and reduced food intake. · While the word *hunger* is often used to describe the physical discomfort experienced by individuals who have consumed insufficient amounts of food, it is more commonly used on a global level to describe a shortage of available food. · Clinical measurements of nutritional status (such as anthropometry) are not always useful indicators of food insecurity. Instead, the prevalence of food insecurity in U.S. households is typically assessed using data regarding food availability and accessibility. · Many individual and socioeconomic factors can predispose a person or family to food insecurity, but poverty is the most telling risk factor of all.

food desert An environment that lacks access to affordable and/or nutritious foods.

food security A condition whereby a person is able to access sufficient amounts of nutritious food.

low food security A condition characterized by reduced food quality, variety, and/or desirability, but not reduced food intake.

very low food security A condition characterized by disrupted eating patterns and reduced food intake caused by a lack of food access and availability.

15-2 What Are the Consequences of Food Insecurity?

· Although food insecurity does not typically lead to starvation or nutrient deficiencies in the United States, it still represents a major public health concern. · There are numerous consequences of food insecurity, many of which have been studied most extensively in women and children. · Older adults are also at high risk for food insecurity, though they often experience it differently than children, young adults, and other population groups. · Food assistance programs that help at-risk individuals obtain food include the Supplemental Nutrition Assistance Program (SNAP); the Special Supplemental Nutrition Program for Women, Infants, and Children (WIC); the School Breakfast and National School Lunch Programs; food recovery programs; and community food banks and kitchens. · Some programs are federally funded, whereas others are community efforts staffed by volunteers.

food bank An organization that collects donated foods and distributes them to local food pantries, shelters, and soup kitchens.

food kitchen A program that prepares and serves meals to members of the community who are in need.

food pantry A program that provides canned, boxed, and sometimes fresh foods directly to individuals in need.

food recovery program A program that collects and redistributes discarded food that would have otherwise gone to waste.

National School Lunch Program A federally funded program that provides nutritionally balanced meals either free of charge or at a reduced cost to school-age children at lunch time.

School Breakfast Program A federally funded program that provides nutritionally balanced breakfasts either free of charge or at a reduced cost to school-age children.

Supplemental Nutrition Assistance Program (SNAP) A federally funded program, formerly known as the Food Stamp Program, that helps low-income households pay for food.

15-3 What Causes Worldwide Hunger and Malnutrition?

· Because poverty is more prevalent in developing countries than in industrialized ones, food insecurity tends to be most ubiquitous and severe in nations with low *per capita* incomes. · There is sufficient food in the world to feed and nourish all its inhabitants. · Worldwide food insecurity is usually caused by a combination of poverty, inadequate food distribution, political instability, urbanization, population growth, gender inequities, and other factors. · Although there are many consequences of food insecurity in poor and developing countries, perhaps the most important and devastating consequence is malnutrition. · Consequences of food insecurity in poorer regions of the world include greater incidence of low birth weight, neonatal death, inhibited growth, vitamin A deficiency, iron deficiency, and iodine deficiency. · Beyond its effects on the health of the individual, malnutrition can harm whole societies. · Though the United Nations serves many purposes, combating international hunger is among its most important efforts. · The United Nations Children's Fund (UNICEF) helps combat international hunger by working toward a common goal of improving the quality of life for people in the world's poorest countries. · Conditional cash-transfer programs may be one way to encourage sound nutritional and educational behaviors and simultaneously alleviate poverty. · The U.S. Peace Corps offers opportunities to make a difference in the lives of others, helping address the problems of food insecurity and malnutrition in many parts of the world. · Heifer International® is a humanitarian effort with a global commitment to foster environmentally sound farming methods that combat both hunger and environmental concerns. · Heifer's living loans program has fostered a living cycle of sustainability for over 65 years.

conditional cash-transfer program An initiative that offers financial reimbursement for individuals or communities that work to improve quality of life.

urbanization A population shift whereby large numbers of people move from rural to urban regions within a country.

U.S. Peace Corps A federally funded volunteer program that promotes world peace and friendship worldwide.

FIGURE 15.3 UNITED NATION'S 2015 MILLENNIUM DEVELOPMENT GOALS

1 ERADICATE EXTREME POVERTY AND HUNGER

2 ACHIEVE UNIVERSAL PRIMARY EDUCATION

3 PROMOTE GENDER EQUALITY AND EMPOWER WOMEN

4 REDUCE CHILD MORTALITY

5 IMPROVE MATERNAL HEALTH

6 COMBAT HIV/AIDS, MALARIA AND OTHER DISEASES

7 ENSURE ENVIRONMENTAL SUSTAINABILITY

8 GLOBAL PARTNERSHIP FOR DEVELOPMENT

As the era of these Millennium Development goals comes to an end in 2016, the UN will launch its 2030 Agenda for Sustainable Development.

15-4 What Can You Do to Alleviate Food Insecurity? • It is only by considering the complexity of food insecurity that the relative importance of each contributing factor can be addressed and effective solutions developed. • Leading world health experts agree that improving food availability and access must be a global priority. • Although malnutrition is a direct consequence of insufficient dietary intake, the ultimate causes of malnutrition often have more to do with economic and societal circumstances. • Although the problem of food insecurity may seem staggering both at home and abroad, it is important to remember that there is much an individual can do to take action against hunger. • Working alone or collectively toward the elimination of hunger and malnutrition is a worthwhile and noble personal and/or professional priority. • The Academy of Nutrition and Dietetics (AND) and American Society for Nutrition (ASN) are two of many professional organizations that work to alleviate world hunger by challenging their members to take action.

The Dietary Reference Intakes (DRIs) include four dietary reference standards used to assess and plan dietary intake: Estimated Average Requirements (EARs), Recommended Dietary Allowances (RDAs), Adequate Intake (AI) levels, and Tolerable Upper Intake Levels (ULs). Acceptable Macronutrient Distribution Ranges (AMDRs) and Estimated Energy Requirements (EERs) are also used to calculate and maintain ideal energy and macronutrient intakes. Use the tables below to assess the adequacy of your own intake levels.

ESTIMATED ENERGY REQUIREMENTS (EERs), RECOMMENDED DIETARY ALLOWANCES (RDAs), AND ADEQUATE INTAKE (AI) LEVELS FOR ENERGY AND THE MACRONUTRIENTS

Age (yr)	Reference BMI (kg/m²)	Reference height, cm (in)	Reference weight, kg (lb)	Water^a AI (L/day)	Energy EER^b (kcal/day)	Carbohydrate RDA (g/day)	Total fiber AI (g/day)	Total fat AI (g/day)	Linoleic acid AI (g/day)	Linolenic acid^c AI (g/day)	Protein RDA (g/day)^d	Protein RDA (g/kg/day)
Males												
0–0.5	—	62 (24)	6 (13)	0.7^e	570	60	—	31	4.4	0.5	9.1	1.52
0.5–1	—	71 (28)	9 (20)	0.8^f	743	95	—	30	4.6	0.5	11	1.2
1–3^g	—	86 (34)	12 (27)	1.3	1,046	130	19	—	7	0.7	13	1.05
4–8^g	15.3	115 (45)	20 (44)	1.7	1,742	130	25	—	10	0.9	19	0.95
9–13	17.2	144 (57)	36 (79)	2.4	2,279	130	31	—	12	1.2	34	0.95
14–18	20.5	174 (68)	61 (134)	3.3	3,152	130	38	—	16	1.6	52	0.85
19–30	22.5	177 (70)	70 (154)	3.7	3,067^h	130	38	—	17	1.6	56	0.8
31–50	22.5	177 (70)	70 (154)	3.7	3,067^h	130	38	—	17	1.6	56	0.8
>50	22.5	177 (70)	70 (154)	3.7	3,067^h	130	30	—	14	1.6	56	0.8
Females												
0–0.5	—	62 (24)	6 (13)	0.7^e	520	60	—	31	4.4	0.5	9.1	1.52
0.5–1	—	71 (28)	9 (20)	0.8^f	676	95	—	30	4.6	0.5	11	1.2
1–3^g	—	86 (34)	12 (27)	1.3	992	130	19	—	7	0.7	13	1.05
4–8^g	15.3	115 (45)	20 (44)	1.7	1,642	130	25	—	10	0.9	19	0.95
9–13	17.4	144 (57)	37 (81)	2.1	2,071	130	26	—	10	1.0	34	0.95
14–18	20.4	163 (64)	54 (119)	2.3	2,368	130	26	—	11	1.1	46	0.85
19–30	21.5	163 (64)	57 (126)	2.7	2,403^i	130	25	—	12	1.1	46	0.8
31–50	21.5	163 (64)	57 (126)	2.7	2,403^i	130	25	—	12	1.1	46	0.8
>50	21.5	163 (64)	57 (126)	2.7	2,403^i	130	21	—	11	1.1	46	0.8
Pregnancy												
1st trimester				3.0	+0	175	28	—	13	1.4	+25	1.1
2nd trimester				3.0	+340	175	28	—	13	1.4	+25	1.1
3rd trimester				3.0	+452	175	28	—	13	1.4	+25	1.1
Lactation												
1st 6 months				3.8	+330	210	29	—	13	1.3	+25	1.3
2nd 6 months				3.8	+400	210	29	—	13	1.3	+25	1.3

Note: For all nutrients, values set for infants are AIs. A dash (—) indicates that a value has not been established.

^{a}The water AI includes drinking water, water in beverages, and water in foods. In general, drinking water and other beverages contribute 70–80 percent, while foods contribute the remainder.

^{b}An EER represents the average dietary energy intake that will maintain energy balance in a healthy person of a given sex, age, weight, height, and physical activity level. Values listed are based on an active person at the reference height and weight and at the midpoint ages for each group until age 19. The values listed for pregnancy and lactation represent energy needed in addition to nonpregnant/nonlactating values.

^{c}Linolenic acid referred to in this table and text is α-linolenic acid, an essential omega-3 fatty acid.

^{d}Values listed are based on reference body weights.

^{e}Assumed to be from human milk.

^{f}Assumed to be from human milk and complementary foods and beverages. This includes approximately 0.6 L (3 cups) as total fluid including formula, juices, and drinking water.

^{g}For energy, the age groups for young children are 1–2 years and 3–8 years.

^{h}For males, subtract 10 kilocalories per day for each year of age above 19.

^{i}For females, subtract 7 kilocalories per day for each year of age above 19.

Source: Adapted from the Dietary Reference Intakes series, National Academies Press. Copyright 1997, 1998, 2000, 2001, 2002, 2004, 2005 by the National Academy of Sciences.

RECOMMENDED DIETARY ALLOWANCES (RDAs) AND ADEQUATE INTAKE (AI) LEVELS FOR VITAMINS

Age (yr)	Thiamin RDA (mg/day)	Riboflavin RDA (mg/day)	Niacin RDA (mg/day)[a]	Biotin AI (μg/day)	Pantothenic acid AI (mg/day)	Vitamin B_6 RDA (mg/day)	Folate RDA (μg/day)[b]	Vitamin B_{12} RDA (μg/day)	Choline AI (mg/day)	Vitamin C RDA (mg/day)	Vitamin A RDA (μg/day)[c]	Vitamin D RDA (μg/day)[d]	Vitamin E RDA (mg/day)[e]	Vitamin K AI (μg/day)
Infants														
0–0.5	0.2	0.3	2	5	1.7	0.1	65	0.4	125	40	400	10	4	2.0
0.5–1	0.3	0.4	4	6	1.8	0.3	80	0.5	150	50	500	10	5	2.5
Children														
1–3	0.5	0.5	6	8	2	0.5	150	0.9	200	15	300	15	6	30
4–8	0.6	0.6	8	12	3	0.6	200	1.2	250	25	400	15	7	55
Males														
9–13	0.9	0.9	12	20	4	1.0	300	1.8	375	45	600	15	11	60
14–18	1.2	1.3	16	25	5	1.3	400	2.4	550	75	900	15	15	75
19–30	1.2	1.3	16	30	5	1.3	400	2.4	550	90	900	15	15	120
31–50	1.2	1.3	16	30	5	1.3	400	2.4	550	90	900	15	15	120
51–70	1.2	1.3	16	30	5	1.7	400	2.4	550	90	900	15	15	120
>70	1.2	1.3	16	30	5	1.7	400	2.4	550	90	900	20	15	120
Females														
9–13	0.9	0.9	12	20	4	1.0	300	1.8	375	45	600	15	11	60
14–18	1.0	1.0	14	25	5	1.2	400	2.4	400	65	700	15	15	75
19–30	1.1	1.1	14	30	5	1.3	400	2.4	425	75	700	15	15	90
31–50	1.1	1.1	14	30	5	1.3	400	2.4	425	75	700	15	15	90
51–70	1.1	1.1	14	30	5	1.5	400	2.4	425	75	700	15	15	90
>70	1.1	1.1	14	30	5	1.5	400	2.4	425	75	700	20	15	90
Pregnancy														
≤18	1.4	1.4	18	30	6	1.9	600	2.6	450	80	750	15	15	75
19–30	1.4	1.4	18	30	6	1.9	600	2.6	450	85	770	15	15	90
31–50	1.4	1.4	18	30	6	1.9	600	2.6	450	85	770	15	15	90
Lactation														
≤18	1.4	1.6	17	35	7	2.0	500	2.8	550	115	1,200	15	19	75
19–30	1.4	1.6	17	35	7	2.0	500	2.8	550	120	1,300	15	19	90
31–50	1.4	1.6	17	35	7	2.0	500	2.8	550	120	1,300	15	19	90

Note: For all nutrients, values for infants are AIs.

[a]Niacin recommendations are expressed as niacin equivalents (NEs), except for recommendations for infants younger than 6 months, which are expressed as preformed niacin.
[b]Folate recommendations are expressed as dietary folate equivalents (DFEs).
[c]Vitamin A recommendations are expressed as retinol activity equivalents (RAEs).
[d]Vitamin D recommendations are expressed as cholecalciferol.
[e]Vitamin E recommendations are expressed as α-tocopherol.

RECOMMENDED DIETARY ALLOWANCES (RDAs) AND ADEQUATE INTAKE (AI) LEVELS FOR MINERALS

Age (yr)	Sodium AI (mg/day)	Chloride AI (mg/day)	Potassium AI (mg/day)	Calcium RDA (mg/day)	Phosphorus RDA (mg/day)	Magnesium RDA (mg/day)	Iron RDA (mg/day)	Zinc RDA (mg/day)	Iodine RDA (μg/day)	Selenium RDA (μg/day)	Copper RDA (μg/day)	Manganese AI (mg/day)	Fluoride AI (mg/day)	Chromium AI (μg/day)	Molybdenum RDA (μg/day)
Infants															
0–0.5	120	180	400	200	100	30	0.27	2	110	15	200	0.003	0.01	0.2	2
0.5–1	370	570	700	260	275	75	11	3	130	20	220	0.6	0.5	5.5	3
Children															
1–3	1,000	1,500	3,000	700	460	80	7	3	90	20	340	1.2	0.7	11	17
4–8	1,200	1,900	3,800	1,000	500	130	10	5	90	30	440	1.5	1	15	22
Males															
9–13	1,500	2,300	4,500	1,300	1,250	240	8	8	120	40	700	1.9	2	25	34
14–18	1,500	2,300	4,700	1,300	1,250	410	11	11	150	55	890	2.2	3	35	43
19–30	1,500	2,300	4,700	1,000	700	400	8	11	150	55	900	2.3	4	35	45
31–50	1,500	2,300	4,700	1,000	700	420	8	11	150	55	900	2.3	4	35	45
51–70	1,300	2,000	4,700	1,000	700	420	8	11	150	55	900	2.3	4	30	45
>70	1,200	1,800	4,700	1,200	700	420	8	11	150	55	900	2.3	4	30	45

RECOMMENDED DIETARY ALLOWANCES (RDAs) AND ADEQUATE INTAKE (AI) LEVELS FOR MINERALS

Age (yr)	Sodium AI (mg/day)	Chloride AI (mg/day)	Potassium AI (mg/day)	Calcium RDA (mg/day)	Phosphorus RDA (mg/day)	Magnesium RDA (mg/day)	Iron RDA (mg/day)	Zinc RDA (mg/day)	Iodine RDA (μg/day)	Selenium RDA (μg/day)	Copper RDA (μg/day)	Manganese AI (mg/day)	Fluoride AI (mg/day)	Chromium AI (μg/day)	Molybdenum RDA (μg/day)
Females															
9–13	1,500	2,300	4,500	1,300	1,250	240	8	8	120	40	700	1.6	2	21	34
14–18	1,500	2,300	4,700	1,300	1,250	360	15	9	150	55	890	1.6	3	24	43
19–30	1,500	2,300	4,700	1,000	700	310	18	8	150	55	900	1.8	3	25	45
31–50	1,500	2,300	4,700	1,000	700	320	18	8	150	55	900	1.8	3	25	45
51–70	1,300	2,000	4,700	1,200	700	320	8	8	150	55	900	1.8	3	20	45
>70	1,200	1,800	4,700	1,200	700	320	8	8	150	55	900	1.8	3	20	45
Pregnancy															
≤18	1,500	2,300	4,700	1,300	1,250	400	27	12	220	60	1,000	2.0	3	29	50
19–30	1,500	2,300	4,700	1,000	700	350	27	11	220	60	1,000	2.0	3	30	50
31–50	1,500	2,300	4,700	1,000	700	360	27	11	220	60	1,000	2.0	3	30	50
Lactation															
≤18	1,500	2,300	5,100	1,300	1,250	360	10	13	290	70	1,300	2.6	3	44	50
19–30	1,500	2,300	5,100	1,000	700	310	9	12	290	70	1,300	2.6	3	45	50
31–50	1,500	2,300	5,100	1,000	700	320	9	12	290	70	1,300	2.6	3	45	50

TOLERABLE UPPER INTAKE LEVELS (ULs) FOR VITAMINS AND MINERALS

Age (yr)	Niacin (mg/day)[a]	Vitamin B₆ (mg/day)	Folate (μg/day)[a]	Choline (mg/day)	Vitamin C (mg/day)	Vitamin A (μg/day)[b]	Vitamin D (μg/day)	Vitamin E (mg/day)[c]	Sodium (mg/day)	Chloride (mg/day)	Calcium (mg/day)	Phosphorus (mg/day)
Infants												
0–0.5	—	—	—	—	—	600	25	—	—[e]	—[e]	1,000	—
0.5–1	—	—	—	—	—	600	38	—	—[e]	—[e]	1,500	—
Children												
1–3	10	30	300	1,000	400	600	63	200	1,500	2,300	2,500	3,000
4–8	15	40	400	1,000	650	900	75	300	1,900	2,900	2,500	3,000
Adolescents												
9–13	20	60	600	2,000	1,200	1,700	100	600	2,200	3,400	3,000	4,000
14–18	30	80	800	3,000	1,800	2,800	100	800	2,300	3,600	3,000	4,000
Adults												
19–70	35	100	1,000	3,500	2,000	3,000	100	1,000	2,300	3,600	2,000–2,500	4,000
>70	35	100	1,000	3,500	2,000	3,000	100	1,000	2,300	3,600	2,000	3,000
Pregnancy												
≤18	30	80	800	3,000	1,800	2,800	100	800	2,300	3,600	3,000	3,500
19–50	35	100	1,000	3,500	2,000	3,000	100	1,000	2,300	3,600	2,500	3,500
Lactation												
≤18	30	80	800	3,000	1,800	2,800	100	800	2,300	3,600	3,000	4,000
19–50	35	100	1,000	3,500	2,000	3,000	100	1,000	2,300	3,600	2,500	4,000

TOLERABLE UPPER INTAKE LEVELS (ULs) FOR VITAMINS AND MINERALS

Age (yr)	Magnesium (mg/day)[d]	Iron (mg/day)[b]	Zinc (mg/day)	Iodine (μg/day)	Selenium (μg/day)	Copper (μg/day)	Manganese (mg/day)	Fluoride (mg/day)	Molybdenum (μg/day)	Boron (mg/day)	Nickel (mg/day)
Infants											
0–0.5	—	40	4	—	45	—	—	0.7	—	—	—
0.5–1	—	40	5	—	60	—	—	0.9	—	—	—
Children											
1–3	65	40	7	200	90	1,000	2	1.3	300	3	0.2
4–8	110	40	12	300	150	3,000	3	2.2	600	6	0.3
Adolescents											
9–13	350	40	23	600	280	5,000	6	10	1,100	11	0.6
14–18	350	45	34	900	400	8,000	9	10	1,700	17	1.0
Adults											
19–70	350	45	40	1,100	400	10,000	11	10	2,000	20	1.0
>70	350	45	40	1,100	400	10,000	11	10	2,000	20	1.0
Pregnancy											
≤18	350	45	34	900	400	8,000	9	10	1,700	17	1.0
19–50	350	45	40	1,100	400	10,000	11	10	2,000	20	1.0
Lactation											
≤18	350	45	34	900	400	8,000	9	10	1,700	17	1.0
19–50	350	45	40	1,100	400	10,000	11	10	2,000	20	1.0

NOTE: ULs were not established for vitamins and minerals not listed and for those age groups listed with a dash (—) because of a lack of data, not because these nutrients are safe to consume at any level of intake. All nutrients can have adverse effects when intakes are excessive.

[a]ULs for niacin and folate apply to synthetic forms obtained from supplements, fortified foods, or a combination of the two.
[b]The UL for vitamin A applies to preformed vitamin A only.
[c]The UL for vitamin E applies to any form of supplemental α-tocopherol, fortified foods, or a combination of the two.
[d]The UL for magnesium applies to synthetic forms obtained from supplements or drugs only.
[e]Source of intake should be from human milk (or formula) and food only.
Source: Adapted from the Dietary Reference Intakes series, National Academies Press. Copyright 1997, 1998, 2000, 2001, 2002, 2004, 2005, and 2011 by the National Academy of Sciences.

ACCEPTABLE MACRONUTRIENT DISTRIBUTION RANGES (AMDRs)

Macronutrient	Range (percent of energy)		
	Children, 1–3 years	Children, 4–18 years	Adults, 19+ years
Fat	30–40	25–35	20–35
ω-6 polyunsaturated acids[a] (linoleic acid)	5–10	5–10	5–10
ω-3 polyunsaturated fatty acids[a] (linolenic acid)	0.6–1.2	0.6–1.2	0.6–1.2
Carbohydrate	45–65	45–65	45–65
Protein	5–20	10–30	10–35

[a]Approximately 10 percent of the total can come from long-chain ω-3 or ω-6 fatty acids.
Source: Adapted from Institute of Medicine. Dietary reference intakes for energy, carbohydrate, fiber, fat, fatty acids, cholesterol, protein, and amino acids (macronutrients). Washington, DC: National Academies Press, 2005.

EATING ON A BUDGET—THE 3 P'S

PLAN
- Plan meals and snacks for the week according to an established budget.
- Find quick and easy recipes online.
- Include meals that will "stretch" expensive food items (stews, casseroles, stir-fried dishes).
- Make a grocery list.
- Check for sales and coupons in the local paper or online and consider discount stores.
- Ask about a loyalty card at your grocery store.

PURCHASE
- Buy groceries when you are not hungry and when you are not too rushed.
- Stick to the grocery list and stay out of the aisles that don't contain items on your list.
- Buy store brands if cheaper.
- Find and compare unit prices listed on shelves to get the best price.
- Purchase some items in bulk or as family packs which usually cost less.
- Choose fresh fruits and vegetables in season; buy canned vegetables with less salt.
- Pre-cut fruits and vegetables, individual cups of yogurt, and instant rice and hot cereal are convenient, but usually cost more than those that require a bit more prep time.
- Good low–cost items available all year include:
  - Protein—beans (garbanzo, black, and cannellini)
  - Vegetables—carrots, greens, and potatoes
  - Fruits—apples, and bananas

PREPARE
- Some meal items can be prepared in advance; pre-cook on days when you have time.
- Double or triple up on recipes and freeze meal-sized containers of soups and casseroles or divide into individual portions.
- Try a few meatless meals by substituting with beans and peas or try "no-cook" meals like salads.
- Incorporate leftovers into a subsequent meal.
- Be creative with a fruit or vegetable and use it in different ways during the week.

Source: Adapted from U.S. Department of Agriculture, Center for Nutrition Policy and Promotion. DG TipSheet September 2011. Available from: http://www.choosemyplate.gov/sites/default/files/printablematerials/PlanPurchasePrepare.pdf.

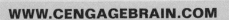

Estimated Energy Requirement (EER) Calculations and Physical Activity (PA) Values

An Estimated Energy Requirement (EER) value represents the average energy intake needed for a healthy person to maintain weight. EER values vary by age, sex, weight, height, and physical activity (PA) level. Note that this is different from the other DRI reference values, which only vary by life stage and sex. EERs are calculated using relatively simple mathematical equations. Using the EER equations for adult men and women of healthy weight below, you can calculate your own EER in kilocalories per day (kcal/day). The lower your PA value, the less active you are, and consequently the lower your EER. Note that at every age, active individuals need more energy than do their sedentary counterparts.

ESTIMATED ENERGY REQUIREMENT (EER) CALCULATIONS

Age Group	Equations for Estimated Energy Requirement (EER; kcal/day)[a]
0–3 months	$[89 \times \text{weight (kg)} - 100] + 175$ kcal
4–6 months	$[89 \times \text{weight (kg)} - 100] + 56$ kcal
7–12 months	$[89 \times \text{weight (kg)} - 100] + 22$ kcal
13–36 months	$[89 \times \text{weight (kg)} - 100] + 20$ kcal
3–8 years (male)	$88.5 - [61.9 \times \text{age (y)}] + \text{PA} \times [26.7 \times \text{weight (kg)} + 903 \times \text{height (m)}] + 20$ kcal
3–8 years (female)	$135.3 - [30.8 \times \text{age (y)}] + \text{PA} \times [10.0 \times \text{weight (kg)} + 934 \times \text{height (m)}] + 20$ kcal
9–18 years (male)	$88.5 - [61.9 \times \text{age (y)}] + \text{PA} \times [26.7 \times \text{weight (kg)} + 903 \times \text{height (m)}] + 25$ kcal
9–18 years (female)	$135.3 - [30.8 \times \text{age (y)}] + \text{PA} \times [10.0 \times \text{weight (kg)} + 934 \times \text{height (m)}] + 25$ kcal
19+ years (male)	$662 - [9.53 \times \text{age (y)}] + \text{PA} \times [15.91 \times \text{weight (kg)} + 539.6 \times \text{height (m)}]$
19+ years (female)	$354 - [6.91 \times \text{age (y)}] + \text{PA} \times [9.36 \times \text{weight (kg)} + 726 \times \text{height (m)}]$
Pregnancy	
14–18 years	
1st trimester	Adolescent EER + 0
2nd trimester	Adolescent EER + 340 kcal
3rd trimester	Adolescent EER + 452 kcal
19–50 years	
1st trimester	Adult EER + 0
2nd trimester	Adult EER + 340 kcal
3rd trimester	Adult EER + 452 kcal
Lactation	
4–18 years	
1st 6 months postpartum	Adolescent EER + 330 kcal
2nd 6 months postpartum	Adolescent EER + 400 kcal
19–50 years	
1st 6 months postpartum	Adult EER + 330 kcal
2nd 6 months postpartum	Adult EER + 400 kcal
Overweight or Obese[b]	
3–18 years (male)	$114 - [50.9 \times \text{age (y)}] + \text{PA} \times [19.5 \times \text{weight (kg)} + 1{,}161.4 \times \text{height (m)}]$
3–18 years (female)	$389 - [41.2 \times \text{age (y)}] + \text{PA} \times [15.0 \times \text{weight (kg)} + 701.6 \times \text{height (m)}]$
19+ years (male)	$1{,}086 - [10.1 \times \text{age (y)}] + \text{PA} \times [13.7 \times \text{weight (kg)} + 416 \times \text{height (m)}]$
19+ years (female)	$448 - [7.95 \times \text{age (y)}] + \text{PA} \times [11.4 \times \text{weight (kg)} + 619 \times \text{height (m)}]$

[a] "PA" stands for the physical activity values appropriate for the age and physiological state. These can be found in the next table.
[b] Body mass index (BMI) ≥ 25 kg/m²; values represent estimated total energy expenditure (TEE; kcal/day) for weight maintenance; weight loss can be achieved by a reduction in energy intake and/or an increase in energy expenditure.

PHYSICAL ACTIVITY (PA) VALUES

Age Group (sex)	PA Level[a]	PA Value	Age Group (sex)	PA Level[a]	PA Value
3–8 years (male)	Sedentary	1.00	3–8 years (female)	Sedentary	1.00
	Low active	1.13		Low active	1.16
	Active	1.26		Active	1.31
	Very active	1.42		Very active	1.56
3–18 years (overweight male)	Sedentary	1.00	3–18 years (overweight female)	Sedentary	1.00
	Low active	1.12		Low active	1.18
	Active	1.24		Active	1.35
	Very active	1.45		Very active	1.60
9–18 years (male)	Sedentary	1.00	9–18 years (female)	Sedentary	1.00
	Low active	1.13		Low active	1.16
	Active	1.26		Active	1.31
	Very active	1.42		Very active	1.56
19+ years (male)	Sedentary	1.00	19+ years (female)	Sedentary	1.00
	Low active	1.11		Low active	1.12
	Active	1.25		Active	1.27
	Very active	1.48		Very active	1.45
19+ years (overweight/obese male)	Sedentary	1.00	19+ years (overweight/obese female)	Sedentary	1.00
	Low active	1.12		Low active	1.16
	Active	1.39		Active	1.27
	Very active	1.59		Very active	1.44

[a]*Sedentary* activity level is characterized by no physical activity aside from that needed for independent living. *Low active* level is characterized by walking 1.5–3 miles/day at 2–4 mph (or equivalent) in addition to the light activity associated with typical day-to-day life. People who are *active* walk 3–10 miles/day at 2–4 mph (or equivalent) in addition to the light activity associated with typical day-to-day life. *Very active* individuals walk 10 or more miles/day at 2–4 mph (or equivalent) in addition to the light activity associated with typical day-to-day life.

[b]Body mass index (BMI) ≥ 25 kg/m².

Source: Institute of Medicine. Dietary Reference Intakes for energy, carbohydrate, fiber, fat, fatty acids, cholesterol, protein, and amino acids (macronutrients). Washington, DC: National Academies Press, 2005.

COMMONLY USED WEIGHTS, MEASURES, AND METRIC CONVERSION FACTORS

Length
1 meter (m) = 39 in, 3.28 ft, or 100 cm.
1 centimeter (cm) = 0.39 in, 0.032 ft, or 0.01 m.
1 inch (in) = 2.54 cm, 0.083 ft, or 0.025 m.
1 foot (ft) = 30 cm, 0.30 m, or 12 in.

Temperature

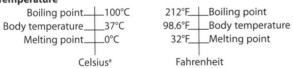

Boiling point — 100°C   212°F — Boiling point
Body temperature — 37°C   98.6°F — Body temperature
Melting point — 0°C   32°F — Melting point

Celsius[a]   Fahrenheit

· To find degrees Fahrenheit (°F) when you know degrees Celsius (°C), multiply by 1.8 and then add 32.
· To find degrees Celsius (°C) when you know degrees Fahrenheit (°F), subtract 32 and then multiply by 0.56.

Volume
1 liter (L) = 1,000 mL, 0.26 gal, 1.06 qt, 2.11 pt, or 34 oz.
1 milliliter (mL) = 1/1,000 L or 0.03 fluid oz.

[a]Also known as *centigrade*.

Volume (continued)
1 gallon (gal) = 128 oz, 16 c, 3.78 L, 4 qt, or 8 pt.
1 quart (qt) = 32 oz, 4 c, 0.95 L, or 2 pt.
1 pint (pt) = 16 oz, 2 c, 0.47 L, or 0.5 qt.
1 cup (c) = 8 oz, 16 tbsp, 237 mL, or 0.24 L.
1 ounce (oz) = 30 mL, 2 tbsp, or 6 tsp.
1 tablespoon (tbsp) = 3 tsp, 15 mL, or 0.5 oz.
1 teaspoon (tsp) = 5 mL or 0.17 oz.

Weight
1 kilogram (kg) = 1,000 g, 2.2 lb, or 35 oz.
1 gram (g) = 1/1,000 kg, 1,000 mg, or 0.035 oz.
1 milligram (mg) = 1/1,000 g or 1,000 μg.
1 microgram (μg) = 1/1,000 mg.
1 pound (lb) = 16 oz, 454 g, or 0.45 kg.
1 ounce (oz) = 28 g or 0.062 lb.

Energy
1 kilojoule (kJ) = 0.24 kcal, 240 calories, or 0.24 Calories.
1 kilocalorie (kcal) = 4.18 kJ, 1,000 calories, or 1 Calorie.

Summary of the 2015 Dietary Guidelines for Americans

Put forth by the U.S. Department of Agriculture and the U.S. Department of Health and Human Services, the 2015 Dietary Guidelines for Americans encompasses two overarching goals: (1) to help individuals maintain energy balance over time and (2) to help Americans choose nutrient-dense foods and beverages. Underlying these broad goals is the proposition that nutrient needs should be met primarily through the consumption of foods, not supplements. The current version of the Dietary Guidelines (its 8th iteration) goes a step further than previous editions by acknowledging that "everyone has a role in the movement to make America healthy." By working together to enact policies, programs, and partnerships that strengthen America's overall health, organizations and individuals can improve the health of the current generation and ensure better health for generations to come.

TABLE 2.3 2015 DIETARY GUIDELINES FOR AMERICANS

Guideline	Justifications and Recommendations
Follow a healthy eating pattern across the lifespan.	• All food and beverage choices matter. • Choose a healthy eating pattern at an appropriate calorie level to help achieve and maintain a healthy body weight, support nutrient adequacy, and reduce the risk of chronic disease.
Focus on variety, nutrient density, and amount.	• To meet nutrient needs within calorie limits, choose a variety of nutrient-dense foods across and within all food groups in recommended amounts.
Limit calories from added sugars and saturated fats and reduce sodium intake.	• Consume an eating pattern low in added sugars, saturated fats, and sodium. • Cut back on foods and beverages higher in these components to amounts that fit within healthy eating patterns.
Shift to healthier food and beverage choices.	• Choose nutrient-dense foods and beverages across and within all food groups in place of less healthy choices. • Consider cultural and personal preferences to make these shifts easier to accomplish and maintain.
Support healthy eating patterns for all.	• Everyone has a role in helping to create and support healthy eating patterns in multiple settings nationwide, from home to school to work to communities.

Source: Adapted from U.S. Department of Agriculture and U.S. Department of Health and Human Services. 2015–2020 Dietary Guidelines for Americans, 2015–2020. 8th edition. Washington, DC Government Printing Office. December 2015.

TABLE 2.4 AMOUNTS OF EACH FOOD GROUP RECOMMENDED (PER DAY) FOR EACH OF THE 2015 DIETARY GUIDELINES HEALTHY DIETARY PATTERNS AND MYPLATE FOOD GUIDANCE SYSTEM

Food Group	Healthy Food Pattern[a] U.S.-Style	Mediterranean-Style	Vegetarian	Dietary Significance
Vegetables	1–2½ cups	1–2½ cups	1–2½ cups	Vegetables are rich sources of potassium, vitamin C, folate, dietary fiber, and vitamins A and E.
Fruits	1–2 cups	1–2½ cups	1–2 cups	Fruits are good sources of folate, vitamin C, vitamin A, and fiber.
Grains	3–6 ounces	3–6 ounces	3–6½ ounces	Grains are a major source of B vitamins, iron, magnesium, selenium, energy, and dietary fiber.
Dairy	2–3 cups	2 cups	2–3 cups	Dairy products are major sources of calcium, potassium, vitamin D, and protein.
Protein foods	2–5½ ounces	2–6½ ounces	1–3½ ounces	Protein foods are rich sources of protein, magnesium, iron, zinc, B vitamins, vitamin D, energy, and potassium.
Oils	15–27 grams	15–27 grams	15–27 grams	Polyunsaturated oils, like those found in vegetable oils and oily fish, are recommended.
Limit on calories for other uses (% of calories)[b]	150–270 kcal (8–15%)	100–260 kcal (8–15%)	170–290 kcal (11–19%)	

[a]Recommended amounts depend on age, sex, and physical activity level. Personalized recommendations can be generated at the MyPlate website (http://www.choosemyplate.gov).
[b]Assumes food choices to meet food group recommendations are in nutrient-dense forms. Calories from added sugars, added refined starches, solid fats, alcohol, and/or to eat more than the recommended amount of nutrient-dense foods are accounted for under this category.

Based on the ratio of weight to height, body mass index (BMI) is a better indicator of obesity than weight alone. For these reasons, BMI has become the standard for gauging if a person is overweight or obese. This table can help you determine whether your BMI is characteristic of being underweight, healthy weight, overweight, or obese. Commonly used units and abbreviations, also listed below, provide easy reference to selected nutritional conventions.

WEIGHT CLASSIFICATIONS USING BODY MASS INDEX (BMI)

Use this chart to calculate your BMI. Locate your weight on the bottom of the chart and your height on the left of the chart. The number located at the intersection of these two values is your BMI.

Height	120	130	140	150	160	170	180	190	200	210	220	230	240	250
4'6"	29	31	34	36	39	41	43	46	48	51	53	56	58	60
4'8"	27	29	31	34	36	38	40	43	45	47	49	52	54	56
4'10"	25	27	29	31	34	36	38	40	42	44	46	48	50	52
5'0"	23	25	27	29	31	33	35	37	39	41	43	45	47	49
5'2"	22	24	26	27	29	31	33	35	37	38	40	42	44	46
5'4"	21	22	24	26	28	29	31	33	34	36	38	40	41	43
5'6"	19	21	23	24	26	27	29	31	32	34	36	37	39	40
5'8"	18	20	21	23	24	26	27	29	30	32	34	35	37	38
5'10"	17	19	20	22	23	24	26	27	29	30	32	33	35	36
6'0"	16	18	19	20	22	23	24	26	27	28	30	31	33	34
6'2"	15	17	18	19	21	22	23	24	26	27	28	30	31	32
6'4"	15	16	17	18	20	21	22	23	24	26	27	28	29	30
6'6"	14	15	16	17	19	20	21	22	23	24	25	27	28	29
6'8"	13	14	15	17	18	19	20	21	22	23	24	25	26	28

Height in feet and inches (left axis) — Weight in pounds (bottom axis)

Key

BMI (kg/m^2)	Classification
<18.5	Underweight
18.5–24.9	Healthy weight
25.0–29.9	Overweight
≥30	Obese

Source: Centers for Disease Control and Prevention. Overweight and obesity: Defining overweight and obesity. Available at: http://www.cdc.gov/obesity/defining.html.

COMMONLY USED NUTRITION-RELATED ABBREVIATIONS

Abbreviation	Definition	Abbreviation	Definition
AI	Adequate Intake	FDA	U.S. Food and Drug Administration
AMDR	Acceptable Macronutrient Distribution Range	GI	Gastrointestinal
ATP	Adenosine triphosphate	HDL	High-density lipoprotein
BMI	Body mass index	LDL	Low-density lipoprotein
BMR	Basal metabolic rate	MUFA	Monounsaturated fatty acid
CDC	U.S. Centers for Disease Control and Prevention	PUFA	Polyunsaturated fatty acid
CVD	Cardiovascular disease	RDA	Recommended Dietary Allowance
DNA	Deoxyribonucleic acid	SFA	Saturated fatty acid
DRI	Dietary Reference Intake	TEE	Total energy expenditure
DV	Daily value	UL	Tolerable Upper Intake Level
EAR	Estimated Average Requirement	USDA	U.S. Department of Agriculture
EER	Estimated Energy Requirement		